11th International Conference on Clustering Aspects of Nuclear Structure and Dynamics (CLUSTER'16)

Journal of Physics: Conference Series Volume 863

Napoli, Italy
23 – 27 May 2016

ISBN: 978-1-5108-4420-9
ISSN: 1742-6588

Printed from e-media with permission by:

Curran Associates, Inc.
57 Morehouse Lane
Red Hook, NY 12571

Some format issues inherent in the e-media version may also appear in this print version.

Printed by Curran Associates, Inc. (2017)

For permission requests, please contact the Institute of Physics
at the address below.

Institute of Physics
Dirac House, Temple Back
Bristol BS1 6BE UK

Phone: 44 1 17 929 7481
Fax: 44 1 17 920 0979

techtracking@iop.org

Additional copies of this publication are available from:

Curran Associates, Inc.
57 Morehouse Lane
Red Hook, NY 12571 USA
Phone: 845-758-0400
Fax: 845-758-2633
Email: curran@proceedings.com
Web: www.proceedings.com

Table of contents

Volume 863

11th International Conference on Clustering Aspects of Nuclear Structure and Dynamics

23–27 May 2016, Complesso dei Ss. Marcellino e Festo Conference Center, Napoli, Italy

Accepted papers received: 14 June 2017
Published online: 4 July 2017

Preface

011001
OPEN ACCESS
11th International Conference on Clustering Aspects of Nuclear Structure and Dynamics

011002
OPEN ACCESS
List of committees

011003
OPEN ACCESS
Conference Programme

011004
OPEN ACCESS
List of participants

011005
OPEN ACCESS
Peer review statement

Introduction

012001
OPEN ACCESS
Opening address

E Vardaci and I Lombardo.....1

012002
OPEN ACCESS
Recent Experimental Results on Nuclear Cluster Physics

C Beck.....3

012003
OPEN ACCESS
Recent results on the cluster structure of light nuclei

F Iachello.....9

012004
OPEN ACCESS
An Overview of the History of Cluster Conferences

H Horiuchi.....15

Cluster Theory

012005
OPEN ACCESS
Alpha-Particle Condensate Structure of the Hoyle State: where do we stand?

P Schuck, Y Funaki, H Horiuchi, G Röpke, A Tohsaki and T Yamada.....19

012006
OPEN ACCESS
Alpha Decay Width of ^{212}Po from a quartetting wave function approach

G Röpke, P Schuck, Y Funaki, H Horiuchi, Z Ren, A Tohsaki, C Xu, T Yamada and B Zhou.....24

012007
OPEN ACCESS
Alpha clustering near nuclear surface and harmonic-oscillator excitations

W Horiuchi and Y Suzuki.....29

012008
OPEN ACCESS
No-core Symplectic Model: Exploiting Hidden Symmetry in Atomic Nuclei

J P Draayer, K D Launey, T Dytrych, A C Dreyfuss, R B Baker and M E Miora.....34

012009
OPEN ACCESS
Geometric symmetries in light nuclei

R Bijker.....40

012010
OPEN ACCESS
Structure of 4n nuclei in a formalism of quartets

M Sambataro and N Sandulescu.....44

012011
OPEN ACCESS
Preformation of alpha-clusters in alpha-decay

R G Lovas.....48

012012
OPEN ACCESS
Alpha clustering and nuclear molecules in ^{8}Be, ^{12}C, ^{16}O, ^{20}Ne, ^{24}Mg, and ^{32}S

G Royer and P Eudes.....52

012013
OPEN ACCESS
A new generator coordinate method to describe gas-like states

R Imai and M Kimura.....56

012014
OPEN ACCESS
Study of similarity and dissimilarity between ^{10}Be and ^{9}Li nuclei with microscopic structure and reaction models

T Furumoto, T Suhara and N Itagaki.....59

012015
OPEN ACCESS
Radiative Capture Processes in Multi-Size-Scale Algebraic Version of Resonating Group Model

A S Solovyev, S Yu Igashov and Yu M Tchuvil'sky.....62

012016
OPEN ACCESS
Constructing realistic alpha cluster channels

K Kravvaris and A Volya.....65

012017
OPEN ACCESS
Semimicroscopic algebraic description of α clustering in Ne isotopes

Géza Lévai.....68

12C: Experiments and Theory

012018
OPEN ACCESS
Experimental progress in clarifying the excitation spectrum of ^{12}C

O S Kirsebom.....71

012019
OPEN ACCESS
Measurement of the 3-α decay from the Hoyle and the broad 10 MeV states in ^{12}C

M Itoh, S Ando, T Aoki, H Arikawa, S Ezure, K Harada, T Hayamizu, T Inoue, T Ishikawa, K Kato, H Kawamura, Y Sakemi and A Uchiyama.....74

012020
OPEN ACCESS
α-cluster breaking effects on cluster structures in ^{12}C

T Suhara and Y Kanada-En'yo.....79

012021
OPEN ACCESS
Nonlocalized clustering and evolution of cluster structure in nuclei

H Horiuchi.....83

012022
OPEN ACCESS
Clustering effects in fusion evaporation reactions with light even-even N=Z nuclei

L Morelli, M D'Agostino, M Bruno, C Frosin, F Gulminelli, F Gramegna, M Cinausero, T Marchi, D Fabris, M Degerlier, G Casini, S Barlini, M Bini, G Pasquali, A Olmi, S Piantelli, S Valdré, G Pastore, N Gelli and E Vardaci.....87

Clusters in Light Nuclei

012023
OPEN ACCESS
Beautiful clusters – boron-9: simulations and decay

Tz Kokalova Wheldon and C Wheldon.....91

012024
OPEN ACCESS
Probing asymmetric clusters by using isoscalar monopole and dipole transitions

M Kimura, Y Chiba and Y Taniguchi.....96

012025
OPEN ACCESS
Study on α-cluster levels in non-4n nuclei using low-energy RI beams

H Yamaguchi, D Kahl, S Hayakawa, Y Sakaguchi, K Abe, T Nakao, T Suhara, N Iwasa, A Kim, D H Kim, S M Cha, M S Kwag, J H Lee, E J Lee, K Y Chae, Y Wakabayashi, N Imai, N Kitamura, P Lee, J Y Moon, K B Lee, C Akers, H S Jung, N N Duy, L H Khiem and C S Lee.....101

012026
OPEN ACCESS
^{15}O+α resonant elastic scattering to study cluster states in ^{19}Ne

D Torresi, C Wheldon, Tz Kokalova, S Bailey, A Boiano, C Boiano, M Cavallaro, S Cherubini, A Di Pietro, J P Fernandez Garcia, M Fisichella, T R Glodariu, J Grebosz, M La Cognata, M La Commara, M Lattuada, M Mazzocco, D Mengoni, C Parascandolo, D Pierroutsakou, G Pizzone, C Signorini, C Stefanini, L Stroe, C Spitaleri, E Strano and M Zadro.....105

012027
OPEN ACCESS
Some Evidence of the Cluster Struture Inside of ^{9}Be

S M Lukyanov, A S Denikin, M A Naumenko, V Burjan, V Kroha, J Mrazek, K Mendibayev, I Sivaček, V Glagolev, Š Piskoř, Yu E Penionzhkevich, N K Skobelev, Yu G Sobolev, W H Trzaska and K Kuterbekov.....109

012028
OPEN ACCESS
Evidence of enhanced radius of Hoyle rotational state in ^{12}C inelastic scattering

M Ito.....113

012029
OPEN ACCESS
Study of the cluster structure of ^{10}Be and ^{16}C neutron-rich isotopes by means of intermediate energies breakup reactions

D Dell'Aquila, L Acosta, L Auditore, G Cardella, E De Filippo, S De Luca, L Francalanza, B Gnoffo, G Lanzalone, I Lombardo, N S Martorana, A Pagano, E V Pagano, M Papa, S Pirrone, G Politi, L Quattrocchi, F Rizzo, E Rosato, P Russotto, A Trifirò, M Trimarchi, G Verde and M Vigilante.....117

012030
OPEN ACCESS
Cluster structure effects of projectile in $^{7}Li+^{208}Pb$ reaction

P K Rath, Md Ashaduzzaman, E Vardaci, A Di Nitto, G La Rana, F Davide, A Pulcini, D Quero, M Mazzocco, D Pierroutsakou, M Cinausero, G Prete, M La Commara, C Parascandolo, M Romoli and N Gelli.....121

012031
OPEN ACCESS
Coulomb breakup of the Borromean nucleus ^{22}C

E C Pinilla and P Descouvemont.....125

012032
OPEN ACCESS
Disentangling unclear nuclear breakup channels of beryllium-9 using the three-axis Dalitz plot

R Smith, M Freer, C Wheldon, N Curtis, S Almaraz-Calderon, A Aprahamian, N I Ashwood, M Barr, B Bucher, P Copp, M Couder, X Fang, G Goldring, F Jung, Tz Kokalova, S R Lesher, W Lu, J D Malcolm, A Roberts, W P Tan and V A Ziman.....128

012033
OPEN ACCESS
Structure and decay pattern of linear-chain states in neutron-rich Carbon isotopes

T Baba and M Kimura.....131

Cluster Effects in Nuclear Reactions and Decay

012034
OPEN ACCESS
Influence of Nuclear Cluster Structure in Proton-Induced Pre-Equilibrium Composite Particle Emission

A A Cowley.....134

012035
OPEN ACCESS
Inclusive breakup of Borromean nuclei

M S Hussein, B V Carlson and T Frederico.....138

012036
OPEN ACCESS
Probing surface distributions of a clusters in ^{20}Ne via α-transfer reaction

T Fukui, Y Taniguchi, T Suhara, Y Kanada-En'yo and K Ogata.....142

012037
OPEN ACCESS
Photodisintegration cross section of ^{9}Be up to E_{γ} = 16 MeV in the α + α + n three-body model

Y Kikuchi, M Odsuren, T Myo and K Katō.....145

012038
OPEN ACCESS
A Virtual State in a Two-Body System Using the Complex Scaling Method

M Odsuren, Y Kikuchi, T Myo, M Aikawa, G Khuukhenkhuu and K. Katō.....148

012039
OPEN ACCESS
A Theoretical Study of Deuteron-induced Surrogate Reactions

B V Carlson, R Capote and M Sin.....151

012040
OPEN ACCESS
Isomeric cross section ratios for the reactions $^{86(87)}$Sr(α,n(2n))^{89mg}Zr

T V Chuvilskaya.....154

012041
OPEN ACCESS
Microscopic time-dependent description of alpha-cluster transfer and incomplete fusion in reactions with ^{12}C and ^{16}O near Coulomb barrier

V V Samarin.....157

012042
OPEN ACCESS
Four- and three-body breakup mechanism of ^{6}Li elastic scattering

S Watanabe, T Matsumoto, K Ogata and M Yahiro.....160

Superheavy Nuclei and Cluster Decay

012043
OPEN ACCESS
Fusion-fission of superheavy nuclei and clustering phenomena

M G Itkis, I M Itkis, G N Knyazheva and E M Kozulin.....163

012044
OPEN ACCESS
Pre-Scission Model Calculation of Fission Fragment Mass and Total Kinetic Energy Distributions for Even-Even Fm, No and Rf Isotopes

N Carjan, F A Ivanyuk and Yu Ts Oganessian.....168

012045
OPEN ACCESS
Observation of shape isomers states in fission fragments

D V Kamanin, Yu V Pyatkov, A A Alexandrov, I A Alexandrova, N Mkaza, V Malaza, E A Kuznetsova, A O Strekalovsky, O V Strekalovsky and V E Zhuchko.....173

012046
OPEN ACCESS
Collinear cluster tri-partition – the brightest observations and their treating

Yu V Pyatkov, D V Kamanin, J E Lavrova, N Mkaza, V Malaza and A O Strekalovsky.....177

012047
OPEN ACCESS
Fission dynamics at high excitation energies investigated in complete kinematics measurements

J L Rodríguez-Sánchez, J Benlliure, J Taïeb, H Álvarez-Pol, L Audouin, Y Ayyad, G B´elier, G Boutoux, E Casarejos, A Chatillon, D Cortina-Gil, T Gorbinet, A Heinz, A Kelić-Heil, B Laurent, J F Martin, C Paradela, E Pellereau, B Pietras, D Ramos, C Rodríguez-Tajes, H Simon and J Vargas.....181

Nuclear Molecules

012048
OPEN ACCESS
A novel visage of molecular resonances in $^{12}C + ^{12}C$

E Uegaki and Y Abe.....184

012049
OPEN ACCESS
Experimental investigations of the sub-Coulomb $^{12}C+^{12}C$ and $^{12}C+^{16}O$ reactions

S Courtin, G Fruet, D G Jenkins, C L Jiang, M Heine, D Montanari, D Santiago-Gonzalez, M Avila-Coronado, L G Morris, A Goasduff, E Rehm, B Back, D Bourgin and C Beck.....189

012050
OPEN ACCESS
$^{16}O + ^{16}O$ molecular structures of superdeformed states in S isotopes

Y Taniguchi.....194

012051
OPEN ACCESS
Search for highly excited states in ^{28}Si

D Montanari, S Courtin, D G Jenkins, C Diget, N Yavuzkanat, R Neveling, J P Mira, F Nemulodi, F D Smit, I Usman, P Papka, J. A Swartz, J J van Zyl and N Orce.....198

012052
OPEN ACCESS
High spin structures in the A $\approx$ 40 mass region: from superdeformation to extreme deformation and clusterization (an example of ^{28}Si)

A V Afanasjev and D Ray.....202

012053
OPEN ACCESS
Asymmetric cluster structure and isoscalar monopole/dipole transitions of ^{28}Si

Y Chiba, Y Taniguchi and M Kimura.....205

Clusters in Heavy Ion Collisions and Nuclear Matter

012054
OPEN ACCESS
Alpha-particle clustering in excited alpha-conjugate nuclei

B Borderie, Ad R Raduta, G Ademard, M F Rivet, E De Filippo, E Geraci, N Le Neindre, R Alba, F Amorini, G Cardella, M Chatterjee, D Guinet, P Lautesse, E La Guidara, G Lanzalone, G Lanzano, I Lombardo, O Lopez, C Maiolino, A Pagano, M Papa, S Pirrone, G Politi, F Porto, F Rizzo, P Russotto and J P Wieleczko.....208

012055
OPEN ACCESS
From real clusters to experimental nuclear matter phase diagram

L G Moretto and L Phair.....213

012056
OPEN ACCESS
Cluster production and nuclear dynamics

M Papa.....217

012057
OPEN ACCESS
Pre-equilibrium emission and clustering in medium-mass nuclei: ^{46}Ti from $^{16}O + ^{30}Si$, $^{18}O + ^{28}Si$, $^{19}F + ^{27}Al$

M Cicerchia, T Marchi, F Gramegna, M Cinausero, J Mabiala, D Fabris, A Caciolli, G Collazzuol, D Mengoni, M Degerlier, L Morelli, M Bruno, M D'Agostino, S Barlini, M Bini, G Pasquali, S Piantelli, G Casini, G Pastore, D Gruyer, P Ottanelli, S Valdrè, N Gelli, A Olmi, G Poggi, E Vardaci, I Lombardo, D Dell'Aquila, S Leoni, N Cieplicka and B Fornal.....221

012058
OPEN ACCESS
Production of exotic nuclei in heavy ion collisions at Fermi energies

R Ogul, N Buyukcizmeci and A Ergun.....224

012059
OPEN ACCESS
Probing the Symmetry Term of the Nuclear Equation of State at High Baryonic Densities

P Russotto, L Acosta, M Adamczyk, A Al-Ajlan, M Al-Garawi, S Al-Homaidhi, F Amorini, L Auditore, T Aumann, Y Ayyad, Z Basrak, J Benlliure, M Boisjoli, K Boretzky, J Brzychczyk, A Budzanowski, C Caesar, G Cardella, P Cammarata, Z Chajecki, M Chartier, A Chbihi, M Colonna, M D Cozma, B Czech, E De Filippo, M Di Toro, M Famiano, S Gannon, I Gašparić, L Grassi, C Guazzoni, P Guazzoni, M Heil, L Heilborn, R Introzzi, T Isobe, K Kezzar, M Kiš, A Krasznahorkay, S Kupny, N Kurz, E La Guidara, G Lanzalone, P Lasko, A Le Fèvre, Y Leifels, R C Lemmon, QF Li, I Lombardo, J Łukasik, W G Lynch, P Marini, Z Matthews, L May, T Minniti, M Mostazo, A Pagano, E V Pagano, M Papa, P Pawłowski, S Pirrone, G Politi, F Porto, W Reviol, F Riccio, F Rizzo, E Rosato, D Rossi, S Santoro, D G Sarantites, H Simon, I Skwirczynska, Z Sosin, L Stuhl, W Trautmann, A Trifirò, M Trimarchi, M B Tsang, G Verde, M Veselsky, M Vigilante, Yongjia Wang, A Wieloch, P Wigg, J Winkelbauer, H H Wolter, P Wu, S Yennello, P Zambon, L Zetta and M Zoric.....227

012060
OPEN ACCESS
Dynamical Dipole mode in heavy-ion fusion reactions

C Parascandolo, D Pierroutsakou, R Alba, A Del Zoppo, C Maiolino, D Santonocito, C Agodi, V Baran, A Boiano, M Colonna, R Coniglione, E De Filippo, M Di Toro, U Emanuele, F Farinon, A Guglielmetti, M La Commara, B Martin, C Mazzocchi, M Mazzocco, C Rizzo, M Romoli, C Signorini, R Silvestri, F Soramel, E Strano, D Torresi, A Trifirò and M Trimarchi.....230

012061
OPEN ACCESS
In medium fragment break-up of projectile in $^{58}Ar+^{36}Ni$ central collisions

L Francalanza.....233

012062
OPEN ACCESS
N/Z effect on reaction mechanisms cross sections in the $^{78}Kr + ^{40}Ca$ and $^{86}Kr + ^{48}Ca$ collisions at 10 AMeV

B Gnoffo, S Pirrone, G Politi, M La Commara, J P Wieleczko, E De Filippo, P Russotto, M Trimarchi, M Vigilante, G Ademard, L Auditore, C Beck, I Bercenau, E Bonnet, B Borderie, G Cardella, A Chibihi, M Colonna, D Dell'Aquila, S De Luca, A D'Onofrio, J D Frankland, G Lanzalone, P Lautesse, D Lebhertz, N Le Neidre, I Lombardo, N S Martorana, K Mazurek, S Norella, A Pagano, E V Pagano, M Papa, E Piasecki, F Porto, L Quattrocchi, F Rizzo, G Spadaccini, A Trifirò and G Verde.....236

012063
OPEN ACCESS
α clustering and its connection to the $E1$ response of heavy nuclei

M Spieker, S Pascu and A Zilges.....239

012064
OPEN ACCESS
The effect of clusters on fragment emission mechanism

S Manna, T K Rana, C Bhattacharya, S Bhattacharya, S Kundu, K Banerjee, P Roy, R Pandey, V Srivastava, A Chaudhuri, T Roy, T K Ghosh, G Mukherjee, J K Meena, S K Pandit, K Mahata, A Shrivastava and V Nanal.....242

Clusters and Correlations

012065
OPEN ACCESS
Tensor correlations in light nuclei with tensor-optimized antisymmetrized molecular dynamics

T Myo, H Toki, K Ikeda, H Horiuchi and T Suhara.....245

012066
OPEN ACCESS
Covalent Binding with Neutrons on the Femto-scale

W von Oertzen, Y Kanada-En'yo and M Kimura.....249

012067
OPEN ACCESS
Proton and neutron pair correlations in ^{10}B

H Morita and Y Kanada-En'yo.....254

012068
OPEN ACCESS
Two-neutron clustering aspects in the transitions induced by the ^{13}C(^{18}O,^{16}O)^{15}C reaction at 84 MeV incident energy

D Carbone, J L Ferreira, F Cappuzzello, J Lubian, C Agodi, M Cavallaro, A Gargano, S M Lenzi, R Linares and G Santagati.....257

012069
OPEN ACCESS
Deuteron-Like Correlation for the $T = 0$ Channel in ^{18}F Studied within the Continuum Contributions

H Masui and M Kimura.....260

012070
OPEN ACCESS
High multiplicity α-particle breakup measurements to study α-condensate states

J Bishop, Tz Kokalova, M Freer, M Assie, L Acosta, S Bailey, G Cardella, N Curtis, E De Filippo, D Dell'Aquila, S De Luca, L Francalanza, B Gnoffo, G Lanzalone, I Lombardo, N Martorana, S Norella, A Pagano, E V Pagano, M Papa, S Pirrone, G Politi, F Rizzo, P Russotto, L Quattrocchi, R Smith, I Stefan, A Trifirò, M Trimarchì, G Verde, M Vigilante and C Wheldon.....263

012071
OPEN ACCESS
Study of resonances produced in light nuclei through two and multi particle correlations

L Quattrocchi, L Acosta, F Amorini, A Anzalone, L Auditore, I Berceanu, G Cardella, A Chbihi, E De Filippo, S De Luca, D Dell' Aquila, L Francalanza, B Gnoffo, A Grzeszczuk, G Lanzalone, I Lombardo, I Martel, N S Martorana, T Minniti, S Norella, A Pagano, E V Pagano, M Papa, S Pirrone, G Politi, F Porto, F Rizzo, E Rosato, P Russotto, A Trifirò, M Trimarchi, G Verde, M Veselsky and M Vigilante.....266

Clusters and Nuclear Astrophysics

012072
OPEN ACCESS
Clusterization of light nuclei and the Trojan Horse Method

A Tumino, C Spitaleri, S Cherubini, G D'Agata, L Guardo, M Gulino, I Indelicato, M La Cognata, L Lamia, R G Pizzone, S M R Puglia, G G Rapisarda, S Romano, M L Sergi and R Spartá.....269

012073
OPEN ACCESS
Measurement of the cross section for the $^4He(\alpha, n)^7Be$ reaction as a possible solution to the cosmological lithium problem

T Kawabata, T Furuno, M Ichikawa, N Iwasa, Y Kanada-En'yo, A Koshikawa, S Kubono, E Miyawaki, T Morimoto, M Murata, T Nanamura, S Nishimura, Y Shikata, Y Takahashi, T Takeda, M Tsumura and K Watanabe.....274

012074
OPEN ACCESS
Capture reactions into borromean two-proton systems at rp-waiting points

D Hove, E Garrido, A S Jensen, H O U Fynbo, N T Zinner and D V Fedorov.....278

012075
OPEN ACCESS
Search for the rare γ-decay mode in ^{12}C

M Tsumura, T Kawabata, T Furuno, A Koshikawa, M Murata, T Morimoto, S Adachi, A Tamii, K Hatanaka, T Ito, J Zenihiro, S Kubono, M Itoh, Y Matsuda, Y Maeda, S Sakaguchi, H Akimune, H Fujimura, I Ou, T Hashimoto and C Iwamoto.....282

012076
OPEN ACCESS
Active target MAIKo to investigate cluster structures in unstable nuclei

T Furuno, T Kawabata, M Murata, S Adachi, Y Ayyad, H Baba, T Baba, M Cavallaro, T Hashimoto, Y Ishii, R Kobayakawa, Y Matsuda, Y Matsuoka, T Morimoto, T Nanamura, S Beceilo-Novo, H J Ong, A Sakaue, J Tanaka, I Tanihata, D T Trong, M Tsumura and H D Watanabe.....286

012077
OPEN ACCESS
The measurement of ^{4}He photodisintegration with MAIKo active target

M Murata, S Adachi, H Akimune, T Furuno, T Kawabata and M Tsumura.....289

Summary and Conclusions

012078
OPEN ACCESS
Closing remarks from a theoretical perspective

Y Kanada-En'yo.....292

012079
OPEN ACCESS
Future Challenges for Nuclear Clustering

M Freer.....298

IOP Conf. Series: Journal of Physics: Conf. Series **863** (2017) 011001 doi:10.1088/1742-6596/863/1/011001

Preface

The 11th edition of the International Conference on Clustering Aspects of Nuclear Structure and Dynamics (CLUSTER'16) was held in Napoli, Italy, on May 23-27 2016. All the Conference Sessions took place in the magnificent *Complesso Monumentale dei Ss. Marcellino e Festo*, located in the Historical Centre of Naples. This is one of the most prestigious building complexes of the Federico II University of Naples, the main home institution of the organizers, together with the Istituto Nazionale di Fisica Nucleare (Naples division). This building is 500 years old; in the XVIII century, it was reshaped by Luigi Vanvitelli, a famous architect of the pre-Neoclassical period in Italy, designer of the prestigious Royal Palace in the near town of Caserta. The site of Plenary Sessions was the *Church of Ss. Marcellino e Festo*, where all the participants were surrounded by the Baroque frescoes by Belisario Corenzio and, among the others, the paintings of Battistello Caracciolo, Massimo Stanzione, Giuseppe Simonelli and Francesco De Mura, important artists of the late Baroque period in South Italy. The sites of Parallel Sessions were two halls in the arcade of the S. Marcellino cloister: this gave the opportunity to the participants to enjoy the beauty of the architecture of the cloister with its fountains, gardens, sculptures.

As organizers of the Conference, our main aims were: (1) to provide an excellent programme, with the expectation to be a reference point for the Nuclear Cluster Physics in the next four years; (2) to assure a very relaxing stay to the participants, allowing them to explore the bounty of artistic, and also culinary, masterpieces that Naples offers to its guests. The first point was assured by all the conveners that, with their excellent talks, gave a very precise and complete overview of the most recent achievements on Nuclear Cluster Physics, both from the experimental and theoretical point of view. We are indebted with the International Advisory Committee for their suggestions of excellent invited speakers and with all the conveners for their contributions to the discussion.

The Conference had a very rich social programme. We remember the intense emotions at the Welcome Cocktail, held in the marvellous *Basilica of S. Giovanni Maggiore*, during the Concert of Classical and Neapolitan music. In the same way, it is impossible for us to forget the evening (*na sera 'e maggio*) of the Social Dinner in St. Chiara cloister, with the flawless talk of Prof. Hisashi Horiuchi on the history of Cluster Conference and the atmosphere and tastes of the dinner. We hope also that the social excursion at the Capodimonte Royal Palace will remain in the hearts and minds of the conveners.

We had 54 talks in plenary session and 40 talks presented in three different parallel sessions. More than 130 physicists participated in the conference, from many countries around the world. We were honoured to have among the participants several Directors of Physics Departments and of eminent Scientific Institutions in the world.

In the Programme the talks are arranged following the session structure of the Conference. In this list invited speakers are marked by asterisks. Plenary sessions contain also the invited talks. The proceedings of 80 talks have been submitted by authors for publication, and are reported in this volume. The papers are arranged according to their subjects, without considering whether they were delivered in a plenary or in a parallel session. We thank Profs. C. Beck, F. Iachello, M. Freer and Y. Kanada-En'yo for delivering excellent introductory and summary talks of the Conference. We want to dedicate this volume of proceedings to the memory of Prof. Paulo R. S. Gomes. He unfortunately passed away in Naples on the last day of the conference. He was a world reknown expert in the field of nuclear reactions with light ions and an excellent physicist. We express once more our sincere condolences to his family and colleagues.

After several months from this event, we believe this *Neapolitan edition* was successful in continuing the high scientific and cultural tradition of the past Cluster Conferences; on our side, we will never forget all the phases of conference preparations and all the moments of conference.

We want to acknowledge the support of the following institutions:

IOP Conf. Series: Journal of Physics: Conf. Series **863** (2017) 011001 doi:10.1088/1742-6596/863/1/011001

- Departments of Physics, Geology and Political Sciences of the Federico II University of Naples;
- Istituto Nazionale di Fisica Nucleare – Sezione di Napoli;
- StudioCongress organizzazione eventi;
- Fondazione Ordine degli Ingegneri della Provincia di Napoli;
- CAEN s.p.a., Viareggio, Italy

and the fundamental help of the members of the Local Organizing Committee during all the phases of conference organization.

Finally, we are pleased to announce that, as decided by the International Advisory Committee, the 12th edition of the International Conference on Clustering Aspects of Nuclear Structure and Dynamics will be held in Brasil in 2020.

E. Vardaci
I. Lombardo

https://agenda.infn.it/conferenceDisplay.py?ovw=True&confId=9919

List of committees

International Advisory Committee

Y. Abe (*Osaka, Japan*)
D. Baye (*Bruxelles, Belgium*)
C. Beck (*Strasbourg, France*)
S. Bhattacharya (*Kolkata, India*)
A. Bracco (*Milano, Italy*)
J. Cseh (*Debrecen, Hungary*)
M. Dasgupta (*Canberra, Australia*)
A. Fomichev (*Dubna, Russia*)
M. Freer (*Birmingham, United Kingdom*)
M. Fujiwara (*Osaka, Japan*)
M. Gai (*Yale and UConn, United States*)
M. Harakeh (*Groningen, Nederlands*)
H. Horiuchi (*Osaka, Japan*)
M. Hussein (*Sao Paulo, Brasil*)
N. Itagaki (Kyoto, *Japan*)
M.G. Itkis (*Dubna, Russia*)
B. Jonson (*Goteborg, Sweden*)
T. Kajino (*Tokyo, Japan*)
Y. Kanada-En'yo (*Kyoto, Japan*)
N. Lo Iudice (*Napoli, Italy*)
M. Milin (*Zagreb, Croatia*)
A. Moro (*Sevilla, Spain*)
T. Neff (*Darmstadt, Germany*)
Y.Ts. Oganessian (*Dubna, Russia*)
N. Orr (*Caen, France*)
P. Papka (*Stellenbosch, South Africa*)
D. Poenaru (*Bucarest, Romania*)
Z. Ren (*Nanjing, China*)
G. Rogachev (*College Station, United States*)
P. Schuck (*Orsay, France*)
A. Tumino (*Enna, Italy*)
E. Vardaci (*Napoli, Italy*)
F.S. Zhang (*Beijing, China*)

Local Organizing Committee

E. Vardaci (*Chairman*), I. Lombardo (*Scientific Secretary*), Md. Ashaduzzaman, F. Davide, B. De Canditiis, D. Dell'Aquila, G. De Gregorio, A. Di Leva, L. Francalanza, M. La Commara, G. La Rana, R. La Torre, L. Lista, C. Parascandolo, D. Pierroutsakou, A. Porrino, D. Quero, P. Rath, M. Romoli, G. Spadaccini, M. Vigilante.

IOP Conf. Series: Journal of Physics: Conf. Series **863** (2017) 011003 doi:10.1088/1742-6596/863/1/011003

Cluster '16

The 11th International Conference on Clustering Aspects of Nuclear Structure and Dynamics

CONFERENCE PROGRAMME

Monday 23.05.2016

9:00 – 10:00 *Registration at the Conference Desk*

10:00 – 10:30 Welcome address *G. Trombetti, L. Merola, G. La Rana, P. Cappelletti, E. Vardaci*

PLE- Session 1: State of the Art: Experiment (Chair: E. Vardaci)

10:30 – 11:10 C. Beck* *Recent experimental results on nuclear cluster physics*

11:10 – 11:30 COFFEE BREAK

PLE- Session 2: State of the Art: Theory – Part 1 (Chair: H. Horiuchi)

11:30 – 12:10 F. Iachello* *Recent theoretical results in nuclear cluster physics*

12:10 – 12:35 R. Bijker *Geometric symmetries in light nuclei*

12:35 – 13:00 M. Sambataro *Structure of 4n nuclei in a formalism of quartets*

13:00 – 13:25 J. Cseh *Clustering and quarteting in light nuclei*

13:25 – 15:00 LUNCH

PLE-Session 3: Clusters and Nuclear Astrophysics (Chair: M. Gai)

15:00 – 15:30 A. Tumino* *Clusterization of Light Nuclei and the Trojan Horse Method*

15:30 – 16:00 P. Descouvemont* *Microscopic description of ^{6}He elastic scattering on heavy targets*

16:00 – 16:25 T. Kawabata *Measurement of the cross section for the ^{4}He(^{4}He, n)^{7}Be reaction as a possible solution to the cosmological lithium problem*

16:25 – 16:50 D. Hove *Structure and decay at rapid proton capture waiting points*

16:50 – 17:15 COFFEE BREAK

IOP Conf. Series: Journal of Physics: Conf. Series **863** (2017) 011003 doi:10.1088/1742-6596/863/1/011003

Cluster '16

The 11th International Conference on Clustering Aspects of Nuclear Structure and Dynamics

PAR-1.A: Parallel Session A (Chair: G. Rogachev)

17:15 – 17:35	R. Smith	*Spectroscopy of the Borromean ^{9}Be and ^{9}B Mirror Nuclei*
17:35 – 17:55	T. Furumoto	*Study of similarity and dissimilarity between ^{10}Be and ^{9}Li nuclei with microscopic structure and reactions models*
17:55 – 18:15	T. Baba	*Structure and decay pattern of linear chain states in neutron-rich Carbon isotopes*
18:15 – 18:35	D. Dell'Aquila	*Study of the cluster structure of ^{10}Be and ^{16}C neutron-rich isotopes by means of intermediate energies breakup reactions*

PAR-1.B: Parallel Session B (Chair: P. Gomes)

17:15 – 17:35	S. Koyama	*Cluster structure on neutron rich nuclei ^{16}C*
17:35 – 17:55	H. Morita	*Proton and Neutron Pair Correlations in ^{10}B*
17:55 – 18:15	R.D. Lasseri	*Impact of 4-body correlations on clustering*
18:15 – 18:35	K. Kravvaris	*Microscopic configuration interaction studies of nuclear clustering*

PAR-1.C: Parallel Session C (Chair: M. La Commara)

17:15 – 17:35	A.V. Afanasjev	*High spin structures in the A≈ 40 mass region: from superdeformation to extreme deformation and clusterization*
17:35 – 17:55	T. Fukui	*Probing surface distribution of α-cluster in ^{20}Ne via α-transfer Reaction*
17:55 – 18:15	Y. Kikuchi	*Photodisintegration cross section of ^{9}Be up to E_γ = 16 MeV in the $\alpha + \alpha + n$ three-body model*
18:15 – 18:35	T. Furuno	*An active target MAIKo to investigate cluster structures in unstable nuclei*

20:00 – 23:00 **WELCOME COCKTAIL AND CONCERT** (Arcibasilica di S. Giovanni Maggiore a Pignatelli)

IOP Conf. Series: Journal of Physics: Conf. Series **863** (2017) 011003 doi:10.1088/1742-6596/863/1/011003

Cluster '16

The 11th International Conference on Clustering Aspects of Nuclear Structure and Dynamics

Tuesday 24.05.2016

PLE-Session 4: ^{12}C: Experiments and Theory – Part 1 (Chair: W. von Oertzen)

9:00 – 9:30	O. S. Kirsebom*	*Experimental progress on clarifying the excitation spectrum of ^{12}C*
9:30 – 10:00	M. Gai*	*Current and Future Tests of the Algebraic Cluster Model in Carbon Isotopes*
10:00 – 10:25	M. Tsumura	*Search for the rare γ-decay mode in ^{12}C*
10:25 – 10:50	T. Suhara	*α-cluster breaking effects on cluster structures in ^{12}C*

10:50 – 11:30 COFFEE BREAK

PLE-Session 5: ^{12}C: Experiments and Theory – Part 2 (Chair: R. Lovas)

11:30 – 12:00	J. Draayer*	*No Core Symplectic Model - Exploiting Hidden Symmetry in Atomic Nuclei*
12:0 0 – 12:30	M. Itoh*	*Measurement of the 3-α decay from the Hoyle and the broad 10 MeV states in ^{12}C*
12:30 – 12:55	H. Horiuchi	*Nonlocalized clustering and evolution of cluster structure in nuclei*
12:55 – 13:20	L. Morelli	*Clustering effects in fusion evaporation reactions with light even-even N=Z nuclei*

13:20 – 15:00 LUNCH

PLE-Session 6: Clusters in Heavy Ion Collisions and Nuclear Matter (Chair: A. Bracco)

15:00 – 15:30	B. Borderie*	*Alpha-particle clustering in Excited alpha-conjugate nuclei*
15:30 – 16:00	L. Moretto*	*From the cluster emission to the experimental liquid–vapor phase diagram of bulk, symmetric, uncharged nuclear matter*
16:00 – 16:25	S. Typel	*A comparison of equation of state models with different cluster dissolution mechanisms*
16:25 – 16:50	M. Papa	*Effective interaction and cluster production in Nuclear Dynamics*

16:50 – 17:15 COFFEE BREAK

IOP Conf. Series: Journal of Physics: Conf. Series **863** (2017) 011003 doi:10.1088/1742-6596/863/1/011003

Cluster '16

The 11th International Conference on Clustering Aspects of Nuclear Structure and Dynamics

PAR-2.A: Parallel Session A (Chair: I. Lombardo)

17:15 – 17:35	J. Bishop	*Searching for α-gas states in ^{28}Si with CHIMERA and FARCOS*
17:35 – 17:55	L. Quattrocchi	*Study of resonances produced in light nuclei through two and multi-particle correlations.*
17:55 – 18:15	J.Rodriguez-Sanchez	*Fission dynamics at high excitation energies investigated in complete kinematic measurements*
18:15 – 18:35	Z.H. Yang	*Study on the cluster structure in nuclei by using direct reactions*
18:35 – 18:55	L. Francalanza	*Projectile's in medium break-up in $^{36}Ar+^{58}Ni$ central collisions*

PAR-3.B: Parallel Session B (Chair: F. Cappuzzello)

17:15 – 17:35	V.V. Samarin	*Microscopic time-dependent description of alpha-cluster transfer and incomplete fusion in reactions with ^{12}C and ^{16}O near Coulomb barrier*
17:35 – 17:55	D. Carbone	*Two-neutron clustering aspects in the transitions induced by the $^{13}C(^{18}O,^{16}O)^{15}C$ reaction at 84 MeV incident energy*
17:55 – 18:15	H. Masui	*Deuteron-Like Correlation for the T = 0 Channel in ^{18}F Studied within the Continuum Contributions*
18:15 – 18:35	E.C. Pinilla	*Coulomb breakup of the Borromean nucleus ^{22}C*
18:35 – 18:55	Yu. M. Tchuvil'sky	*Exchange-Induced Effective Interaction in Cluster Reaction and Decay Channels*

PAR-4.C: Parallel Session C (Chair: S. Dmitriev)

17:15 – 17:35	M. Odsuren	*A Virtual State in a Two-Body System Using the Complex Scaling Method*
17:35 – 17:55	B.V. Carlson	*A Theoretical Study of Deuteron-induced Surrogate Reactions*
17:55 – 18:15	T.V.Chuvilskaya	*Isomeric cross section ratios for the reactions $^{86(87)}Sr(\alpha,n(2n))89^{mg}Zr$*
18:15 – 18:35	S. Bailey	*Extracting the Spectral Signature of Alpha Clustering in Medium Mass Nuclei*
18:35 – 18:55	S. Manna	*The effect of clusters on fragment emission mechanism*
18:55 – 19:15	B. Gnoffo	*N/Z effects on the reaction mechanisms cross sections in the $^{78}Kr+^{40}Ca$ and $^{86}Kr+^{48}Ca$ collisions at 10 AMeV*

[19:25 – 19:45]: IAC Council for Organization of **Cluster 2020** Conference

IOP Conf. Series: Journal of Physics: Conf. Series **863** (2017) 011003 doi:10.1088/1742-6596/863/1/011003

Cluster '16

The 11th International Conference on Clustering Aspects of Nuclear Structure and Dynamics

Wednesday 25.05.2016

<u>PLE-Session 7</u>: Clusters in Light Nuclei – Part 1 (Chair: M. Fujiwara)

9:00 – 9:30	Tz. Kokalova Wheldon*	*Beautiful clusters*
9:30 – 10:00	Y. Funaki*	*Cluster structures in the excited states above the Hoyle state*
10:00 – 10:25	H. Yamaguchi	*Study on α-cluster levels in non-4n nuclei using low-energy RI beams*
10:25 – 10:50	D. Torresi	*$^{15}O+\alpha$ resonant elastic scattering to studying cluster states in ^{19}Ne*

<u>10:50 – 11:15</u> <u>COFFEE BREAK</u>

<u>PLE-Session 8</u>: Clusters and Correlations (Chair: G. Verde)

11:15 – 11:45	S. Shimoura*	*Tetraneutron states populated via $^4He(^8He,^8Be)$ reaction*
11:45 – 12:15	L. Grigorenko*	*High-precision studies of exotic structure and decay effects in the dripline nuclei within three-body core+N+N model*
12:15 – 12:40	T. Myo	*Tensor correlations in light nuclei with tensor-optimized antisymmetrized molecular dynamics*
12:40 – 13:05	W. von Oertzen	*Covalent Binding by Neutrons on the Femto-scale*

<u>13:05 – 14:45</u> <u>LUNCH</u>

<u>PLE-Session 9</u>: State of the Art: Theory – Part 2 (Chair: N. Lo Iudice)

14:45 – 15:15	P. Schuck*	*The α-particle condensate structure of the Hoyle and analogous states. Where do we stand after 15 years?*
15:15 – 15:45	G. Ropke*	*Alpha Decay Width of ^{212}Po from a quartetting wave function approach*
15:45 – 16:15	E. Khan*	*Crystal, cluster and quantum liquid nuclear states*
16:15 – 16:40	T. Otsuka	*Ab initio Monte Carlo Shell Model calculations on Be and C isotopes*

<u>16:40 – 17:00</u> <u>COFFEE BREAK</u>

IOP Conf. Series: Journal of Physics: Conf. Series **863** (2017) 011003 doi:10.1088/1742-6596/863/1/011003

Cluster '16

The 11th International Conference on Clustering Aspects of Nuclear Structure and Dynamics

PAR-3.A: Parallel Session A (Chair: M. Papa)

17:00 – 17:20	M. Cicerchia	*Pre-equilibrium emission and clustering in medium-mass nuclei: ^{46}Ti from ^{16}O + ^{30}Si, ^{18}O + ^{28}Si, ^{19}F + ^{27}Al*
17:20 – 17:40	R. Ogul	*Production of exotic nuclei in heavy ion collisions at Fermi energies*
17:40 – 18:00	P. Russotto	*Probing the Symmetry Term of the Nuclear Equation of State at High Baryonic Densities*
18:00 – 18:20	C. Parascandolo	*Evidence of Dynamical Dipole Mode excitation in the ^{40,48}Ca + ^{152,144}Sm fusion evaporation and fission reactions at 11MeV/A*

PAR-3.B: Parallel Session B (Chair: C. Beck)

17:00 – 17:20	S. Watanabe	*Four- and three-body breakup mechanism of ^{6}Li elastic scattering*
17:20 – 17:40	P. Rath	*Cluster structure effect of projectile in ^{7}Li+^{208}Pb reaction*
17:40 – 18:00	M. Spieker	*α-clustering and its connection to the E1 response of heavy nuclei*
18:00 – 18:20	M. Murata	*The measurement of ^{4}He Photodisintegration with MAIKo active target*

PAR-3.C: Parallel Session C (Chair: M. Kimura)

17:00 – 17:20	Y. Chiba	*Asymmetric cluster structure and isoscalar monopole/dipole transitions of ^{28}Si*
17:20 – 17:40	R. Imai	*A new generator coordinate method to describe gas-like states*
17:40 – 18:00	G. Lévai	*Semimicroscopic algebraic description of α clustering in Ne isotopes*
18:00 – 18:20	A. Solovyev	*Radiative Capture Processes in Multi-Size-Scale Algebraic Version of Resonating Group Model*

19:45 – 23:00 KEYNOTE TALK AND SOCIAL DINNER (Reale Basilica di S. Chiara ed annesso Monastero)

19:45 – 20:15	H. Horiuchi*	*An overview of the History of Cluster Conferences*
20:20 – 23:00	All participants	*Social dinner at St. Chiara Cloister*

Cluster '16

The 11th International Conference on Clustering Aspects of Nuclear Structure and Dynamics

Thursday 26.05.2016

PLE-Session 10: Superheavy Nuclei and Cluster Decay (Chair: D. Poenaru)

9:00 – 9:30	M. Itkis*	*Fusion-Fission of superheavy nuclei and clustering phenomena*
9:30 – 10:00	N. Carjan*	*Symmetric and Super-Symmetric Spontaneous Fission of Very Heavy Elements: Challenges and Prospects*
10:00 – 10:25	D.M. Kamanin	Observation *of shape isomer states in fission fragments*
10:25 – 10:50	Yu.V. Pyatkov	*Collinear cluster tri-partition - the brightest observations and their treating*

10:50 – 11:15 COFFEE BREAK

PLE-Session 11: Clusters in Light Nuclei – Part 2 (Chair: Y. Abe)

11:15 – 11:45	M. Kimura*	*Probing asymmetric clusters by isoscalar monopole/dipole transitions*
11:45 – 12:15	G. Rogachev*	*Quantifying clustering near* α*-threshold*
12:15 – 12:40	S.V. Lukyanov	*Some Insights into* ^{5}*He-Cluster Structure of* ^{9}*Be*
12:40 – 13:05	F. M. Marqués	*Exploring the Structure of the Most Neutron-rich Boron and Carbon Isotopes*
13:05 – 13:30	M. Ito	*Spatial measures of the nuclear reactions*

13:30 – 15:30 LUNCH

15:30 – 18:00	All participants	*Excursion at the Capodimonte Royal Palace*

IOP Conf. Series: Journal of Physics: Conf. Series **863** (2017) 011003 doi:10.1088/1742-6596/863/1/011003

Cluster '16

The 11th International Conference on Clustering Aspects of Nuclear Structure and Dynamics

Friday 27.05.2016

PLE-Session 12: Nuclear Molecules (Chair: M. Freer)

9:00 – 9:30	E. Uegaki*	*A Novel Visage of Molecular Resonances in $^{12}C + ^{12}C$*
9:30 – 10:00	S. Courtin*	*Experimental investigations of the sub-Coulomb $^{12}C+^{12}C$ and $^{12}C+^{16}O$ reactions*
10:00 – 10:25	Y. Taniguchi	*$^{16}O + ^{16}O$ molecular structures of superdeformed states in S isotopes*
10:25 – 10:50	D. Montanari	*Search for highly-excited superdeformed states in ^{28}Si*

10:50 – 11:15 COFFEE BREAK

PLE-Session 13: State of the Art: Theory – Part 3 (Chair: S. Typel)

11:15 – 11:45	W. Horiuchi*	*Alpha-clustering near nuclear surface and harmonic-oscillator excitations*
11:45 – 12:10	R. Lovas	*Preformation of alpha clusters in alpha decay*
12:10 – 12:35	G. Royer	*Alpha clustering and nuclear molecules in ^{8}Be, ^{12}C, ^{16}O, ^{20}Ne, ^{24}Mg, and ^{32}S*
12:35 – 13:00	M. S. Hussein	*Theories of Inclusive Non-Elastic Breakup and Incomplete Fusion of Weakly-Bound Two-Cluster Projectiles Revisited: Critical Assessment*

13:00 – 15:00 LUNCH

PLE-Session 14: Cluster Effects in Nuclear Reactions and Decay (Chair: M. S. Hussein)

15:00 – 15:25	P.R.S. Gomes	*Breakup of the weakly bound $^{6,7}Li$ two-cluster nuclei at near barrier energies*
15:25 – 15:50	A.A. Cowley	*Influence of Nuclear Cluster Structure in Proton-Induced Pre-Equilibrium Composite Particle Emission*

PLE-Session 15: Summary and Conclusions (Chair: D. Baye)

15:50 – 16:30	Y. Kanada En'yo*	*Closing Remarks: Theoretical Perspectives*
16:30 – 17:10	M. Freer*	*Closing Remarks: Experimental Perspectives*
17:10	E. Vardaci, I. Lombardo	*Closing Remarks and Acknowledgments*

IOP Conf. Series: Journal of Physics: Conf. Series **863** (2017) 011004 doi:10.1088/1742-6596/863/1/011004

List of participants

Y. Abe
Research Center for Nuclear Physics, Osaka University, Ibaraki, Osaka 567-0047, Japan
abey@rcnp.osaka-u.ac.jp

A. V. Afanasjev
Mississippi State University, P.O. Drawer 5167, Mississippi State, MS 39762
Anatoli.Afanasjev@gmail.com

Md. Ashaduzzaman
Univ. Napoli Federico II & INFN-*Sezione di Napoli, Via Cintia, M.S. Angelo, Edificio 6, 80126 Napoli, Italy*
ashad@na.infn.it

T. Baba
Department of Physics, Hokkaido University, North 10, West 8, Kita-ku, Sapporo 060-0810, Japan
baba@nucl.sci.hokudai.ac.jp

S. Bailey
School of Physics and Astronomy, University of Birmingham, Edgbaston, Birmingham, B15 2TT, UK
s.c.bailey@pgr.bham.ac.uk

D. Baye
Université Libre de Bruxelles, Bâtiment NO, niveau N6, CP229, boulevard du Triomphe, 1050 Bruxelles
dbaye@ulb.ac.be

C. Beck
IPHC & Université de Strasbourg, 23, rue du Loess, Bat. 20(2), BP 28, F-67037 Strasbourg Cedex 2 France
christian.beck@iphc.cnrs.fr

R. Bijker
Instituto de Ciencias Nucleares, UNAM, AP 70-543, 04510 México DF, México
bijker@nucleares.unam.mx

J. Bishop
School of Physics and Astronomy, University of Birmingham, Edgbaston, Birmingham, B15 2TT, UK
j.bishop.2@pgr.bham.ac.uk

B. Borderie
IPN Orsay – IN2P3, 15 rue Georges Clemenceau - 91406 Orsay, France
borderie@ipno.in2p3.fr

A. Bracco
Dipartimento di Fisica, Università di Milano, Via Celoria 16, 20133 Milano, Italy
Angela.Bracco@mi.infn.it

F. Cappuzzello
Dip. Fisica e Astronomia, Univ. di Catania, Via Santa Sofia, 64, 95123, Catania, Italy
cappuzzello@lns.infn.it

D. Carbone
INFN – Laboratori Nazionali del Sud, Via Santa Sofia, 64, 95123, Catania, Italy
carboned@lns.infn.it

N. Carjan
Joint Institute for Nuclear Research, FLNR, 141980 Dubna, Moscow Region, Russia
carjan@in2p3.fr

IOP Conf. Series: Journal of Physics: Conf. Series **863** (2017) 011004 doi:10.1088/1742-6596/863/1/011004

B. V. Carlson
Inst. Tecnológico Aero., P. Mar. E. Gomes, 50, Vila das Acácias, 12228-900, São José dos Campos/SP - Brasil
brett@ita.br

Y. Chiba
Department of Physics, Hokkaido University, North 10, West 8, Kita-ku, Sapporo 060-0810, Japan
chiba@nucl.sci.hokudai.ac.jp

T. V. Chuvilskaya
Skobeltsyn Institute of Nuclear Physics, Lomonosov Moscow State University, 119991, Moscow, Russia
tatchuv@anna19.sinp.msu.ru

M. Cicerchia
Dip. di Fisica G. Galilei, Università degli Studi di Padova, via Marzolo, 8, 35131 Padova
Magda.Cicerchia@lnl.infn.it

S. Courtin
IPHC & Université de Strasbourg, 23, rue du Loess, Bat.20(2), BP 28, F-67037 Strasbourg Cedex 2 France
sandrine.courtin@iphc.cnrs.fr

A. A. Cowley
Dep. Physics, Stellenbosch University Private Bag X1, Matieland, Stellenbosch 7602, South Africa
aac@sun.ac.za

J. Cseh
MTA ATOMKI, *Debrecen, Bem tér 18/c, 4026, Hungary*
cseh@atomki.mta.hu

B. De Canditiis
Univ. Napoli Federico II & INFN-*Sezione di Napoli, Via Cintia, M.S. Angelo, Edificio 6, 80126 Napoli, Italy*
bart.deca@na.infn.it

D. Dell'Aquila
Univ. Napoli Federico II & INFN-*Sezione di Napoli, Via Cintia, M.S. Angelo, Edificio 6, 80126 Napoli, Italy*
dellaquila@na.infn.it

G. De Gregorio
Univ. Napoli Federico II & INFN-*Sezione di Napoli, Via Cintia, M.S. Angelo, Edificio 6, 80126 Napoli, Italy*
degregorio@na.infn.it

P. Descouvemont
Université Libre de Bruxelles, Bâtiment NO, niveau N6, CP229, boulevard du Triomphe, 1050 Bruxelles
pdesc@ulb.ac.be

S. Dmitriev
Joint Institute for Nuclear Research, FLNR JINR, 141980 Dubna, Moscow region, Russia
dmitriev@jinr.ru

J. P. Draayer
Louisiana State University, Dep. Phys. & Astr., 221-E Nicholson Hall, Tower Dr. Baton Rouge, LA 70803-4001, US
draayer@sura.org

L. Francalanza
INFN-*Sezione di Napoli, Via Cintia, M.S. Angelo, Edificio 6, 80126 Napoli, Italy*
francalanza@na.infn.it

M. Freer
School of Physics and Astronomy, University of Birmingham, Edgbaston, Birmingham, B15 2TT, UK
M.Freer@bham.ac.uk

IOP Conf. Series: Journal of Physics: Conf. Series **863** (2017) 011004 doi:10.1088/1742-6596/863/1/011004

M. Fujiwara
Research Center for Nuclear Physics, Osaka University, Suita, Osaka 567-0047, Japan
fujiwara@rcnp.osaka-u.ac.jp

T. Fukui
Nuclear Data Center, Japan Atomic Energy Agency, Tokai, Ibaraki 319-1195, Japan
fukui.tokuro@jaea.go.jp

Y. Funaki
RIKEN 2-1, *Hirosawa, Wako, Saitama 351-0198 Japan*
funaki@riken.jp

T. Furumoto
Graduate School of Education, Yokohama National University, Yokohama 240-8501, Japan
furumoto@ichinoseki.ac.jp

T. Furuno
Dep. of Physics, Kyoto University, Kitashirakawa-Oiwakecho, Sakyo-ku, Kyoto 606-8502, Japan
furuno@scphys.kyoto-u.ac.jp

M. Gai
LNS at Avery Point, University of Connecticut, 1084 Shennecossett Rd., Groton, CT 06340, US
moshe.gai@uconn.edu

B. Gnoffo
INFN - *Sez. Catania, Via Santa Sofia, 64, 95123, Catania, Italy*
brunilde.gnoffo@ct.infn.it

P. R. S. Gomes†
Universidade Federal Fluminense, R. Miguel de Frias, 9 - Icaraí, Niterói - RJ, Brasil
paulogom@if.uff.br

L. Grigorenko
Joint Institute for Nuclear Research, FLNR JINR, 141980 Dubna, Moscow region, Russia
lgrigorenko@yandex.ru

H. Horiuchi
Research Center for Nuclear Physics, Osaka University, Suita, Osaka 567-0047, Japan
horiuchi@rcnp.osaka-u.ac.jp

W. Horiuchi
Department of Physics, Hokkaido University, North 10, West 8, Kita-ku, Sapporo 060-0810, Japan
whoriuchi@nucl.sci.hokudai.ac.jp

D. Hove
Department of Physics and Astronomy, Aarhus University, DK-8000 Aarhus C, Denmark
dennish@phys.au.dk

M. S. Hussein
Univ. de São Paulo, Rua do Matão, Travessa R, 187 - Cidade Universitária, 05508-900 - Sao Paulo, SP - Brasil
hussein@if.usp.br

F. Iachello
Center for Theoretical Physics, Yale University, P.O. Box 208120, New Haven, CT 06520-8120, US
francesco.iachello@yale.edu

R. Imai
Department of Physics, Hokkaido University, North 10, West 8, Kita-ku, Sapporo 060-0810, Japan
yosuke@nucl.sci.hokudai.ac.jp

M. Itkis
Joint Institute for Nuclear Research, FLNR JINR, 141980 Dubna, Moscow region, Russia
itkis@jinr.ru

M. Ito
Department of Pure and Applied Physics, Kansai University, 3-3-35 Yamatecho Suita, Osaka, Japan
itomk@kansai-u.ac.jp

M. Itoh
Cyclotron and Radioisotope Center (CYRIC), Tohoku University, Sendai, Miyagi 980-8578, Japan
itoh@cyric.tohoku.ac.jp

D. Kamanin
Joint Institute for Nuclear Research, FLNR JINR, 141980 Dubna, Moscow region, Russia
kamanin@jinr.ru

Y. Kanada En'yo
Dep. of Physics, Kyoto University, Kitashirakawa-Oiwakecho, Sakyo-ku, Kyoto 606-8502, Japan
yenyo@ruby.scphys.kyoto-u.ac.jp

K. Kato
Department of Physics, Hokkaido University, North 10, West 8, Kita-ku, Sapporo 060-0810, Japan
kato@nucl.sci.hokudai.ac.jp

T. Kawabata
Dep. of Physics, Kyoto University, Kitashirakawa-Oiwakecho, Sakyo-ku, Kyoto 606-8502, Japan
kawabata@scphys.kyoto-u.ac.jp

E. Khan
IPN Orsay – IN2P3, 15 rue Georges Clemenceau - 91406 Orsay, France
khan@ipno.in2p3.fr

M. Kimura
Department of Physics, Hokkaido University, North 10, West 8, Kita-ku, Sapporo 060-0810, Japan
masaaki@nucl.sci.hokudai.ac.jp

O. S. Kirsebom
Department of Physics and Astronomy, Aarhus University, DK-8000 Aarhus C, Denmark
oliver.kirsebom@gmail.com

S. Koyama
Department of Physics, University of Tokyo, 7-3-1 Hongo, Bunkyo, Tokyo 113-0033, Japan
koyama@nucl.phys.s.u-tokyo.ac.jp

K. Kravvaris
Department of Physics, Florida State University, 600 W. College Avenue, Tallahassee, FL 32306, US
kk11f@my.fsu.edu

M. La Commara
Univ. Napoli Federico II & INFN-*Sezione di Napoli, Via Cintia, M.S. Angelo, Edificio 6, 80126 Napoli, Italy*
Marco.LaCommara@na.infn.it

G. La Rana
Univ. Napoli Federico II & INFN-*Sezione di Napoli, Via Cintia, M.S. Angelo, Edificio 6, 80126 Napoli, Italy*
larana@na.infn.it

R. La Torre
Univ. Napoli Federico II & INFN-*Sezione di Napoli, Via Cintia, M.S. Angelo, Edificio 6, 80126 Napoli, Italy*
rlatorre@na.infn.it

R. D. Lasseri
IPN Orsay – IN2P3, 15 rue Georges Clemenceau - 91406 Orsay, France
lasseri@ipno.in2p3.fr

G. Lévai
MTA ATOMKI, *Debrecen, Bem tér 18/c, 4026, Hungary*
levai@atomki.mta.hu

N. Lo Iudice
Univ. Napoli Federico II & INFN-*Sezione di Napoli, Via Cintia, M.S. Angelo, Edificio 6, 80126 Napoli, Italy*
loiudice@na.infn.it

I. Lombardo
Univ. Napoli Federico II & INFN-*Sezione di Napoli, Via Cintia, M.S. Angelo, Edificio 6, 80126 Napoli, Italy*
ivlombardo@na.infn.it

R. G. Lovas
MTA *Institute for Nuclear Research, PO Box 51, H-4001 Debrecen, Hungary*
rgl@atomki.hu

S. V. Lukyanov
Joint Institute for Nuclear Research, FLNR JINR, 141980 Dubna, Moscow region, Russia
lukyan@jinr.ru

S. Manna
Variable Energy Cyclotron Centre, 1/AF Bidhan Nagar, Kolkata - 700064, India
smanna@vecc.gov.in

F. M. Marqués
LPC *Caen* – ENSICAEN, *6, Bd. du Maréchal Juin, 14050 Caen Cedex CEDEX, France*
marques@lpccaen.in2p3.fr

H. Masui
Information Processing Center, Kitami Institute of Technology, Kitami 090-8507, Japan
hgmasui@mail.kitami-it.ac.jp

D. Montanari
IPHC & Université de Strasbourg, 23, rue du Loess, Bat. 20(2), BP 28, F-67037 Strasbourg Cedex 2 France
daniele.montanari@iphc.cnrs.fr

L. Morelli
Dip. Fisica e Astronomia, Univ. Bologna, Via Irnerio, 46, 40126 Bologna, Italy
luca.morelli@bo.infn.it

L.G. Moretto
College of Chemistry, 420 Latimer Hall, University of California, Berkeley, CA 94720-1460, US
lgmoretto@lbl.gov

H. Morita
Dep. of Physics, Kyoto University, Kitashirakawa-Oiwakecho, Sakyo-ku, Kyoto 606-8502, Japan
morita.hiroyuki.47a@st.kyoto-u.ac.jp

M. Murata
Dep. of Physics, Kyoto University, Kitashirakawa-Oiwakecho, Sakyo-ku, Kyoto 606-8502, Japan
m.murata@scphys.kyoto-u.ac.jp

T. Myo
Faculty of Engineering, Osaka Institute of Technology, 5-16-1 Omiya, Asahi-ku, Osaka, 535-8585 Japan
myo@ge.oit.ac.jp

M. Odsuren
Nuclear Research Center, National University of Mongolia, Ulan Bator 14200, Mongolia
odsurenn@gmail.com

R. Ogul
Selçuk University, Department of Physics, 42075 Kampus, Konya, Turkey
rogul@selcuk.edu.tr

T. Otsuka
Department of Physics, University of Tokyo, Hongo, Bunkyo-ku, Tokyo 1130033, Japan
otsuka@phys.s.u-tokyo.ac.jp

M. Papa
INFN – Sezione di Catania, Via Santa Sofia, 64, 95123, Catania, Italy
papa@ct.infn.it

C. Parascandalo
INFN-*Sezione di Napoli, Via Cintia, M.S. Angelo, Edificio 6, 80126 Napoli, Italy*
concetta.parascandalo@na.infn.it

D. Pierroutsakou
INFN-*Sezione di Napoli, Via Cintia, M.S. Angelo, Edificio 6, 80126 Napoli, Italy*
pierroutsakou@na.infn.it

E. C. Pinilla
Universidad Nacional de Colombia, Dep. de Fisica, Carrera 45 N° 26-85, Bogotà D.C., Colombia
ecpinillab@unal.edu.co

D. Poenaru
Horia Hulubei National Institute of Physics and Nuclear Engineering, Strada Reactorului 30, Măgurele, Romania
poenaru@nipne.ro

Yu. V. Pyatkov
Joint Institute for Nuclear Research, FLNR JINR, 141980 Dubna, Moscow region, Russia
yvp_nov@mail.ru

D. Quero
Univ. Napoli Federico II & INFN-*Sezione di Napoli, Via Cintia, M.S. Angelo, Edificio 6, 80126 Napoli, Italy*
quero@na.infn.it

L. Quattrocchi
INFN – Sezione di Catania, Via Santa Sofia, 64, 95123, Catania, Italy
lquattrocchi@unime.ut

P. K. Rath
Univ. Napoli Federico II & INFN-*Sezione di Napoli, Via Cintia, M.S. Angelo, Edificio 6, 80126 Napoli, Italy*
rath@na.infn.it

J. L. Rodrìguez-Sànchez
Dep. de Física de Partículas, Univ. de Santiago de Compostela E-15782 Santiago de Compostela, Spain
joseluisrodriguez88@gmail.com

G. Roepke
Institut für Physik, Universitaet Rostock, Albert-Einstein-Straße 23-24, 18059 Rostock, Germania
gerd.roepke@uni-rostock.de

G. Rogachev
Dep. of Physics and Astronomy, Mitchell Physics Building, 4242 TAMU, College Station, TX 77843, US
rogachev@tamu.edu

M. Romoli
INFN-*Sezione di Napoli, Via Cintia, M.S. Angelo, Edificio 6, 80126 Napoli, Italy*
romoli@na.infn.it

G. Royer
Dep. de Physique, 2 rue de la Houssiniere BP92208, 44322 Nantes Cedex 3, France
royer@subatech.in2p3.fr

P. Russotto
INFN – Sezione di Catania, Via Santa Sofia, 64, 95123, Catania, Italy
russotto@lns.infn.it

V. V. Samarin
Joint Institute for Nuclear Research, FLNR JINR, 141980 Dubna, Moscow region, Russia
samarin@jinr.ru

M. Sambataro
INFN – Sezione di Catania, Via Santa Sofia, 64, 95123, Catania, Italy
michelangelo.sambataro@ct.infn.it

P. Schuck
IPN Orsay – IN2P3, 15 rue Georges Clemenceau - 91406 Orsay, France
schuck@ipno.in2p3.fr

S. Shimoura
Center for Nuclear Study, the University of Tokyo, 2-1 Hirosawa, Wako, Saitama Prefecture 351-0106, Japan
shimoura@cns.s.u-tokyo.ac.jp

R. Smith
School of Physics and Astronomy, University of Birmingham, Edgbaston, Birmingham, B15 2TT, UK
r.smith.3@pgr.bham.ac.uk

G. Spadaccini
Univ. Napoli Federico II & INFN-*Sezione di Napoli, Via Cintia, M.S. Angelo, Edificio 6, 80126 Napoli, Italy*
spadac@na.infn.it

A. Solovyev
Dukhov Research Institute of Automatics (VNIIA), *22, ul. Sushchevskaya, Moscow 127055, Russia*
alexander.solovyev@mail.ru

M. Spieker
Institute for Nuclear Physics, University of Cologne, Zülpicher Strasse 77, 50937 Cologne, Germany
spieker@ikp.uni-koeln.de

T. Suhara
Matsue College of Technology, 14-4 Nishiikumacho, Matsue, Shimane Prefecture 690-0865, Japan
suhara@matsue-ct.ac.jp

Y. Taniguchi
Nihon Inst. of Medical Science, 1276 Shimogawara, Moroyama, Iruma District, Saitama Prefecture 350-0435, Japan
yasutaka@nims.ac.jp

Yu. M. Tchuvil'sky
Skobeltsyn Institute of Nuclear Physics, Lomonosov Moscow State University, 119991, Moscow, Russia
tchuvl@nucl-th.sinp.msu.ru

D. Torresi
School of Physics and Astronomy, University of Birmingham, Edgbaston, Birmingham, B15 2TT, UK
d.torresi@bham.ac.uk

IOP Conf. Series: Journal of Physics: Conf. Series **863** (2017) 011004 doi:10.1088/1742-6596/863/1/011004

A. Trifirò
Dipartimento di Scienze Mat., Inf., Fis., Terra, Univ. Messina, Via Stagno d'Alcontres, 98166, Messina, Italy
atrifiro@unime.it

M. Tsumura
Dep. of Physics, Kyoto University, Kitashirakawa-Oiwakecho, Sakyo-ku, Kyoto 606-8502, Japan
tsumura@scphys.kyoto-u.ac.jp

A. Tumino
Università degli Studi di Enna "Kore" - Cittadella Universitaria - 94100 Enna, Italy
aurora.tumino@unikore.it

S. Typel
GSI Helmholtzzentrum fuer Schwerionenforschung, Planckstraße 1, 64291 Darmstadt, Germany
s.typel@gsi.de

E. Uegaki
Graduate School of Engineering Science, Akita University, Akita 010-8502, Japan
uegaki@phys.akita-u.ac.jp

E. Vardaci
Univ. Napoli Federico II & INFN-*Sezione di Napoli, Via Cintia, M.S. Angelo, Edificio 6, 80126 Napoli, Italy*
vardaci@na.infn.it

G. Verde
INFN – Sezione di Catania, Via Santa Sofia, 64, 95123, Catania, Italy
verde@ipno.in2p3.fr

M. Vigilante
Univ. Napoli Federico II & INFN-*Sezione di Napoli, Via Cintia, M.S. Angelo, Edificio 6, 80126 Napoli, Italy*
vigilant@na.infn.it

W. Von Oertzen
Helmholtz-Zentrum Berlin, Hahn-Meitner Platz, 14109 Berlin, Germany
oertzen@helmholtz-berlin.de

S. Watanabe
RIKEN 2-1, *Hirosawa, Wako, Saitama 351-0198 Japan*
shin.watanabe.vf@riken.jp

H. Yamaguchi
Center for Nuclear Study, the University of Tokyo, 2-1 Hirosawa, Wako, Saitama Prefecture 351-0106, Japan
yamag@cns.s.u-tokyo.ac.jp

Z. H. Yang
RIKEN 2-1, *Hirosawa, Wako, Saitama 351-0198 Japan*
yzh0505@pku.edu.cn

F. S. Zhang
College of Nuclear Science and Technology, Beijing Normal University, 100875 Beijing, China
fszhang@bnu.edu.cn

Peer review statement

All papers published in this volume of *Journal of Physics: Conference Series* have been peer reviewed through processes administered by the proceedings Editors. Reviews were conducted by expert referees to the professional and scientific standards expected of a proceedings journal published by IOP Publishing.

IOP Conf. Series: Journal of Physics: Conf. Series **863** (2017) 012001 doi:10.1088/1742-6596/863/1/012001

Opening address

E Vardaci and I Lombardo on behalf of the organizers
Dipartimento di Fisica, Università di Napoli Federico II and INFN Sezione di Napoli, Italy
E-mail: Emanuele.Vardaci@na.infn.it

Dear Ladies and Gentlemen,

welcome to Napoli and to the *11th International Conference on Clustering Aspects of Nuclear Structure and Dynamics.* We are deeply honored to host this event here in Napoli and to have all of you here. This welcome is jointly forwarded by the Rector of the University of Napoli Prof. G. Manfredi, the Head of the Physics Department Prof. L. Merola, the Director of the Napoli Division of the INFN Prof. G. La Rana, and the head of the Geology Department Prof. P. Cappelletti who is also in charge of preserving this magnificent monumental complex.

With the help of the International Advisory Committee we have prepared a very intense and ambitious scientific program. With the support of the Local Organizing Committee, we have decided to organize this event in the heart of the historical center of the town of Napoli. We have allocated accommodations around the historical center to make it possible to all the participants to arrive by foot to the conference site and in the meanwhile to breathe the emotions, the colors and the tastes of this amazing town.

We would like to express our gratitude to the Rector Prof. Gaetano Manfredi for having enthusiastically granted us the use of this historical complex to host this international event. We would like to thank the people in charge responsible of this complex Dott. C. Montola, dott.ssa M. Prece, and the security service headed by sig. A. Savarese. Furthermore, we would like to express our gratitude to the Questore di Napoli (superintendent) dott. A. De Iesu and the Comandante Provinciale Carabinieri di Napoli Gen. B. Antonio de Vita for having organized a capillary security service to watch over all of us.

As you may have already seen, the social program is located in several places around the conference site, but still in the historical center. All of them are reachable by foot. We have allocated time intervals between events long enough to allow everyone to walk comfortably and arrive to the specific event on time.

One last word is about the two logos. The short title of the conference, CLUSTER16, takes advantage of the Gulf of Napoli to represent the initial. The splendid sea overlooked by the Vesuvio vulcano was an irresistible temptation. The second logo depicts a very common scene in one of the typical small streets of this dense town, called *vicoli*, where kids play soccer having not much room anywhere else (as you well know, a sport that carries the full passion of Neapolitans). The scene, common for Neapolitans, of few kids grabbing the ball to rip it off to the others reminded us the idea of how cluster particles can stay together: clusters all want the same object. For instance, it can be a neutron, or another cluster itself, just like kids. Indeed also the morphology of Napoli itself recalls the idea of clusterized nuclei. We may say that Napoli is a strongly clusterized town.

Clusters in Napoli

Last but not least, we would like to thank the StudioCongress Agency,and in particular Nicoletta de Pertis and Antonella Catocci, for having preciously helped us in the organization of this event. Furthermore, we are grateful to the sponsors Fondazione Ordine degli Ingegneri della Provincia di Napoli and CAEN s.p.a., Viareggio (Italy).

We now thank all of you once more for being here. We declare open the Conference. Let's start with the first session.

Emanuele Vardaci and Ivano Lombardo

IOP Conf. Series: Journal of Physics: Conf. Series **863** (2017) 012002 doi:10.1088/1742-6596/863/1/012002

Recent Experimental Results on Nuclear Cluster Physics

C Beck

Département de Recherches Subatomiques, Institut Pluridisciplinaire Hubert Curien, IN_2P_3-CNRS and Université de Strasbourg - 23, rue du Loess BP 28, F-67037 Strasbourg Cedex 2, France

E-mail: `christian.beck@iphc.cnrs.fr`

Abstract. Knowledge on nulear cluster physics has increased considerably since the pioneering discovery of $^{12}C+^{12}C$ resonances half a century ago and nuclear clustering remains one of the most fruitful domains of nuclear physics, facing some of the greatest challenges and opportunities in the years ahead. The occurrence of "exotic" shapes and/or Bose-Einstein α condensates in light $N=Z$ α-conjugate nuclei is investigated. Evolution of clustering from stability to the drip-lines examined with clustering aspects persisting in light neutron-rich nuclei is consistent with the extension of the "Ikeda-diagram" to non α-conjugate nuclei.

1. Introduction

Among the greatest challenges in nuclear science was the understanding of the clustered structure of nuclei from both the experimental and theoretical perspectives [1, 2, 3, 4, 5, 6, 7, 8]. Progress in physics of nuclear molecules and nuclear clustering were facing some of the greatest challenges and opportunities in the years ahead. Besides the well known series of Cluster conferences [9, 10, 11], a series of workshops on the State-Of-The-Art in Nuclear Cluster Physics was started citeStrasbourg,Brussels,Yokohama. The first one was held in Strasbourg in 2008 [12], the second one in Brussels in 2010 [13] and the last one in Yokohama in 2014 [14]. Fig. 1 (taken from the cover of Ref. [14]) summarizes the different types of clustering that were discussed during the last two or three decades [8, 15]. Most of these exotic structures were investigated in an experimental context by using either some new approaches or developments of older methods [16].

Starting in the 1960s the search for resonant structures in the excitation functions for various combinations of light α-conjugate ($N=Z$) nuclei in the energy regime from the Coulomb barrier up to regions with high excitation energies (E_x=20$-$50 MeV) remains a subject of contemporary debate [1, 8]. These resonances were interpreted in terms of nuclear molecules [1]. The question of how quasimolecular resonances may reflect continuous transitions from scattering states in the ion-ion potential to true cluster states in the compound systems was still unresolved in the 1990s [1]. In many cases, these resonant structures were associated with strongly-deformed shapes and with α-clustering phenomena [6, 8, 17, 18], predicted from the Nilsson-Strutinsky approach, the cranked α-cluster model [6, 17], or other mean-field calculations [18, 19]. In light α-like nuclei clustering is observed as a general phenomenon not only at high excitation energies close to the α-decay thresholds [6, 8, 17, 20] but may also play a key role at low excitation

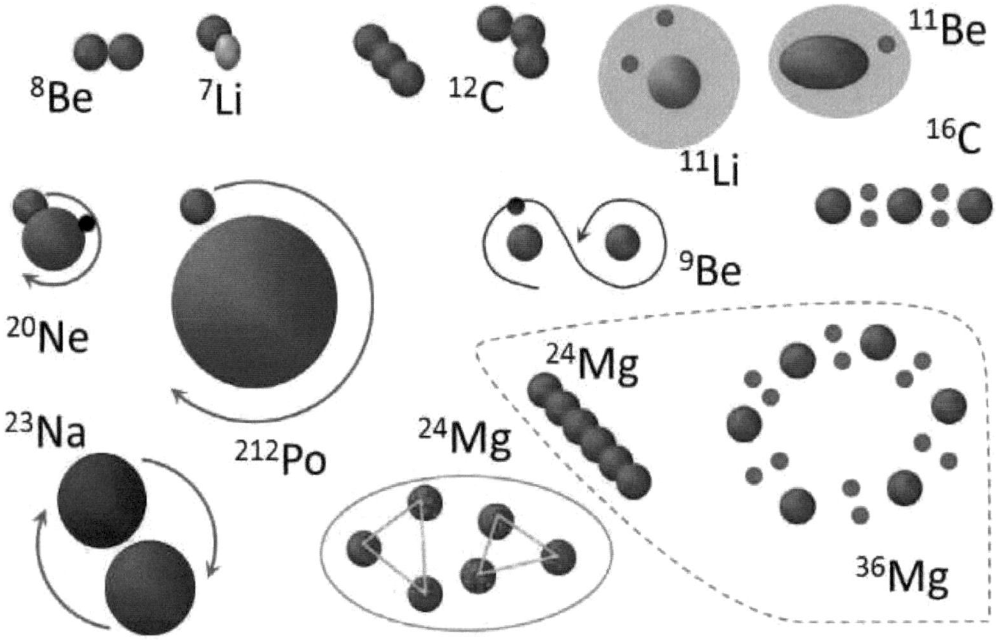

Figure 1. Different types of nuclear clusters discussed in recent years [8, 15, 36, 37].

energies and for the ground states. This exotic behavior was perfectly illustrated by the famous "Ikeda-diagram" for N=Z nuclei in 1968 [21], which was modified and recently extended by von Oertzen [22, 23] for neutron-rich nuclei, as shown in the right panel of Fig. 2. Despite the early inception of cluster studies, it is only recently that radioactive ion beams experiments, with great help from advanced theoretical works, enabled new generation of studies, in which data with variable excess neutron numbers or decay thresholds are compared to predictions with least or no assumptions of cluster cores. For instance, new experimental approaches for the tetraneutron system at RIKEN [24] challenge the most advanced theoretical models [25, 26]. On the other hand, predicted but elusive phenomena, such as molecular orbitals or linear chain structures, are now gradually coming to light from experimental data [6, 8].

Clustering is a general feature [6, 8, 23] not only observed in light neutron-rich nuclei [27], but also in halo nuclei such as ^{11}Li [28] or ^{14}Be [29], for instance. The problem of cluster formation has also been treated extensively for very heavy systems by R.G. Gupta [19], by D. Poenaru, V. Zagrebaev and W. Greiner [30, 31] and by C. Simenel [32] suggesting that giant molecules and collinear ternary fission may co-exist [33]. Finally, signatures of α-clustering have also been predicted and/or discovered in light nuclei surviving from intermediate-energy [34] to ultrarelativistic-energy [35] nuclear collisions. The topic of nuclear clusters benefits of intense theoretical activity [7] where new experimental information has come to light very recently [6, 8]. Several status reports were given in conferences and their written contributions can be found in their respective proceedings [15, 36, 37].

2. Renewed interest in the spectroscopy of light α-like nuclei: α condensates ?

The renewed interest in the ^{12}C nucleus [38] was mainly focused to a better understanding of the nature of the so called "Hoyle" state [39] that can be described in terms of a bosonic

condensate, a cluster state and/or a α-particle gas [40, 41, 42]. Much experimental progress has been achieved recently as far as the spectroscopy of ^{12}C near and above the α-decay threshold is concerned [38, 43, 44]. More particularly, the 2_2^+ "Hoyle" rotational excitation in ^{12}C has been observed [43]. The ^{12}C(α,α)^{12}C* reaction [44] populates a new state compatible with an equilateral triangle configuration of three α particles. Still, the structure of the "Hoyle" state remains controversial as experimental results of its direct decay into three α particles are found to be in disagreement [38, 45].
In the study of Bose-Einstein Condensation (BEC), the α-particle states in light N=Z nuclei [40, 41, 42], are of great importance. At present, the search for an experimental signature of BEC in ^{16}O is of highest priority. A state with the structure of the "Hoyle" state [39] in ^{12}C coupled to an α particle is predicted in ^{16}O at about 15.1 MeV (the 0_6^+ state), the energy of which is $\approx$ 700 keV above the 4α-particle breakup threshold [46]. However, any state in ^{16}O equivalent to the "Hoyle" state [39] in ^{12}C is most certainly going to decay exclusively by particle emission with very small γ-decay branches, thus, very efficient particle-γ coincidence techniques will have to be used in the near future to search for them. BEC states are expected to decay by α emission to the "Hoyle" state and could be found among the resonances in α-particle inelastic scattering on ^{12}C decaying to that state. In 1967, Chevallier et al. [47] could excite these states in an α-particle transfer channel leading to the ^{8}Be–^{8}Be final state and proposed that a structure corresponding to a rigidly rotating linear arrangement of four alpha particles may exist in ^{16}O. A more sophisticated experimental setup was used recently [48]: although the excitation function is generally in good agreement with the previous results [47] a phase shift analysis of the angular distributions does not provide evidence to support the reported hypothesis of a 4α-chain state configuration. Tetrahedral symmetries are predicted to occur in ^{16}O [49]. Experimental investigations are still underway to understand the nuclear structure of high-spin states of ^{16}O, ^{20}Ne and heavier nuclei [50, 51]. The search for exotic chain-like structures remains an exciting prospect but, up to now, tentative evidence of chain states have been unsubstantiated and the view is that such structure have not yet been definitively observed.

3. Alpha clustering, deformations and α condensates in heavier nuclei

The relashionship between α-clustering, nuclear molecules and superdeformation (SD) [8, 18, 52, 53] is of particular interest, since nuclear shapes with major-to-minor axis ratios of 2:1 are typical ellipsoidal elongations for light nuclei (corresponding to a quadrupole deformation parameter $\beta_2 \approx 0.6$). Furthermore, the structure of possible octupole-unstable 3:1 nuclear shapes (hyperdeformation (HD) with $\beta_2 \approx 1.0$) has also been discussed for actinide nuclei in terms of clustering phenomena [8]. Typical examples for the link between quasimolecular bands and extremely deformed (SD/HD) shapes can be found in the literature for $A = 20 - 60$ α-conjugate N=Z nuclei [8, 18, 36, 52, 54, 55, 56, 57, 58, 59, 60, 61, 62, 63, 64, 65, 66, 67]. Highly deformed shapes and SD rotational bands have been discovered in several light α-like nuclei, such as ^{36}Ar and ^{40}Ca by using γ-ray spectroscopy techniques [57, 59]. In particular, the extremely deformed rotational SD band in ^{36}Ar [57] is comparable in shape to the quasimolecular bands observed in both ^{12}C+^{24}Mg and ^{16}O+^{20}Ne reactions [8, 58, 64]. Ternary clusterizations are also predicted theoretically, but were not found experimentally in ^{36}Ar so far [8, 58]. On the other hand, ternary fission of ^{56}Ni, related to its HD shapes, was identified from out-of-plane angular correlations measured in the ^{32}S+^{24}Mg reaction with the Binary Reaction Spectrometer at the VIVITRON Tandem facility [67]. This possibility [8, 67] is not limited to light N=Z compound nuclei, true ternary fission [31, 33, 68] can also occur for very heavy [33, 68] and superheavy [69] nuclei. The next natural question to be addressed is whether dilute-gas-like BEC structures [40, 41, 42] also exist in medium-mass α-conjugate nuclei as predicted by several theoretical investigations [70, 71, 72]. Several recent undergoing experiments indicate that it might be the case at least for ^{24}Mg [73], and much work is in progress in this field [8, 41].

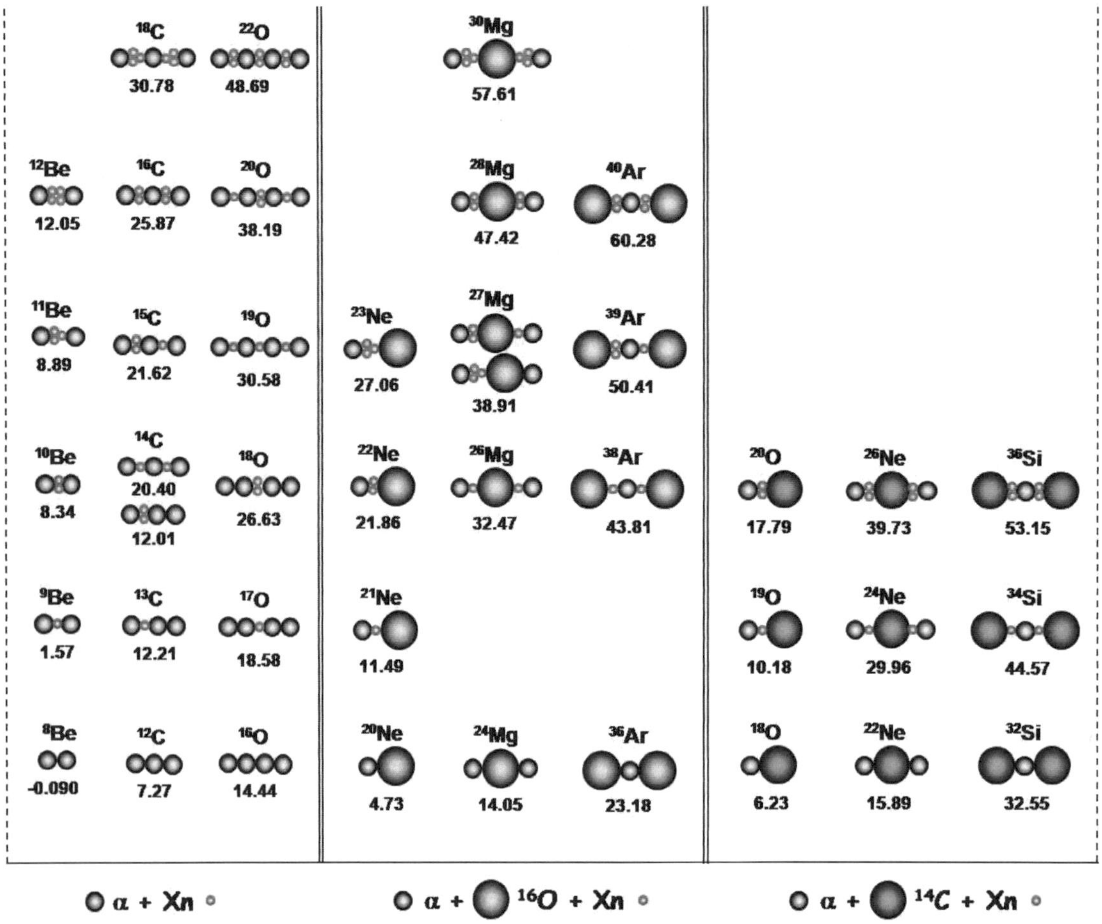

Figure 2. Schematic illustration of the structures of molecular shape isomers in light neutron-rich isotopes of nuclei consisting of α-particles, ^{16}O- and ^{14}C-clusters plus some covalently bound neutrons (Xn means X neutrons). The so called "Extended Ikeda-Diagram" [22] with α-particles (left panel) and ^{16}O-cores (middle panel) can be generalized to ^{14}C-cluster cores (right panel). The lowest line of each configuration corresponds to parts of the original "Ikeda-Diagram" [21].

4. Clustering in light neutron-rich nuclei

Clustering is a general phenomenon [8] observed also in nuclei with extra neutrons as illustrated in Fig. 2 by the "Ikeda-Diagram" [21] modified and extended by von Oertzen [22]. With additional neutrons, specific molecular structures appear with binding effects based on covalent molecular neutron orbitals. In these diagrams α-clusters and ^{16}O-clusters are the main ingredients. Actually, the ^{14}C nucleus may play a similar role in clusterisation as the ^{16}O nucleus does. Both of them have similar properties as a cluster: i) closed neutron p-shells, ii) first excited states well above $E^* = 6$ MeV, and iii) high binding energies for α-particles. A general picture of clustering and molecular configurations in light nuclei can be drawn from investigation of the oxygen isotopes with $A \geq 17$. Recent results on the even-even oxygen isotopes i.e. ^{18}O [74, 75] and ^{20}O [23] as well as very striking cluster states found in odd-even oxygen isotopes such as ^{17}O [76] and ^{19}O [23] have been obtained. Therefore, the "Ikeda-Diagram" [21] and the "Extended Ikeda-Diagram" consisting of ^{16}O cluster cores with covalently bound neutrons [22] is being further revised to include also the ^{14}C cluster cores as illustrated in Fig. 2 (right panel).

5. Concluding remarks

Marked progress is being accomplished in many traditional or novel subjects of nuclear cluster physics. High-precision spectroscopy techniques enable us to uncover important parts of the complete spectroscopy of the "Hoyle" state in ^{12}C. Thus, the origin of carbon for life is likely to be understood in the very near future with answer to the question of the "Hoyle" state structure. The connection of α-clustering and quasimolecular resonances with α condensates in very light nuclei and with extreme deformations (SD, HD, ...) in heavier nuclei as investigated by more and more sophisticated experimental devices is discussed in this introductory talk. Some neutron-rich nuclei displaying very well defined quasimolecular bands in agreement with theoretical predictions justify the "Extended Ikeda-Diagram" to be generalized by the inclusion of the ^{14}C cluster as a core, similarly to the ^{16}O one. The developments in these selected subjects of nuclear clusters physics show the importance of clustering among the basic modes of motion of nuclear many-body systems.

Acknowledgments

This article is dedicated to the memory of my friends Paulo Gomes, Alex Szanto de Toledo and Valery Zagrebaev. I would like to thank the CLUSTERS '16 organizers for giving me the opportunity to introduce this conference. Christian Caron (Springer) is acknowledged for initiating the series of three volumes of *Lecture Notes in Physics* entitled "Clusters in Nuclei".

References

[1] Greiner W, Park Y Jae, and Scheid W 1995 *Nuclear Molecules* (Singapore: World Scientific)
[2] Beck C 2010 *Clusters in Nuclei, Vol. 1*, ed. Springer Verlag Berlin Heidelberg ; *Lecture Notes in Physics* **818**
[3] Beck C 2012 *Clusters in Nuclei, Vol. 2*, ed. Springer Verlag Berlin Heidelberg; *Lecture Notes in Physics* **848**
[4] Beck C 2014 *Clusters in Nuclei, Vol. 3*, ed. Springer Verlag Berlin Heidelberg; *Lecture Notes in Physics* **875**
[5] Delion D S 2010 *Theory of Cluster Emission* (Berlin Heidelberg: Springer); *Lecture Notes in Physics* **819**
[6] Freer M 2014 *Clustering in Light Nuclei* (Berlin Heidelberg: Springer) *Lecture Notes in Physics* **879**
[7] Funaki Y, Horiuchi H, and Tohsaki A 2015 *Prog. Part. Nucl. Phys.* **82** 78
[8] Beck C 2017 *International Journal of Modern Physics* E Special-Topics Issue on *Nuclear Particle Correlations and Cluster Physics*; 2016 arXiv:**1604.02281**
[9] Proceedings of the 8th International Conference on Clustering Aspects of Nuclear Structure and Dynamics (CLUSTERS '03) 24-29 November 2003, Nara, Japan, ed. Ikeda K, Tanihata I and Horiuchi H 2004 *Nucl. Phys.* A **738**
[10] Proceedings of the 9th International Conference on Clustering Aspects of Nuclear Structure and Dynamics (CLUSTERS '07) 3-7 September 2007, Stratford upon Avon, UK, ed. Freer M 2008 *J. Phys.: Conference Series* **111**
[11] Proceedings of the 10th International Conference on Clustering Aspects of Nuclear Structure and Dynamics (CLUSTERS '12) 24-28 September 2012, Debrecen, Hungary, ed. Lovas R, Dombradi Z, Kruppa A, and Levai G 2013 *J. Phys.: Conference Series* **436**
[12] Proceedings of the Workshop on the State of the Art in Nuclear Cluster Physics (SOTANCP 2008) 13-16 May 2008, Strasbourg, France, ed. Beck C, Dufour M, and Schuck P 2008 Special issue of *International Journal of Modern Physics* E **17**
[13] Proceedings of the International Workshop on State of the Art in Nuclear Cluster Physics (SOTANCP 2010) 25-28 May 2010, Brussels, Belgium, ed. Descouvemont P, Dufour M, and Sparenberg J M 2010 Special issue of *International Journal of Modern Physics* E **20**
[14] Proceedings of the International Workshop on State of the Art in Nuclear Cluster Physics (SOTANCP 2014) 26-30 May 2014, Yokohama, Japan, ed. Yamada T and Kanada-En'yo Y 2014 *J. Phys.: Conference Series* **569** (2014)
[15] Beck C 2015 *J. Phys.: Conference Series* **569** 012002
[16] Papka P and Beck C 2012 Clusters in Nuclei - Vol.2 ed. Beck C, *Lecture Notes in Physics* **848** 299
[17] Freer M 2007 *Rep. Prog. Phys.* **70** 2149
[18] Horiuchi H 2010 Clusters in Nuclei - Vol.1, ed. Beck C, *Lecture Notes in Physics* **818** 57
[19] Gupta R K 2010 Clusters in Nuclei - Vol.1, ed. Beck C, *Lecture Notes in Physics* **818** 232
[20] von Oertzen W, Freer M and Kanada-En'yo Y 2006 *Phys. Rep.* **432** 43
[21] Horiuchi H and Ikeda K 1968 *Prog. Theor. Phys.* **40** 277

[22] von Oerzten W 2001 *Eur. Phys. J.* A **11** 403
[23] von Oerzten W and Milin M, 2014 Clusters in Nuclei - Vol.3, ed. Beck C, *Lecture Notes in Physics* **875** 147
[24] Kisamori K *et al* 2016 *Phys. Rev. Lett.* **116** 052501; Orr N 2016 *Physics* **9** 14
[25] Hiyama E *et al* 2016 *Phys. Rev.* C **93** 044004
[26] Bertulani C and Zelevinsky V 2016 *Nature* **532** 448
[27] Kanada-En'yo Y *et al* 2010 Clusters in Nuclei - Vol.1, ed. Beck C, *Lecture Notes in Physics* **818** 129
[28] Ikeda K *et al.* 2010 Clusters in Nuclei - Vol.1, ed. Beck C, *Lecture Notes in Physics* **818** 165
[29] Nakamura N and Kondo Y 2012 Clusters in Nuclei - Vol.2, ed. Beck C, *Lecture Notes in Physics* **848** 67
[30] Poenaru D N and Greiner W 2010 Clusters in Nuclei -Vol.1, ed. Beck C, *Lecture Notes in Physics* **818** 1
[31] Zagrebaev V and Greiner W 2010 Clusters in Nuclei - Vol.1, ed. Beck C, *Lecture Notes in Physics* **818** 267
[32] Simenel C 2014 Clusters in Nuclei - Vol.3, ed. Beck C, *Lecture Notes in Physics* **875** 95
[33] Kamanin D V and Pyatkov Y 2014 Clusters in Nuclei - Vol.3, ed. Beck C, *Lecture Notes in Physics* **875** 183
[34] Borderie B *et al* 2016 *Phys. Lett.* B **755** 475
[35] Zarubin P L 2014 Clusters in Nuclei - Vol.3, ed. Beck C, *Lecture Notes in Physics* **875** 51
[36] Beck C 2014 Proceedings of the First International African Symposium on Exotic Nuclei (IASEN '13) 2-6 December 2013, Cape Town, South Africa, in "Exotic Nuclei", ed. E. Cherepanov, Y. Penionzkhevich, R. Bark and J. Cornell (World Scientific) p.165; arXiv:**1402.3758**
[37] Beck C 2016 Proceedings of the First International Symposium on "New Horizons in Fundamental Physics - From Neutrons Nuclei via Superheavy Elements and Supercritical Fields to Neutron Stars and Cosmic Rays, Makutsi Safari Farm, South Africa, November 23-29, 2015; arXiv:**1603.02574**
[38] Freer M and Fynbo H O U 2014 *Prog. Part. Nucl. Phys.* **71** 1
[39] Hoyle F 1954 *Astrophys. J. Suppl. Ser.* **1** 121
[40] Tohsaki A *et al* 2001 *Phys. Rev. Lett.* **87** 192501
[41] von Oertzen W 2010 Clusters in Nuclei - Vol.1, ed. Beck C, *Lecture Notes in Physics* **818** 102
[42] Yamada T *et al* 2012 Clusters in Nuclei - Vol.2, ed. Beck C, *Lecture Notes in Physics* **848** 229
[43] Zimmerman W R *et al* 2013 *Phys. Rev. Lett.* **110** 152502
[44] Marin-Lambarri D J *et al* 2014 *Phys. Rev. Lett.* **113** 012502
[45] Itoh *et al* 2014 *Phys. Rev. Lett.* **113** 102501
[46] Funaki Y *et al* 2008 *Phys. Rev. Lett.* **101** 082502
[47] Chevallier P *et al* 1967 *Phys. Rev.* **160** 827
[48] Curtis N *et al* 2013 *Phys. Rev.* C **88** 064309
[49] Robson D 1979 *Phys. Rev. Lett.* **42** 876; Bijker R and Iachello F 2014 *Phys. Rev. Lett.* **112** 152501
[50] Kokalova Tz *et al* 2013 *Phys. Rev.* C **87** 057309
[51] Swartz J A *et al* 2015 *Phys. Rev.* C **91** 034317
[52] Beck C 2004 *Nucl. Phys.* A **738** 24; *Int. J. Mod. Phys.* E **13** 9; 2011 *Acta Physica Polonica* B **42** 747
[53] Ray D and Afanasjev A V 2016 *Phys. Rev.* C **94** 014310
[54] Jenkins D G 2013 Clusters in Nuclei - Vol.3, ed. Beck C, *Lecture Notes in Physics* **875** 25
[55] Jenkins D G and Courtin S 2015 *J. Phys. G: Nucl. Part. Phys.* **42** 034010
[56] Pandit D *et al* 2010 *Phys. Rev.* C **81** 061302; 2005 *Phys. Rev.* C **72** 021601
[57] Svensson C E *et al* 2000 *Phys. Rev. Lett.* **85** 2693
[58] Beck C *et al* 2009 *Phys. Rev.* C **80** 034604; 2013 *J. Phys.: Conferences Series* **436** 012014
[59] Ideguchi E *et al* 2001 *Phys. Rev. Lett.* **87** 222501
[60] Rousseau M *et al* 2002 *Phys. Rev.* C **66** 034612
[61] Salsac M D *et al* 2008 *Nucl. Phys.* A **801** 1
[62] Di Nitto A *et al* 2016 *Phys. Rev.* C **93** 044602
[63] Nouicer R *et al* 1999 *Phys. Rev.* C **60** 041303(R)
[64] Sciani W *et al* 2009 *Phys. Rev.* C **80** 034319; Cseh J *et al* 2009 *Phys. Rev.* C **80** 034320
[65] Beck C *et al* 2001 *Phys. Rev.* C **63** 014607
[66] Bhattacharya C *et al* 2002 *Phys. Rev.* C **65** 014611
[67] von Oertzen W *et al* 2008 *Eur. Phys. J.* A **36** 279
[68] Pyatkov Y F *et al* 2010 *Eur. Phys. J.* A **45** 29
[69] Zagrebaev V I *et al* 2010 *Phys. Rev.* C **81** 044608
[70] Yamada T and Schuck P 2004 *Phys. Rev.* C **69** 024309
[71] Ebran J P *et al* 2012 *Nature* **487** 341
[72] Girod M and Schuck P 2013 *Phys. Rev. Lett.* **111** 132503
[73] Morelli L *et al* 2014 *J. Phys.: (London)* G **41** 075107; 2016 *J. Phys. (London)* G **43** 045110
[74] von Oertzen W *et al* 2010 *Eur. Phys. J.* A **43** 17
[75] Avila M *et al* 2014 *Phys. Rev.* C **90** 024327
[76] Milin M *et al* 2009 *Eur. Phys. J.* A **41** 335

IOP Conf. Series: Journal of Physics: Conf. Series **863** (2017) 012003 doi:10.1088/1742-6596/863/1/012003

Recent results on the cluster structure of light nuclei

F Iachello
Center for Theoretical Physics, Sloane Laboratory,
Yale University, New Haven, Connecticut 06520-8120, USA

E-mail: **francesco.iachello@yale.edu**

Abstract. A recently developed model (ACM) is introduced and applied to the study of kα structures with k=2 (^{8}Be), k=3 (^{12}C) and k=4 (^{16}O). Evidence for Z_2 (k=2), D_3 (k=3) and T_d (k=4) symmetry is presented. An extension (ACFM) of the model to kα+x(neutrons, protons) structures is briefly mentioned and applied to the study of ^{9}Be (k=2,x=1).

1. Introduction

The cluster structure of light nuclei has been the subject of many investigations since the seminal work of Wheeler [1]. Further work of Dennison [2] and others, culminated in the study of Brink [3] who suggested specific geometric configurations for kα nuclei (k=2-7) based on 'microscopic' calculations, the so-called Brink-Bloch model. Since then, many theoretical approaches have been introduced and used to study cluster properties of light nuclei. Recently, renewed interest in clustering has arisen from new experiments, especially in ^{12}C, where additional states, predicted by cluster models have been observed [4]. A review of recent experimental results is given in These Proceedings [5].

Open questions in cluster physics are: (i) what are the signatures of specific geometric configurations, (ii) are these configurations rigid or soft, (iii) how far clustering extends in excitation energy, and (iv) how far clustering extends in mass number. In order to answer these questions, especially the first two, we have recently developed an algebraic approach to clustering based on the algebraic theory of molecules introduced in 1981 [6]. For kα structures, the approach amounts to a bosonic quantization of the Jacobi variables in terms of the algebra of U(3k-2), where k is the number of constituents. Some results for k=2-4, will be reported in this contribution.

2. Algebraic cluster model (ACM)

This model describes cluster configurations of any type in terms of the Lie algebra U(3k-2). An explicit construction of the algebra has been completed for k=2,3,4 and is summarized in Table 1.

k	Nucleus	U(3k-2)	Jacobi variables	Discrete symmetry
2α	^{8}Be	U(4) [6,7]	$\boldsymbol{\rho}$	Z_2
3α	^{12}C	U(7) [8,9]	$\boldsymbol{\rho}, \boldsymbol{\lambda}$	D_3
4α	^{16}O	U(10) [10,11,12]	$\boldsymbol{\rho}, \boldsymbol{\lambda}, \boldsymbol{\eta}$	T_d

Table 1. Algebraic description of kα structures

A major advantage of ACM is that it is possible to derive analytical results that provide benchmarks for experiments.

2.1. *Energy formulas*. For rigid structures, explicit expressions for energy levels in terms of the angular momentum L and the vibrational quantum numbers v_i (i=1,2,...), in a semi-classical approximation to ACM, are:

$$E(v,L) = E_0 + \omega\left(v + \frac{1}{2}\right) + BL(L+1)$$

$$E(v_1,v_2,L,K) = E_0 + \omega_1\left(v_1 + \frac{1}{2}\right) + \omega_2(v_2+1) + BL(L+1) + (C-B)\left(K \mp 2\ell_2\right)^2 \quad (1)$$

$$E(v_1,v_2,v_3,L) = E_0 + \omega_1\left(v_1 + \frac{1}{2}\right) + \omega_2(v_2+1) + \omega_3\left(v_3 + \frac{3}{2}\right) + BL(L+1)$$

for 2α (Z_2), 3α (D_3) and 4α (T_d) respectively, where the ω's, B and C are parameters.
Spectra are characterized by representations of the discrete group G and consist in a set of rotation-vibration bands, with specific values of the angular momentum and parity. Representations can be labeled either by G or by the isomorphic group S_n, the permutation group. The conversion from G to S_n is: G=Z_2 ~ S_2~ P, A≡[2]; G=D_3 ~S_3 , A≡[3], E≡ [21]; G=T_d ~ S_4 , A≡ [4], F≡ [31], E≡ [22].
For 2α with Z_2 symmetry, the spectrum consists of a set of rotation-vibration bands with A symmetry and $L^P = 0^+, 2^+, 4^+, \ldots$
For 3α with D_3 symmetry, the spectrum consists of a set of rotation-vibration bands with A and E symmetry as shown in Fig.1. For A-representations, the rotational band has $L^P = 0^+, 2^+, 3^-, 4^\pm, \ldots$, while for E-representations the rotational band has $L^P = 1^-, 2^\pm, 3^\pm, \ldots$. Note the unusual angular momentum and parity content of the rotational bands and the *parity doubling*.

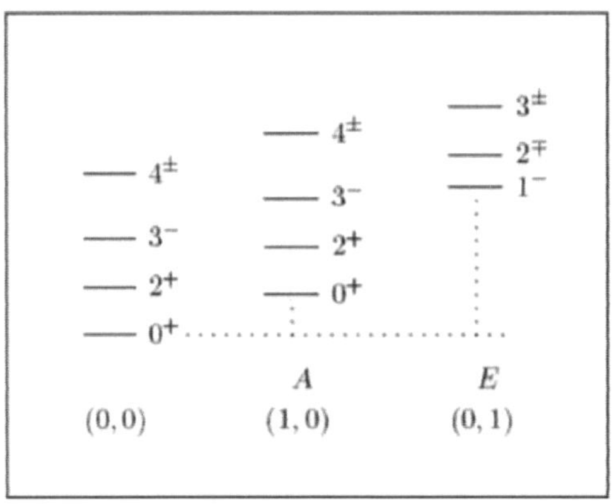

Figure 1. Allowed states of 3α with D_3 symmetry

For 4α with T_d symmetry the spectrum consists of a set of rotation-vibration bands with A, E and F symmetry as shown in Fig.2. For A-representations, the rotational band has $L^P = 0^+, 3^-, 4^+, 6^\pm, \ldots$, while for E-representations it has $L^P = 2^\pm, 4^\pm, 5^\pm, \ldots$ and for F-representations $L^P = 1^-, 2^+, 3^\pm, \ldots$. Note again the unusual angular momentum and parity content and the *parity doubling*.

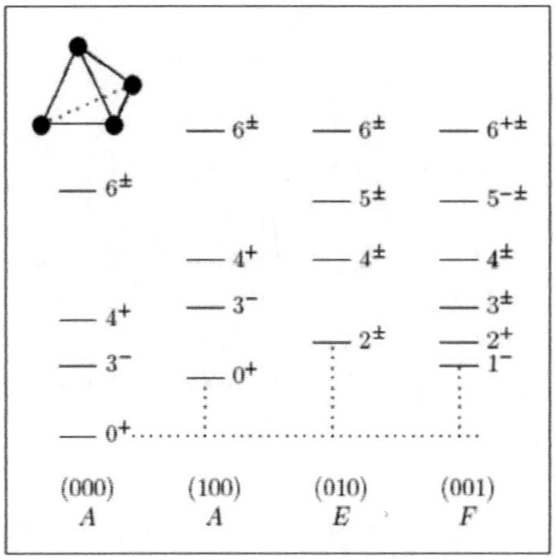

Figure 2. Allowed states of 4α with T_d symmetry

The occurrence of D_3 symmetry in ^{12}C has been confirmed by recent experiments [4, 13] and is reported in These Proceedings [14]. The occurrence of T_d symmetry in ^{16}O was discussed long ago by Robson [15], and it has been emphasized recently in [12, 16]. The evidence is shown in Fig.3.

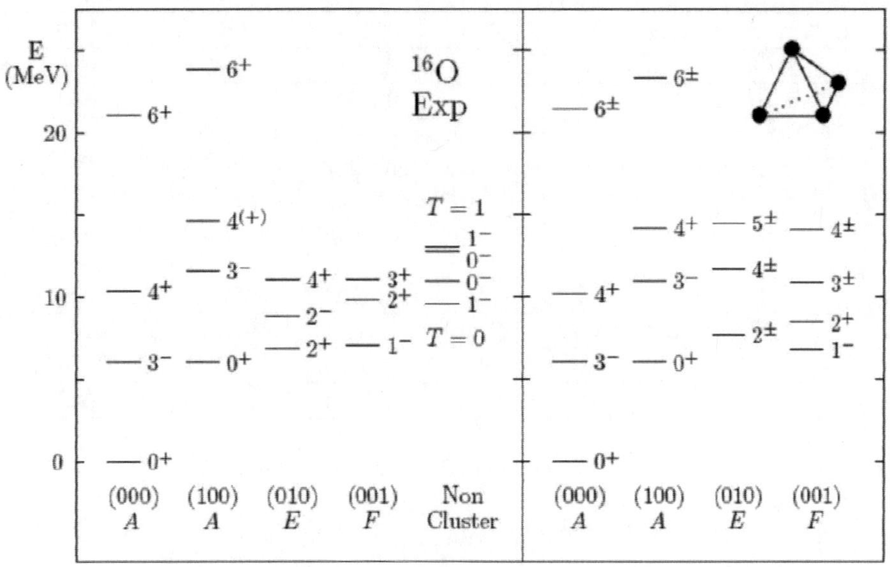

Figure 3. Comparison between theory and experiment in ^{16}O

2.2. *Electromagnetic transition rates.* In the rigid limit, explicit expressions for B(EL) values along the ground state band can be derived. They are:

$$B(EL;0\rightarrow L)=\left(\frac{Ze\beta^{L}}{2}\right)^{2}\frac{2L+1}{4\pi}\left[2+2P_{L}(-1)\right]$$

$$B(EL;0\rightarrow L)=\left(\frac{Ze\beta^{L}}{3}\right)^{2}\frac{2L+1}{4\pi}\left[3+6P_{L}(-\frac{1}{2})\right] \qquad (2)$$

$$B(EL;0\rightarrow L)=\left(\frac{Ze\beta^{L}}{4}\right)^{2}\frac{2L+1}{4}\left[4+12P_{L}(-\frac{1}{3})\right]$$

for 2α , 3α and 4α respectively. Here β is the displacement of each α particle from the center of mass and Z is the total charge. These expressions are compared with experiment in Tables 2, 3 and 4 for ^{8}Be (k=2), ^{12}C (k=3) and ^{16}O (k=4). The measured B(EL) values provide strong evidence for close-packed clustering, Z_2, D_3, T_d, in the ground state bands of ^{8}Be, ^{12}C, ^{16}O.

$B(EL;L^P\rightarrow 0^+)$	Th	Exp	$E(L^P)$	Th	Exp
$B(E2;2\rightarrow 0)$	20.4	21±2.3 (23.0(2.5) W.u.)	$E(2^+)$	3060	3030
$B(E4;4\rightarrow 0)$	326.1		$E(4^+)$	10200	11350

Table 2. B(EL) values and energies in ^{8}Be compared to those expected from Z_2 symmetry. B(EL) values in e^2fm^{2L} and E in keV. The theoretical energies are calculated using E(keV)=510L(L+1). The experimental value for B(E2; $2^+\rightarrow 0^+$) is estimated from radiative capture [17]

$B(EL;L^P\rightarrow 0^+)$	Th	Exp	$E(L^P)$	Th	Exp
$B(E2;2^+\rightarrow 0^+)$	9.3	7.6±0.4 (4.65(26) W.u.)	$E(2^+)$	4400	4439
$B(E3;3^-\rightarrow 0^+)$	84	103±17 (12(2) W.u.)	$E(3^-)$	9640	9641
$B(E4;4^+\rightarrow 0^+)$	68		$E(4^+)$	14670	14080

Table 3. B(EL) values and energies in ^{12}C compared with those expected from D_3 symmetry. B(EL) values in e^2fm^{2L} and E in keV. The theoretical energies are calculated using E(keV)=730L(L+1). The value of β=1.9fm is estimated from the elastic form factor measured in electron scattering

$B(EL;L^P\rightarrow 0^+)$	Th	Exp	$E(L^P)$	Th	Exp
$B(E3;3^-\rightarrow 0^+)$	181	205±10 (13.6(7)W.u.)	$E(3^-)$	6132	6130
$B(E4;4^+\rightarrow 0^+)$	338	378±133 (3.7(1.3)W.u.)	$E(4^+)$	10220	10356
$B(E6;6^+\rightarrow 0^+)$	8245		$E(6^+)$	21462	21052

Table 4. B(EL) values and energies in ^{16}O compared with those expected from T_d symmetry. B(EL) values in e^2fm^{2L} and E in keV. The theoretical energies are calculated from E(keV)=511L(L+1). The value of β=2.0fm is extracted from the elastic form factor in electron scattering

2.3. *Form factors in electron scattering.* In the rigid limit and semi-classical approximation, the form factors are given by $F_L(0\rightarrow L)=c_L j_L(q\beta)$. A discussion of these form factors and comparison with experiment is given in the accompanying contribution of Bijker [18].

3. Current work

An important question is to what extent clustering persists when additional particles are added to kα structures. To this end we have initiated a study of kα+x neutrons (protons) structures (ACFM model). These structures were suggested by von Oertzen [19] and have been the subject of many investigations. The ACM offers a simple way to calculate single particle levels in a field with discrete symmetry on top of which rotation-vibration bands are built. Calculations are currently being done [20] for ^{9}Be, ^{9}B

(2α+1neutron, proton); ^{13}C, ^{13}N (3α+1neutron, proton); and ^{17}O, ^{17}F (4α+1 neutron, proton). Preliminary results for cluster rotational bands in ^{9}Be are shown in Fig.4. It should be noted that no additional parameter is required in the calculation except for the value β=2.6fm. This value is larger than the value β=2.0fm obtained from the moment of inertia of ^{8}Be due to the finite extent of the nucleon-alpha interaction.

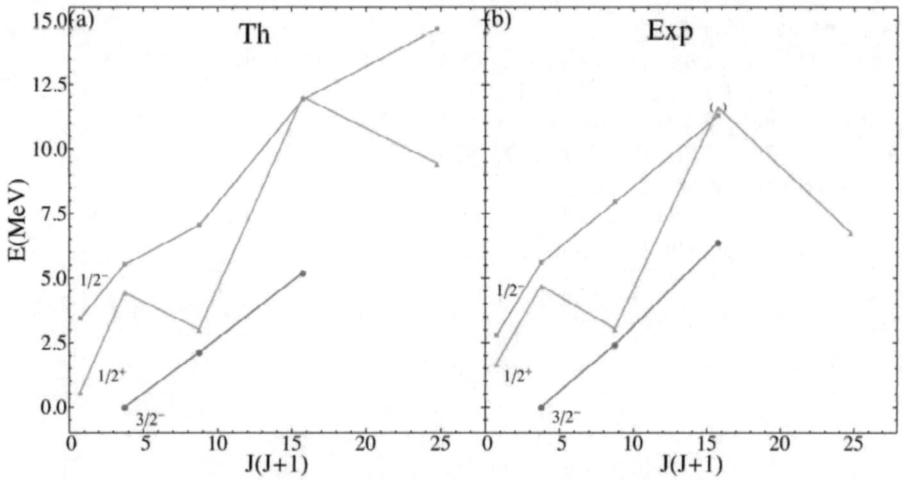

Figure 4. Cluster rotational bands in ^{9}Be

4. Summary and conclusions

A model based on symmetry has been developed (ACM) within which one can derive analytic expressions for energy levels, electromagnetic transition rates and form factors in kα nuclei. Evidence has been presented for k=2, 3, 4 (^{8}Be, ^{12}C, ^{16}O). Clustering appears to be robust for the ground state band and somewhat soft for vibrational bands. A model (ACFM) is currently being developed to derive analytic expressions for energy levels and electromagnetic transition rates in kα+x (neutrons, protons). Evidence has been presented for k=2, x=1. The development of ACM and ACFM opens the way for a renewed study of clustering in light nuclei.

Current answers to the questions posed in the introduction are: (i) signatures of clustering are properties of spectra and electromagnetic transition rates, especially *parity doubling, unusual structures of rotational bands, and enhanced E2, E3, E4, E5, ... transition rates*; (ii) clustering appears to extend at least to angular momentum J=6 in k=2, 3, 4 systems and up to excitation energies of the order of 25MeV; (iii) clustering appears to extend to A=16; (iv) clustering appears to survive the coupling of fermions, at least up to x=1.

Acknowledgements

This work was supported in part by U.S.D.O.E. Grant DE-FG02-91ER40608.

References

[1] Wheeler JA 1937 *Phys. Rev.* **52** 1083
[2] Dennison DM 1954 *Phys. Rev.* **96** 378
[3] Brink DM 1965 Proc. Int. School of Physics Enrico Fermi, Course XXXVI, 247
Brink DM, Friedrich H, Weiguny A and Wong CW 1970 *Phys. Lett.* B **33** 143
[4] Marin-Lambarri *et al* 2014 *Phys. Rev. Lett.* **113** 012502
[5] Beck C 2016 These Proceedings
[6] Iachello F 1981 *Chem. Phys. Lett.* **78** 581

IOP Conf. Series: Journal of Physics: Conf. Series **863** (2017) 012003 doi:10.1088/1742-6596/863/1/012003

[7] Iachello F 1981 *Phys. Rev.* C **23** 2778 (R)
[8] Bijker R and Iachello F 2000 *Phys. Rev.* C **61** 067305
[9] Bijker R and Iachello F 2002 *Ann. Phys. (N.Y.)* **298** 334
[10] Bijker R 2010 *AIP Conf. Proc.* **1323** 28
[11] Bijker R 2012 *J. Phys.: Conf. Series* **380** 012003
[12] Bijker R and Iachello F 2014 *Phys. Rev. Lett.* **112** 152501
[13] Freer M *et al* 2011 *Phys. Rev.* C **83** 034314
[14] Gai M 2016 These Proceedings
[15] Robson D 1978 *Nucl. Phys.* A **308** 381; Robson D 1979 *Phys. Rev. Lett.* **42** 876
Robson D 1982 *Phys Rev.* C **25** 1108 (R)
[16] Bijker R and Iachello F 2016 in preparation
[17] Daker DM *et al* 2013 *Phys. Rev. Lett.* **111** 062502
[18] Bijker R 2016 These Proceedings
[19] von Oertzen W 1996 *Z. Phys.* A **354** 37; von Oertzen W 1997 *Z. Phys.* A **357** 355
[20] Della Rocca V and Iachello F 2016 in preparation

IOP Conf. Series: Journal of Physics: Conf. Series **863** (2017) 012004 doi:10.1088/1742-6596/863/1/012004

An Overview of the History of Cluster Conferences

H Horiuchi

Research Center for Nuclear Physics, Osaka University, Ibaraki 567-0047, Japan

E-mail: horiuchi@rcnp.osaka-u.ac.jp

Abstract. An overview is given on the historical development of the cluster conference series which started at Bohum in 1969. I start with the discussion of the philosophy of Karl Wildermuth and then I make a review on the main subjects and topics in cluster conferences. Since the cluster dynamics is a main nuclear dynamics together with the mean-field dynamics, we see that development of the cluster conference has been along with the rises of many new subjects in nuclear physics itself. Examples of them include superheavy nuclei, nuclear astrophysics, neutron-rich nuclei, cluster-gas states and ab initio calculations. Finally I discuss that more activities in and attention to cluster physics are seen in recent days.

1. Introduction

According to the request of the organizer to me to give a short review of the history of the cluster conferences, I try to review the main subjects or topics of clustering phenomena and theories discussed in the cluster conferences since the first conference at Bochum. The list of cluster conferences is as follows:

1. Bochum, Germany (1969)
2. College Park, Maryland, United States (1975)
3. Winnipeg, Manitoba, Canada (1978)
4. Chester, United Kingdom (1984)
5. Kyoto, Japan, (1988)
6. Strasbourg, France (1994)
7. Rab, Croatia (1999)
8. Nara, Japan (2003)
9. Stratford-upon-Avon, United Kingdom (2007)
10. Debrecen, Hungary (2012)

I was absent from three conferences, Bochum, Maryland, and Strasbourg.

I first discuss the philosophy of Karl Wildermuth who started this conference series. His philosophy is based on the duality property of the shell-model wave function of the nuclear ground state and this duality character of the ground state has been recently revisited. Next I review that the development of the cluster conference has been along with the rises of many new subjects and topics in nuclear physics itself. Examples are molecular resonances, superheavy nuclei, nuclear astrophysics, fragmentation in nucleus-nucleus collisions, neutron-rich nuclei, cluster-gas states, and ab initio calculations. Finally I discuss the present status of nuclear cluster physics that more activities in and attention to cluster physics are seen in recent days.

IOP Conf. Series: Journal of Physics: Conf. Series **863** (2017) 012004 doi:10.1088/1742-6596/863/1/012004

2. Duality nature of the ground state and coexistence of cluster dynamics and mean-field dynamics

The first cluster conference at Bochum was held about 10 years after Karl Wildermuth had started his microscopic cluster model study of nuclear structure in 1958 [1]. At the year 1958, the shell model had already been well established because it was proposed about 10 years ago by Mayer and Jensen at 1949. Wildermuth aimed to discuss the cluster physics which coexists with mean-field physics. Wildermuth's cluster model was based on the duality property of the shell-model wave function of the ground state which means that the ground-state wave function can be rewritten in the form of the cluster-model wave function. A typical example is the closed-shell wave function of the ^{16}O ground state which is equivalent to ^{12}C + α cluster-model wave function:

$$\det|(0s)^4(0p)^{12}| = c_L \mathcal{A}\left[R_{(4-L)/2,L}(r_{C-\alpha}, 3\nu)[Y_L(\widehat{r}_{C-\alpha})\phi_L(^{12}\mathrm{C})]_{J=0}\phi(\alpha)\right] g(\boldsymbol{X}_G, 16\nu), \qquad (1)$$

where $R_{nL}(r)$ is radial H.O. function, $g(r,\gamma) = (2\gamma/\pi)^{3/4}\exp(-\gamma r^2)$ and $\boldsymbol{X}_G$ is the total center-of-mass coordinate. The duality character of ground state shell-model wave functions was known by many people who studied microscopic cluster-model wave functions in 1950's. For example, Perring and Skyrme wrote at the beginning of the abstract of their paper [2] "It is shown that it is possible to write down α-particle wave functions for the ground states of ^{8}Be, ^{12}C, and ^{16}O, which become, when antisymmetrized, identical with shell-model wave functions."

Wildermuth considered that the formation of cluster states is due to the activation of cluster degrees of freedom embedded in the ground state. So his philosophy can be shown as in Fig.1.

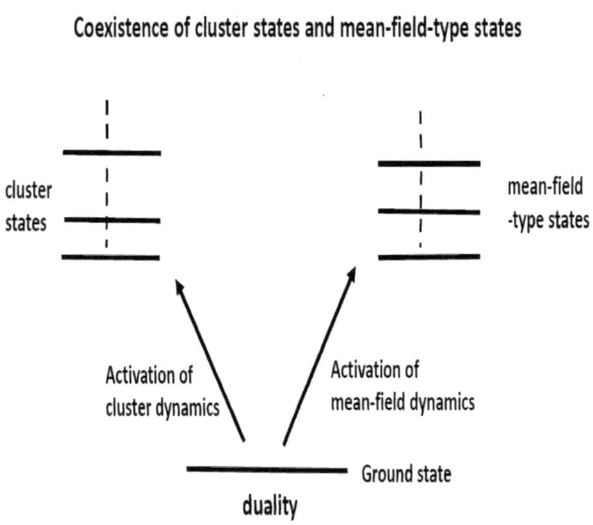

Figure 1. Coexistence of cluster states and mean-field-type states due to the duality of the ground state. Ground state is a compact shell-model state but it has degrees of freedom of clustering dynamics together with that of mean-field dynamics. Activation of cluster degrees of freedom yields cluster states while that of mean-field degrees of freedom yields mean-field-type excited states.

Recently the duality of the ground state has been revisited. It has been recognized that observed large values of $E0$ transitions between cluster states and the ground state in many nuclei can be explained very naturally by the duality property of the ground state. Large observed value means that it has the magnitude of single-particle value in spite of the fact that the cluster state has many-particle many-hole configurations in the shell-model language. But according to the Wildermuth's philosophy, the cluster state is formed by the excitation of the single degree of freedom which is the inter-cluster relative motion embedded in the ground state. Thus it is quite natural that the $E0$ transition between cluster state and the ground state has the magnitude of single-particle value.

IOP Conf. Series: Journal of Physics: Conf. Series **863** (2017) 012004 doi:10.1088/1742-6596/863/1/012004

3. Development of the cluster conference along with the rises of many new subjects in nuclear physics

At the time of Bochum conference the Brink model was already accepted as a powerful cluster model [3]. The Brink wave function has the form which is an antisymmetrized wave function of crystalline structure of clusters. When inter-cluster distance becomes zero, Brink wave function is equivalent to SU(3) shell-model wave function, which is nothing but the duality character of the shell-model wave function. As is well known, the RGM (resonating group method) calculation is now made by using the Brink wave function.

The orthogonality condition model (OCM) which has been one of the most important methods of cluster model was proposed around 1970. It is the point cluster model with orthogonalities of inter-cluster wave functions to the Pauli-forbidden states. At the second cluster conference at Maryland in 1975, D. Brink reviewed in detail in his introductory talk the results of $^{12}C + \alpha$ OCM by Y. Suzuki which is a masterpiece of the OCM approach to clustering.

The molecular resonance has been an important subject in cluster physics and many discussions were made in cluster conferences especially at Manitoba, Chester, Kyoto, and Strasbourg. At Chester conference in 1984, A. Bromley gave an introductory talk. In his talk, the Ikeda diagram was displayed for the first time in this series of cluster conferences. It was 16 years later after the publication of the diagram in 1968 whose presentation owed much on the observation of molecular resonances. An important recent study of the molecular resonance is about its possible relation with the superdeformed state. For example it was proposed that the $^{16}O + {}^{16}O$ molecular states are generated as excited states of the ^{32}S superdeformed states which have the duality of cluster and mean-field structures.

At Kyoto conference in 1988 I gave an introductory talk under the session-chairmanship of K. Wildermuth. One of the subjects discussed by me was nuclear astrophysics especially "nucleosynthesis and cluster state". As is well known now, nuclear reactions of nucleosynthesis are largely influenced by the cluster states near the cluster-breakup threshold energies as seen in Ikeda diagram. Other subjects which I discussed included "cluster decay and fission" and "intermediate-energy heavy-ion collisions".

The fragmentation reaction in nucleus-nucleus collisions has been also an important subject in nuclear cluster physics. In cluster conferences especially at Strasbourg (1994), Rab (1999), Nara (2003), many discussions were made and, in the case of theoretical studies, research reports by the use of BUU (Boltzmann-Uehling-Uhlenbeck), QMD (quantum molecular dynamics), and AMD (antisymmetrized molecular dynamics) methods were presented. A very important harvest was obtained from the subject of the fragmentation reaction to the subject of nuclear structure study. It is the use of the AMD method as a novel method of nuclear structure. The AMD method enables the study of clustering without assuming the existence of clusters. The FMD (Fermionic molecular dynamics) also participated in the nuclear structure study.

The advent of beams of unstable nuclei has been extending the nuclear physics to wide region of nuclear chart beyond the stable nucleus region. This new trend of nuclear physics which started from around 1990 in a large scale immediately affected the nuclear cluster physics. Novel types of cluster structure in neutron-rich nuclei have been extensively discussed as is well known. At Rab conference in 1999 I gave a summary talk in which I gave a conjecture on clustering near neutron dripline as shown in Fig.2.

At Nara conference in 2003, two important novel subjects were discussed extensively; they are "cluster-gas states" and "ab initio calculations of cluster structure". The cluster-gas structure is a novel type of nuclear structure and can be investigated spectroscopically, which is in sharp contrast to nucleon gas. The starting point of this subject was the discovery that the Hoyle state (0_2^+) of ^{12}C has a structure of 3α condensate-like structure. As for the ab initio calculation, it is well known that in late 1990's the quantum Monte Carlo calculation reported that the exact calculation with realistic nuclear force yielded the 2α dumbbell structure for the ground band

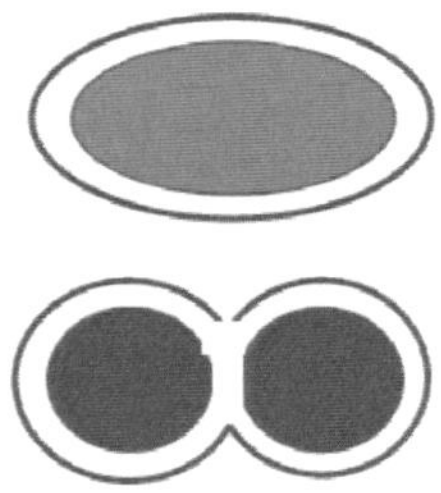

Figure 2. Possible cluster structure near neutron dripline with clusterized core surrouded by skin neutrons. Because of the constancy of thickness ($\approx$ 1 fm) of neutron skin which is due to the incapability of selfbinding of neutrons, clusterized core can sustain more valence neutrons than ordinary deformed core. The core can have three-cluster structure.

states of ^{8}Be. What is remarkable in the status of present-day ab initio calculations is that an important check of the ability of an ab initio method is to examine whether the Hoyle state is well reproduced or not.

4. More activities in and attention to cluster physics

In 2008, a new series of cluster workshop was inaugurated, which is named SOTANCP (state of the art in nuclear cluster physics). The first workshop was held at Strasbourg (France) in 2008, the second was at Brussels (Belgium) in 2010, and the third was at Yokohama (Japan) in 2014. This clearly shows that in recent days we have more activities in and attention to cluster physics. I show in Fig.3 that the above trend in recent days is well reflected in the number of citations of the old paper of Ikeda diagram. [4].

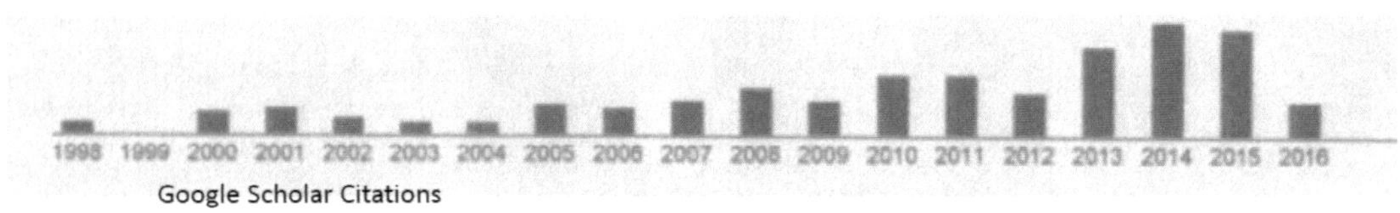

Figure 3. Number of citations of the old paper of Ikeda diagram published in 1968. We can see that this paper is now cited more than before. Data is taken from Google Scholar Citations.

References

[1] Wildermuth K and Kanellopoulos Th 1958 *Nucl. Phys.* **7** 150; 1958/59 *Nucl. Phys.* **9** 449
[2] Perring J K and Skyrme T H R 1956 *Proc. Phys. Soc.* **69** 600
[3] Brink D M 1966 *Proc. Int. Sch. Phys. Enrico Fermi* **36** 247
[4] Ikeda K, Takigawa N and Horiuchi H 1968 *Prog. Theor. Phys. Supplement* Extra Number 464

IOP Conf. Series: Journal of Physics: Conf. Series **863** (2017) 012005 doi:10.1088/1742-6596/863/1/012005

Alpha-Particle Condensate Structure of the Hoyle State: where do we stand?

P Schuck[1,2,3], Y Funaki[4], H Horiuchi[5], G Röpke[6], A Tohsaki[5] and T Yamada[7]

[1]Institut de Physique Nucléaire, CNRS, UMR8608,
Orsay, F-91406, France
[2]Université Paris-Sud, Orsay, F-91505, France
[3]Laboratoire de Physique et de Modélisation des Milieux Condensés, CNRS et Université Joseph Fourier, UMR5493, 25 Av. des Martyrs, BP 166, F-38042 Grenoble Cedex 9, France
[4]The Institute of Physical and Chemical Research (RIKEN), Wako,
Saitama, 351-0198, Japan
[5]Research Center for Nuclear Physics (RCNP), Osaka University,
Ibaraki, Osaka 567-0047, Japan
[6]Institut für Physik, University of Rostock, Universitätsplatz 1,
18051 Rostock, Germany
[7]Laboratory of Physics, Kanto Gakuin University,
Yokohama 236-8501, Japan

Abstract. The present understanding of the structure of the Hoyle state in ^{12}C is reviewd. It is pointed out that a crucial test of any theory is the good reproduction of the experimental results for the inelastic form factor from ground to the Hoyle state. The performances of the so-called THSR wavwe function are outlined confirming the α particle condensation hypothesis proposed 15 years back in [1].

1. Introduction

One of the most amazing phenomena in quantum many-particle systems is the formation of quantum condensates. At present, the formation of condensates is of particular interest in strongly coupled fermion systems in which the crossover from Bardeen-Cooper-Schrieffer (BCS) pairing to Bose-Einstein condensation (BEC) may be investigated. Among very different quantum systems, nuclear matter is especially well suited for the study of correlation effects in a quantum liquid. In [2], the possibility of α particle (quartet) condensation in infinite matter was investigated. It was found that quartetting is possible at low densities, below about a fifth of saturation density. At higher densities, around the point where the chemical potential μ turns from negative (binding) to positive, the condensation breaks down. This is contrary to ordinary pairing which can exist for considerably positive μ values, depending only on the range of the pairing force. The reason for this strong qualitative difference between the two cases is explained in [3].

The question then arises whether in analogy to pairing, also for quartetting exist nuclei where this phenomenon is born out. In [1], we found that such a possibility very likely exists in lighter self-conjugate nuclei for excitation energies around the α disintegration threshold. In this contribution, we want to discuss the successes and eventual failures of this idea which was proposed 15 years back.

2. The THSR approach and the Hoyle state

The Hoyle state is the first excited 0^+ state in ^{12}C at 7.65 MeV. This state is one of the most fameous states in nuclear physics because without its existence life on earth would be absent in its present form. Indeed, since ^{8}Be is unstable, the stellar production of Carbon in the universe would be lower by a huge factor without the existence of the Hoyle state. It is just at the right energy to allow for the so-called triple α reaction $\alpha+\alpha+\alpha \to^8$ Be$+\alpha \to^{12}$C* to become strongly accelerated.

For the microscopic description of the Hoyle state several approaches have been put forward in the past [4, 5, 6, 7, 8, 9, 10]. However, only the following, so-called THSR wave function (according to the authors Tohsaki, Horiuchi, Schuck, Roepke) which was proposed in [1] concentrates on the α particle condensation aspect (the spin-isospin part is not written out)

$$\Psi_{\rm THSR} \propto \mathcal{A}\psi_1\psi_2\psi_3 \equiv \mathcal{A}|B\rangle \tag{1}$$

with

$$\psi_i = e^{-((\mathbf{R}_i-\mathbf{X}_G)^2)/B^2}\phi_{\alpha_i} \tag{2}$$

and

$$\phi_{\alpha_i} = e^{-\sum_{k<l}(\mathbf{r}_{i,k}-\mathbf{r}_{i,l})^2/(8b^2)} \tag{3}$$

In (1) the $\mathbf{R}_i$ are the c.o.m. coordinates of α particle 'i' and $\mathbf{X}_G$ is the total c.o.m. coordinate of ^{12}C. $\mathcal{A}$ is the antisymmetrizer of the twelve nucleon wave function with ϕ_{α_i} the intrinsic translational invariant wave function of the α-particle 'i'. The whole 12 nucleon wave function in (1) is, therefore, translationally invariant. The special Gaussian form given in Eqs. (2), (3) was chosen in [1] to ease the variational calculation. The condensate aspect lies in the fact that (1) is a (antisymmetrized) product of three times the same α-particle wave function and is, thus, analogous to a number projected BCS wave function in the case of pairing. This twelve nucleon wave function has two variational parameters, b and B. It possesses the remarkable property that for $B = b$ it is a pure harmonic oscillator Slater determinant (this aspect of (1) is explained in [11, 12]) whereas for $B \gg b$ the α's are at low density so far apart from one another that the antisymmetrizer can be dropped and, thus, (1) becomes a simple product of three α particles, all in identical 0S states, that is, a pure condensate state. The minimization of the energy with a Hamiltonian containing a nucleon-nucleon force determined earlier independently [13] allows to obtain a reasonable value for the ground state energy of ^{12}C. Variation of energy under the condition that (1) is orthogonal to the previously determined ground state allows to calculate the first excited 0^+ state, i.e., the Hoyle state. While the size of the individual α particles remains very close to their free space value ($b \simeq 1.37$ fm), the variationally determined B parameter takes on about three times this value. This entails a quite enhanced value of the rms radius of 3.83 fm of the Hoyle state with repect to the one of the ground state (2.4 fm). This gives a volume (density) of the Hoyle state about a factor 3-4 larger (smaller) than for the ground state. In such a large volume the α's have space to develope themselves what is not the case in the ground state where they overlap strongly.

Still the question may be asked: is the Hoyle state closer to a Slater determinant or to an α condensate? A precise answer is obtained from the calculation of the bosonic occupation numbers which have been obtained in three different works [14, 15, 10] with very similar results. The ones of [15] are displayed in Fig.1. We see that the distribution in the ground state is more or less equipartitioned and compatible with the SU3 shell model theory whereas the distribution

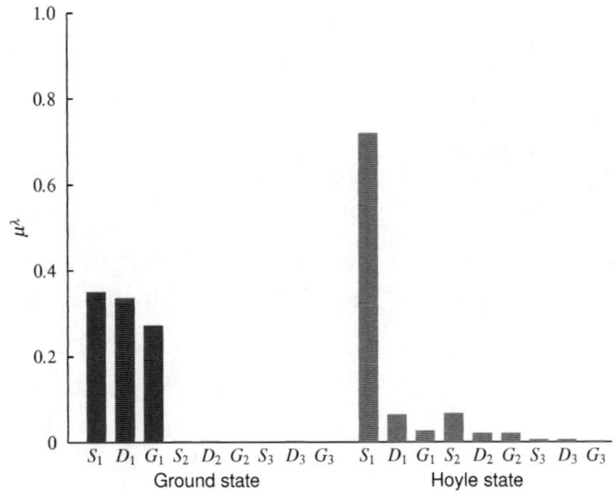

Figure 1. α particle occupation numbers in the ground state (left) and in the Hoyle state (right) [15].

of the Hoyle state has an overwhelming contribution of over 70% of the α's being in the lowest 0S state, all other contributions are down by a factor of at least ten. We, therefore, can say that the three α particles in the Hoyle state occupy with their c.o.m. motion to a large fraction the same 0S orbit meaning that, indeed, the Hoyle state can be considered to within good approximation as a condensate of 3 α particles. However, the Pauli principle is still active and antisymmetrisation scatters the α's out of the condensate about 30% of the time. It can be mentioned here that this number is very similar concerning good single particle states in odd nuclei where the fermionic occupation numbers also are in the range of 70-80%.

We mentioned that the Hoyle state has an extended volume being by a factor 3-4 larger than the one of the ground state. How to prove this? It turns out that the inelastic form factor, measured by inelastic electron scattering, is very sensitive to the size of the Hoyle state [16]. Increasing artificially the size of the Hoyle state by about 20% reduces the form factor globally by a factor of two. The fact that the THSR theory reproduces very precisely the experimental values of the inelastic form factor, see Fig.2 without any adjustable parameter can be considered as a great achievement and gives large credit to the picture that in the Hoyle state three α particles are well born out moving almost independently in their proper mean field. In the same figure we show recent Green's function Monte Carlo (GFMC) results [17] which also also reproduce the inelastic form factor very nicely. In the insert of the GFMC-panel, we see that the rather precise experimental transition radius of 5.29 ± 0.14 fm^2 given in [8] is better reproduced than with the THSR approach which yields an about 20% too large value. On the other hand, the GFMC approach gives the position of the Hoyle state about 2.5 MeV too high whereas with the THSR wave function the experimental value of 7.65 MeV is quite well reproduced with no adjustable parameter.

3. A brief account of the situation in ^{16}O

The situation in ^{16}O is quite a bit more complicated than in ^{12}C. The fact is that between the 4α threshold and the ground state, there are several 0^+ states which can be interpreted as $\alpha+^{12}$C cluster configurations. In Fig. 3, we show the result of a calculation with the so-called Orthogonal Condition Model (OCM) method [18].

We see that there is a very nice one to one correspondence between the first six calculated 0^+ states and experiment. In regard of the complexity of the situation the agreement between

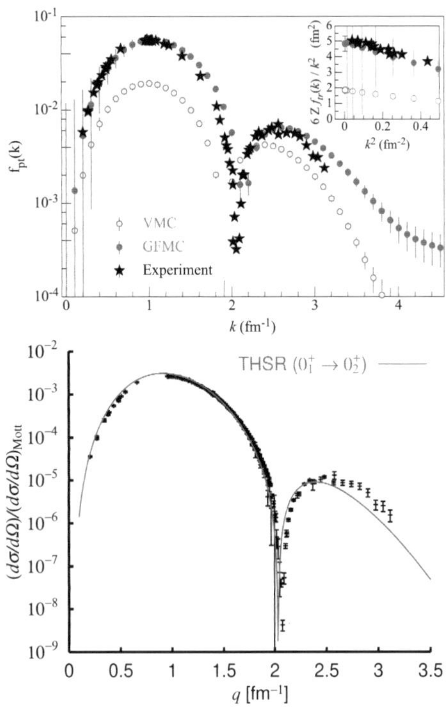

Figure 2. Inelastic form factors from GFMC [17], upper panel, and THSR [16], lower panel. The THSR result cannot be distinguished from the one of [5] on the scale of the figure meaning that also the approach of [5] implicitly contains the α condensation aspect (this, by the way, is also the case with the approach in [6]).

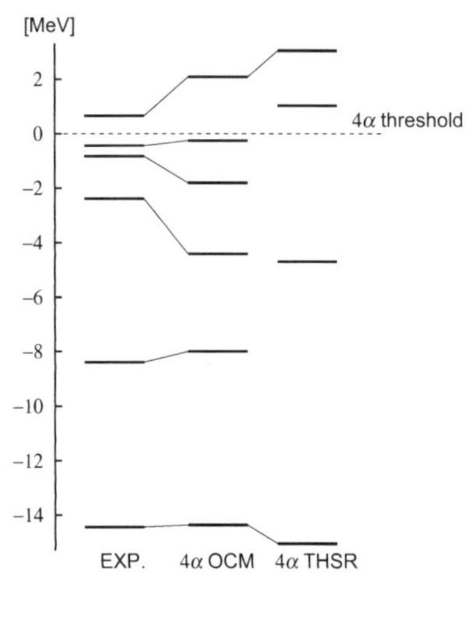

Figure 3. Spectrum of 0^+ states in ^{16}O from the OCM and THSR approaches [18, 1].

both can be considered as very satisfactory. Only the highest state was identified with the 4α condensate state. The four other excited 0^+ states are $\alpha+^{12}$C configurations. For example the 5-th 0^+ state is interpreted as an α orbiting in a higher nodal S-wave around the ground state of ^{12}C. The 4-th 0^+ state contains an α orbiting in a P-wave around the first 1^- state in ^{12}C. In the 3-rd 0^+ state the α is in a D-wave coupled to the 2_1^+ state of ^{12}C and in the 2-nd 0^+ state the α is in a 0S-wave and the ^{12}C in its ground state. The single parameter THSR calculation can only reproduce correctly the ground state and the α condensate state (0_6^+). By construction it cannot describe $\alpha+$ ^{12}C configurations. So, the two intermediate states give some sort of average picture of the four α plus ^{12}C configurations. One would have to employ a more general ansatz like in ([19] to cope with the situation. Work in this direction is in progress. The 0_6^+ state is theoretically identified as the α-condensate state from the overlap squared $|\langle 0_6^+|\alpha +^{12} \mathrm{C}(0_i^+)\rangle|^2$ [20].

4. Discussion and Conclusions

In this contribution, we concentrated on the α particle condensation aspect of the Hoyle state introduced with the THSR wave function 15 years back [1]. This approach reproduces all known experimental results of the Hoyle state without any adjustable parameter and, thus, gives credit to the condensation scenario. This, despite of the fact that its direct experimental verification is difficult. Indeed, while pairing induces clear signs of superfluidity in rotating nuclei, no analogous effects have been detected so far from quartetting. However, several experiments are under way or planned concerning the Hoyle state and analogous states in ^{16}O or even heavier nuclei, what may shed further light on the situation in the near future [21, 22]. A major issue in this respect is the understanding not only of the Hoyle state but of excited states thereof. The 0_3^+ and 0_4^+ have been identified experimentally recently and have been interpreted as α gas states with one α in a higher nodal S-state and a linear chain state, respectively, see [19] and references therein for discussions. Also the structure of the second 2^+ state is strongly debated. It is considered either as a member of a rotational band with the Hoyle state as band head [23] or more as a nodal excitation of one of the α's into a D-wave [19]. Further experimental and theoretical studies are necessary to elucidate the situation. With respect to the excited Hoyle states an interesting paper has appeared recently [24] where the authors explain with a single adjustable parameter very well the Hoyle spectrum on grounds that the Hoyle state is a Bose condensate with broken U1 symmetry (particle number). However, also this approach is not well tested and needs further work.

References

[1] Tohsaki A, Horiuchi H, Schuck P and Röpke G 2001 *Phys. Rev. Lett.* **87** 192501
[2] Röpke G, Schnell A, Schuck P and Noziéres P 1998 *Phys. Rev. Lett.* **80** 3177
[3] Sogo T, Lazauskas R, Röpke G and Schuck P 2009 *Phys. Rev.* C **79** 051301
[4] Horiuchi H 1974 *Progr. Theor. Phys.* **51** 1266; idem 1975 **53** 447
[5] Kamimura M 1981 *Nucl. Phys.* A **351** 456
[6] Uegaki E, Okabe S, Abe Y and Tanaka H 1977 *Progr. Theor. Phys.* **57** 1262; *idem* 1978 **59** 1031
[7] Kanada-En'yo Y 1998 *Phys. Rev. Lett.* **81** 5291; 2007 *Progr. Theor. Phys.* **117** 655
[8] Chernykh M, Feldmeier H, von Neumann-Cosel P and Richter A 2007 *Phys. Rev. Lett.* **98** 032501
[9] Lazauskas R and Dufour M 2011 *Phys. Rev. Lett.* **84** 064318
[10] Ishikawa S 2014 *Phys. Rev.* C **90** 061604(R) and private communication.
[11] Bayman B F and Bohr A 1958 *Nucl. Phys.* **9** 596
[12] Yamada T, Funaki Y, Horiuchi H, Ikeda K and Tohsaki A 2008 *Prog. Theor. Phys.* **120** 1139
[13] Tohsaki A *Phys. Rev.* C **49** 1814
[14] Matsumara M and Suzuki Y *Nucl. Phys.* A **739** 238
[15] Yamada T and Schuck P 2005 *Eur. Phys. J.* A **26** 185
[16] Funaki Y, Tohsaki A, Horiuchi H, Schuck P and Röpke G *Eur. Phys. J.* A **28** 259
[17] Carlson J, Gandolfi S, Pederiva F, Pieper S C, Schiavilla R, Schmidt K E and Wiringa R B 2015 *Rev. Mod. Phys.* **87** 1067
[18] Funaki Y, Yamada T, Horiuchi H, Röpke G and Schuck P 2008 *Phys. Rev. Lett.* **101** 082502
[19] Funaki Y *Phys. Rev.* C **92** 021302
[20] Yamada Y, Funaki Y, Horiuchi H, Röpke G and Schuck P 2011 arXiv:1103.3940; "Cluster in Nuclei (Lecture Notes in Physics)-Vol.2-", ed. by C. Beck, (Springer -Verlag, Berlin)
[21] Natowitz J, private communication.
[22] Freer M, private communication.
[23] Marin-Lambarri D J, Bijker R, Freer M, Gai M, Kokalova Tz, Parker D J and Wheldon C 2014 *Phys. Rev. Lett.* **113** 012502
[24] Nakamura Y, Takahashi J, Yamanaka Y and Ohkubo S 2016 *Phys. Rev.* C **94** 014314

IOP Conf. Series: Journal of Physics: Conf. Series **863** (2017) 012006 doi:10.1088/1742-6596/863/1/012006

Alpha Decay Width of ^{212}Po from a quartetting wave function approach

G Röpke[1,2], P Schuck[3,4,5], Y Funaki[6], H Horiuchi[7], Z Ren[8], A Tohsaki[9], C Xu[10], T Yamada[11] and B Zhou[12]

[1] Institut für Physik, Universität Rostock, D-18051 Rostock, Germany
[2] National Research Nuclear University (MEPhI), 115409 Moscow, Russia
[3] Institut de Physique Nucléaire, CNRS, UMR8608, Orsay, F-91406, France
[4] Université Paris-Sud, Orsay, F-91505, France
[5] Laboratoire de Physique et de Modélisation des Milieux Condensés, CNRS et Université Joseph Fourier, UMR5493, 25 Av. des Martyrs, BP 166, F-38042 Grenoble Cedex 9, France
[6] The Institute of Physical and Chemical Research (RIKEN), Wako 351-0198, Japan
[7] Research Center for Nuclear Physics (RCNP), Osaka University, Osaka 567-0047, Japan
[8] Department of Physics, Nanjing University, Nanjing 210093, China
[9] Research Center for Nuclear Physics (RCNP), Osaka University, Osaka 567-0047, Japan
[10] Department of Physics, Nanjing University, Nanjing 210093, China
[11] Laboratory of Physics, Kanto Gakuin University, Yokohama 236-8501, Japan
[12] Faculty of Science, Hokkaido University, Sapporo 060-0810, Japan

E-mail: gerd.roepke@uni-rostock.de

Abstract. An effective α particle equation is derived for cases where an α particle is bound to a doubly magic nucleus. As an example, we consider ^{212}Po with the α on top of the ^{208}Pb core. The fully quantal solution of the problem is inspired by the THSR (Tohsaki-Horiuchi-Schuck-Röpke) wave function concept that has been successfully applied to light nuclei. Shell model calculations are improved by including four-particle (α-like) correlations that are of relevance when the matter density becomes low. In the region where the α-like cluster penetrates the core nucleus, the intrinsic bound state wave function transforms at a critical density into an unbound four-nucleon shell model state. We present a microscopic calculation of both α-cluster preformation probability and decay width in typical α-emitter ^{212}Po. Using actually measured density distribution of the ^{208} Pb core, the calculated alpha decay width of ^{212}Po agrees fairly well with the measured one. Applications to other nuclei (^{20}Ne) are discussed.

1. Introduction

The radioactive α decay is known for a long time in nuclear physics. In contrast to extensive and precise experimental data, see [1] and refs. given there, the theoretical description needs some semiempirical assumptions at present. Whereas the tunneling of an α particle across the Coulomb barrier is well described in quantum physics, the problem in understanding the α decay within a microscopic approach is the preformation of the α cluster in the decaying nucleus.

Recently many attempts are performed to investigate α-like correlations in nuclear systems. For nuclear matter, it has been shown that α-like correlations are possible in the low-density region, but are dissolved because of the Pauli blocking around nucleon density $n_B = 0.03$ fm^{-3} [2]. Such low-density nuclear systems occur, for instance, in excited selfconjugate light nuclei

(^{8}Be, ^{12}C, ^{16}O, ^{20}Ne). Four-nucleon correlations have been identified successfully within the THSR approach [3], see also [4]. Recently, α-like correlations have been investigated also in the outer tails of the density of a nucleus where the nucleon density is below 0.03 fm^{-3} [5, 6].

2. Four nucleons moving in a mean field

We consider a quantum many-particle approach. The propagation of a quartet $(n_\uparrow, n_\downarrow, p_\uparrow, p_\downarrow)$ is described by a four-particle Green function which leads to the in-medium Schrödinger equation

$$[E_4 - \hat{h}_1 - \hat{h}_2 - \hat{h}_3 - \hat{h}_4]\Psi_4(\mathbf{r}_1\mathbf{r}_2\mathbf{r}_3\mathbf{r}_4) = \int d^3\mathbf{r}_1'\, d^3\mathbf{r}_2' \langle \mathbf{r}_1\mathbf{r}_2|B\, V_{N-N}|\mathbf{r}_1'\mathbf{r}_2'\rangle \Psi_4(\mathbf{r}_1'\mathbf{r}_2'\mathbf{r}_3\mathbf{r}_4)$$
$$+ \int d^3\mathbf{r}_1'\, d^3\mathbf{r}_3' \langle \mathbf{r}_1\mathbf{r}_3|B\, V_{N-N}|\mathbf{r}_1'\mathbf{r}_3'\rangle \Psi_4(\mathbf{r}_1'\mathbf{r}_2\mathbf{r}_3'\mathbf{r}_4) + \text{four further permutations.} \quad (1)$$

The single-nucleon operator $\hat{h}_1$ contains the kinetic energy and the potential energy of an external mean field, produced by the other nucleons which are not under consideration. The six nucleon-nucleon interaction terms contain besides the nucleon-nucleon potential V_{N-N} also the blocking operator B which can be given in quasi-particle state representation. For the first term on the r.h.s. of Eq. (1), the expression $B(1,2) = [1 - f_1(\hat{h}_1) - f_2(\hat{h}_2)]$ results. The phase space occupation $f_\nu(\hat{h}) = \sum_n^{\text{occ.}} |n,\nu\rangle\langle n,\nu|$ indicates the phase space which is occupied and therefore according to the Pauli principle not available for an interaction process. In Ref. [5], the Tamm-Dancoff expression $[1 - f_1(\hat{h}_1)][1 - f_2(\hat{h}_2)]$ is used which neglects pairing.

It is advantageous to introduce new variables, the c.o.m. position $\mathbf{R} = \sum_i^4 \mathbf{r}_i/4$ and the relative coordinates $\mathbf{s}_j$, $j = 1 \ldots 3$, in particular Jacobian coordinates. It has been shown [5] that we can subdivide the wave function $\Psi(\mathbf{R}, \mathbf{s}_j)$ into two parts in an unique way,

$$\Psi(\mathbf{R}, \mathbf{s}_j) = \varphi^{\text{intr}}(\mathbf{s}_j, \mathbf{R})\, \Phi(\mathbf{R}). \quad (2)$$

A coupled system of wave equations is obtained describing the c.o.m. motion

$$-\frac{\hbar}{8m}\nabla_R^2\Phi(\mathbf{R}) - \frac{\hbar}{4m}\int ds_j \varphi^{\text{intr},*}(\mathbf{s}_j,\mathbf{R})[\nabla_R\varphi^{\text{intr}}(\mathbf{s}_j,\mathbf{R})][\nabla_R\Phi(\mathbf{R})]$$
$$-\frac{\hbar}{8m}\int ds_j \varphi^{\text{intr},*}(\mathbf{s}_j,\mathbf{R})[\nabla_R^2\varphi^{\text{intr}}(\mathbf{s}_j,\mathbf{R})]\Phi(\mathbf{R}) + \int dR'\, W(\mathbf{R},\mathbf{R}')\Phi(\mathbf{R}') = E\,\Phi(\mathbf{R}) \quad (3)$$

and a similar equation for the intrinsic motion.

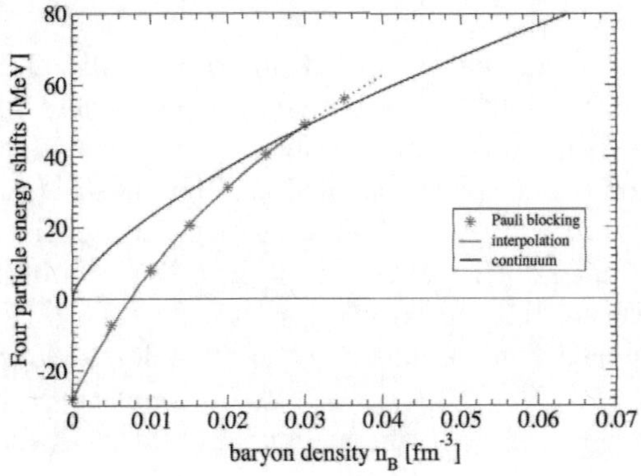

Figure 1. Pauli blocking: four-nucleon continuum edge (full line) and α-like bound state (stars) as function of n_B (homogeneous nuclear matter). The bound state energy of the α particle is shifted with increasing density n_B of the surrounding matter and merges with the four-nucleon continuum $\mu_4 = 4E_F = 2\hbar^2(3\pi^2 n_B/2)^{2/3}/m$ at the critical density $n_B^{\text{cr}} \approx 0.03$ fm^{-3}.

The c.o.m. potential

$$W(\mathbf{R},\mathbf{R}') = \int ds_j\, ds_j'\, \varphi^{\text{intr},*}(\mathbf{s}_j,\mathbf{R}) \left[T[\nabla_{s_j}]\delta(\mathbf{R}-\mathbf{R}')\delta(\mathbf{s}_j-\mathbf{s}_j') + BV(\mathbf{R},\mathbf{s}_j;\mathbf{R}',\mathbf{s}_j')\right] \varphi^{\text{intr}}(\mathbf{s}_j',\mathbf{R}')$$

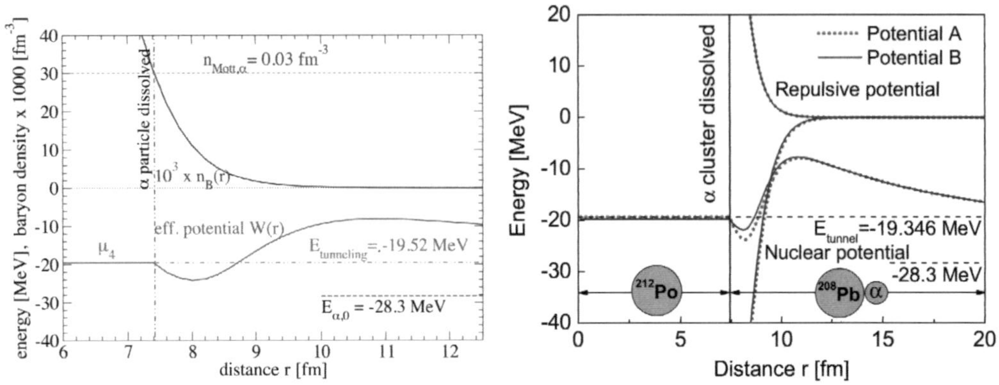

Figure 2. Potential for the c.o.m. motion of the four-nucleon system on top of the Pb core. Left: core nucleus density and $r_{\rm cr}$, Thomas-Fermi condition (Potential A). Right: different contributions to the c.o.m. potential (A and B).

contains the external (mean-field) contribution $V^{\rm mf}_\tau(r)$ and the intrinsic energy of the α-like cluster. Neglecting gradient terms in (3), the main problem to solve the intrinsic energy is the Pauli blocking which is in general non-local.

Within a local-density approximation, results for homogeneous nuclear matter can be used; see Fig. 1. The intrinsic wave function with lowest energy changes from an α-like bound state to a product of quasi-free single nucleon states at the critical density $n_B^{\rm cr} \approx 0.03$ fm^{-3}.

3. Results for ^{212}Po

Following [5, 6], we consider ^{212}Po which decays into a ^{208}Pb core and an α particle, half-life 0.299 μs and decay energy $Q_\alpha = 8954.13$ keV. The core nucleons are distributed with the baryon density $n_B(r)$ [7] and produce a mean field $V^{\rm mf}_\tau(r)$ acting on the two neutrons ($\tau = n$) and two protons ($\tau = p$) moving on top of the lead core. In addition, $n_B(r)$ is responsible for the Pauli blocking shift shown in Fig. 1. The critical density $n_B = 0.03$ fm^{-3} occurs at $r_{\rm cr} = 7.4383$ fm. This means that α-like clusters can exist only at distances $r > r_{\rm cr}$, for smaller values of r the intrinsic wave function is characterized by the uncorrelated motion. The mean-field potential of the ^{208}Pb core is obtained as the M3Y double-folding potential, see [8], in addition to the Coulomb interaction.

An important issue is that we select four nucleons which are described microscopically and can form an α like bound state. The interaction with the ^{208}Pb core is taken as mean field so that correlations between the nucleons of the core and the four quartet nucleons are neglected. Furthermore, we will not give a microscopic description of the core nucleons (e.g., Thomas-Fermi or shell model calculations) but consider both $n_B(r)$ and $V^{\rm mf}_\tau(r)$ as phenomenological input.

The local c.o.m. potential $W(r)$ is shown in Fig. 2. To derive the the M3Y double-folding potential, a parameterized nucleon-nucleon effective interaction $v(s) = c\exp(-4s)/(4s) - d\exp(-2.5s)/(2.5s)$ has been used, describing a short-range repulsion (c) and a long-range attraction (d); s denotes the nucleon-nucleon distance. At $r = r_{\rm cr}$, the four nucleons are at the Fermi energy μ_4, see Fig. 1. Within the Thomas-Fermi model, the chemical potential μ_4 for the free nucleon gas inside the core nucleus is constant, not depending on r.

Solving a Schrödinger equation describing the c.o.m. motion with the potential $W(r)$, the energy eigenvalue E_4 of the quartet is obtained. The c.o.m. wave function is found, the preformation factor is calculated from the probability to find the quartet c.o.m. position at $r > r_{\rm cr}$. In that region, α particles are formed. The tunneling rate and the half-life are obtained

IOP Conf. Series: Journal of Physics: Conf. Series **863** (2017) 012006 doi:10.1088/1742-6596/863/1/012006

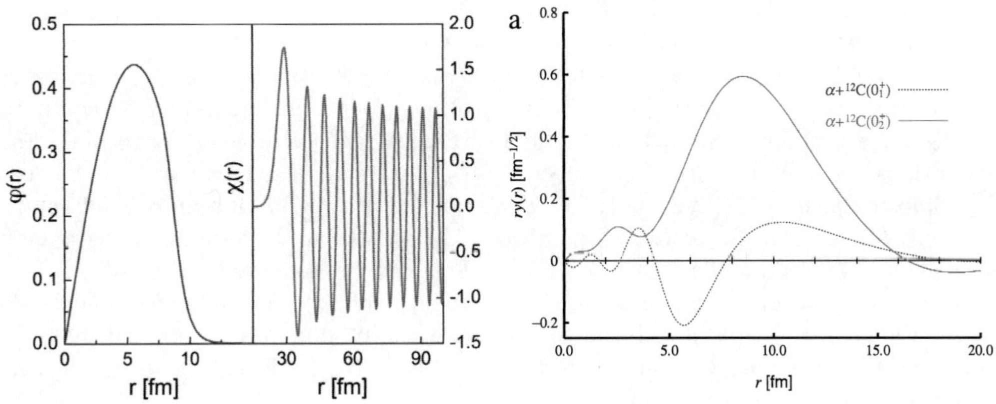

Figure 3. Left panel: c.o.m. wave function for the α particle in ^{212}Po [6]. Right: c.o.m. wave function for the α particle in ^{16}O [4].

in the standard way, see [6].

Table 1. The calculated preformation probability and decay half-life of ^{212}Po using different sets of effective c.o.m. potentials.

Pot.	c [MeV fm]	d [MeV fm]	E_{tunnel} [MeV]	Fermi en. μ_4[MeV]	$E_{\text{tunnel}} - \mu_4$ [MeV]	Pref. fact. P_α	Decay half-life $T_{1/2}$[s]
A	13866.30	4090.51	-19.346	-19.346	0	0.367	2.91×10^{-8}
B	11032.08	3415.56	-19.346	-19.771	0.425	0.142	2.99×10^{-7}

Two sets of parameters c, d have been used as given in Tab. 1. For both sets A, B, the parameter values c, d reproduce the measured tunneling energy $E_4 = -19.52$ MeV. Furthermore, set A fulfills the condition $E_4 = \mu_4$ of the Thomas-Fermi model. The result for the decay half-life $T_{1/2}$ is too short. Instead, set B is constructed to reproduce the measured decay half-life. Then, the position of E_4 is above the Fermi energy μ_4 of the core nuclei. Note that a gap is expected to occur in a shell model when adding the four nucleons on top of the double magic core nucleus.

4. THSR and ^{20}Ne

Several approximations have been made which should be discussed and improved, if possible. We neglected gradient terms so that our approach is close to the local-density approximation. Within a rigorous approach, the c.o.m. potential $W(\mathbf{r}, \mathbf{r}')$ is non-local. A full treatment of the inhomogeneous case relevant for finite nuclei, which includes the gradient terms and the non-local potentials, is a future goal, presently not in reach.

Furthermore the simple TF model for the core nucleus has to be improved by shell model calculations. To be consistent, also few-particle correlations for the core nucleons have to be considered, not only for the quartet on top of the core.

The THSR wave function concept is a systematic approach to treat few-body correlations between all nucleons. It has been successfully applied to light nuclei. Our quartetting wave function approach is inspired by the THSR concept so that we have to compare with respect to the treatment of few-nucleon correlations going beyond an uncorrelated quasiparticle model. However, because of complexity THSR calculations are restricted to small numbers of nucleons.

At present, THSR calculations have been performed up to ^{20}Ne [9, 10]. Calculations for large nuclei such as ^{212}Po are not in reach.

The nucleus ^{20}Ne can also be considered as a quartet on top of a double magic nucleus ^{16}O as core nucleus. The investigation of this nucleus opens the possibility to compare our quartetting wave function approach with the THSR approach. Problems concerning different approximations inherent in both approaches can be analyzed. First exploratory calculations using the known density distribution of ^{16}O give an value $r_{\rm cr} = 3.302$ fm, an α-like bound state can exist only for $r > r_{\rm cr}$. To calculate the mean-field potential we have to fit the parameter c, d occurring in the M3Y-type effective interaction. This can be done using the difference of the binding energy of ^{20}Ne and ^{16}O (binding energy of the α particle). Assuming the Thomas-Fermi model condition $E_4 = \mu_4$ discussed above, the c.o.m. wave function can be calculated. The corresponding rms radii can be compared with the respective empirical values of both nuclei. Preliminary calculations give reasonable results.

Of interest is the comparison of the c.o.m. wave function for the quartet in the different approaches, see Fig. 3. The typical behavior of the quartetting wave function approach gives a nearly linear increase up to $r_{\rm cr}$, after this a continuous decrease. THSR calculations show also a nearly linear increase, the deviations are related to the internal structure of the core. The comparison with THSR calculations would lead to a better understanding of the microscopic calculations, in particular the c.o.m. potential, the c.o.m. wave function, and the preformation factor.

Note that the position of the chemical potential determines the preformation probability. If $\mu_4 < E_{\rm tunnel}$, the preformation probability becomes smaller. If $\mu_4 > E_{\rm tunnel}$, the preformation probability becomes larger. In the latter case, the states at the Fermi surface are also correlated. This is the case, e.g., for the Hoyle state where all core nucleons are found in correlated states.

5. Conclusions

- α-like clustering occurs in nuclear systems in the region of low density. Examples are the density tails near the surface of heavy nuclei, e.g. ^{212}Po.
- A consistent quantum statistical approach is needed to describe the formation of four-nucleon correlations and its dissolution at increasing nucleon density. We present an approach where a quartet moves under the influence of the core nucleus which is treated as a mean field.
- A simple description of the core nucleus is possible within a local density approach. It has to be improved taking into account density gradients and non-locality of Pauli blocking. An essential issue for future research is the treatment of correlations within the core nucleus.
- The investigation of ^{20}Ne where the full treatment of four-nucleon correlations within the THSR approach has been done, would be of high interest to discuss α preformation and to describe the α-like particle on top of the ^{16}O core [9].

References

[1] Delion D S, Liotta R J, and Wyss R 2015 *Phys. Rev.* C **92** 051301(R)
[2] Röpke G, Schnell A, Schuck P and Nozières P 1998 *Phys. Rev. Lett.* **80** 3177
[3] Tohsaki A, Horiuchi H, Schuck P and Röpke G 2001 *Phys. Rev. Lett.* **87** 192501
[4] FunakiY, Horiuchi H and Tohsaki A 2015 *Progr. Part. Nucl. Phys.* **82** 78
[5] Röpke G, Schuck P, Funaki Y, Horiuchi H, Zhongzhou Ren, Tohsaki A, Chang Xu, Yamada T, and Bo Zhou 2014 *Phys. Rev.* C **90** 034304
[6] Xu C *et al* 2016 *Phys. Rev.* C **93** 011306(R)
[7] Tarbert C M *et al* 2014 *Phys. Rev. Lett.* **112** 242502
[8] Misicu S and Esbensen H 2007 *Phys. Rev.* C **75** 034606; **76** 054609
[9] Bo Zhou *et al* 2012 *Phys. Rev.* C **86** 014301
[10] Bo Zhou *et al* 2014 *Phys. Rev.* C **89** 034319

IOP Conf. Series: Journal of Physics: Conf. Series **863** (2017) 012007 doi:10.1088/1742-6596/863/1/012007

Alpha clustering near nuclear surface and harmonic-oscillator excitations

Horiuchi W[1] and Suzuki Y[2,3]

[1]Department of Physics, Hokkaido University, Sapporo 060-0810, Japan
[2]Department of Physics, Niigata University, Niigata 950-2181, Japan
[3]RIKEN Nishina Center, Wako 351-0198, Japan

E-mail: whoriuchi@nucl.sci.hokudai.ac.jp

Abstract. We quantify how it is difficult to describe an alpha(α)-cluster state with single-particle harmonic-oscillator (HO) bases in the low-lying ^{16}O states by counting the number of HO quanta of ^{12}C+n+n+p+p five-body wave functions. We also discuss how many HO quanta are needed for describing a localized α cluster near the nuclear surface towards understanding of the shell and cluster coexistence in heavier nuclei.

1. Introduction

One of the characteristic features of an atomic nuclear system is coexistence and competition of the shell and cluster structure. Especially, describing a localized α clustering near the nuclear surface is essential to understand excitations of light nuclei as well as α-decay phenomena in heavy nuclei. Since these two different aspects are often studied with the different models, a unified description is desired for deep understanding of the nuclear clustering phenomena.

A simultaneous description of the ground and the first excited 0^+ states in ^{16}O is one of the most famous and long-standing problems in nuclear theory. Calculations based on nuclear shell model have still been difficult to describe them even in present-day large-scale calculations [1, 2]. Among many efforts for resolving the problem, Ref. [3, 4] showed that both the ground and first excited 0^+ states could be understood naturally if ^{12}C+α cluster structure was considered.

Recently we showed that both the 0^+ states were described very well with a ^{12}C+n+n+p+p five-body model [5]. Five-body dynamics with Pauli constraint from the ^{12}C core was fully taken into account by using explicitly correlated Gaussian bases with use of stochastic sampling for huge number of variational parameters. We concluded that, without an assumption of the α cluster, the first excited state had well-developed ^{12}C+α cluster structure.

In this contribution, we quantitatively discuss how it is difficult to describe the nuclear α-clustering near nuclear surface by means of the HO excitations. In the next section, we briefly explain the explicitly correlated Gaussian method. The results of the ^{12}C+n+n+p+p five-body model calculations are reviewed in Sec. 3. With counting the number of HO quanta, we quantify the difficulty of representing the cluster state with the shell-model basis. In Sec. 3, we discuss how many HO quanta are needed to have an α-cluster near the nuclear surface by using a simple α-cluster model wave function. Conclusions and future perspectives are given in Sec. 4.

2. Correlated Gaussian method

In order to study the nuclear clustering and shell structure in a unified manner, here we take a powerful and a flexible few-body approach, that is, correlated Gaussian (CG) method [6, 7, 8]. Many examples have confirmed that the method can describe complex few-body dynamics, e.g., describing different types of structure and correlated motion of the particles, and accurately describing the tail of the wave function in the asymptotic region. The method has been applied not only to nuclear physics but also other quantum mechanical fields (See recent review [9]). Our strategy to understand the coexistence of the cluster and shell structure is the following: First, we choose degrees of freedom of interest and determine the Hamiltonian. Two-body interactions are chosen consistently with properties of the two-body system, e.g., energy, phase-shift. Once the Hamiltonian is determined, the many-body Schrödinger equation is accurately solved without recourse to any many-body approximation. Appearance and distortion of clusters are naturally taken into account within the degrees of freedom chosen in the calculation.

A nuclear system is specified by the angular momentum J, its z component M_J, parity π, and isospin quantum numbers T, M_T. The total wave function is expressed by a combination of various components of orbital and spin angular momenta. Here the basis function we use takes a general form in the LS coupling scheme

$$\Phi^{\pi}_{(LS)JM_J,TM_T} = \mathcal{A}\left\{[\phi^{\pi}_{L} \times \chi_S]_{JM_J}\, \eta_{TM_T}\right\}, \tag{1}$$

where $\mathcal{A}$ is the antisymmetrizer for nucleons, and the symbol $[\phi^{\pi}_{L} \times \chi_S]_{JM_J}$ stands for angular momentum coupling. The total spin (isospin) function χ_S (η_T) is constructed by a successive coupling of all spin (isospin) functions of a nucleon.

For the spatial part of the basis function, ϕ^{π}_{L}, we use the CG. Let $\boldsymbol{x}=(\boldsymbol{x}_i)$ denote a set of the Jacobi coordinates excluding the center-of-mass coordinate. We express ϕ^{π}_{L} as a combination of CG [6, 7] and the two global vectors [8]

$$\phi^{\pi}_{(L_1L_2)LM_L}(A, u_1, u_2) = \exp(-\tilde{\boldsymbol{x}}A\boldsymbol{x})[\mathcal{Y}_{L_1}(\tilde{u}_1\boldsymbol{x}) \times \mathcal{Y}_{L_2}(\tilde{u}_2\boldsymbol{x})]_{LM_L} \tag{2}$$

with a solid harmonic $\mathcal{Y}_\ell(\boldsymbol{r}) = r^\ell Y_\ell(\hat{\boldsymbol{r}})$. Here $\tilde{\boldsymbol{x}}A\boldsymbol{x} = \sum_{i,j} A_{ij}\boldsymbol{x}_i\cdot\boldsymbol{x}_j$ and $\tilde{u}_i\boldsymbol{x} = \sum_k (u_i)_k\boldsymbol{x}_k$, where the positive-definite symmetric matrix A, the vectors of u_1 and u_2 are variational parameters. It is noted that all coordinates are explicitly correlated through them. Both u_1 and u_2 define the global vectors, $\tilde{u}_1\boldsymbol{x}$ and $\tilde{u}_2\boldsymbol{x}$, that are responsible for the rotational motion of the system. With the two global vectors, any L^π states except for 0^- can be constructed with a suitable choice of L_1 and L_2. An advantage of the CG is that it keeps its functional form under a linear transformation of the coordinates [6, 7], which is a key for describing both the cluster and shell structure in a unified manner.

3. α clustering near the nuclear surface

3.1. The low-lying 0^+ states of ^{16}O

A ^{12}C$+n+n+p+p$ five-body model successfully describes the ground and first excited 0^+ states of ^{16}O [5]. Here we review Ref. [5] and discuss how the cluster state is represented by the HO excitations. The five-body system is characterized by the Hamiltonian which consists of an NN potential for valence nucleons and a ^{12}C-nucleon (CN) potential. A central Minnesota (MN) potential [10] is employed as the NN potential. To fine tune the binding energy of the α particle, the potential strengths are multiplied by 0.9814. Symmetrized Woods-Saxon and its derivative forms are assumed for the central and spin-orbit CN potential, respectively. Their parameters are determined so as to reproduce the low-lying spectrum of ^{13}C with $1/2^-$, $1/2^+$, and $5/2^+$. Our CN potential is deep enough to accommodate some redundant or Pauli-forbidden states. To eliminate such states we impose the orthogonality constraint for the relative

IOP Conf. Series: Journal of Physics: Conf. Series **863** (2017) 012007 doi:10.1088/1742-6596/863/1/012007

motion of the valence nucleons, which is practically achieved by adding a pseudo potential to the Hamiltonian [11]. Here we assume the HO wave functions of $0s_{1/2}$ and $0p_{3/2}$ as the occupied orbits of ^{12}C. The HO frequency is set to be 16.0 MeV, which reproduces the size of the ^{12}C ground state. The core excitation of ^{12}C may be included, for example, based on a collective model [12]. However, for the sake of simplicity, the core excitation is ignored in this calculation.

We express the wave function in terms of a linear combination of many basis states of Eq. (1). Each basis element contains so many variational parameters that discretizing them on grids leads to an enormous dimension of at least 10^{10}. The stochastic variational method [6, 7] provides us with an efficient way to obtain an accurate solution with reducing the number of bases. We test a number of candidate bases with stochastic sampling and select the best one among them, and increase the basis dimension one by one until the convergence is reached.

Energy convergence is very slow and more than 9000 basis states were required [5]. This number is very large compared to a four-nucleon system with the same MN interaction that requires only few tens of basis states [6]. The present problem belongs to a class of quantum few-body problem with orthogonality constraints, which often appears in atomic and subatomic physics when the system contains composite particles [7]. Solving such a problem is quite challenging. In the present case, the most basis states are used for eliminating the forbidden states. Since the valence nucleons are strongly correlated, the basis states may have a large overlap with the forbidden states, and thus many basis states are needed to get a converged solution. After the convergence two 0^+ states appear below the ^{12}C+α threshold and their energies are both remarkably close to experiment [5].

For quantitative discussion of the clustering degree, we calculate the ^{12}C+α spectroscopic amplitude of ^{16}O (See Fig. 3 of Ref. [5]). For the 0_1^+ state, the highest peak is located below the touching distance of ^{12}C+α (~ 5.9 fm). In such a case, the α particle is strongly distorted by the core nucleus. In fact, a spectroscopic factor (SF), is small (0.105), whereas the ^{12}C+α-cluster-model calculation gives 0.300 [3]. In this study, we do not assume the α cluster, and thus the distortion of the α cluster shows up naturally. In the 0_2^+ state, the amplitude is much larger than that of the 0_1^+ state and peaked at the touching distance of ^{12}C+α, showing very long tail which suggests the well developed α-cluster structure. The SF is 0.680 which is in agreement with the value of the ^{12}C+α calculation, 0.679. The phase space of the first excited state of ^{16}O is exhausted by the ^{12}C+α cluster component.

Why is it so difficult to describe the 0_2^+ state with the ordinary shell model truncation? To answer this, it is useful to calculate the probability distribution of the total HO quanta Q in our wave functions (See Fig. 4 of Ref. [5]). The oscillator frequency is set to be the same as the forbidden states in the ^{12}C core. For the ground state, the distribution starts with $Q = 4$, which may correspond to the configuration of $(0p_{1/2})_\nu^2(0p_{1/2})_\pi^2$. The distribution converges rapidly with increasing Q. Since a few major shell configurations, $Q \leq 8$, are required to cover more than 90% of the model spaces, the state may be described with the shell model picture.

As predicted by the cluster model [3, 4], the distribution of the first excited state of ^{16}O is quite different from the ground state. Since the ground state exhausts the $Q = 4$ component, the component for the excited state becomes small approximately 10%. The HO quanta are again peaked at $Q = 12$ and widely distributed showing still undiminished probability even at $Q = 30$. One needs at least $Q = 26$ to cover more than 90% of our model space. Therefore, it is practically difficult to describe such a state by the HO basis with the usual major shell truncation.

The α-cluster structure is a key to understand $N = Z$ and doubly closed nuclei. Interestingly, ^{4}He itself also exhibits the coexistence of shell and cluster structure in its spectrum [13]. The first excited state is 0^+ as like ^{16}O and parity inverted partners are actually observed. Similarly to the ^{16}O case, if one considers the cluster excitation of $3N + N$ (^{3}He+n and ^{3}H+p), the first excited state can be regarded as the excitation of the relative motion function between $3N$ and

N. Differently from the case of ^{16}O, the two cluster have the spin and isospin degrees of freedom, and thus seven negative parity states are possible having $3N+N$ with the P-wave relative motion, excluding the center-of-mass excitation with $T=0$ state. In fact, the occupation probability of the HO quanta for those states show extended distributions indicating the $(3N+N)$-cluster structure [14]. For the 0_2^+ state, the high Q component is used for describing the S-wave relative motion wave function between $3N$ and N. Inversion doublet partners also need high Q components in order to describe the asymptotic behavior of the P-wave relative motion wave function.

3.2. HO components of α-cluster wave function near nuclear surface

Let us consider more about the role of the HO components to describe such cluster states with a simple α-cluster wave function. Detailed discussions are given in the recent review book [15]. We assume that a $(0s)^4$-α cluster located at position $\boldsymbol{s}$ is represented by the product of the Gaussian-wave packet with width parameter $\nu^{-1/2}$

$$\phi_{\boldsymbol{s}}^{\nu}(\boldsymbol{r}) = \left(\frac{2\nu}{\pi}\right)^{3/4} \exp(-\nu(\boldsymbol{r}-\boldsymbol{s})^2). \tag{3}$$

The spin- and isospin-wave functions are totally antisymmetrized. Therefore, the spacial part of the wave function is totally symmetric. The probability that $\phi_{\boldsymbol{s}}^{\nu}$ contains shell model orbits with the orbital angular momentum l is calculated by

$$P_\eta(l) = \sum_{m=-l}^{l} \frac{1}{4\pi}\int d\hat{s}\int_0^\infty dr r^2 |\langle Y_{lm}(\hat{\boldsymbol{r}})|\phi_{\boldsymbol{s}}^{\nu}(\boldsymbol{r})\rangle|^2 = (2l+1)i_l(\eta)e^{-\eta}, \tag{4}$$

where $i_l(z)$ is the modified spherical Bessel function, and $\eta=\nu s^2$ is a dimensionless quantity which can be a measure of spatial localization of the cluster. In case that the cluster size is the free-α particle ($\nu = 0.26$ fm^{-2}), η quickly increases, while if ν is taken to be the standard value, $\nu = 0.48A^{-1/3}$ fm^{-2}, η gradually increases. Now we put the free-α particle around the nuclear surface at $|\boldsymbol{s}| = 1.1A^{1/3}$ fm. In cases of ^{16}O ($\eta = 2.0$) and ^{212}Po ($\eta = 11$), cumulative sum $\sum_{l=0}^{l_{\rm max}} P_\eta(l)$ exceeds 80% at $l_{\rm max}$ =3 and 6 for $\eta = 2.0$ and 11, respectively.

Applying this analysis to the $(0s)^4$-α-cluster wave function, we can discuss partial-wave components of the wave function [15]. For ^{16}O case, $l_{\rm max} = 2$ is required for satisfying about 90% of the probability following the distribution in Eq. (4). The most of HO quanta are used to describe the spatial extent of the ^{12}C+α S-wave relative motion wave function rather than increasing the partial angular momentum. This information will be a guide to a theoretical description of the α clustering. In case of ^{212}Po, a famous α emitter, higher partial wave ($l_{\rm max} = 7$) is required to describe the surface clustering around the ^{208}Pb core.

4. Summary and future perspectives

Applying the ^{12}C+n+n+p+p five-body model to the low-lying spectrum of ^{16}O, we succeeds in describing both the shell-model and cluster configurations in a unified scheme. By counting the number of the harmonic-oscillator (HO) quanta of the realistic five-body wave functions as well as the simple α-cluster ones, we quantify how it is difficult to describe the cluster configurations with the HO shell-model basis.

It is interesting to study the α clustering in the other spin-parity states of ^{16}O as important inputs for astrophysical reactions. A description of the α decay of ^{212}Po has of particular interest as a typical phenomena of α-clustering near the nuclear surface. Those projects are underway and will be reported, hopefully, in the next Cluster conference.

Acknowledgments
This work was in part supported by JSPS KAKENHI Grant number JP15K05072.

References

[1] Włoch M, Dean D J, Gour J R, Hjorth-Jensen M, Kowalski, Papenbrock T and Piecuch P 2005 *Phys. Rev. Lett.* **94** 212501
[2] Maris P, Vary J P and Shirokov A M 2009 *Phys. Rev.* C **79** 014308
[3] Suzuki Y 1976 *Prog. Theor. Phys.* **55** 1751
[4] Suzuki Y 1976 *Prog. Theor. Phys.* **56** 111
[5] Horiuchi W and Suzuki Y 2014 *Phys. Rev.* C **89** 011601(R)
[6] Varga K and Suzuki Y 1995 *Phys. Rev.* C **52** 2885
[7] Suzuki Y and Varga K 1998 *Stochastic Variational Approach to Quantum-Mechanical Few-Body Problems, Lecture Notes in Physics Monographs* Vol. m54 (Berlin: Springer)
[8] Suzuki Y, Horiuchi W, Orabi M and Arai K 2008 *Few-Body Syst.* **42** 33
[9] Mitroy J, Bubin S, Horiuchi W, Suzuki Y, Adamowicz L, Cencek W, Szalewicz K, Komasa J, Blume D and Varga K 2013 *Rev. Mod. Phys.* **85** 693
[10] Thompson D R, LeMere M and Tang Y C 1977 *Nucl. Phys.* **A286** 53
[11] Kukulin V I and Pomerantsev V N 1978 *Ann. Phys. (NY)* **111** 330
[12] Horiuchi W and Suzuki Y 2014 *Few-Body Syst.* **55** 121
[13] Horiuchi W and Suzuki Y 2008 *Phys. Rev.* C **78** 034305
[14] Horiuchi W and Suzuki Y 2014 *Phys. Rev.* C **90** 034001
[15] Suzuki Y and Horiuchi W 2016 *Emergent phenomena in Atomic Nuclei from Large-scale Modeling: a Symmetry-guided Perspective* ed K D Launey (World Scientific) Chapter 7 173-197, ISBN: 978-981-3146-04-4

No-core Symplectic Model: Exploiting Hidden Symmetry in Atomic Nuclei

J P Draayer[1], K D Launey[1], T Dytrych[1,2], A C Dreyfuss[1], R B Baker[1] and M E Miora[3]

[1] Dept. of Physics & Astronomy, Louisiana State University, Baton Rouge, LA 70803, USA
[2] Nuclear Physics Institute, 250 68 Řež, Czech Republic
[3] Rollins College, Winter Park, FL 32789, USA

Abstract. We report on recent developments within the framework of the no-core symplectic shell model (NCSpM) that complements the no-core shell model (Navrátil, Vary, and Barrett) by exploiting the algebraic features of the symplectic shell model (Rowe and Rosensteel) while also allowing for high-performance computing applications, but in highly truncated, physically relevant subspaces of the complete space. The leading symplectic symmetry typically accounts for 70% to 90% of the structure of the low-lying states, a result that is only moderately dependent on the details of the selected inter-nucleon interaction. Examples for ^{6}Li, ^{12}C, ^{16}O, and ^{20}Ne are shown to illustrate the efficacy the NCSpM, and as well the strong overlap with cluster-like and pairing configurations that dominate the dynamics of low-lying states in these nuclei.

1. Introduction

The no-core shell model (NCSM) is a complete hamiltonian-based microscopic theory for studying the structure of atomic nuclei [1, 2]. The simplest implementation of the NCSM ignores all symmetries except for time reversal, parity and translational invariance; it is ideally suited for straightforward high performance computing (HPC) applications. The symplectic shell model [3, 4], developed by Rowe and Rosensteel, which predates the NCSM, establishes an algebraic framework that enables a partitioning of the complete space into physically relevant subspaces that respect rotational invariance and exposes important dynamical symmetries. In particular, in one limit the symplectic model reduces to Bohr-Mottelson theory [5], while in another it reduces to the Elliott model [6, 7].

That symplectic Sp(3, ℝ) symmetry is an important symmetry of atomic nuclei can be traced to the fact that, within a microscopic framework that relates to particle position and momentum coordinates, it naturally describes rotations and vibrations of equilibrium deformation (or a set of a reasonable number of deformed configurations). The existence of the symmetry and its slight symmetry breaking has been recently confirmed in light nuclei from first principles in the *ab initio* symmetry-adapted no-core shell model (SA-NCSM) [8, 9]. The SA-NCSM capitalizes on exact as well as partial symmetries that underpin the structure of nuclei and provides remarkable insight into how simple symmetry patterns emerge in the many-body nuclear dynamics from first principles. Complementary to this, a fully microscopic no-core symplectic shell model (NCSpM) [10], by employing the symplectic symmetry and schematic interactions, can reach model spaces beyond the reach of any *ab initio* theory, thereby making possible to probe the

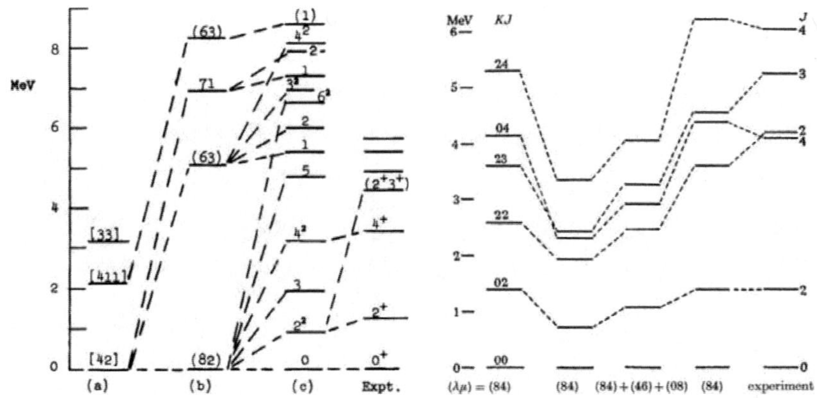

Figure 1. Elliott's SU(3) model applied to *sd*-shell nuclei. Left panel: Spectrum of ^{22}Ne (a) with a Majorana potential, (b) with the addition of the second-order SU(3) Casimir invariant, C_2^{su3}, and (c) with the Majorana potential plus an attractive $Q \cdot Q$ interaction [or (b) with the addition of L^2]. Figure taken from [15]. Right panel: Spectrum of ^{24}Mg with a Gaussian central force. Figure taken from [7]. The vertical axis in both figures represents energy in MeV. Note the importance of the most deformed SU(3) configuration (8 2) in ^{22}Ne and (8 4) in ^{24}Mg for reproducing the experimental low-lying states.

physics, e.g., of the challenging ^{12}C Hoyle state and its rotational band, as well as negative-parity states, together with highly deformed intermediate-mass nuclei. The NCSpM reveals the nature of collectivity with outcomes that corroborate the results of earlier studies [4, 11, 12]. It provides a description of bound states in terms of a relatively small fraction of the complete space when the latter is expressed in an $(LS)J$ coupling scheme with the spatial configurations further organized into irreducible representations (irreps) of SU(3) and then into irreps of the embedding Sp(3, $\mathbb{R}$). The SU(3) plays a key role as shown by the seminal work of Elliott [6, 7] (Fig. 1). The SU(3)-symmetry dominance has been also observed in heavier nuclei, where pseudo-spin symmetry and its pseudo-SU(3) complement have been shown to play a similar role in accounting for deformation in the upper *pf* and lower *sdg* shells, and in particular, in strongly deformed nuclei of the rare-earth and actinide regions [13], as well as in many other studies (e.g., [14]).

Concurrently to collective approaches, particle-driven models have been developed, following the success of the independent-particle model of Mayer and Jensen [16, 17]. One of the most successful is the *ab initio* NCSM discussed above. It adopts the harmonic oscillator (HO) single-particle basis characterized by the $\hbar\Omega$ oscillator strength and limited by the $N_{\max}$ cutoff, which is defined as the maximum number of HO quanta allowed in a many-body basis state above the minimum for a given nucleus. It divides the space in "horizontal" HO shells dictated by particle-hole excitations (complementary to a "vertical" space organization in the symplectic model dictated by collectivity). The NCSM has achieved remarkable descriptions of low-lying states of *s*- and *p*-shell light nuclei [1, 18], and is further augmented by several techniques, such as NCSM/RGM [19], Importance Truncation NCSM [20] and Monte Carlo NCSM [21]. This supports and complements results of other first-principle approaches, such as Green's function Monte Carlo (GFMC) [22], Coupled-cluster (CC) method [23], In-medium SRG [24], and Lattice Effective Field Theory (EFT) [25], along with the symmetry-adapted no-core shell model (SA-NCSM) [8] that combines both particle-driven and collective concepts.

2. Exploiting hidden symmetries

2.1. Simple pattern formation from first principles

The *ab initio* symmetry-adapted no-core shell model (SA-NCSM) [8] adopts the first-principle concept and utilizes many-particle basis states that are organized with respect to the physically

IOP Conf. Series: Journal of Physics: Conf. Series **863** (2017) 012008 doi:10.1088/1742-6596/863/1/012008

relevant, deformation-related $\mathrm{SU}(3)_{(\lambda\,\mu)} \overset{\kappa}{\supset} \mathrm{SO}(3)_L$ subgroup chain (for reviews, see [9, 26]), that is, it is a no-core shell model with SU(3)-coupled basis states (hence, the SA-NCSM results obtained in a complete $N_{\max}$ space are equivalent to the $N_{\max}$ NCSM results). This allows the full model space to be down-selected to the physically relevant space.

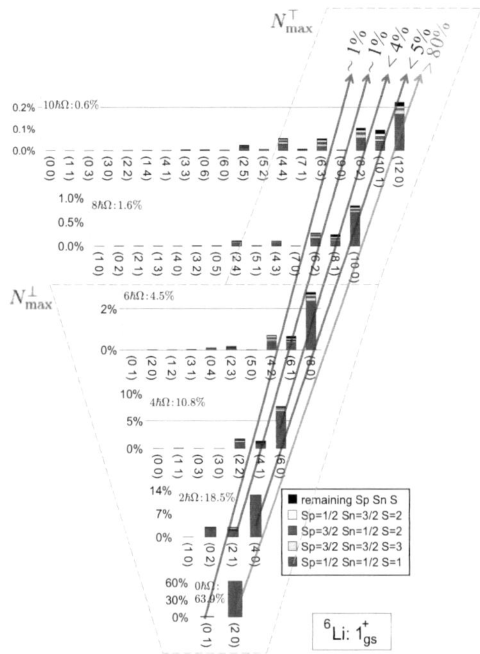

Figure 2. Probability distributions for proton, neutron, and total intrinsic spin components $(S_p S_n S)$ across the Pauli-allowed deformation-related $(\lambda\,\mu)$ values for the 1^+ ground state of ^{6}Li, calculated in 12 HO shells with the JISP16 bare interaction ($\hbar\Omega = 20$ MeV). The concentration of strengths to the far right demonstrates the dominance of collectivity. This supports a *symmetry-guided* model space selection, which implies inclusion of the full space up through $N^{\perp}_{\max}$, and a subset of deformation/spin configurations beyond this, up through $N^{\top}_{\max}$, labeled as $\langle N^{\perp}_{\max}\rangle N^{\top}_{\max}$. The projection onto symplectic vertical slices (with probability $\geq 1\%$) is schematically illustrated for ^{6}Li by arrows and clearly reveals the preponderance of a single symplectic irrep (vertical cone). Figure adapted from Ref. [8].

The *ab initio* SA-NCSM results for *p*-shell nuclei reveal a dominance of configurations of large deformation in the $0\hbar\Omega$ subspace. For example, the *ab initio* $N_{\max} = 10$ SA-NCSM results with the bare JISP16 realistic interaction [27] (similarly, for the bare N^{3}LO realistic interaction [28]) for the 1^+ ground state (*g.st.*) and its rotational band for ^{6}Li (similarly for ^{6}He, ^{8}Be, and ^{12}C) reveal the dominance of the leading $0\hbar\Omega$ (2 0) irrep and its symplectic excitations (Fig. 2). The outcome points to a remarkable feature common to the low-lying states of nuclei that has heretofore gone unrecognized in other first-principle studies; namely, the emergence, without a priori constraints, of simple orderly patterns that favor strongly deformed configurations and low spin values. This feature confirms the dominant role the SU(3) and $\mathrm{Sp}(3,\mathbb{R})$ symmetries play in nuclear dynamics and is central to expanding the reach of first-principle studies to heavier nuclei [29]. For example, model spaces that expand up to 20 HO shells for ^{12}C are feasible and en route to be employed. This allows the SA-NCSM to advance an extensible microscopic framework for studying nuclear structure and reactions that capitalizes on advances being made in *ab initio* methods while exploiting symmetries found to dominate the dynamics.

2.2. The elusive Hoyle state and negative-parity states in ^{12}C

The Hoyle state was predicted based on observed abundances of heavy elements in the universe [32] and has attracted much recent attention both in theory (e.g., see [25,33,34]) and experiment (e.g., [35–37]). The findings of the no-core symplectic shell model (NCSpM) inform key features of the primary physics responsible for the emergent phenomena of large deformation and alpha-cluster substructures in ^{8}Be and ^{12}C [10,38], as well as enhanced collectivity in intermediate-mass nuclei, such as ^{20}O, ^{20,22,24}Ne, ^{20,22}Mg, and ^{24}Si [39] (Fig. 3).

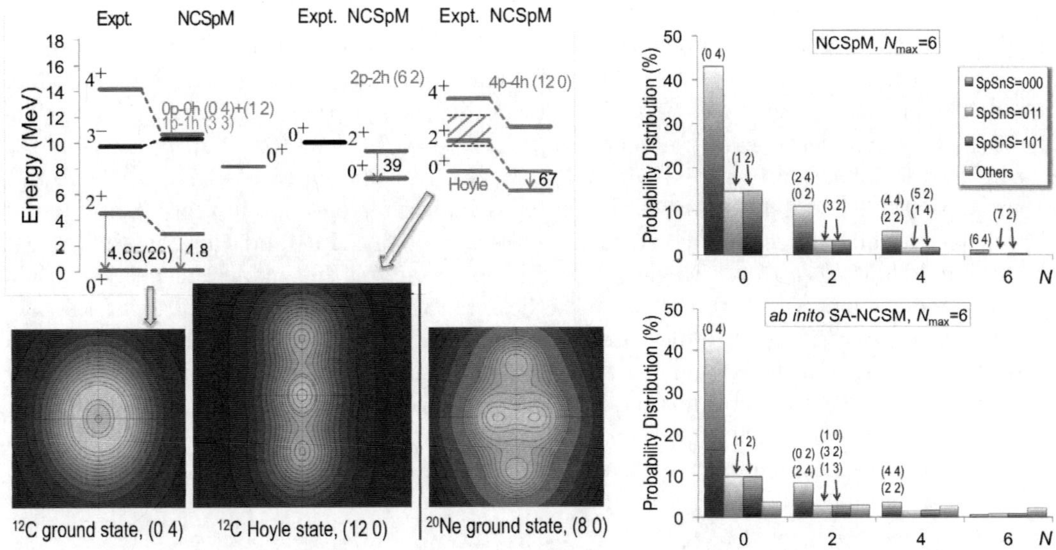

Figure 3. Left: ^{12}C energy spectrum (in MeV) and $B(E2)$ rates (in W.u.) calculated by the NCSpM using the schematic interaction of Eq. (1) with the bare JISP16 NN and 5 symplectic irreps (shown above each band) expanded up to $N_{\max} = 20$, and compared to experiment ("Expt.") [30]. One-body density profiles *in the intrinsic frame* reveal a torus-like shape for the ^{12}C 0^+_{gs} and overlapping clusters in the Hoyle state. Right: Probability distribution across total HO quanta excitations for the ^{12}C 0^+_{gs} – close similarity is observed for NCSpM (top) and SA-NCSM (bottom) for $N_{\max} = 6$ and $\hbar\Omega = 18$ MeV [10]. Dominant ($\geq 1\%$) SU(3) modes are also shown. Very similar results are obtained for 2^+_1 and 4^+_1 [31].

The NCSpM is a fully microscopic no-core shell model that can be employed in model spaces beyond current NCSM limits. It uses a symplectic Sp(3, $\mathbb{R}$) basis but is not limited to Sp(3, $\mathbb{R}$)-preserving interactions. The NCSpM employed within a full model space up through a given $N_{\max}$ coincides with the NCSM for the same $N_{\max}$ cutoff. However, in the case of the NCSpM, the symplectic irreps divide the space into 'vertical slices' that are comprised of basis states of a definite deformation $(\lambda\,\mu)$. Hence, the model space can be reduced to only a few important configurations that are chosen among all possible Sp(3, $\mathbb{R}$) irreps within the $N_{\max}$ model space. We employ a Hamiltonian with an effective interaction derived from the long-range expansion of the two-body central nuclear force,

$$H_{\text{eff}} = H_0 + \frac{\chi}{2}\frac{\left(e^{-\gamma Q\cdot Q} - 1\right)}{\gamma} + V^{\text{SB}}_{NN}, \tag{1}$$

where V^{SB}_{NN} is a symmetry-breaking term applied to symplectic bandheads (we adopt either a spin-orbit term, $-\kappa \sum_{i=1}^{A} l_i \cdot s_i$, or a realistic interaction, such as the bare interactions, JISP16 [27] or chiral N3LO [28] NN). The Sp(3, $\mathbb{R}$)-preserving part of the Hamiltonian includes: the spherical HO potential, which together with the kinetic energy yields the HO Hamiltonian, $H_0 = \sum_{i=1}^{A}\left(\frac{\mathbf{p}_i^2}{2m} + \frac{m\Omega^2\mathbf{r}_i^2}{2}\right)$, and the important $\frac{1}{2}Q\cdot Q = \frac{1}{2}\sum_s q_s \cdot \left(\sum_t q_t\right)$ interaction, which realizes the physically relevant interaction of each particle with the total quadrupole moment of the nuclear system (for a valence shell and very small γ, the Elliott model is recovered). The considerable average contribution, $\langle Q\cdot Q\rangle$, of $Q\cdot Q$ is removed [40]. We take the coupling constant χ to be proportional to $\hbar\Omega$ and, to leading order, to decrease with the total number of HO excitations, as shown by Rowe [41] based on self-consistent arguments.

The NCSpM outcome reveals a quite remarkable agreement with the experiment and *ab initio* results (Fig. 3) [10]. Energies and eigenstates for ^{12}C were calculated for $\hbar\Omega = 18$ MeV given by the empirical estimate $\approx 41/A^{1/3} = 17.9$ MeV. The results are shown for $N_{\max} = 20$, which we found sufficient for convergence. This $N_{\max}$ model space is further reduced by selecting the five most relevant symplectic irreps, extended up through $N_{\max} = 20$ (6.6×10^3 positive-parity basis states). In comparison to experiment (Fig. 3), the outcome for $\gamma = -1.71 \times 10^{-4}$ reveals that the lowest 0^+, 2^+, and 4^+ states of the 0p-0h symplectic irreps closely reproduce the *g.st.* rotational band, while the lowest 0^+ states of the $4\hbar\Omega$ 4p-4h (12 0) and the $2\hbar\Omega$ 2p-2h (6 2) irreps are found to lie close to the Hoyle state and the 10-MeV 0^+ resonance, respectively. The NCSpM successfully reproduces the 3^- state by the $1\hbar\Omega$ (3 3), as well as other observables, such as mass rms radii, electric quadrupole moments and $B(E2)$ transition strengths (Fig. 3). The model is also applicable to the low-lying states of ^{8}Be and *sd*-shell nuclei without any adjustable parameters [38,39]. This suggests that the fully microscopic NCSpM model has indeed captured an important part of the underlying physics and informs key features of the interaction and structure primarily responsible for the formation of such simple patterns.

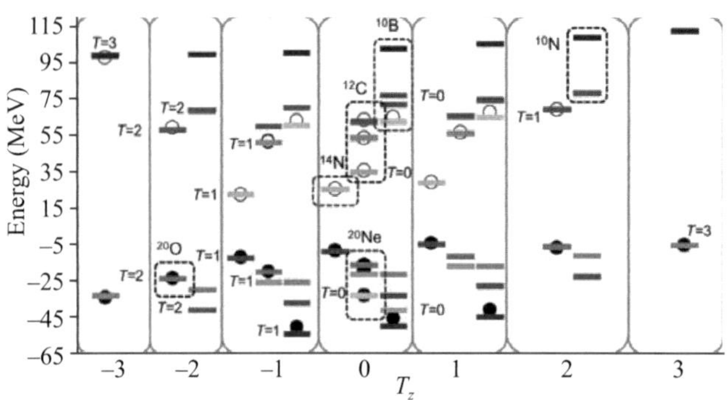

Figure 4. Exact solutions (lines) for a pairing Hamiltonian with experimentally deduced non-degenerate single-particle energies, like-nucleon and proton-neutron isovector pairing, proton-neutron isoscalar term, and a Coulomb potential, as compared to experimental 0^+ states (circle).

In addition, for the nuclei under consideration, we explore the formation of like-nucleon and proton-neutron (*pn*) isovector pairs together with *pn* isoscalar correlations, of importance to two-nucleon correlations in break-up reaction channels. These investigations are carried forward in the framework of exact pairing plus a mean field with single-particle energies (s.p.e.) that are empirically derived (in general, they can also be determined in a mean-field approach, including density functional theory for heavy nuclei, and in a spherical shell model). The method provides, for the first time, exact solutions when non-degenerate s.p.e. and the challenging *pn* correlations are considered, based on recent mathematical developments [42,43]. Results for light up through medium-mass nuclei ($10 \le A \le 62$) are found to remarkably agree with experiment [44] (see Fig. 4 for selected *p*- and *sd*-shell nuclei including the deformed ^{20}Ne). The outcome suggests that enhanced deformation does not preclude pair formation – indeed, collective modes appear to remain dominant even in the presence of pairing (see also [45,46]), while the most noticeable effect of the latter is typically in reducing the moment of inertia.

In short, we have demonstrated the emergence of orderly patterns in nuclei from first principles, with associated hidden symmetry exploited, in turn, in the NCSpM. We have illustrated the efficacy of the NCSpM to gain new insights into the microscopic structure of nuclei, as well as into cluster-like and pairing correlations.

Acknowledgments

We thank D. Rowe, G. Rosensteel, and C. Bahri, together with the PetaApps Collaboration, J. P. Vary, P. Maris, U. Catalyurek, and M. Sosonkina, for useful discussions. This work was supported by the U.S. NSF (OCI-0904874, ACI -1516338), the U.S. DOE (DE-SC0005248),

SURA (Southeastern Universities Research Association), Louisiana State University (through its direct support of J.P.D.'s work with SURA), and the Czech Science Foundation under Grant No. 16-16772S. This work benefitted from HPC resources provided by Blue Waters, the Louisiana Optical Network Initiative, as well as LSU (`www.hpc.lsu.edu`).

References

[1] Navrátil P, Vary J P and Barrett B R 2000 *Phys. Rev. Lett.* **84** 5728
[2] Barrett B, Navrátil P and Vary J 2013 *Prog. Part. Nucl. Phys.* **69** 131
[3] Rosensteel G and Rowe D J 1977 *Phys. Rev. Lett.* **38** 10
[4] Rowe D J 1985 *Reports on Progr. in Phys.* **48** 1419
[5] Bohr A and Mottelson B R 1969 *Nuclear Structure* vol 1 (Benjamin, New York)
[6] Elliott J P 1958 *Proc. Roy. Soc.* A **245** 128
[7] Elliott J P and Harvey M 1962 *Proc. Roy. Soc.* A **272** 557
[8] Dytrych T, Launey K D, Draayer J P, Maris P, Vary J P *et al.* 2013 *Phys. Rev. Lett.* **111** 252501
[9] Launey K D, Dytrych T and Draayer J P 2016 *Prog. Part. Nucl. Phys.* **89** 101 (invited review)
[10] A C Dreyfuss K D Launey T D, Draayer J P and Bahri C 2013 *Phys. Lett. B* **727** 511
[11] Bahri C and Rowe D J 2000 *Nucl. Phys.* A **662** 125
[12] Draayer J, Weeks K and Rosensteel G 1984 *Nucl. Phys.* **A419** 1
[13] Bahri C, Draayer J and Moszkowski S 1992 *Phys. Rev. Lett.* **68** 2133
[14] Zuker A, Retamosa J, Poves A and Caurier E 1995 *Phys. Rev.* C **52** R1741
[15] Harvey M 1968 *Adv. Nucl. Phys.* **1** 67
[16] Mayer M G 1949 *Phys. Rev.* **75** 1969
[17] Haxel O, Jensen J H D and Suess H E 1949 *Phys. Rev.* **75** 1766
[18] Maris P, Shirokov A M and Vary J P 2010 *Phys. Rev.* C **81** 021301(R)
[19] Navrátil P, Quaglioni S, Stetcu I and Barrett B R 2009 *J. Phys.* G **36** 083101
[20] Roth R and Navrátil P 2007 *Phys. Rev. Lett.* **99** 092501
[21] Abe T, Maris P, Otsuka T, Shimizu N, Utsuno Y and Vary J 2012 *Phys. Rev.* C **86** 054301
[22] Pieper S C, Varga K and Wiringa R B 2002 *Phys. Rev.* C **66** 044310
[23] Wloch M, Dean D J, Gour J R, Hjorth-Jensen M, Kowalski K *et al.* 2005 *Phys. Rev. Lett.* **94** 212501
[24] Tsukiyama K, Bogner S K and Schwenk A 2011 *Phys. Rev. Lett.* **106** 222502
[25] Epelbaum E, Krebs H, Lee D and Meissner U G 2011 *Phys. Rev. Lett.* **106** 192501
[26] Dytrych T, Sviratcheva K D, Draayer J P, Bahri C and Vary J P 2008 *J. Phys.* G **35** 123101
[27] Shirokov A, Vary J, Mazur A and Weber T 2007 *Phys. Lett. B* **644** 33
[28] Entem D R and Machleidt R 2003 *Phys. Rev.* C **68** 041001
[29] Dytrych T, Launey K D and Draayer J P 2014 *Symmetry-adapted no-core shell model* (McGraw-Hill)
[30] Launey K D, Dreyfuss A C, Draayer J P, Dytrych T and Baker R B 2014 *J. Phys.: Conf. Series* **569** 012061
[31] Dreyfuss A C, Launey K D, Dytrych T, Draayer J P, Baker R B *et al.* 2016 *submitted to Phys. Rev.* C
[32] Hoyle F 1954 *Astrophys. J. Suppl. Ser.* **1** 121
[33] Chernykh M, Feldmeier H, Neff T, von Neumann-Cosel P and Richter A 2007 *Phys. Rev. Lett.* **98** 032501
[34] Khoa D T, Cuonga D C and Kanada-En'yo Y 2011 *Phys. Letts.* B **695** 469
[35] Fynbo H O U *et al.* 2005 *Nature* **433** 136
[36] Freer M *et al.* 2009 *Phys. Rev.* C **80** 041303
[37] Zimmerman W R *et al.* 2013 *Phys. Rev. Lett.* **110** 152502
[38] Launey K D, Dytrych T, Draayer J P, Tobin G K, Ferriss M C *et al.* 2013 Symmetry-adapted No-core Shell Model for Light Nuclei *Proceedings of the 5th Int.l Conf. Fission and properties of neutron-rich nuclei, ICFN5, November 4 - 10, 2012, Sanibel Island, Florida* ed Hamilton J H and Ramayya A V (World Scientific, Singapore) p 29
[39] Tobin G K, Ferriss M C, Launey K D, Dytrych T, Draayer J P and Bahri C 2014 *Phys. Rev.* C **89** 034312
[40] Rosensteel G and Draayer J P 1985 *Nucl. Phys.* A **436** 445
[41] Rowe D J 1967 *Phys. Rev.* **162** 866
[42] Guan X, Launey K D, Xie M, Bao L, Pan F and Draayer J P 2012 *Phys. Rev.* C **86** 024313
[43] Pan F and Draayer J P 2002 *Phys. Rev.* C **66** 044314
[44] Miora M E, Launey K D, Kekejian D, Pan F and Draayer J P 2015 Exact Solutions for Pairing Correlations Among Protons and Neutrons *APS Conference, DNP, Santa Fe, NM, 28-31 October 2015* `http://meetings.aps.org/Meeting/DNP15/Session/EA.80`
[45] Bahri C, Escher J and Draayer J 1995 *Nucl. Phys.* A **592** 171
[46] Pan F, Ding X, Launey K D, Li H, Xu X and Draayer J P 2016 *Nucl. Phys.* A **947** 234

IOP Conf. Series: Journal of Physics: Conf. Series **863** (2017) 012009 doi:10.1088/1742-6596/863/1/012009

Geometric symmetries in light nuclei

R Bijker

Instituto de Ciencias Nucleares, Universidad Nacional Autónoma de México, A.P. 70-543, 04510 México, D.F., México

E-mail: bijker@nucleares.unam.mx

Abstract. The algebraic cluster model is is applied to study cluster states in the nuclei ^{12}C and ^{16}O. The observed level sequences can be understood in terms of the underlying discrete symmetry that characterizes the geometrical configuration of the α-particles, *i.e.* an equilateral triangle for ^{12}C, and a regular tetrahedron for ^{16}O. The structure of rotational bands provides a fingerprint of the underlying geometrical configuration of α-particles.

1. Introduction

Ever since the early days of nuclear physics the structure of ^{12}C has been extensively investigated both experimentally and theoretically [1, 2, 3, 4]. In recent years, the measurement of new rotational excitations of both the ground state [5, 6, 7] and the Hoyle state [8, 9, 10, 11] has generated a lot of renewed interest to understand the structure of ^{12}C and that of α cluster nuclei in general. Especially the (collective) nature of the 0^+ Hoyle state at 7.65 MeV which is of crucial importance in stellar nucleosynthesis to explain the observed abundance of ^{12}C, has presented a challenge to nuclear structure calculations, such as α-cluster models [12], antisymmetrized molecular dynamics [13], fermionic molecular dynamics [14], BEC-like cluster model [15], (no-core) shell models [16, 17], *ab initio* calculations based on lattice effective field theory [18, 19], and the algebraic cluster model [7, 20, 21].

In this contribution, I discuss some properties of the α-cluster nuclei ^{12}C and ^{16}O in the framework of the algebraic cluster model.

2. Algebraic Cluster Model

The Algebraic Cluster Model (ACM) describes the relative motion of the n-body clusters in terms of a spectrum generating algebra of $U(\nu+1)$ where $\nu = 3(n-1)$ represents the number of relative spatial degrees of freedom. For the two-body problem the ACM reduces to the $U(4)$ vibron model [22], for three-body clusters to the $U(7)$ model [20, 23] and for four-body clusters to the $U(10)$ model [21, 24]. In the application to α-cluster nuclei the Hamiltonian has to be invariant under the permuation group S_n for the n identical α particles. Since one does not consider the excitations of the α particles themselves, the allowed cluster states have to be symmetric under the permutation group.

The potential energy surface corresponding to the S_n invariant ACM Hamiltonian gives rise to several possible equilibrium shapes. In addition to the harmonic oscillator (or $U(3n-3)$ limit) and the deformed oscillator (or $SO(3n-2)$ limit), there are other solutions which are of special interest for the applications to α-cluster nuclei. These cases correspond to a geometrical

IOP Conf. Series: Journal of Physics: Conf. Series **863** (2017) 012009 doi:10.1088/1742-6596/863/1/012009

Table 1. Algebraic Cluster Model for two-, three- and four-body clusters

	2α	3α	4α
ACM	$U(4)$	$U(7)$	$U(10)$
Point group	$\mathcal{C}_2$	$\mathcal{D}_{3h}$	$\mathcal{T}_d$
Geometry	Linear	Triangle	Tetrahedron
G.s. band	0^+	0^+	0^+
	2^+	2^+	
		3^-	3^-
	4^+	$4^\pm$	4^+
		5^-	
	6^+	$6^{\pm+}$	$6^\pm$

configuration of α particles located at the vertices of an equilateral triangle for ^{12}C and of a regular tetrahedron for ^{16}O. Even though they do not correspond to dynamical symmetries of the ACM Hamiltonian, one can still obtain approximate solutions for the rotation-vibration spectrum

$$E = \begin{cases} \omega_1(v_1 + \frac{1}{2}) + \omega_2(v_2 + 1) + \kappa\, L(L+1) & \text{for } n = 3 \\ \omega_1(v_1 + \frac{1}{2}) + \omega_2(v_2 + 1) + \omega_3(v_3 + \frac{3}{2}) + \kappa\, L(L+1) & \text{for } n = 4 \end{cases}$$

The rotational structure of the ground-state band depends on the point group symmetry of the geometrical configuration of the α particles and is summarized in Table 1.

The triangular configuration with three α particles has point group symmetry $\mathcal{D}_{3h}$ [20]. Since $\mathcal{D}_{3h} \sim \mathcal{D}_3 \times P$, the transformation properties under $\mathcal{D}_{3h}$ are labeled by parity P and the representations of $\mathcal{D}_3$ which is isomorphic to the permutation group S_3. The corresponding rotation-vibration spectrum is that of an oblate top: v_1 represents the vibrational quantum number for a symmetric stretching A vibration, v_2 denotes a doubly degenerate E vibration. The rotational band structure of ^{12}C is shown in the left panel of Fig. 1.

The tetrahedral group $\mathcal{T}_d$ is isomorphic to the permutation group S_4. In this case, there are three fundamental vibrations: v_1 represents the vibrational quantum number for a symmetric stretching A vibration, v_2 denotes a doubly degenerate E vibration, and v_3 a three-fold degenerate F vibration. The right panel of Fig. 1 shows the rotational band structure of ^{16}O.

3. Electromagnetic transitions

For transitions along the ground state band the transition form factors are given in terms of a product of a spherical Bessel function and an exponential factor arising from a Gaussian distribution of the electric charges, $\mathcal{F}(0^+ \rightarrow L^P; q) = c_L\, j_L(q\beta)\, e^{-q^2/4\alpha}$ [20]. The charge radius can be obtained from the slope of the elastic form factor in the origin $\langle r^2 \rangle^{1/2} = \sqrt{\beta^2 + 3/2\alpha}$. The transition form factors depend on the parameters α and β which can be determined from the first minimum in the elastic form factor and the charge radius.

The transition probabilities $B(EL)$ along the ground state band can be extracted from the form factors in the long wavelength limit

$$B(EL; 0^+ \rightarrow L^P) = \frac{(Ze)^2}{4\pi}\, c_L^2\, \beta^{2L} ,$$

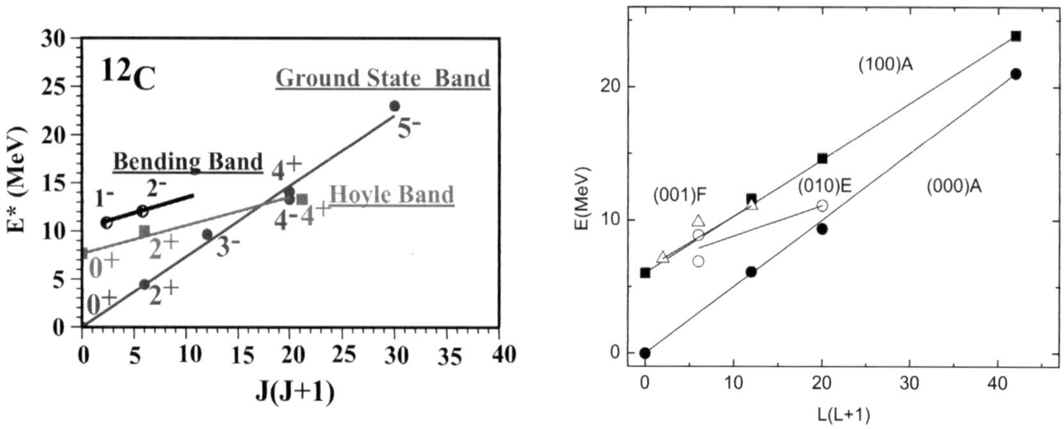

Figure 1. (Color online) Rotational band structure of the ground-state band, the Hoyle band (or A vibration) and the bending vibration (or E vibration) in ^{12}C (left) [7], and the ground-state band (closed circles), the A vibration (closed squares), the E vibration (open circles) and the F vibration (open triangles) in ^{16}O (right) [21].

Table 2. $B(EL)$ values in ^{12}C (top) and ^{16}O (bottom).

^{12}C	Th	Exp		Ref
$B(E2; 2_1^+ \to 0_1^+)$	8.4	7.6 ± 0.4	e^2fm^4	[25, 26, 27]
$B(E3; 3_1^- \to 0_1^+)$	44	103 ± 17	e^2fm^6	[25, 26, 27]
$B(E4; 4_1^+ \to 0_1^+)$	73		e^2fm^8	[25, 26, 27]
$M(E0; 0_2^+ \to 0_1^+)$	0.4	5.5 ± 0.2	fm^2	[25, 26, 27]
^{16}O	Th	Exp		Ref
$B(E3; 3_1^- \to 0_1^+)$	215	205 ± 10	e^2fm^6	[28]
$B(E4; 4_1^+ \to 0_1^+)$	425	378 ± 133	e^2fm^8	[28]
$B(E6; 6_1^+ \to 0_1^+)$	9626		e^2fm^{12}	[28]
$M(E0; 0_2^+ \to 0_1^+)$	0.54	3.55 ± 0.21	fm^2	[28]

with

$$c_L^2 = \begin{cases} \frac{2L+1}{3}\left[1 + 2P_L(-\frac{1}{2})\right] & \text{for } n = 3 \\ \\ \frac{2L+1}{4}\left[1 + 3P_L(-\frac{1}{3})\right] & \text{for } n = 4 \end{cases}$$

The good agreement for the $B(EL)$ values for the ground band in Table 2 shows that both in ^{12}C and in ^{16}O the positive and negative parity states merge into a single rotational band. Moreover, the large values of $B(EL; L_1^P \to 0_1^+)$ indicate a collectivity which is not predicted for simple shell model states. The large deviation for the $E0$ between the first excited 0^+ (Hoyle) state and the ground state indicates that the 0_2^+ state cannot be interpreted as a simple vibrational excitation of a rigid triangular (^{12}C) or tetrahedral (^{16}O) configuration, but rather corresponds to a more floppy configuration with large rotation-vibration couplings. A more detailed study of the electromagnetic properties of α-cluster nuclei in the ACM for non-rigid configurations is in progress.

4. Summary and conclusions

In this contribution, the cluster states in ^{12}C and ^{16}O were interpreted in the framework of the ACM as arising from the rotations and vibrations of a triangular and tetrahedral configuration of α particles, respectively. In both cases, the ground state band consist of positive and negative parity states which coalesce to form a single rotational band. This interpretation is validated by the observance of strong $B(EL)$ values. The rotational sequences can be considered as the fingerprints of the underlying geometric configuration (or point-group symmetry) of α particles.

For the Hoyle band in ^{12}C there are several interpretations for the geometrical configuration of the three α particles. In order to determine whether the geometrical configuration of the α-particles for the Hoyle band is linear, bent or triangular, the measurement of a possible 3^- Hoyle state is crucial, since its presence would indicate a triangular configuration, just as for the ground state band.

Finally, the results presented here for ^{12}C and ^{16}O emphasize the occurrence of α-cluster states in light nuclei with $\mathcal{D}_{3h}$ and $\mathcal{T}_d$ point group symmetries, respectively.

Acknowledgments

This work was supported in part by research grant IN107314 from PAPIIT-DGAPA.

References

[1] Wheeler J A 1937 *Phys. Rev.* **52** 1083
[2] Hafstad L R and Teller E 1938 *Phys. Rev.* **54** 681
[3] Von Oertzen W, Freer M and Kanada-En'yo Y 2006 *Phys. Rep.* **432** 43
[4] Freer M and Fynbo H O U 2014 *Prog. Part. Nucl. Phys.* **78** 1
[5] Freer M *et al* 2007 *Phys. Rev.* C **76** 034320
[6] Kirsebom O S *et al* 2010 *Phys. Rev.* C **81** 064313
[7] Marín-Lámbarri D J, Bijker R, Freer M, Gai M, Kokalova T, Parker D J and Wheldon C 2014 *Phys. Rev. Lett.* **113** 012502
[8] Itoh M *et al* 2011 *Phys. Rev.* C **84** 054308
[9] Freer M *et al* 2012 *Phys. Rev.* C **83** 034314
[10] Zimmerman W R *et al* 2013 *Phys. Rev. Lett.* **110** 152502
[11] Freer M *et al* 2007 *Phys. Rev.* C **76** 034320
[12] Robson D 1978 *Nucl. Phys.* A **308** 381
[13] Kanada-En'yo Y 2007 *Prog. Theor. Phys.* **117** 655
[14] Chernykh M, Feldmeier H, Neff H, Von Neumann-Cosel P and Richter A 2007*Phys. Rev. Lett.* **98** 032501
[15] Funaki Y, Horiuchi H, Von Oertzen W, Ropke G, Schuck P, Tohsaki A and Yamada T 2009 *Phys. Rev.* C **80** 64326
[16] Roth R, Langhammer J, Calci A, Binder S and Navrtil P *Phys. Rev. Lett.* **107** 072501
[17] Dreyfuss A C, Launey K D, Dytrych T, Draayer J P and Bahri C 2013 *Phys. Lett.* B **727** 511
[18] Epelbaum E, Krebs H, Lee D and Meissner U G 2011 *Phys. Rev. Lett.* **106** 192501
[19] Epelbaum E, Krebs H, Lähde T, Lee D and Meissner U G 2012 *Phys. Rev. Lett.* **109** 252501
[20] Bijker R and Iachello F 2000 *Phys. Rev.* C **61** 067305
Bijker R and Iachello F 2002 *Ann. Phys. (N.Y.)* **298** 334
[21] Bijker R and Iachello F 2014 *Phys. Rev. Lett.* **112** 152501
[22] Iachello F 1981 *Chem. Phys. Lett.* **78** 581
[23] Bijker R, Iachello F and Leviatan A 1994 *Ann. Phys. (N.Y.)* **236** 69
Bijker R, Iachello F and Leviatan A 2000 *Ann. Phys. (N.Y.)* **284** 89
[24] Bijker R 2010 *AIP Conference Proceedings* **1323** 28
Bijker R 2012 *J. Phys.: Conf. Series* **380** 012003
[25] Ajzenberg-Selove F 1990 *Nucl. Phys.* A **506** 1
[26] Reuter W, Fricke G, Merle K and Miska H 1982 *Phys. Rev.* C **26** 806
[27] Strehl P and Schucan Th H 1968 *Phys. Lett.* B **27** 641
[28] Tilley D R, Weller H R and Cheves C M 1993 *Nucl. Phys.* A **564** 1

IOP Conf. Series: Journal of Physics: Conf. Series **863** (2017) 012010 doi:10.1088/1742-6596/863/1/012010

Structure of $4n$ nuclei in a formalism of quartets

M Sambataro[1] **and N Sandulescu**[2]

[1] Istituto Nazionale di Fisica Nucleare - Sezione di Catania, Via S. Sofia 64, I-95123 Catania, Italy

[2] National Institute of Physics and Nuclear Engineering, P.O. Box MG-6, Magurele, Bucharest, Romania

E-mail: michelangelo.sambataro@ct.infn.it, sandulescu@theory.nipne.ro

Abstract. We show that low-energy spectra of $4n$ nuclei in the *sd* shell can be described with high accuracy in a formalism of four-body correlated structures (quartets). The states of all $N \geq Z$ nuclei belonging to the $A = 24$ isobaric chain are represented as a superposition of two-quartet states, with quartets being characterized by isospin T and angular momentum J. These quartets are assumed to be those describing the lowest states in ^{20}Ne (T_z=0), ^{20}F (T_z=1) and ^{20}O (T_z=2). We find, in particular, that the spectrum of the self-conjugate nucleus ^{24}Mg can be well reproduced in terms of T=0 quartets only and that, among these, the J=0 quartet plays by far the leading role in the structure of the ground state. These results highlight the importance of quartetting in the structure of $4n$ nuclei.

1. Introduction

Self-conjugate nuclei possess the unique property of carrying an equal number of protons and neutrons distributed over the same single particle orbits. In these nuclei, owing to the charge independence of the nuclear interaction, the isovector proton-neutron (pn) pairing is expected to come into play on equal footing as the like-particle proton-proton and neutron-neutron pairing of the more common $N > Z$ nuclei. In addition, pn pairing is also expected to occur in an isoscalar form. The competition between these two types of pn pairing in $N = Z$ nuclei has been matter of great debate in recent years [1].

We have recently carried out an analysis of the pn pairing in even-even $N = Z$ nuclei both in the isovector and in the isoscalar channels [2, 3, 4, 5]. This analysis has evidenced, on the one hand, that a description of the ground state correlations induced by this interaction in terms of a condensate of collective pairs (of various form [5]) is not satisfactory and, on the other hand, that these correlations can be accounted for to a high degree of precision by approximating the ground state as a product of identical $T = 0$ quartets. $T = 0$ quartets are four-body correlated structures formed by two protons and two neutrons that, in the case of a spherical mean field, are also characterized by a total angular momentum $J = 0$. We have also explored a more sophisticated approximation which consists in letting the quartets to be all distinct and we have verified that it leads to basically exact results in the case of pn isovector pairing in deformed systems [3]. In all cases the quartets have been constructed variationally for each nucleus.

The pn pairing is for sure a key ingredient of the nuclear force for $N = Z$ nuclei but it is nonetheless only a part of it. In the presence of a full Hamiltonian, other quartets are reasonably expected to come into play besides the $T = 0, J = 0$ ones emerging from the analysis of the pn pairing. In the following we will describe how the approach of Refs. [2, 3, 4, 5] has been extended to include quartets with arbitrary values of isospin and angular momentum and to treat realistic interactions. As an application of the method we will provide a description of some *sd* shell nuclei.

The criterion adopted for the selection of the quartets has been that of choosing, as representative of the quartets with a given isospin T, those describing the lowest levels with that isospin in nuclei with four active particles outside the inert core of reference [6]. For applications within the *sd* shell, the inert core is represented by ^{16}O and the nuclei which have therefore been considered for the definition of the quartets are $^{20}_{10}\mathrm{Ne}_{10}$, $^{20}_{9}\mathrm{F}_{11}$ and $^{20}_{8}\mathrm{O}_{12}$. The lowest states of these nuclei are characterized by T =0, 1 and 2, respectively. Each of these states therefore identifies a quartet with given T,J (and a projection $T_z = (N-Z)/2$). We have carried out SM calculations for these three nuclei and, as initial sets of quartets, we have selected those formed by the lowest 6 states in ^{20}Ne ($0 \leq J \leq 6$), the lowest 5 states in ^{20}F ($1 \leq J \leq 5$) and the lowest 9 states in ^{20}O ($0 \leq J \leq 4$). The mixing amplitudes defining each collective quartet have resulted from these SM calculations. As it will be seen below, we have also explored reductions of these sets to identify the most relevant quartets in the structure of the nuclei under investigation. Throughout these calculations we have employed the USDB interaction [7].

2. Formalism

We work in a spherically symmetric mean field and label the single-particle states by $i \equiv \{n_i, l_i, j_i\}$, where the standard notation for the orbital quantum numbers is used. The quartet creation operator is defined as

$$Q^{+}_{\alpha,JM,TT_z} = \sum_{i_1 j_1 J_1 T_1} \sum_{i_2 j_2 J_2 T_2} C^{(\alpha)}_{i_1 j_1 J_1 T_1, i_2 j_2 J_2 T_2} \times \left[[a^{+}_{i_1} a^{+}_{j_1}]^{J_1 T_1} [a^{+}_{i_2} a^{+}_{j_2}]^{J_2 T_2} \right]^{JT}_{MT_z}, \tag{1}$$

where a^{+}_{i} creates a fermion in the single particle state i and $J(T)$ and $M(T_z)$ denote, respectively, the total angular momentum(isospin) and the relative projections. No restrictions on the intermediate couplings J_1T_1 and J_2T_2 are introduced in the calculations. In order to generate the spectra of $4n$ nuclei we perform configuration interaction calculations in spaces built in terms of selected sets of the above quartets.

3. Results

We start our analysis by examining the self-conjugate nucleus ^{24}Mg. In Fig. 1, we compare the experimental spectrum of this nucleus with that resulting from a shell model (SM) calculation and with the spectrum obtained in the quartet model (QM) when all the selected T=0, 1, 2 quartets are taken into account. The quartet approach is seen to reproduce well both the SM ground state correlation energy and the SM excited states up to an energy of about 9 MeV. The SM is in turn able to fit well the experimental levels.

Having verified that the selected sets of T=0, 1, 2 quartets are sufficient to describe the low-energy spectrum of ^{24}Mg, it becomes of interest to investigate the role of the different quartets. We begin by focusing on isospin. In Fig. 1, on the right hand side, we show the theoretical spectrum obtained in the quartet formalism when only T=0 quartets are retained. On top of the lowest levels we also show the overlaps with the corresponding SM eigenstates. One can see that this spectrum does not exhibit relevant differences with respect to the full QM calculation and that the above overlaps are pretty large, what confirms the good quality of the QM(T=0) wave functions. This result provides a clear evidence of the marginal role played by the T=1 and T=2 quartets in the structure of these states.

As a next step, we concentrate on the ground state by employing only $T = 0$ quartets. In Fig. 2, we show how the error in the correlation energy of this state, relative to the SM result,

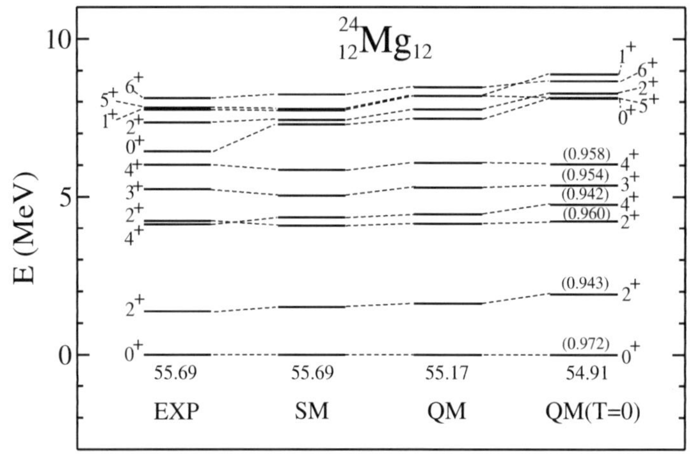

Figure 1. Spectrum of ^{24}Mg obtained in the quartet model (QM) compared to experimental data (EXP) and shell model (SM) results. QM(T=0) denotes the results obtained only with T=0 quartets; the numbers on top of these levels are the overlaps between the QM and SM eigenfunctions. The number below each spectrum gives the ground state correlation energy, namely the difference between the total ground state energy and the energy in the absence of interaction.

varies by reducing, one quartet at a time and starting from the highest one in energy, the set of T=0 quartets. The correlation energy remains basically unchanged up to the point where only the lowest J=0,2,4 quartets are left. From this point on further reductions in the set of quartets lead to significant variations in the energy. The overlaps between the SM and QM ground states which are shown in the same figure exhibit a behavior which is consistent with that of the energies. Thus, these calculations indicate that the T=0 quartets with J=0,2,4 play a major role in the ground state of ^{24}Mg. Among these quartets, the T=0, J=0 quartet is by far the one which contributes most to the correlation energy since, as it can be seen in Fig. 2, an approximation in terms of only this quartet accounts for about 94% of the total energy. A very similar conclusion has emerged also from an analysis of the ground state of the three-quartet system ^{28}Si [6].

So far we have discussed only self-conjugate nuclei. The range of nuclei which are accessible in terms of a set of T=0,1,2 quartets is, however, much broader. Limiting ourselves to the case of 8 active particles (two quartets) outside the ^{16}O core, the whole isobaric chain of nuclei with A=24 is within reach. We have investigated to what extent the spectra of these nuclei can be described in terms of the full set of T=0,1,2 quartets employed in the case of ^{24}Mg. The results can be seen in Ref. [6] where we compare the low-energy spectra of $^{24}_{11}Na_{13}$, $^{24}_{10}Ne_{14}$, $^{24}_{9}F_{15}$ and $^{24}_{8}O_{16}$ obtained in the quartet model with the SM results and with the experimental data. As one can see, for all these nuclei the quartet formalism generates spectra which agree well with the SM ones. It is worth stressing that for the nuclei shown in Fig. 3 the quartets are built not only by two protons and two neutrons, as in the case of $N = Z$ nuclei, but also by one proton and three neutrons and by four neutrons. The case of quartets built by four like-particles had

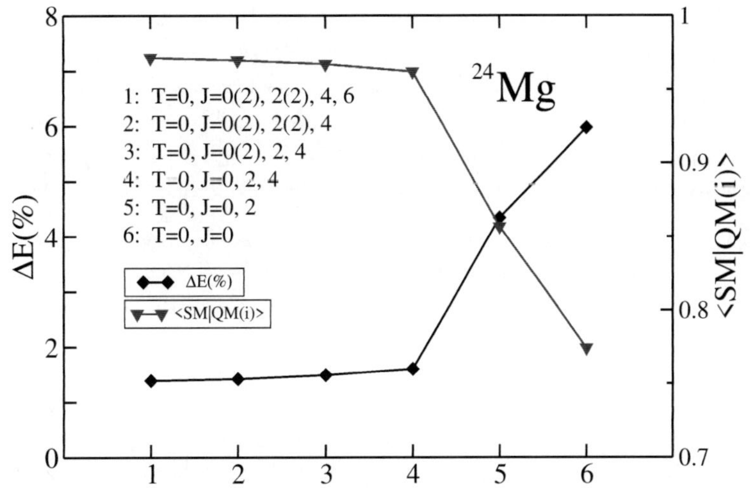

Figure 2. Relative errors (with respect to the SM value) in the ground state correlation energy of ^{24}Mg obtained within the QM with the sets of quartets indicated in the figure. For each set we also show the overlaps between SM and QM wave functions.

been already discussed in Ref. [8] in relation with a treatment of the pairing Hamiltonian.

4. Conclusions

We have provided a description of $4n$ nuclei in a formalism of quartets, i.e., four-body correlated structures characterized by total isospin T and total angular momentum J. We have analyzed in particular the whole isobaric chain of $A = 24$ nuclei. For all these nuclei the quartet formalism has provided a description of the low-energy spectra comparable in accuracy with that of shell model calculations. This fact confirms the importance of quartet degrees of freedom in any type of $4n$ nuclei and validates the present quartet formalism as the appropriate tool for treating them.

Acknowledgments

This work was supported by the Romanian National Authority for Scientific Research, CNCS UEFISCDI, Project Number PN-II-ID-PCE-2011-3-0596.

References

[1] Frauendorf S and Macchiavelli A 2014 *Progr. Part. Nucl. Phys.* **78** 24
[2] Sandulescu N, Negrea D, Dukelsky J and Johnson C W 2012 *Phys. Rev.* C **85** 061303(R); Sandulescu N, Negrea D and Johnson C W 2012 *Phys. Rev.* C **86** 041302(R); Sandulescu N and Negrea D 2014 *Phys. Rev.* C **90** 024322
[3] Sambataro M and Sandulescu N 2013 *Phys. Rev.* C **88** 061303(R)
[4] Sambataro M, Sandulescu N and Johnson C W 2015 *Phys. Lett.* B **740** 137
[5] Sambataro M and Sandulescu N 2016 *Phys. Rev.* C **93** 054320
[6] Sambataro M and Sandulescu N 2015 *Phys. Rev. Lett.* **115** 112501
[7] Brown B A and Richter W A 2006 *Phys. Rev.* C **74** 034315
[8] Sambataro M and Sandulescu N 2013 *J. Phys.* G: *Nucl. Phys.* **40** 055107

IOP Conf. Series: Journal of Physics: Conf. Series **863** (2017) 012011 doi:10.1088/1742-6596/863/1/012011

Preformation of alpha-clusters in alpha-decay

R G Lovas

MTA Institute for Nuclear Research, PO Box 51, H–4001 Debrecen, Hungary

Abstract. Comprehensive cluster-model analyses of all decay cases show that the probability of clustering in most α-decaying states is close to unity. Although early shell-model results for the preformation probability grossly undershot this value and the α-decay width accordingly, more recent sophisticated microscopic structure calculations are in agreement with this observation, except that of a complex-energy shell model. I review the relavant shell model calculations in an attempt to pinpoint the crucial ingredients, which are likely to be responsible for this disagreement.

1. Introduction

Before 1975 it was generally assumed that α preformation in the parent nucleus is meagre. This belief was primarily based on contemporary shell-model calculations. But with the shell-model bases enlarged, the estimates for clustering increased [1, 2]. In particular, the ^{208}Pb+α component in the ^{212}Po g.s. has been found to be 0.3. At the same time, the widths of all favoured α-decays were reproduced in an extreme cluster model, providing evidence for strong clustering in α-decaying states [3, 4]. The consistency of appreciable clustering with realistic width has been confirmed by a nuclear-matter-motivated approach [5]. Thus there seemed to be an accord among various results.

In the last cluster conference [6], however, it was reported that the decay width of ^{212}Po is reproduced in a shell model that yields a much smaller value for clustering (about 1%) than the the classical microscopic calculations (> 0.1). This contradiction requires a review of the intricacies and pitfalls in the advanced shell models. My attention is focussed on the ^{212}Po$\to^{208}$Pb+α system, in which ^{212}Po is assumed to consist of a passive ^{208}Pb core and four active valence nucleons.

2. Clustering and decay properties

In a state $\Psi^{\rm P}$ of a parent nucleus P the formation of a daughter+α-particle (D+α) pair, described by wave functions $\Phi^{\rm D}$ and Φ^{α}, is characterized by $g(\mathbf{R}) = \langle \mathcal{A}\{\Phi^{\rm D}\Phi^{\alpha}\delta(\mathbf{R}-\mathbf{R}_{\alpha\rm D})\}|\Psi^{\rm P}\rangle$, where the interfragment antisymmetrizing operator $\mathcal{A}$ acts on both the intrinsic coordinates and the relative coordinate $\mathbf{R}_{\alpha\rm D}$ of the fragments. This formation amplitude and the spectroscopic factor $S = \int d\mathbf{R}|g(\mathbf{R})|^2$ had been attributed probability meaning. It was recognized by Fliessbach [7] that this cannot be correct. He introduced a proper probability amplitude $G(\mathbf{R})$ and probability $\mathcal{S}$ as

$$G(\mathbf{R}) = \mathcal{N}^{-1/2}g(\mathbf{R}), \qquad \mathcal{S} = \int d\mathbf{R}|G(\mathbf{R})|^2,$$

where $\mathcal{N}$ is a resonating-group-like norm operator, whose kernel is

$$N(\mathbf{R},\mathbf{R}') = \langle \mathcal{A}\{\Phi^{\rm D}\Phi^{\alpha}\delta(\mathbf{R}-\mathbf{R}_{\alpha\rm D})\}|\mathcal{A}\{\Phi^{\rm D}\Phi^{\alpha}\delta(\mathbf{R}'-\mathbf{R}_{\alpha\rm D})\}\rangle.$$

For a pure cluster state, $\mathcal{S} = 1$ as it should be, while $S \neq 1$. In fact, for a heavy two-cluster system, $S \ll \mathcal{S} < 1$ since the Pauli suppression in $g(\mathbf{R})$ is 'overdosed'. (For recent examples, see Ref. [8].) The former misinterpretation of S as a clustering probability contributed to the underrating of clustering. For a spherical nucleus the decay width is usually calculated from the tail of the amplitude $g(\mathbf{R})$ or $G(\mathbf{R})$. The decay width is then $\Gamma = 2P(R_0)(\hbar^2/2\mu R_0)g^2(R_0)$, where $g(R)$ is the radial factor of $g(\mathbf{R}) = (g(R)/R)Y_{lm}(\hat{R})$, l is the relative angular momentum, μ is the reduced mass and P is the penetrability. It is important that $R_0 = |\mathbf{R}_0|$ should be in the asymptotic region, i.e., it should be larger than the range of $\mathcal{A}$ and of the nuclear force between the fragments. Then in that region $g(\mathbf{R}_0) = G(\mathbf{R}_0)$, and P is the Coulomb penetrability.

3. Improved shell models

The standard shell model is distinguished from other microscopic models by its being unbiassed for clustering. Its bases consist of all configurations of the lowest-lying s.p. states. The low-lying s.p. states, however, do not stretch out far from the centre, which makes it difficult to produce a $g(R)$ that is realistic in the asymptotic region.

3.1. Cluster-configuration shell model (CCSM)

If the core is taken infinitely heavy, the shell model basis can be complemented by states of the form of $\mathcal{A}\{\Phi^\alpha_\mathbf{s}\Phi^\mathrm{D}\}$, where $\Phi^\alpha_\mathbf{s}$ is a Slater determinant of shifted 0s orbits $\phi^{(\beta)}_\mathbf{s}(\mathbf{r}) \propto \exp[-\frac{1}{2}\beta(\mathbf{r}-\mathbf{s})^2]\chi$, where χ is a spin-isospin state. A c.m. factor can then be separated from the determinant $\Phi^\alpha_\mathbf{s}$

$$\mathcal{A}\{\Phi^\alpha_\mathbf{s}\Phi^\mathrm{D}\} \propto \mathcal{A}\{\exp[-2\beta(\mathbf{R}_{\alpha\mathrm{D}} - \mathbf{s})^2]\Phi^\alpha\Phi^\mathrm{D}\},$$

where Φ^α is an intrinsic alpha-particle wave function. This shows that the shifted Gaussian orbits $\phi^{(\beta)}_\mathbf{s}$ give rise to basis elements that contain two-cluster states. (These are to be projected to angular momentum 0.) The antisymmetrization eliminates the components of the s.p. Gaussians that are not orthogonal to the orbits occupied in the core. This makes it possible to get rid of the core so as to incorporate the core effects in a projection operator $\mathcal{P}$, which projects off all the core orbits [9].

The model with one major valence shell (538 basis states) and 30 to 40 shifted Gaussians with well-positioned centres $\mathbf{s}$ and $R_0 > 10$ fm yield Γ=1.2$\times10^{-15}$ MeV to 1.6$\times10^{-15}$ MeV, thus reproducing the experimental value of Γ=1.5$\times10^{-15}$ MeV.

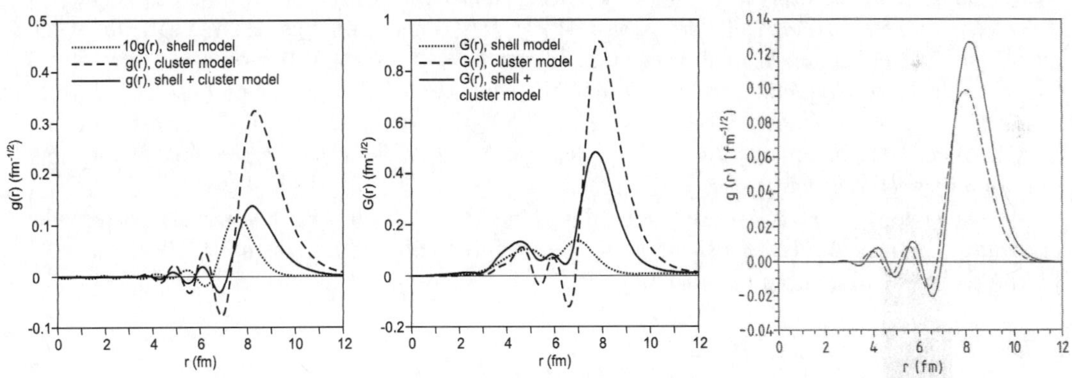

Figure 1. Left and middle panels: $g(r)$ and $G(r)$ for the ^{212}Po=^{208}Pb+α decay in the thee models given in the legends [10]. Right panel: $g(r)$ in the SSM (dashed line; see Sect. 3.2) and in the CCSM (full line) [11]

The $g(r)$ and $G(r)$ produced by the shell model with and without cluster configurations and by the cluster configurations alone are shown on the left and middle panels of Fig. 1. The $g(R)$ of the CCSM is an order of magnitude larger and stretches out more than that of the ordinary shell model. All $g(R)$ oscillate around 0 like ordinary wave functions, with 11 or 12 nodes. The curves $G(R)$, on the contrary, have arkward shapes, except for that of the pure cluster model. This bears out the tendency observed in other works [1, 12, 13, 14]. To explain this behaviour, let us expand the amplitudes in terms of the complete orthonormal set of the norm-operator eigenfunctions $\{\varphi_\nu\}$ [15]:

$$g^{\rm sh}(\mathbf{R}) = \sum_\nu n_\nu^{1/2}\langle\Xi_\nu|\Phi^{\rm sh}\rangle\varphi_\nu(\mathbf{R}), \quad G^{\rm sh}(\mathbf{R}) = \sum_\nu \langle\Xi_\nu|\Phi^{\rm sh}\rangle\varphi_\nu(\mathbf{R}),$$

$$g^{\rm cl}(\mathbf{R}) = \sum_\nu n_\nu\langle\varphi_\nu|\phi^{\alpha \rm D}\rangle\varphi_\nu(\mathbf{R}), \quad G^{\rm cl}(\mathbf{R}) = \sum_\nu n_\nu^{1/2}\langle\varphi_\nu|\phi^{\rm rel}\rangle\varphi_\nu(\mathbf{R}).$$

Here $\phi^{\alpha\rm D}$ stands for the relative-motion function in the cluster-model or in the cluster-model term of the CCSM wave function, n_ν are the eigenvalues of the norm operator ($\mathcal{N}\varphi_\nu = n_\nu\varphi_\nu$), Ξ_ν are normalized two-cluster states in which the relative motion function is φ_ν (that is $\Xi_\nu = n_\nu^{-1/2}\mathcal{A}\{\Phi^{\rm D}\Phi^\alpha\varphi_\nu\}$), and $\Phi^{\rm sh}$ is the shell-model wave function or the shell-model term of the CCSM. Both $\langle\Xi_\nu|\Phi^{\rm sh}\rangle$ and $\langle\varphi_\nu|\phi^{\rm rel}\rangle$ are overlaps between normalized functions, and, for small ν values, both the bra and the ket functions are appreciable in the nuclear volume. The first eigenvalue $n_0 \approx 10^{-7}$ and $n_\nu \longrightarrow 1$ monotonously (see the left panel of Fig. 2). Since φ_ν have $\nu = 0, 1, 2, \ldots$ internal nodes and $n_\nu < 0.1$ for $\nu \leq 11$, the functions $g^{\rm sh}$, $g^{\rm cl}$ and $G^{\rm cl}$ are approximately orthogonal to all φ_ν up to $\nu = 11$, hence they are bound to have at least 12 nodes. No such statement holds for $G^{\rm sh}$, since its expansion contains no n_ν. That explains the nodal behaviours of the functions in Fig. 1.

3.2. Stochastic shell model (SSM)

A shifted Gaussian basis may be said to be biassed for clustering, although the variational principle underlying the shell model reduces the bias. An unbiassed shell-model of tractable size can be set up by the use of the stochastic variational method with Gaussian orbits [16]. This is a powerful method for the description of few-body systems, such as the five-body system of ^{212}Po=^{208}Pb+p+p+n+n. A Gaussian orbit of parameter μ projected to orbital angular momentum l looks like $u_l^\mu(r) \propto r^l \exp(-\frac{1}{2}\mu r^2)$, and in the basis states the angular momenta are to be coupled according to the following pattern: $[[[l_\pi\frac{1}{2}]_{j_\pi}][l'_\pi\frac{1}{2}]_{j'_\pi}]_J[[l_\nu\frac{1}{2}]_{j_\nu}][l'_\nu\frac{1}{2}]_{j'_\nu}]_J]_0$. These proton ($\pi$) and neutron ($\nu$) quantum numbers should run over all values that yield reasonable configurations over the core, with $J = 0, 1, 2, \ldots$ The trial function is constructed term by term from among many functions of randomly selected parameters $\mu_1, \ldots, \mu_4$. A term is adopted in the basis if it lowers the energy sufficiently when added to the trial function constructed in the previous step [16]. For the case of ^{212}Po [11], convergence was achieved with about 120 configurations and basis dimension about 400, at energy $E = -19.25$ MeV (experiment: $E = -19.35$ MeV). With the same interaction, the CCSM and the conventional shell model produce $E = -19.18$ MeV and $E = -18.96$ MeV, respectively. The SSM α-formation amplitude $g(r)$ is compared with that of the CCSM on the right panel of Fig. 1. There is a factor of 2 between the spectroscopic factors, and a factor of 3 between the widths.

It is remarkable that the SSM nearly reproduces the CCSM result with an ordinary shell-model basis, which is optimized solely for energy, thus is certainly unbiassed for clustering. This nice work [11] has been mostly overlooked hitherto.

3.3. Complex-energy shell model (CSM)

The novelty in this approach is the inclusion, in the basis, of s.p. resonances treated as complex-energy eigenstates with outgoing-wave asymptotics [17] (2 neutron and 11 proton resonances). The width is calculated as $\Gamma = \mathcal{S}\Gamma^{\rm s.p.}$, where $\Gamma^{\rm s.p.}$ is the width in a single-α-particle model. The $G(R)$ implicit in $\mathcal{S}$ is to be calculated via $G(\mathbf{R}) = \mathcal{N}^{-1/2}g(\mathbf{R})$, where $\mathcal{N}^{-1/2} = \sum_\nu |\varphi_\nu\rangle n_\nu^{-1/2}\langle\varphi_\nu|$. (There are no zero eigenvalues.) The resulting $g(R)$ has no internal nodes, while $G(R)$ has 10. Thus these functions do not follow the nodal behaviour shown to be generally valid in Sect. 3.1. They come close to zero at about 10 fm as much as the $g(R)$ and $G(R)$ of the CCSM do at 12 fm, thus their range may be said to be 2 fm shorter, and no wonder that they produce 30 times smaller clustering probability, and the tail of $G(R)$ would predict a width smaller by a factor of 36. This result breaks the trend of $S \ll \mathcal{S}$ as well; in fact, $\mathcal{S} < S$. This suggests that there is something the matter with the values of n_ν.

What is surprising is that, despite the unphysical shapes of $g(R)$ and $G(R)$ and despite their too short tails, the width calculated from $\Gamma = \mathcal{S}\Gamma^{\rm s.p.}$ is perfect. As is shown in Fig. 2b, the eigenvalues strongly depend on the basis used for their calculation. The most inaccurate small n_ν were discarded on the grounds that they are spurious. They are certainly not spurious [18], but their inaccuracy and omission do not influence the tail of $G(R)$, hence the reasonability of the Γ value that is extracted from the tail.

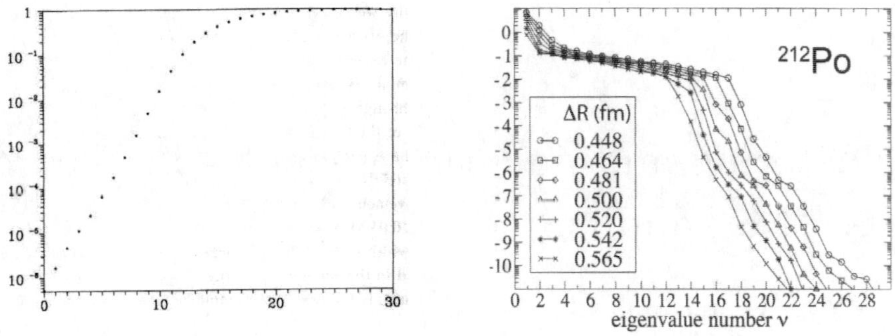

Figure 2. Eigenvalues n_ν of the norm kernel. (a) Left panel: Ref. [9]; (b) right panel: Ref. [6] in reverse sequence; the different symbols distinguish between results with different bases [6].

4. Conclusion

Apart from some aspects of the CSM [6], the shell-model description of α-decay is well understood. The motivation of the CSM was to include the continuum in the basis. The continuum has often been referred to as an important missing ingredient in the usual descriptions of the decay. However, the continuum can be well represented by a set of discrete states. In fact, it is represented by discrete (resonance) states even in Ref. [6]. Moreover, what is included there is not the four-particle continuum but some part of the s.p. continuum. The α-decay width is borne by the tail of the wave function in the four-particle channel, and there is a tremendous difference between the single-nucleon and the four-nucleon continua. The tail behaviour can be made realistic by long-tailed basis functions in the four-particle channel like in the CCSM. The Gaussian wave packets used there are predominantly combinations of four-particle continuum states, thus the continuum states are in fact not missing.

If the formula $\Gamma = \mathcal{S}\Gamma^{\text{s.p.}}$ is used to calculate Γ, it may be enough for the wave function of the single-α-particle model to show the correct tail behaviour, but the clustering probability can only be correct if $G(R)$ is correct at least up to and including the last large peak.

References

[1] Tonozuka I and Arima A 1979 *Nucl. Phys.* A **323** 45
[2] Lovas R G, Liotta R J, Insolia A, Varga K and Delion D S 1998 *Phys. Rep.* **294** 265
[3] Buck B, Merchant A C and Perez S M 1990, *Phys. Rev. Lett.* **65** 2975
[4] Buck B Merchant A C and Perez S M 1993 *At. Data Nucl. Data Tables* **54** 53; Ohkubo S *et al.* 1998 Prog. Theor. Phys. Suppl. **132** 1–228; Buck B, Merchant A C, Perez S M and Seals H E 2005 *J. Phys.* G **31** 1499; Ren Y and Ren Zh 2012 *Phys. Rev.* C **85** 044608; Xu F R, Wang S M, Lin Z J and Pei J C 2013 *J. Phys.: Conf. Ser.* **436** 012064
[5] Xu Ch, Ren Zh, Röpke G, Schuck P, Funaki Y, Horiuchi H, Tohsaki A, Yamada T and Zhou B 2016 *Phys Rev* C **93** 011306(R) Horiuchi H 1977 Prog. Theor. Phys. Suppl. **62** 90
[6] Id Betan R 2013 *J. Phys.: Conf. Ser.* 012061; Id Betan R and Nazarewicz W 2012 *Phys. Rev.* C **86** 034338
[7] Fliessbach T 1975 *Z Phys* A **272** 39
[8] Volya A and Tchuvil'sky Y M 2015 *Phys. Rev.* C **91** 044319
[9] Varga K, Lovas R G and Liotta R J 1992 *Nucl. Phys.* A **550** 421
[10] Varga K, Lovas R G and Liotta R J 1992 *Phys. Rev. Lett.* **69** 37
[11] Varga K and Liotta R J 1994 *Phys. Rev.* C **50** R1292
[12] Fliessbach T and Mang H J 1976 *Nucl. Phys.* A **263** 75
[13] Okabe S 1989 Suppl. J. Phys. Soc. Jpn. **58** 516
[14] Okabe S and Fliessbach T 1984 *Phys. Lett.* B **144** 5
[15] Lovas R G 2016 *Phys. Rev.* C **93** 069801
[16] Varga K, Suzuki Y and Lovas R G 1994 *Nucl. Phys.* A **571** 447
[17] Berggren T 1968 *Nucl. Phys.* A **109** 265
[18] Horiuchi H 1977 Prog. Theor. Phys. Suppl. **62** 90

IOP Conf. Series: Journal of Physics: Conf. Series **863** (2017) 012012 doi:10.1088/1742-6596/863/1/012012

Alpha clustering and nuclear molecules in ^{8}Be, ^{12}C, ^{16}O, ^{20}Ne, ^{24}Mg, and ^{32}S

Royer G and Eudes P

Subatech, Université-IN2P3/CNRS-Ecole des Mines, 44307 Nantes, France

E-mail: royer@subatech.in2p3.fr

Abstract. Within a liquid drop model the energy of the ^{12}C, ^{16}O, ^{20}Ne, ^{24}Mg and ^{32}S 4n-nuclei has been calculated within different configurations of α-molecules : linear chain, triangle, square, tetrahedron, pentagon, trigonal bipyramid, square pyramid, hexagon, octahedron, octagon and cube. The potential barriers governing the decay and entrance channels via α emission or absorption as well as the potential barriers of other possible binary and ternary reactions have been compared. The rms radii of the linear chains do not correspond to the experimental rms radii of the ground states. The binding energies of the three-dimensional molecules at the nascent point of the fragments are higher than the ones of the planar configurations. The A-4 daughter plus α particle configurations have always the lowest potential energy.

1. Introduction

Hydrogen burning in stars leads to a core of helium and, later on, to the nucleosynthesis of n-α nuclei : ^{16}O, ^{20}Ne, ^{24}Mg, ^{28}Si, ^{32}S, ... The cluster-type states coexist with the mean-field-type states [1] in these specific nuclei. In ^{12}C a large amount of 3α cluster wave function is contained in the ground state wave function. Recently, a new high spin 5^- state has been observed. It corresponds to a level of the ground state rotational band of an equilateral triangular spinning top [2]. In ^{16}O, recent studies lead, for the ground state, to a tetrahedral configuration allowing to reproduce the energy spectrum and the electromagnetic properties [3, 4]. Some excited states are due to the mean-field-type excitation mode while other ones are due to the cluster structure of α+^{12}C. For the excited 0_2^+ Hoyle state of ^{12}C and possible excited Hoyle state of ^{16}O, the α condensate character chain has been proposed after comparing a large number of Brink functions with Tohsaki-Horiuchi-Schuck-Röpke wave functions [5, 6].

The generalized liquid drop model (GLDM) describing correctly the fusion [7], alpha emission [8] and fission [9] processes has been used to calculate the L-dependent potential energy of the following α clusters: aligned α chains, isosceles triangle, square, tetrahedron, pentagon, trigonal bipyramid, square pyramid, hexagon, octahedron, octagon and cube [10, 11]. The proximity energy is included in this model to take into account the effects of the nuclear forces between the surfaces in regard in a gap or a neck between nuclei.

2. ^{12}C nucleus

Antisymmetrized Molecular Dynamics and Fermionic Molecular Dynamics models without assuming α clustering led naturally for the different states to triangular α shapes with different angles allowing the reproduction of the low-lying spectrum [12, 13]. Within effective field theory and Monte Carlo lattice calculations it has been demonstrated that the ^{12}C ground state has a compact triangular configuration while the Hoyle state and the second excited state have rather an obtuse triangular shape of alpha particles [14]. These predictions are strengthened by the observation of the high spin 5^- state at 22.4 MeV.

These oblate ternary configurations have been investigated within the GLDM and an isosceles triangular shape characterized by an angle θ. At the contact point the energy of the linear chain of three α particles (θ=180 deg.) is higher than the energy of the equilateral triangular shape, the difference being 7.4 MeV. This is also in favor of the equilateral triangular configuration (the energy being degenerated between 120 and 180 degrees). The experimental rms charge radius of the ground state is $< r^2 >^{1/2}=$ 2.47 fm while, within the GLDM, the rms radius of the triangular shape is 2.43 fm and 3.16 fm for a linear chain at the contact point [10].

3. ^{16}O nucleus

A tetrahedral molecule of alpha particles is predicted for the ground state [3, 4] and a square

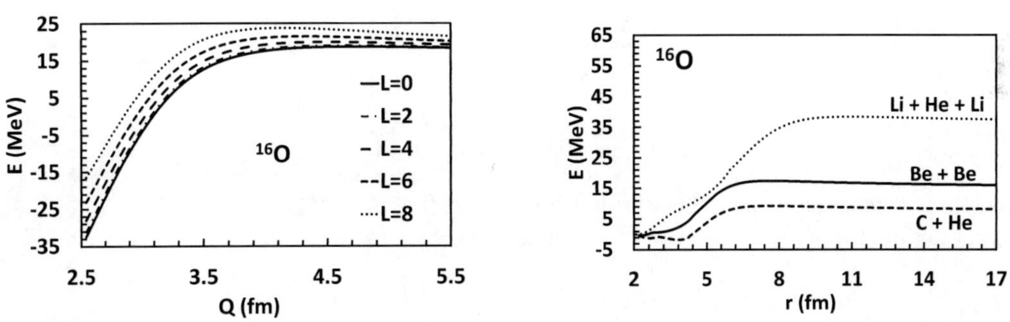

Figure 1. Potential energy of the α-tetrahedron (left part) as functions of the angular momentum (in $\hbar$ unit) and rms radius and (right part) potential energy governing the ^{12}C+^{4}He, ^{8}Be+^{8}Be, and ^{6}Li+^{4}He+^{6}Li nuclear systems versus the distance between the mass centres.

configuration for the first excited spin-0 state. The energies of the tetrahedral shape are given in Fig. 1. The rms radius is 2.54 fm for a tetrahedron, 2.83 fm for a square and 4.15 fm for a prolate linear chain at the contact point between the four α particles. Experimentally the rms charge radius of the ground state is $< r^2 >^{1/2}=$ 2.70 fm excluding probably a linear chain configuration for the ground state. The binding energy is higher for the tetrahedral molecule than for the square shape since for these molecular shapes the proximity energy plays a crucial role. Indeed, the tetrahedron is linked by six bonds and the square by only four bonds. On the contrary, the Coulomb repulsion is higher for the tetrahedral shape. The energy difference between the two configurations is 13.7 MeV close to the $Q_{4\alpha}$ (14.4 MeV), the energy of the 0_6^+ state (15.1 MeV) and 14.0 MeV the energy of a 0^+ state.

The ground state may also be described using double closed shell wave functions but several excited states may also be viewed within the ^{12}C+^{4}He cluster picture. The potential energies of the ^{12}C+^{4}He, ^{8}Be+^{8}Be, and aligned ^{6}Li+^{4}He+^{6}Li systems have been determined assuming a spherical shape for all nuclei (see Fig. 1). The top of the barriers corresponds to separated nuclei maintained in unstable equilibrium. Quasimolecular ^{12}C+^{4}He one-body shapes with a deep neck have roughly the same energy than the spherical compound nucleus.

4. ^{20}Ne nucleus

The trigonal bipyramid, square pyramid and pentagonal molecular configurations have been compared. Their energies are displayed in Fig. 2. The rms radius is 2.76 fm for a trigonal bipyramid, 2.79 fm for a square pyramid and 3.29 fm for a pentagon. The experimental rms charge radius is $< r^2 >^{1/2}= 3.01$ fm, much lower than the rms radius of a prolate chain. The binding energy of the trigonal bipyramid is the highest since the number of bonds is five for the pentagon, eight for the square pyramid and nine for the trigonal bipyramid. At the contact point, the energy difference between the trigonal bipyramid and the square pyramid is 21.3 MeV and 15.3 between the square pyramid and the pentagon, the $Q_{5\alpha}$ value being 19.2 MeV.

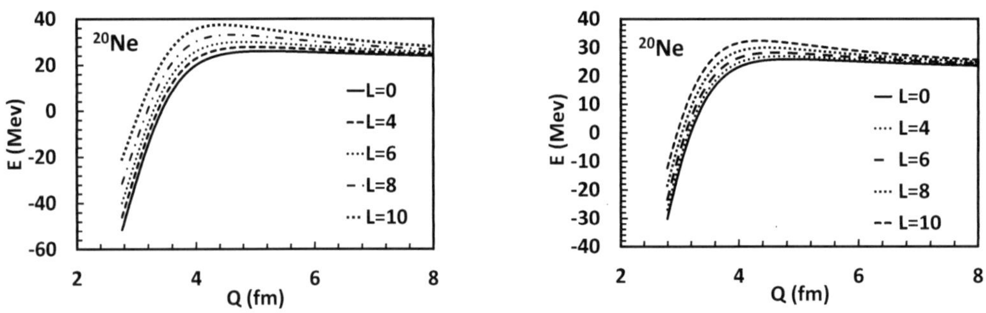

Figure 2. Potential energy of the trigonal bipyramid (on the left) and the square pyramid (on the right) as functions of the angular momentum and rms radius.

The ground state band contains the ^{16}O+^{4}He cluster at most 70% and the potential energies of the ^{16}O+^{4}He, ^{12}C+^{8}Be, ^{10}B+^{10}B, and linear ^{8}Be+^{4}He+^{8}Be systems have been determined. In the ^{16}O+^{4}He channel quasimolecular one-body shapes have roughly the same energy than the spherical nucleus and the minimum has a cluster structure corresponding to the two ^{4}He and ^{16}O nuclei almost in contact.

5. ^{24}Mg nucleus

The energies of the octahedral and hexagonal α-molecules are shown in Fig. 3. The experimental rms charge radius of the ground state is only $< r^2 >^{1/2}= 3.06$ fm. At the contact point the rms radius is 2.85 fm for an octahedron and 3.79 fm for an hexagon excluding the linear chain and planar shape as possible ground state shapes. The binding energy is lower for the hexagonal configuration since there are only six bonds for the hexagon and twelve for the octahedron.

The energies of the ^{16}O+^{8}Be, ^{12}C+^{12}C, ^{8}Be+^{8}Be+^{8}Be, and ^{10}B+^{4}He+^{10}B systems are shown in Fig. 3.

6. ^{32}S nucleus

The octagonal and cubic α-configurations have been studied. The experimental rms charge radius is $< r^2 >^{1/2}= 3.26$ fm while the rms radius is 4.85 fm for an octagon and 3.37 fm for a cube at the contact point which seems to exclude the planar and linear configurations. The binding energy is higher for the cubic configuration than for the octagonal shape since there are twelve bonds for the cube and only eight for the octagon.

The potential energies of the ^{28}Si+^{4}He, ^{24}Mg+^{8}Be, ^{20}Ne+^{12}C, and ^{16}O+^{16}O systems have been also determined. The energy of the one-body ^{28}Si+^{4}He system is relatively constant from

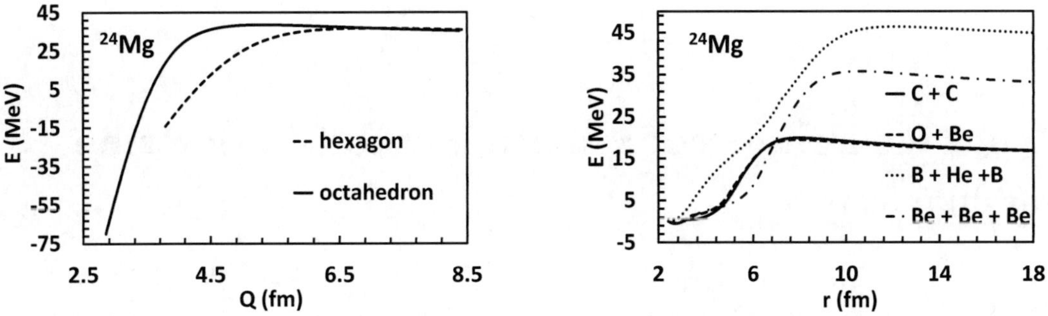

Figure 3. Potential energies of an hexagon and an octahedron from the contact point as a function of the rms radius and potential barriers governing the ^{16}O+^{8}Be, ^{12}C+^{12}C, ^{8}Be+^{8}Be+^{8}Be, and ^{10}B+^{4}He+^{10}B reactions versus the distance between the mass centres.

the spherical nucleus till the contact point allowing the cohabitation of different quasimolecular shapes.

7. Conclusion

Within an α-particle model the energy of the ^{12}C, ^{16}O, ^{20}Ne, ^{24}Mg and ^{32}S nuclei has been calculated within different α-molecular configurations: linear chain, triangle, square, tetrahedron, pentagon, trigonal bipyramid, square pyramid, hexagon, octahedron, octagon, and cube. Within a macroscopic approach the potential barriers governing the entrance and decay channels of these 4n-nuclei via alpha absorption or emision and other possible nuclear systems have also been compared.

The rms radii of the prolate chains seem incompatible with the experimental rms radii of the ground states. The binding energies of the three-dimensional molecules are higher than the binding energies of the planar shapes. The core+α cluster system leads always to the lowest potential barrier.

References

[1] von Oertzen W, Freer M and Kanada-En'yo Y 2006 *Phys. Rep.* **432** 43
[2] Marin-Lambarri D J *et al* 2014 *Phys. Rev. Lett.* **113** 012502
[3] Epelbaum E *et al* 2014 *Phys. Rev. Lett.* **112** 102501
[4] Bijker R and Iachello F 2014 *Phys. Rev. Lett.* **112** 152501
[5] Funaki Y *et al* 2009 *Phys. Rev.* C **80** 064326
[6] Suhara T, Funaki Y, Zhou B, Horiuchi H and Tohsaki A 2014 *Phys. Rev. Lett.* **112** 062501
[7] Royer G and Remaud B 1985 *Nucl. Phys.* A **444** 477
[8] Royer G 2010 *Nucl. Phys.* A **848** 279
[9] Royer G, Jaffré M and Moreau D 2012 *Phys. Rev.* C **86** 044326
[10] Royer G, Escudie A and Sublard B 2014 *Phys. Rev.* C **90** 024607
[11] Royer G, Ramasamy G and Eudes P 2015 *Phys. Rev.* C **92** 054308
[12] Kanada-En'yo Y 2007 *Prog. Theor. Phys.* **117** 655
[13] Chernykh M *et al* 2007 *Phys. Rev. Lett.* **98** 032501
[14] Epelbaum E *et al* 2012 *Phys. Rev. Lett.* **109** 252501

A new generator coordinate method to describe gas-like states

R Imai and M Kimura

Department of Physics, Hokkaido University, 060-0810 Sapporo, Japan.

E-mail: masaaki@nucl.sci.hokudai.ac.jp

Abstract. A new theoretical model which employs the real time as the generator coordinate is developed to describe gas-like cluster states. The model exploits the quantum ergodic nature of the time-evolution of the wave packets. The result of the benchmark calculation shows that the new method describes the dynamics of the 3α system reasonably comparable with or even better than the Tohsaki-Horiuchi-Schuck-Röpke wave function.

1. Introduction

In this decade, the α particle condensates in finite nuclei attract a lot of interest. In the very begging [1, 2], the Hoyle state of ^{12}C was the icon of the α particle condensate. Later, a lot of theoretical and experimental efforts revealed that there are many families of the 3α condensates in ^{12}C and established that the α particle condensate also exists in the 4α system, ^{16}O [3, 4].

Different from the cold atomic systems, the α particle condensates are self-bound systems without the confinement potential, hence the question "how many α particles can form the condensate?" is very interesting and essential. The experiments [5] imply the possible formation of 10 and 15 α particle condensates in the compound states of ^{40}Ca and ^{60}Zn. Theoretically, Yamada and Schuck [6] showed that up to approximately 10α particles can form self-bound condensate. However, their prediction is based on the macroscopic model which approximates α particles as structureless bosons. Therefore, the microscopic model studies which take the Fermi statistics of nucleons into account are highly desirable for more detailed and reliable discussions.

Unfortunately, we immediately find serious difficulty in describing the dynamics of many α particles with the full microscopic models. Imagine that one tries to describe $N\alpha$ particle system by the generator coordinate method (GCM) which superposes basis wave functions Φ_i,

$$\Psi = \sum_i c_i \Phi_i. \tag{1}$$

It is easy to understand that the number of basis wave function Φ_i very quickly increases as the number of α particle increases, which makes the practical numerical calculation impossible. Therefore, the effective method to generate the basis wave function Φ_i is indispensable. Several methods have been suggested, which are based on the random generation of the basis wave function [7] and the imaginary-time development [8]. In this contribution, we suggest an alternative but surprisingly effective method which relies on the ergotic nature of the real-time evolution of the α particle wave packets. We here discuss the result of the benchmark calculation for 3α system.

2. Methodology

The Hamiltonian of the 3α system used in the benchmark calculation is given as,

$$H = \sum_{i=1}^{12} t_i + \sum_{i<j}^{12} v_N(r_{ij}) + \sum_{i<j}^{6} v_C(r_{ij}) - t_{cm}, \tag{2}$$

where t_i and t_{cm} respectively denote the kinetic energies of the nucleons and the center-of-mass. As an effective nucleon-nucleon interaction, we used Volkov No. 2 with a slight modification. In short, the Hamiltonian is common to the other studies using resonating group method (RGM) [9] and Tohsaki-Horiuchi-Schuck-Röpke (THSR) wave function [10].

Our basis wave function for 3α system is the Brink-Bloch wave function,

$$\Phi(\boldsymbol{Z}_1, \boldsymbol{Z}_2, \boldsymbol{Z}_3) = \mathcal{A}\{\, \Phi_\alpha(\boldsymbol{Z}_1)\Phi_\alpha(\boldsymbol{Z}_2)\Phi_\alpha(\boldsymbol{Z}_3)\,\}, \tag{3}$$

$$\Phi_\alpha(\boldsymbol{Z}) = \mathcal{A}\{\, \phi(\boldsymbol{r}_1, \boldsymbol{Z})\chi_{p\uparrow}\phi(\boldsymbol{r}_2, \boldsymbol{Z})\chi_{p\downarrow}\phi(\boldsymbol{r}_3, \boldsymbol{Z})\chi_{n\uparrow}\phi(\boldsymbol{r}_4, \boldsymbol{Z})\chi_{n\downarrow}\,\}, \tag{4}$$

$$\phi(\boldsymbol{r}, \boldsymbol{Z}) = \left(\frac{2\nu}{\pi}\right)^{3/4} \exp\left\{ -\nu\left(\boldsymbol{r} - \frac{\boldsymbol{Z}}{\sqrt{\nu}}\right)^2 + \frac{1}{2}Z^2 \right\}, \tag{5}$$

where $\Phi_\alpha(\boldsymbol{Z})$ denotes the wave function of α particle located at $\boldsymbol{Z}$. Here, the three-dimensional vector $\boldsymbol{Z}$ is complex numbered differently from the ordinary Brink-Bloch wave function, and hence, they are regarded as the dynamical variable. By applying the time-dependent variational principle, one obtains the equation of motion (EoM) for the α particle centroids $\boldsymbol{Z}_1, \boldsymbol{Z}_2$ and $\boldsymbol{Z}_3$,

$$\sum_{j,\sigma} C_{i\rho j\sigma} \frac{dZ_{j\sigma}}{dt} = \frac{\partial \mathcal{H}}{\partial Z^*_{i\rho}}, \quad i = 1, 2, 3, \quad \rho, \sigma = x, y, z, \tag{6}$$

$$\mathcal{H} \equiv \frac{\langle \Phi(\boldsymbol{Z}_1, \boldsymbol{Z}_2, \boldsymbol{Z}_3)|H|\Phi(\boldsymbol{Z}_1, \boldsymbol{Z}_2, \boldsymbol{Z}_3)\rangle}{\langle \Phi(\boldsymbol{Z}_1, \boldsymbol{Z}_2, \boldsymbol{Z}_3)|\Phi(\boldsymbol{Z}_1, \boldsymbol{Z}_2, \boldsymbol{Z}_3)\rangle}, \quad C_{i\sigma j\rho} \equiv \frac{\partial^2 \ln \langle \Phi(\boldsymbol{Z}_1, \boldsymbol{Z}_2, \boldsymbol{Z}_3)|\Phi(\boldsymbol{Z}_1, \boldsymbol{Z}_2, \boldsymbol{Z}_3)\rangle}{\partial Z^*_{i\sigma}\partial Z_{j\sigma}}. \tag{7}$$

We evolve the system by solving the EoM (Eq. (6)). As a result, we obtain the wave function of the 3α system at each time t; $\Phi(\boldsymbol{Z}_1(t), \boldsymbol{Z}_2(t), \boldsymbol{Z}_3(t))$. Despite of its classical-like look, this EoM possesses the following excellent properties [11, 12],

(i) The wave function $\Phi(\boldsymbol{Z}_1(t), \boldsymbol{Z}_2(t), \boldsymbol{Z}_3(t))$ has ergodic nature.

(ii) The ensemble of $\Phi(\boldsymbol{Z}_1(t), \boldsymbol{Z}_2(t), \boldsymbol{Z}_3(t))$ follows *quantum* statistics.

This means that, if time is evolved long enough, the set of wave functions $\Phi(\boldsymbol{Z}_1(t), \boldsymbol{Z}_2(t), \boldsymbol{Z}_3(t))$ spans a good model space for 3α system. Therefore, the superposition of the basis wave function,

$$\Psi = \int_0^T dt\; c(t)\,\{\, \Phi(\boldsymbol{Z}_1(t), \boldsymbol{Z}_2(t), \boldsymbol{Z}_3(t)) + \Phi^*(\boldsymbol{Z}_1(t), \boldsymbol{Z}_2(t), \boldsymbol{Z}_3(t))\,\}, \tag{8}$$

should give a good description for 3α system. Here, the basis wave functions are conjugated to guarantee that Ψ is time even. The coefficients $c(t)$ should be determined by the diagonalization of the Hamiltonian. In other words, by using the EoM, we perform the generator coordinate method by employing the real time t as the generator coordinate.

3. Numerical benchmark

The practical numerical calculation was done as follows. First, we randomly generate the excited 3α wave function by the imaginary-time evolution. The excitation energy is set $E^* \simeq 15$ MeV. Then we solve the EoM and perform the GCM calculation by discretizing the time t in Eq. (8).

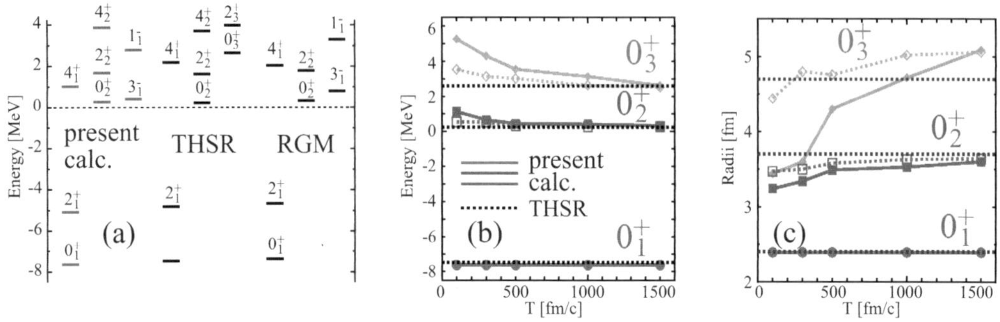

Figure 1. (a) Energy spectra of 3α system calculated by the present method, THSR [10] and RGM [9]. (b) and (c) The convergence of the energies and radii of the eigenstates as functions of the time evolution duration T.

Figure 1 (a) compares the present result with other theoretical calculations which employ the same Hamiltonian. It is clear that our model gives very reasonable result and consistent with the previous methods including the THSR wave function. It is also noted that the radii are also properly described. For example, the radius of the Hoyle state is calculated as 3.66 fm and 3.7 fm by the present model and THSR respectively, while the ordinary RGM yields much smaller radius, 3.47 fm. Thus, the our method can reasonably describe the cluster gas-like states and has the advantage over the ordinary cluster model. Figure 1 (b) and (c) show how the binding energies and radii are converged as the duration T increases. Solid and dashed lines show the results obtained by starting from different initial condition. We see that, except for the 0_3^+ state which has broad width, the calculation converges as T becomes large and the result is independent of the initial condition of the time evolution. Since the extension of this model is rather straightforward, the application to 4α,5α, ... systems looks fascinating and promising.

4. Summary

In summary, we have developed a new theoretical model which employs the real time as the generator coordinate. The model relies on the quantum ergodic nature of the EoM of the wave packets. The result of the benchmark calculation shows that our model can describe the 3α system reasonably.

References

[1] Tohsaki A *et al* 2001 *Phys. Rev. Lett.* **87** 192501
[2] Funaki Y *et al* 2010 *Phys. Rev.* C **82** 024312
[3] Wakasa T *et al* 2007 *Phys. Lett.* B **653** 173
[4] Funaki Y *et al* 2008 *Phys. Rev. Lett.* **101** 082502
[5] Akimune H *et al* 2013 *J. Phys. Conf. Series* **436** 012010
[6] Yamada T and Shuck P 2004 *Phys. Rev.* C **69** 024309
[7] Itagaki N *et al* 2007 *Phys. Rev.* C **75** 037303
[8] Akahori T *et al* 2015 *Phys. Rev.* C **92** 022801(R)
[9] Kamimura M *et al* 1981 *Nucl. Phys.* A **351** 456
[10] Funaki Y *et al* 2015 *Phys. Rev.* C **92** 021302
[11] Schnack J and Feldmeier H 1996 *Nucl. Phys.* A **601** 181
[12] Ono A and Horiuchi H *Phys. Rev.* C **53** 2341

IOP Conf. Series: Journal of Physics: Conf. Series **863** (2017) 012014 doi:10.1088/1742-6596/863/1/012014

Study of similarity and dissimilarity between ^{10}Be and ^{9}Li nuclei with microscopic structure and reaction models

T Furumoto[1], **T Suhara**[2] **and N Itagaki**[3]

[1] Graduate School of Education, Yokohama National University, Yokohama 240-8501, Japan and National Institute of Technology, Ichinoseki College, Ichinoseki, Iwate 021-8511, Japan
[2] Matsue College of Technology, Matsue, Shimane 690-8518, Japan
[3] Yukawa Institute for Theoretical Physics, Kyoto University, Kyoto 606-8502, Japan

E-mail: furumoto-takenori-py@ynu.ac.jp

Abstract. We apply the microscopic nuclear structure and reaction models to the elastic scattering of the ^{10}Be and ^{9}Li nuclei. The states of the ^{10}Be and ^{9}Li nuclei are described as an $\alpha + \alpha(t) + n + n$ four-body system based on the cluster wave function, respectively. The elastic scattering cross section is obtained by the microscopic coupled channel (MCC) calculation with the MPa interaction, which is the latest version of the complex G-matrix interaction derived from the ESC NN interaction model. We compare the channel coupling (CC) effect of the ^{10}Be nucleus with that of the ^{9}Li nucleus in the elastic scattering cross sections calculated with the same incident energy per nucleon. At the backward angles, the CC effect on the elastic scattering cross section for the incident ^{10}Be nucleus is clearly larger than that for the incident ^{9}Li nucleus. We investigate the dissimilarity for the CC effect through the dynamical polarization potential.

1. Introduction

The cluster features of the ^{10}Be and ^{9}Li nuclei have been frequently investigated from both nuclear structure and reaction view points. In terms of the cluster model, the ^{10}Be (^{9}Li) nucleus is described as the $\alpha + \alpha(t) + n + n$ four-body systems, respectively [1, 2, 3, 4]. Recently, their similarity has been discussed from the view point of the nuclear structure as in Refs. [5, 6, 4, 7]. However, their dissimilarity is not clarified in those works. In addition, the similarity and/or dissimilarity are not discussed from the view point of the nuclear reaction, although numerous observations are obtained through the nuclear reaction experiments.

In this paper, we investigate the similarity and/or dissimilarity of the ^{10}Be and ^{9}Li nuclei with the microscopic structure and reaction models. The model consists of the combination of the microscopic cluster model and the microscopic coupled channel (MCC) method. In the microscopic cluster model, the total wave function is composed of superposition of various cluster configurations. The transition densities are derived from the total wave function, and we use them for the elastic cross section calculations in the MCC framework. The channel coupling (CC) effects with the excited states are clearly shown for the ^{10}Be and ^{9}Li scatterings. Here we focus on the dissimilarity of the CC effect in the ^{10}Be and ^{9}Li nuclei.

2. Formalism

We first perform structure calculation and construct the wave functions of ^{10}Be and ^{9}Li based on the stochastic multi-configuration mixing (SMCM) method. In the SMCM method, the total wave function is obtained by superposing many $\alpha+\alpha(t)+n+n$ four-body cluster configurations, which are randomly generated. Next, the differential cross section is obtained by the microscopic coupled-channel (MCC) calculation. Here the coupling potential is obtained by folding the densities of the projectile and target nuclei with the effective NN interaction. The detail of the calculation is presented in Ref. [4] while the NN interaction is replaced by the MPa interaction [8], which is the latest version of the complex G-matrix interaction.

3. Results

For the ^{10}Be nucleus, the 0_1^+, 2_1^+ and 2_2^+ states are obtained as -62.74, -59.49 ($E_x = 3.247$) and -57.46 ($E_x = 5.278$) MeV, respectively. The obtained radius for the ground state is 2.559 fm. The transition strengths from the 2_1^+ and 2_2^+ states to the ground state (B(IS2)) are 43.84 and 0.3551 fm^4, respectively. The obtained values for the ^{10}Be nucleus are comparable to the experimental data [10, 11, 12, 13]. The binding energies, radius of the ground state and transition strengths of the ^{9}Li nucleus are obtained in Ref. [4].

Here, we compare the ^{10}Be nucleus with the ^{9}Li nucleus in the elastic scattering by the ^{12}C target at $E/A = 59.4$ MeV.In the MCC calculation, we take into account the 0_1^+, 2_1^+ and 2_2^+ states for the ^{10}Be nucleus. On the other hand, the $3/2_1^-$, $1/2_1^-$, $5/2_1^-$ and $7/2_1^-$ states are included in our calculation for the ^{9}Li elastic scattering. The transition density for the ^{12}C nucleus is taken from Ref. [9]. For the ^{12}C nucleus, we take into account the excitation of the 2_1^+ ($E_x = 4.44$ MeV), 0_2^+ ($E_x = 7.65$ MeV), 3_1^- ($E_x = 9.64$ MeV) states. In this paper, the result including the full combination of projectile and target excitations is called as full-CC calculation.

The renormalization factor is often introduced to reproduce the data by increasing/reducing the real and imaginary parts of the folding model potential. In this paper, we apply the renormalization factor (N_W) only to the imaginary part of the folding model potential with the complex G-matrix interaction.

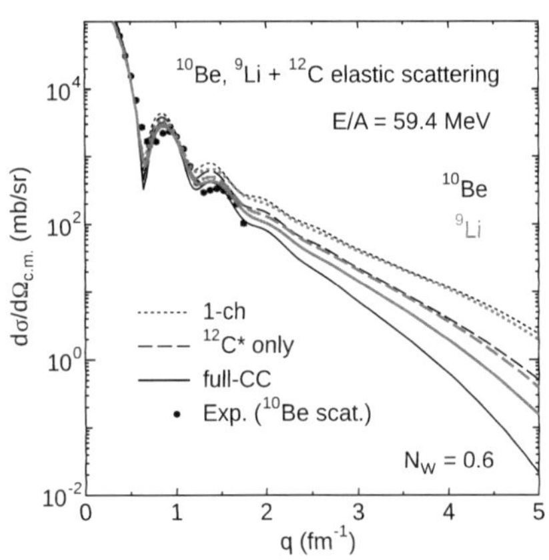

Figure 1. The elastic scattering cross sections for the ^{10}Be + ^{12}C and ^{9}Li + ^{12}C systems at E/A = 59.4 MeV. The experimental data is taken from Ref. [14, 15]

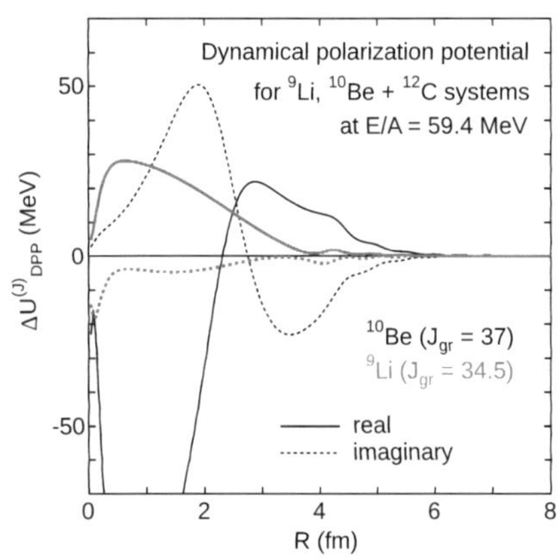

Figure 2. The dynamical polarization potential (DPP) for the ^{10}Be + ^{12}C and ^{9}Li + ^{12}C systems at $E/A = 59.4$ MeV.

Figure 1 shows the calculated elastic cross section for the ^{10}Be + ^{12}C and ^{9}Li + ^{12}C systems at E/A = 59.4 MeV. The filled circles are the experimental data for the ^{10}Be + ^{12}C elastic cross section at $E/A = 59.4$ MeV. The dotted and solid curves are the results by the 1-ch and full-CC calculations, respectively. The dashed curves include the excitation effect of the target nucleus (^{12}C) only. The thin and bold curves are the result of the calculated ^{10}Be and ^{9}Li elastic scatterings, respectively. When we take into account no projectile (^{10}Be and ^{9}Li) excitation effect, the calculated elastic cross sections are closed to each other. The CC effect by the ^{10}Be and ^{9}Li nuclei is clearly seen on the elastic cross section. Their dissimilarity by the CC effect is also clearly seen, although the transition strength of them is not so different.

In order to investigate the reason why the ^{10}Be and ^{9}Li nuclei show the different CC effect, we show the dynamical polarization potential (DPP) in Fig. 2. In the tail region (R =3–6 fm), it is clearly seen that the DPP for the ^{10}Be nucleus is larger than that for the ^{9}Li nucleus. The DPP is known to depend on the coupling potential, elastic wave function, inelastic wave function and the angular momentum algebra as shown in Ref. [16]. Then, we check all of the components of the DPP to clarify the cause of the different CC effect. Finally, we find that the angular momentum algebra has an important role in all of the components which are the elastic wave function, inelastic wave function and DPP. Here, we note that the angular momentum algebra contributes to the DPP not only directly but also through the elastic and inelastic wave functions indirectly.

4. Conclusion

The ^{10}Be (^{9})Li nucleus is constructed by the $\alpha + \alpha(t) + n + n$ four body cluster model with the SMCM method, respectively. With the wave functions, we obtain the ^{10}Be and ^{9}Li elastic cross sections by the ^{12}C target nucleus in the MCC calculation. The ^{10}Be and ^{9}Li elastic cross sections are closed to each other without the CC effect by the ^{10}Be or ^{9}Li nuclei. However, their elastic cross sections give the different CC effect when their excitation effects are taken into account in the MCC calculation. By investigating the DPP, it is found that the different CC effect is caused by the angular momentum algebra.

Acknowledgments

This work was supported by JSPS KAKENHI Grant Numbers 15K17661, 15K17662 and 26400269.

References

[1] Itagaki N and Okabe S 2000 *Phys. Rev.* C **61** 044306
[2] Arai K, Ogawa Y, Suzuki Y and Varga K 2001 *Prog. Theor. Phys. Suppl.* **142** 97
[3] von Oertzen W, Freer M and Kanada-En'yo Y 2006 *Phys. Rep.* **432** 43
[4] Furumoto T, Suhara T and Itagaki N 2013 *Phys. Rev.* C **87** 064320
[5] Suhara T and Kanada-En'yo Y 2010 *Prog. Theor. Phys.* **123** 303
[6] Kanada-En'yo Y and Suhara T 2012 *Phys. Rev.* C **85** 024303
[7] Kanada-En'yo Y 2016 arXiv:1604.01453
[8] Yamamoto Y, Furumoto T, Yasutake N and Rijken Th A 2014 *Phys. Rev.* C **90** 045805
[9] Kamimura M 1981 *Nucl. Phys.* A **351** 456
[10] Tanihata I *et al* 1998 *Phys. Lett.* B **206** 592
[11] Liatard E *et al* 1990 *Europhys. Lett.* **13** 401
[12] Warner R E *et al* 2001 *Phys. Rev.* C **64** 944611
[13] Ajzenberg-Selvo F and Helly J H *et al* 1990 *Nucl. Phys.* A **506** 1
[14] Cortina-Gil M D *et al* 1997 *Phys. Lett.* B **401** 9
[15] Experimental Nuclear Reaction Data (EXFOR) http://www.jcprg.org/exfor/
[16] Furumoto T and Sakuragi Y 2013 *Phys. Rev.* C **87** 014618

Radiative Capture Processes in Multi-Size-Scale Algebraic Version of Resonating Group Model

A S Solovyev[1], S Yu Igashov[1] and Yu M Tchuvil'sky[1,2,3]

[1] Dukhov Research Institute of Automatics (VNIIA), 22, ul. Sushchevskaya, Moscow 127055, Russia
[2] Skobeltsyn Institute of Nuclear Physics, Lomonosov Moscow State University, 1(2), Leninskie gory, GSP-1, Moscow 119991, Russia
[3] Pacific National University, 136, Tikhookeanskaya str., Khabarovsk 680035, Russia

E-mail: alexander.solovyev@mail.ru

Abstract. Novel microscopic approach based on so-called multi-size-scale algebraic version of the resonating group model to describing radiative capture reactions is developed. The main idea of the approach is to expand nuclear system wave functions of discrete spectrum and continuum over basis wave functions of the algebraic version of the resonating group model with different values of oscillator radius. It allows to calculate astrophysical *S*-factor or cross section for radiative capture reaction and at the same time to reproduce exactly the experimental values for breakup thresholds of the final nucleus states into two fragments – colliding nuclei, as well as to obtain corresponding scattering phase shifts fitting existing experimental data. The approach is applied to description of the important mirror $^3H(\alpha,\gamma)^7Li$ and $^3He(\alpha,\gamma)^7Be$ reactions to demonstrate its possibilities. Calculated results are in good agreement with experimental data.

1. Introduction and some details of the approach

Radiative capture reactions are of great interest for nuclear astrophysics. However their cross sections are not available for reliable experimental measurements at low astrophysical energies due to strong Coulomb repulsion. As important examples of such reactions one can be considered the mirror $^3H(\alpha,\gamma)^7Li$ and $^3He(\alpha,\gamma)^7Be$ reactions [1, 2]. Cross section of the latter one in astrophysical relevant energy region is as before inaccessible for experiments and necessary for solving a number of problems [3]. In the current situation theoretical calculations based on microscopic approaches are assumed to be one of the most justified and promising ways to obtain required values of cross sections for the radiative capture reactions at low sub-barrier energies.

The main idea of the algebraic version of the resonating group model (AVRGM) [4] is expansion of relative motion wave function of clusters forming nuclear system onto series over the oscillator basis

$$f_{\nu lm}(\mathbf{q}) = (-1)^{(\nu-l)/2}\left(\frac{2\Gamma((\nu-l+2)/2)}{r_0^3\,\Gamma((\nu+l+3)/2)}\right)^{1/2}\left(\frac{q}{r_0}\right)^{l} L_{(\nu-l)/2}^{(l+1/2)}\left(\frac{q^2}{r_0^2}\right)\exp\left(-\frac{q^2}{2r_0^2}\right)Y_{lm}(\mathbf{n}_{\mathbf{q}}), \quad (1)$$

where Γ, $L_n^{(\beta)}$, Y_{lm} are gamma-function, generalized Laguerre polynomial, and spherical harmonic respectively; ν is number of oscillator quanta; l, m are relative orbital momentum and its projection; $\mathbf{q}$

IOP Conf. Series: Journal of Physics: Conf. Series **863** (2017) 012015 doi:10.1088/1742-6596/863/1/012015

is Jacobi vector of the relative motion of the clusters; r_0 is oscillator radius. As a result, total wave function of two-cluster system can be written as the expansion over so-called AVRGM basis

$$\Psi_{J^\pi Mlsv} = N_{J^\pi lsv}\, \mathrm{A}\left\{ \sum_{m+\sigma=M} C^{JM}_{lm\ s\sigma} \left[\phi^{(1)}_{s_1}\phi^{(2)}_{s_2}\right]_{s\sigma} f_{vlm}(\mathbf{q}) \right\}. \qquad (2)$$

Here $N_{J^\pi lsv}$ is the normalization; A is antisymmetrization operator; $\phi^{(1)}_{s_1}$, $\phi^{(2)}_{s_2}$ are internal wave functions of the clusters with spins s_1, s_2 coupled to the channel spin s with projection σ; J, M are total angular momentum and its projection; π is parity of the system; $C^{JM}_{lm\ s\sigma}$ is Clebsch–Gordan coefficient. In the present work wave functions of translationally invariant oscillator shell model for the lowest states compatible with the Pauli exclusion principle are chosen as the cluster internal wave functions. Oscillator radiuses involved in these functions and (1) are supposed to be equal. In the framework of the usual AVRGM approach applied in our previous works [5, 6] for radiative capture reactions description, wave functions of nuclear system are expanded over unified AVRGM basis (2). More advanced multi-size-scale AVRGM approach developed in this work uses different AVRGM bases at expansion of wave functions for discrete spectrum and continuum. These bases differ by values of the oscillator radius playing a role of scale parameter. When calculating astrophysical S-factor (cross section) for radiative capture reaction, it simultaneously makes possible to get scattering phase shifts fitting ones extracted from experiments and to achieve coincidence between calculated and experimental breakup thresholds into colliding nuclei in entrance channel for states of final nucleus formed as a result of the radiative capture. It is very important to have exact value for those breakup thresholds because they directly enter into expression for the radiative capture cross section [5, 6].

2. Results and conclusion

To demonstrate possibilities and applicability of the approach proposed in the present work the mirror ^{3}H(α, γ)^{7}Li and ^{3}He(α, γ)^{7}Be reactions were adopted for description. Effective modified Hasegawa–Nagata potential [7] was used as nuclear one in calculations. Astrophysical S-factors calculated for the reactions at low energies are shown in figure 1 and 2.

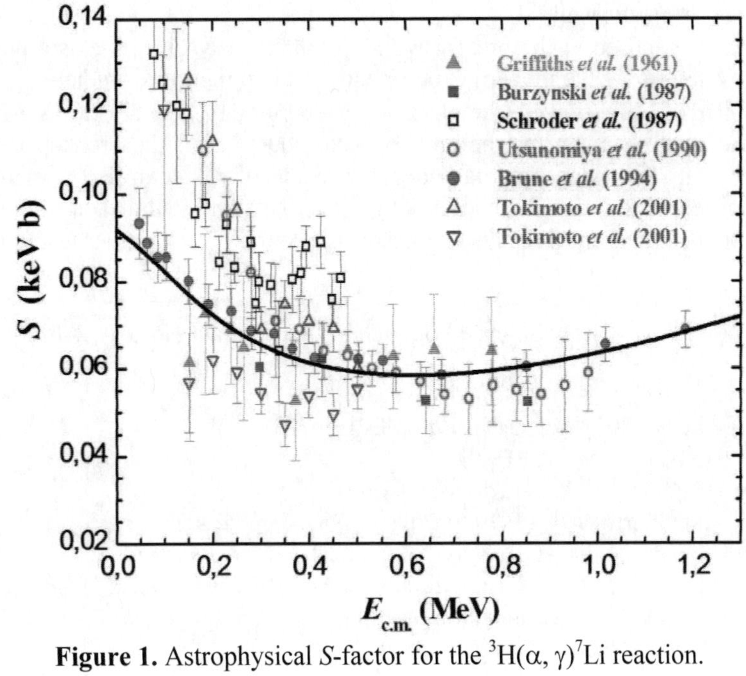

Figure 1. Astrophysical S-factor for the ^{3}H(α, γ)^{7}Li reaction.

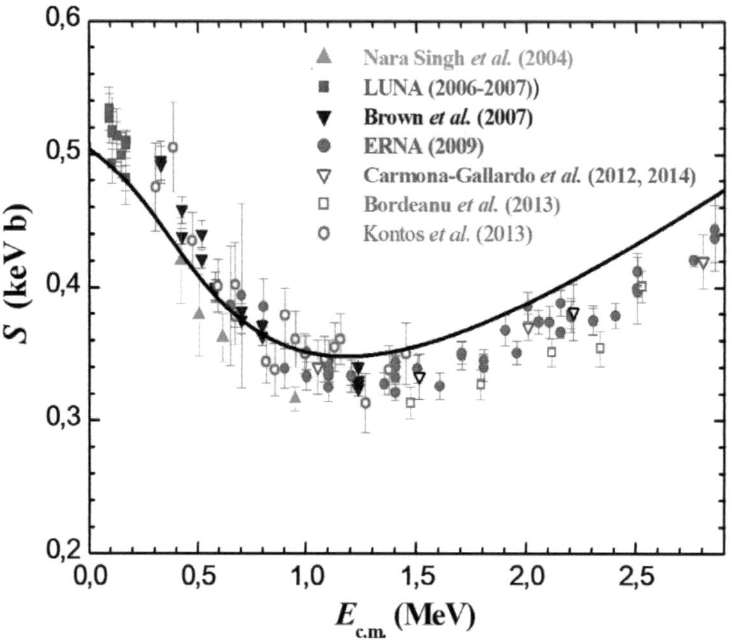

Figure 2. Astrophysical S-factor for the ^{3}He(α, γ)^{7}Be reaction.

As it can be seen from the figures, the calculated results are in good agreement with the modern experimental data (see references cited in [5, 6]). For considered energy ranges the radiative captures in the nuclear ^{4}He + ^{3}H and ^{4}He + ^{3}He systems proceed with formation of ^{7}Li and ^{7}Be nuclei in their ground or first excited states. Calculated breakup thresholds of these states into ^{4}He + ^{3}H and ^{4}He + ^{3}He nuclei for ^{7}Li and ^{7}Be ones respectively coincide with experimental values exactly [8]. Those values are $\varepsilon_0 = 2.467$ MeV, $\varepsilon_1 = 1.989$ MeV for ^{7}Li nucleus and $\varepsilon_0 = 1.586$ MeV, $\varepsilon_1 = 1.157$ MeV for ^{7}Be one. Calculated scattering phase shifts for ^{4}He + ^{3}H and ^{4}He + ^{3}He systems fit data extracted from the experiments well.

Thus the approach proposed in the present work simultaneously gives the astrophysical S-factors energy dependences describing the modern experimental data on the mirror radiative capture reactions ^{3}H(α, γ)^{7}Li and ^{3}He(α, γ)^{7}Be very well, the phase shifts energy dependences fitting data on scattering of the nuclei in entrance channels for the reactions considered, and the breakup thresholds values coinciding exactly with their experimental ones for states of ^{7}Li and ^{7}Be nuclei formed in the reactions. Such unified description of wide enough set of experimental data significant increases reliability of the obtained theoretical results and makes opportunities for further using the approach.

Acknowledgments

The work was supported by Russian Science Foundation (RSF), grant No. 16-12-10048.

References

[1] Adelberger E G et al. 2011 *Rev. Mod. Phys.* **83** 195
[2] Xu Y et al. 2013 *Nucl. Phys.* A **918** 61
[3] Fields B D 2011 *Ann. Rev. Nucl. Part. Sci.* **61** 47
[4] Filippov G F and Okhrimenko I P 1980 *Phys. Atom. Nucl.* **32** 480
[5] Solovyev A S, Igashov S Yu and Tchuvil'sky Yu M 2014 *J. Phys.: Conf. Ser.* **569** 012020
[6] Solovyev A S, Igashov S Yu and Tchuvil'sky Yu M 2015 *Eur. Phys. J.: Web Conf.* **86** 00054
[7] Kanada H, Kaneko T, Nagata S and Nomoto M 1979 *Prog. Theor. Phys.* **61** 1327
[8] Tilley D R et al. 2002 *Nucl. Phys.* A **708** 3

Constructing realistic alpha cluster channels

K Kravvaris and A Volya
Florida State University, Tallahassee, FL 32306, USA

Abstract. We present techniques that allow for α-cluster channels with realistic α-particle wave functions from No Core Shell Model calculations to be constructed. We compare results of several clustering calculations with realistic α wave functions to those assuming a trivial $(0s)^4$ structure.

In this work we report our progress in addressing specific questions related to studies of α clustering from the perspective of the nuclear shell model approach; our discussion follows a series of works in Refs. [1-7], and [8], in particular. Let us start by reviewing several key elements of the shell model approach to clustering. The use of harmonic oscillator (HO) basis, which we adopt for this work, allows for a formal separation of the center-of-mass (CM) degrees of freedom. Having an additional HO confining CM potential allows for factorization of the CM degree of freedom, and the full A-nucleon wave function appears as a product

$$\Psi = \phi_{n\ell m}(\mathbf{R})\,\Psi' \tag{1}$$

of the HO wave function $\phi_{n\ell m}(\mathbf{R})$ that depends only on the CM variables $\mathbf{R}$ and the translationally invariant wave function Ψ', which is a function of relative coordinates only. In our notation n is the number of oscillator quanta. Thus, in eq. (1) the total number of oscillator excitation quanta is shared between the CM and intrinsic degrees of freedom $N = n + N'$.

In the traditional approach to clustering the solutions for parent and daughter systems are both obtained in the form (1) where $n = 0$. It is then assumed than an intrinsic state of the α particle is $(0s)^4$ configuration and thus $N'_\alpha = 0$. This approximation simplifies the approach significantly because then the wave function of the α particle in a given HO CM state, the channel, can be written as

$$\Psi_\alpha = \phi_{n\ell m}(\mathbf{R})\,\Psi'_\alpha = \sum_\eta X^\eta_{n\ell}\,\Phi^\eta_{(n,0):\ell m}. \tag{2}$$

The expansion goes over all possible $A = 4$ nucleon configurations (often also referred to as partitions), labeled by η which have a stretched SU(3) symmetry $(n, 0)$, full permutational symmetry, and spin and isospin quantum numbers $S = T = 0$. The expansion cluster coefficients (CC) $X^\eta_{n\ell}$ are known analytically [3, 8]

$$X^\eta_{n\ell} = \sqrt{\frac{1}{4^n}\,\frac{n!}{\prod_i (n_i!)^{\alpha_i}}\,\frac{4!}{\prod_i \alpha_i!}}. \tag{3}$$

Here α_i refers to the number of particles on an oscillator shell n_i. Then the fractional parentage coefficient between parent and daughter states $\mathcal{F}^\eta_{nL} \equiv \langle \Psi_P | \Phi^{\eta\dagger}_{(n,0):L} | \Psi_D \rangle$ can be evaluated with the standard shell model techniques utilizing operator formalism of the second quantization. The translationally invariant fractional parentage coefficient requires an additional recoil coefficient that emerges as an oscillator bracket from recouping of the center of mass variables of the daughter system and α particle into a relative coordinate and a common CM that coincides with the parent system. The whole procedure and additional renormalization strategy is discussed in Ref. [8].

The purpose of this work is to present an approach that does not require a simple $(0s)^4$ structure for the α particle. We obtain the α wave function, restricted by the maximum total number of HO quanta $N_{\max}$ using the No Core Shell Model (NCSM) scheme, we use interactions from [9]. The realistic α wave function invalidates the expansion in eq. (2), and renders the analytic result in (3) useless. Nevertheless, the proposed new technique is just as effective numerically since it does not require construction of SU(3) operators $\Phi^{\eta}_{(n,0):\ell m}$.

We start with an α particle wave function being calculated using the NCSM approach. Upon application of the Glockner-Lawson procedure this produces a state of the same type as in eq. (1), where realistic α particle is in the CM HO state with $n = 0$. Then using sequential applications of the the CM creation and annihilation operators we obtain states where the α particle is in the CM state (1) with any desired CM HO quantum numbers. The CM creation operators (and annihilation correspondingly) are defined in the usual way as

$$\mathcal{B}^{\dagger}_{m} = \frac{1}{\sqrt{2AM\Omega\hbar}}(AM\Omega R_m - iP_m) = \frac{1}{\sqrt{A}}\sum_{a=1}^{A} b^{\dagger}_{a\,m} \tag{4}$$

where m denotes a specific magnetic projection of vectors and $b^{\dagger}_{am}$ raises the quanta of the a-th particle. The operator is easily constructed using an isoscalar mass-density dipole E1 operator

$$D_m = \sqrt{\frac{4\pi}{3}}\sqrt{\frac{\hbar}{2AM\Omega}}(\mathcal{B}^{\dagger}_{m} + \mathcal{B}_m) \tag{5}$$

and by taking the part that increases the number of quanta.

In order to boost the CM wave function to a particular state with $n = 2p + \ell$ quanta, where p is the number of nodes and ℓ the angular momentum, one can apply various combinations of creation operators $\mathcal{B}^{\dagger}_{m}$. The number of nodes can be increased, keeping rotational quantum numbers unchanged using a scalar combination of two creation operators

$$\mathcal{B}^{\dagger}\cdot\mathcal{B}^{\dagger} \equiv \left(\mathcal{B}^{\dagger}_{+1}\mathcal{B}^{\dagger}_{-1} + \mathcal{B}^{\dagger}_{-1}\mathcal{B}^{\dagger}_{+1} - \mathcal{B}^{\dagger}_{0}\mathcal{B}^{\dagger}_{0}\right), \tag{6}$$

$$\mathcal{B}^{\dagger}\cdot\mathcal{B}^{\dagger}\phi_{n\ell m}(\mathbf{R}) = \sqrt{(n-\ell+2)(n+\ell+3)}/4\;\phi_{n+2\ell m}(\mathbf{R}). \tag{7}$$

In order to increase the angular momentum ℓ while keeping the number of nodes in the wave function fixed one can act with $\mathcal{B}^{\dagger}_{m}$. The simplest strategy is to build an aligned state where $m = \ell$

$$\mathcal{B}^{\dagger}_{+1}\phi_{n\ell\ell}(\mathbf{R}) = \sqrt{\frac{(\ell+1)(n+\ell+3)}{4(2\ell+3)}}\phi_{n+1\ell+1\ell+1}(\mathbf{R}). \tag{8}$$

The CM angular momentum operator is a vector construction from CM quanta creation and annihilation operators $\mathcal{B}^{\dagger}\times\mathcal{B}$. In particular, the usual lowering operator required to obtain the desired magnetic projection is

$$\mathcal{L}_{-}\phi_{n\ell m}(\mathbf{R}) = 4\sqrt{2}\left(\mathcal{B}^{\dagger}_{-1}\mathcal{B}_0 - \mathcal{B}^{\dagger}_{0}\mathcal{B}_{-1}\right)\phi_{n\ell m}(\mathbf{R}) = \sqrt{(l+m)(l-m+1)}\phi_{n\ell m-1}(\mathbf{R}). \tag{9}$$

Configuration	$N_{\max}=0$	$N_{\max}=4$
$(sd)^4$	0.038	0.035
$(p)(sd)^2(pf)$	0.308	0.282
$(p)^2(pf)^2$	0.103	0.094
$(p)^2(sd)(sdg)$	0.154	0.141
$(s)^2(sd)(sdgi)$	0.000	0.005
$(p)(sd)(pf)(sdg)$	0.000	0.009

Table 1: Select configuration content of NCSM wave functions for ^{4}He with $\hbar\Omega$ = 20 MeV boosted by 8 quanta ($L=0$).

$A_P \to A_D$	$N_{\max}=0$	$N_{\max}=6$	Exp.
^{20}Ne→^{16}O	0.755	0.827	1
^{22}Ne→^{18}O	0.481	0.563	0.37
^{24}Mg→^{20}Ne	0.411	0.519	0.66
^{26}Mg→^{22}Ne	0.439	0.548	0.20
^{28}Si→^{24}Mg	0.526	0.575	0.33
^{30}Si→^{26}Mg	0.555	0.600	0.55

Table 2: Spectroscopic Factors for sd shell ground states using SU(3) and realistic α cluster wave functions ($\hbar\Omega$ = 14 MeV).

In Table 1 we compare the weights of select configuration components of α wave functions from NCSM calculations with $N_{\max}=4$ that have been CM-boosted for $n=8$ using the raising operators with the $N_{\max}=0$ case. The $N_{\max}=0$ case corresponds to $(0s)^4$ α configuration and reproduces the squared CC in eq. (3). For configurations with $N=\sum_i n_i\alpha_i=8$ the ratio between the two columns is proportional to the weight of the $(0s)^4$ configuration in the α wave function. This weight depends on the oscillator frequency Ω; here $\hbar\Omega$ = 20MeV. The last two lines in Table 1, show $N=10$ configurations that are only present in a realistic α wave function that is different from $(0s)^4$.

In Table 2 we demonstrate the effect of using a realistic α cluster in spectroscopic calculations. Here, $\hbar\Omega$ = 14 MeV that is a more typical value for the traditional shell model. The spectroscopic factors are calculated using the procedure and *sd* shell model Hamiltonian outlined in [8]. For $N_{\max}=0$ in the *sd* shell the only SU(3) component that contributes is the (8,0) irreducible representation; this is not the case in a more realistic situation with $N_{\max}=6$. Here, channels with $p=4,3,2,1$ all contribute to the $(sd)^4$ valence configuration.

In summary, in this work we put forward a new strategy for clustering calculations using the oscillator-based shell model approach. The strategy is efficient numerically, does not rely on algebraic techniques, and allows to treat realistic wave functions of clusters from NCSM. Select simple examples presented here demonstrate the procedure, show the effects of more complex α wave functions and highlight the limit when our approach reduces to the algebraic method used previously.

Acknowledgments

We thank Yu. M. Tchuvil'sky for collaboration and helpful comments. This material is based upon work supported by the U.S. Department of Energy Office of Science, Office of Nuclear Physics under Grant No. DE-SC0009883

References

[1] Mang H J 1957 *Z. Phys.* **148** 582
[2] Chung W, Vanhienen J, Wildenthal B H and Bennett C L 1978 *Phys. Lett. B* **79** 381
[3] Smirnov Yu F and Tchuvil'sky Yu M 1977 *Phys. Rev. C* **15** 84
[4] Grigorescu M, Brown B A and Dumitrescu O 1993 *Phys. Rev. C* **47** 2666
[5] Tchuvil'sky Yu M, Kurowsky W W, Sakharuk A A and Neudatchin V G 1995 *Phys. Rev. C* **51** 784
[6] Navratil P 2004 *Phys. Rev. C* **70** 054324
[7] Nemetz O F, Neudatchin V G, Rudchik A T, Smirnov Yu F and Tchuvil'sky Yu M 1988 *Nucleon Clusters in Atomic Nuclei and Multi-Nucleon Transfer Reactions* (Kiev: Naukova Dumka) p 295
[8] Volya A and Tchuvil'sky Yu M 2015 *Phys.Rev.C* **91** 044319
[9] Shirokov A M, Vary J P, Mazur A I and Weber T A 2008 *Physics of Atomic Nuclei* **71** 1232

Semimicroscopic algebraic description of α clustering in Ne isotopes

Géza Lévai

Institute for Nuclear Research, Hungarian Academy of Sciences (Atomki), Debrecen, Pf. 51, Hungary 4001

E-mail: `levai@atomki.mta.hu`

Abstract. The α-cluster states of ^{20}Ne, ^{19}Ne and ^{18}Ne have been analyzed as members of a supermultiplet predicted by the $U(4|12)$ cluster supersymmetry scheme. The analysis of $B(E2)$ data and one-nucleon transfer reactions seem to support the validity of this scheme. The α-cluster states of the ^{22}Ne nucleus have been analyzed independently, and were compared with DAMD and GCM calculations. Based on these results, a similar study of ^{21}Ne seems worthwhile.

1. Introduction

Nuclei in the neighborhood of ^{20}Ne are known to possess marked α cluster bands. Among them, some Ne isotopes are in the forefront of investigations: α-clustering in ^{19}Ne is known to play an important role nuclear astrophysical processes in stars [1], while Ne isotopes with $A > 20$ serve as examples for systems with molecular orbitals [2].

The semimicroscopic algebraic cluster model (SACM) [3] has been applied to many p- and sd-shell nuclei assuming core+α or other types of clustering (see e.g. the references of [4]). This model makes heavy use of the SU(3) algebra, which appears in the desciption of both the relative motion and that of the internal cluster structure, where it accounts for the orbital structure in terms of Elliott's SU(3) model. The semimicroscopic nature of the SACM is manifested by the fact that it combines a microscopic model space with effective interactions. Due to the former one the model gives rigorous account of the Pauli exclusion principle between the nucleons, while the latter one allows a relatively simple parametrization of the physical operators.

The systematic application of the SACM to neighbouring core+α-type cluster systems revealed correlations in the model parameters with the mass number [5]. This finding later inspired the extension of the formalism of the SACM to a supersymmetry scheme that allows the unified description of the α-cluster states of neighbouring nuclei [6]. Supersymmetry links bosonic and fermionic degrees of freedom, which appear naturally in the SACM as the excitation of the relative motion, and that of the internal structure of non-closed-shell clusters, respectively. In this scheme the α-cluster states of ^{20}Ne (0 fermion), ^{19}F, ^{19}Ne (1 fermion) and ^{18}O, ^{18}F, ^{18}Ne (2 fermions) belong to the same U(4|12) supermultiplet.

More recently the SACM was appied to describe α-clustering in the ^{22}Ne nucleus [4]. Besides the interpretation of the experimental energy levels and E2 transitions, the results were also compared with the predictions of two other models, the generator-coordinate method (GCM) and the deformed-basis antisymmetrized molecular dynamics (DAMD) approach.

Table 1. α-cluster bands in A=20, 19, and 18 nuclei up to two major shell excitations.

^{16}O+α $T=0$	(^{15}N, ^{15}O)+α $T=\frac{1}{2}$	(^{14}C, ^{14}N, ^{14}O)+α $T=1$ and $T=0$ (^{14}N+α)
8(8,0)	7(6,0)	6(4,0)
9(9,0)	8(8,1), 8(7,0)	7(6,1), 7(5,0)
10(10,0)	9(9,1), 9(8,0)	8(8,2), 8(7,1), 8(6,0)

2. ^{20}Ne, ^{19}Ne and ^{18}Ne: members of a cluster supermultiplet

The algebra chain characterizing the U(4|12) supersymmetry scheme is

$$\begin{aligned} &\mathrm{U}(4|12) \supset \mathrm{U_B}(4) \otimes \mathrm{U_F}(12) \supset \mathrm{SU_B}(3) \otimes \mathrm{SU_F}(3) \otimes \mathrm{U_F^{ST}}(4), \\ &\supset \mathrm{SU}(3) \otimes \mathrm{SU_F^S}(2) \otimes \mathrm{SU_F^T}(2) \supset \mathrm{SO}(3) \otimes \mathrm{SU_F^S}(2) \otimes \mathrm{SU_F^T}(2) \supset \mathrm{SU}(2) \otimes \mathrm{U_F^T}(1) \end{aligned} \quad (1)$$

where B and F refer to *bosonic* and *fermionic* structures, while S and T indicate spin and isospin characterizing the core cluster. The algebra chain (1) contains that of the SACM [3], which starts with $\mathrm{U_B}(4) \otimes \mathrm{SU_F}(3) \otimes \mathrm{U_F^{ST}}(4)$ and breaks down to the same subalgebras.

The $\mathrm{U_B}(4)$ bosonic and $\mathrm{U_F}(12)$ fermionic algebra are generated by the SU(3) coupled, particle number conserving operators $B_{M_L}^{(\lambda,\mu)L}(l,l') = [b^{\dagger(l,0)} \times \tilde{b}^{(0,l')}]_{M_L}^{(\lambda,\mu)L}$, and $A_{M_LM_SM_T}^{(\lambda,\mu)LST} = [a^{\dagger(0,1)\frac{1}{2}\frac{1}{2}} \times \tilde{a}^{(1,0)\frac{1}{2}\frac{1}{2}}]_{M_LM_SM_T}^{(\lambda,\mu)LST}$, and can be embedded into the U(4|12) superalgebra by adding the operators

$$D_{M_Lm_sm_t}^{(\lambda,\mu)Lst}(l') = [a^{\dagger(0,1)st} \times \tilde{b}^{(0,l')}]_{M_Lm_sm_t}^{(\lambda,\mu)Lst} \quad (2)$$

and their inverse, which conserve only the *total* particle number. Here $l, l' = 0$ and 1 refer to σ (monopole) and π (dipole) bosons, while the labels in the $A_{M_LM_SM_T}^{(\lambda,\mu)LST}$ operator characterize the 12 single-particle levels on the p shell in LS coupling [6].

The basis states associated with the chain (1) are labelled by the same quantum numbers as in the SACM [3], plus the fermion number N_F and the total particle number $\mathcal{N} = N_\mathrm{B} + N_\mathrm{F}$ [6]:

$$|\mathcal{N} N_\mathrm{B} n_\pi, (\lambda_\mathrm{F}, \mu_\mathrm{F}); (\lambda, \mu) K_L LSJM_JTM_T\rangle. \quad (3)$$

One important implication of the Pauli blocking between the nucleons of the clusters is that n_π, and thus N_B and $\mathcal{N}$ are restricted from below.

The predictions of the cluster SUSY scheme involving the six nuclei displayed in Table 1 were tested in two ways: comparing the calculated and experimental values for electric quadrupole transition intensities for the six nuclei, and checking the one-nucleon transfer intensities in six known reactions between five of them.

The comparison of the experimental and theoretical B(E2) values was done by adjusting the three model parameters to the most precisely measured transitions in two different ways. First the fitting procedure was carried out for each nucleus separately, then a global fit was done, involving the available electric quadrupole transition data for *all* six nuclei.

There were altogether 52 experimental B(E2) values with well defined mean values and errors, and a few more with lower or upper limits. There were 5, 5 an 2 experimental $B(E2)$ data for ^{20}Ne (in two bands assigned to $(\lambda, \mu) = (8,0)$ and $(9,0)$), ^{19}Ne (with $(6,0)$ and $(8,1)$) and ^{18}Ne (with $(4,0)$). It was found that the parameters determined for the six nuclei individually did not differ much from the parameters extracted from the fit to the experimental data of *all six* nuclei, indicating the approximate validity of the SUSY scheme.

Table 2. α-cluster bands in ^{22}Ne and ^{21}Ne up to two major shell excitations.

n_π	(λ,μ) for ^{22}Ne					(λ,μ) for ^{21}Ne		
8			(8,2)	(6,3)	(4,4)		(8,1)	(6,2)
9		(11,1)	(9,2)	(7,3)	(5,4)	(11,0)	(9,1)	(7,2)
10	(14,0)	(12,1)	(10,2)	(8,3)	(6,4)	(12,0)	(10,1)	(8,2)

In order to calculate the C^2S spectroscopic factors of the one-nucleon transfer reactions, the matrix elements of the SUSY generators (2) were evaluated. Since these operators are the generators of the SUSY algebra, they predict strict selection rules between the states of neighbouring nuclei. Analyzing the six reactions it was found that the average of the experimental C^2S spectroscopic factors for the allowed 28 transitions was typically about an order of magnitude larger than for the 20 forbidden transitions. This finding also seems to support the validity of the U(4|12) supersymmetry scheme [6].

3. ^{22}Ne and ^{21}Ne: candidates for molecular orbital structures

In some $N > Z$ systems the neutrons can be pictured as moving on molecular orbitals around the core+α configuration [2], which, for Neon isotopes is $^{16}\mathrm{O} + \alpha$. This development inspired us to apply the SACM to the ^{22}Ne nucleus [4]. The SACM α-cluster model space of ^{22}Ne and ^{21}Ne is displayed in Table 2.

The low-lying positive-parity part of the experimental ^{22}Ne spectrum is dominated by the 0_1^+ and 2^+ bands, while the 0_2^+ band with large prolate deformation starts at 6.23 MeV. In the SACM these bands correspond to $n_\pi(\lambda,\mu)\kappa$ =8(8,2)0, 8(8,2)2 and 10(14,0)0. For negative parity the 1^-, 0_1^- and 2^- bands are known at low energy, which correspond to $n_\pi(\lambda,\mu)\kappa$ =9(11,1)1, 9(9,2)0 and 9(9,2)2. The energy spectrum was fitted with the SACM Hamiltonian containing five parameters, and good agreement was found for the positive-parity spectrum. The calculated $E2$ transitions agreed well with the handful of known experimental data . The model predicted further bands, but the available experimental data was insufficient for assigning experimental states to the model ones. However, these predictions generally compared well with the results of two microscopic models, the deformed-basis antisymmetrized molecular dynamics (DAMD) [7] approach and the generator-coordinate method (GCM) [8]. The low-lying bands were described by the three models in a qualitatively similar way.

Preliminary studies indicate that the experimental $K^\pi = 3/2^+$ and $1/2^-$ bands of ^{21}Ne may be assigned to the $(\lambda,\mu) = (8,1)$ and (11,0) SU(3) multiplets.

Acknowledgments
Work supported by the Hungarian Scientific Research Fund – OTKA, grant No. K112962.

References

[1] Otani R, Kageyama R, Iwasaki M, Kudo M, Tomita M and Ito M 2014 *Phys. Rev. C* **90** 034316
[2] von Oertzen W and Milin M 2014 *Springer Lecture Notes in Physics* **875** 147
[3] Cseh J and Lévai G 1994 *Ann. Phys. (N.Y.)* **230** 165
[4] Lévai G 2013 *Phys. Rev. C* **88** 014328
[5] Lévai G and Cseh J 1996 *Phys. Lett. B* **381** 1
[6] Lévai G, Cseh J and Van Isacker P 2007 *J. Phys. G: Nucl. Part. Phys.* **34** 1729
[7] Kimura M 2007 *Phys. Rev. C* **75** 034312
[8] Descouvemont P 1988 *Phys. Rev. C* **38** 2397

IOP Conf. Series: Journal of Physics: Conf. Series **863** (2017) 012018 doi:10.1088/1742-6596/863/1/012018

Experimental progress in clarifying the excitation spectrum of ^{12}C

O S Kirsebom

Department of Physics and Astronomy, Aarhus University, DK-8000 Aarhus C, Denmark

E-mail: oliskir@phys.au.dk

Abstract. We briefly review recent progress in clarifying the role of α clustering in ^{12}C and discuss how experimental studies of the $\alpha + \alpha + \alpha$ breakup of excited states in ^{12}C contribute towards this goal. Finally, we show preliminary data from ongoing experimental studies of the $p +^{11}$ B $\rightarrow \alpha + \alpha + \alpha$ reaction at the 5 MV Van dee Graaff accelerator at Aarhus University.

1. Introduction

To what extent can we think of the ^{12}C nucleus as composed of three closely spaced α particles? Recent years have witnessed significant progress towards answering this nearly century-old question. Most notably, microscopic calculations have provided the first theoretical evidence of α clustering; see, e.g., Refs. [1, 2, 3, 4]. While compelling, the theoretical evidence is in itself not conclusive. Ultimately, experimental verification is needed.

The microscopic calculations readily predict excitation spectra and transition strengths which can be checked against experimental data, but since the calculations all employ some type of bound-state approximation they do not make any predictions concerning the couplings to particle channels, such as $\alpha + \alpha + \alpha$ which incidentally is the only particle-decay channel open at low excitation energies. From an experimental point of view this limitation can seem a bit absurd considering that particle emission is the dominant decay mode for all but a few states in ^{12}C. Fortunately, this aspect is receiving growing attention. Computational schemes have been develop to solve the $\alpha+\alpha+\alpha$ three-body problem in the continuum, and these schemes have been employed to determine the final-state momentum distribution of the $^{12}\text{C}^* \rightarrow \alpha + \alpha + \alpha$ breakup process [5, 6] (which can be measured in laboratory experiments) as well as the cross section of the $\alpha + \alpha + \alpha \rightarrow {}^{12}\text{C}+\gamma$ fusion process [7, 8, 9] (which plays a crucial role in astrophysics). Efforts are underway to merge the two approaches (bound-state microscopic and continuum three-body). This will allow particle-decay properties to be predicted from first principles and motivates further experimental studies of the $\alpha + \alpha + \alpha$ breakup of excited states in ^{12}C.

The first experimental studies of the $^{12}\text{C}^* \rightarrow \alpha + \alpha + \alpha$ breakup process date back more than half a century. However, high-statistics, complete-kinematics studies have only become feasible in recent years thanks to the development of compact arrays of segmented silicon detectors. Thanks to the superior quality of the modern data, it has been possible to extract a wealth of new information: A sequential model based on the R-matrix theory [10] has been found to provide a remarkably accurate description of the breakup process [11, 12, 13]. The α-decay branching ratio to the ground state of ^{8}Be has been determined for several states in ^{12}C [13, 14] including the famous Hoyle state [15, 16, 17, 18]. Evidence for hitherto unknown states has been obtained and spin-parities have been constrained [12, 19, 20, 21]. Finally, gamma transitions into unbound states have been measured by an indirect technique [21, 22, 23].

2. Preliminary resutls on the system p+^{11}B

The probes used to populate the excited states in ^{12}C include the β decays of ^{12}B and ^{12}N, the p+^{11}B fusion reaction and reactions such as ^{10}B(^{3}He, p) and ^{11}B(^{3}He, d). Recent work at the 5 MV Van de Graaff accelerator at Aarhus University has focused on the p+^{11}B fusion reaction, which gives access to excited states above the proton-separation energy of 16.0 MeV. The resulting three α particles are detected in coincidence with high resolution and high efficiency using a compact setup consisting of four double-sided silicon detectors [23]. We exploit the complete kinematic information by imposing cuts on total energy and momentum (removes unwanted events such as random coincidences) and perform kinematic fitting to further improve resolution. To visualize the experimental data we use the symmetric Dalitz plot with $(E_1 + 2E_2)/\sqrt{3}$ plotted along the abscissa and E_1 along the ordinate. Here E_i is the kinetic energy of one of the α particles in the centre of mass frame. Figure 1 shows examples of two such Dalitz plots obtained a different beam energies. Note that the results are preliminary. In particular, corrections have not been made for the effects of experimental acceptance.

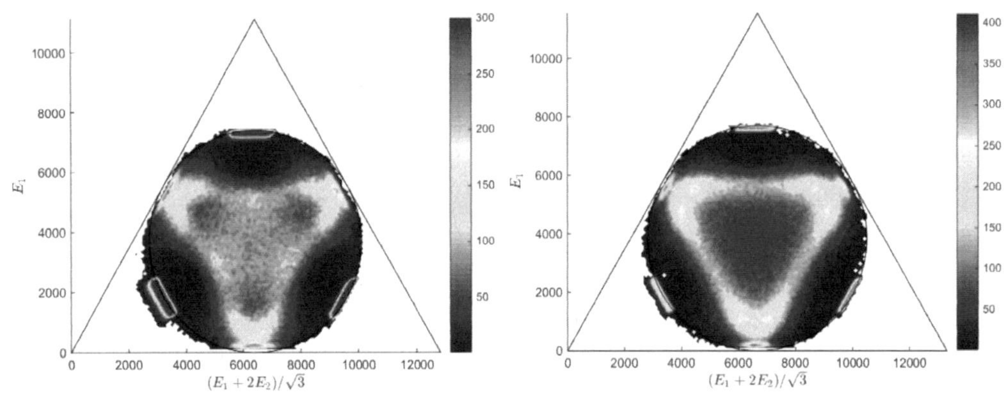

Figure 1. Dalitz plots showing preliminary experimental data for the $p+^{11}\mathrm{B} \rightarrow \alpha + \alpha + \alpha$ reaction at beam energies of 2.63 MeV (left) and 3.12 MeV (right). E_i is the kinetic energy of one of the α particles in the centre of mass frame in units of keV.

The relatively high level density above the proton separation energy makes these studies rather challenging because at a given beam energy we populate a mix of states in ^{12}C with different spin-parities and decay properties. In order to disentangle the various contributions we are currently undertaking a scan of the p+^{11}B fusion reaction in the energy range 0.15–3.5 MeV in steps of about 100 keV, performing complete-kinematics measurements at each energy. We are also working on extending our sequential model to include spin and angular-momentum coupling in the p+^{11}B entrance channel.

Acknowledgments

The author is greatful to Simon P. Pedersen for producing the figures and acknowledges support from the Villum Foundation.

References

[1] Kanada-En'yo Y 2007 *Progr. Theor. Phys.* **117** 655–680

[2] Chernykh M, Feldmeier H, Neff T, von Neumann-Cosel P and Richter A 2007 *Phys. Rev. Lett.* **98** 032501
[3] Epelbaum E, Krebs H, Lähde T A, Lee D and Meißner U G 2012 *Phys. Rev. Lett.* **109** 252501
[4] Lau P H C and Manton N S 2014 *Phys. Rev. Lett.* **113** 232503
[5] Álvarez-Rodríguez R, Jensen A S, Fedorov D V, Fynbo H O U and Garrido E 2007 *Phys. Rev. Lett.* **99** 072503
[6] Álvarez-Rodríguez R, Jensen A S, Garrido E, Fedorov D V and Fynbo H O U 2008 *Phys. Rev.* C **77** 064305
[7] Nguyen N B, Nunes F M, Thompson I J and Brown E F 2012 *Phys. Rev. Lett.* **109** 141101
[8] de Diego R, Garrido E, Fedorov D and Jensen A 2011 *Phys. Lett.* B **695** 324–328
[9] Ishikawa S 2013 *Phys. Rev.* C **87** 055804
[10] Balamuth D P, Zurmühle R W and Tabor S L 1974 *Phys. Rev.* C **10** 975–986
[11] Fynbo H O U, Prezado Y, Bergmann U C, Borge M J G, Dendooven P, Huang W X, Huikari J, Jeppesen H, Jones P, Jonson B, Meister M, Nyman G, Riisager K, Tengblad O, Vogelius I S, Wang Y, Weissman L, Rolander K W and Äystö J 2003 *Phys. Rev. Lett.* **91** 082502
[12] Kirsebom O S, Alcorta M, Borge M J G, Cubero M, Diget C A, Dominguez-Reyes R, Fraile L M, Fulton B R, Fynbo H O U, Hyldegaard S, Jonson B, Madurga M, Martin A M, Nilsson T, Nyman G, Perea A, Riisager K and Tengblad O 2010 *Phys. Rev.* C **81** 064313
[13] Laursen K L, Fynbo H O U, Kirsebom O S, Madsbøl K S and Riisager K *arXiv:1604.01244 [nucl-ex]*
[14] Alcorta M, Borge M J G, Cubero M, Diget C A, Domínguez-Reyes R, Fraile L M, Fulton B R, Fynbo H O U, Galaviz D, Hyldegaard S, Jeppesen H, Jonson B, Kirsebom O S, Madurga M, Maira A, Muñoz-Martín A, Nilsson T, Nyman G, Obradors D, Perea A, Riisager K, Tengblad O and Turrion M 2012 *Phys. Rev.* C **86** 064306
[15] Kirsebom O S, Alcorta M, Borge M J G, Cubero M, Diget C A, Fraile L M, Fulton B R, Fynbo H O U, Galaviz D, Jonson B, Madurga M, Nilsson T, Nyman G, Riisager K, Tengblad O and Turrión M 2012 *Phys. Rev. Lett.* **108** 202501
[16] Manfredi J, Charity R J, Mercurio K, Shane R, Sobotka L G, Wuosmaa A H, Banu A, Trache L and Tribble R E 2012 *Phys. Rev.* C **85** 037603
[17] Rana T K, Bhattacharya S, Bhattacharya C, Kundu S, Banerjee K, Ghosh T K, Mukherjee G, Pandey R, Roy P, Srivastava V, Gohil M, Meena J K, Pai H, Saha A K, Sahoo J K and Saha R M 2013 *Phys. Rev.* C **88** 021601
[18] Itoh M, Ando S, Aoki T, Arikawa H, Ezure S, Harada K, Hayamizu T, Inoue T, Ishikawa T, Kato K, Kawamura H, Sakemi Y and Uchiyama A 2014 *Phys. Rev. Lett.* **113** 102501
[19] Hyldegaard S, Alcorta M, Bastin B, Borge M J G, Boutami R, Brandenburg S, Büscher J, Dendooven P, Diget C A, Van Duppen P, Eronen T, Fox S P, Fraile L M, Fulton B R, Fynbo H O U, Huikari J, Huyse M, Jeppesen H B, Jokinen A S, Jonson B, Jungmann K, Kankainen A, Kirsebom O S, Madurga M, Moore I, Nieminen A, Nilsson T, Nyman G, Onderwater G J G, Penttilä H, Peräj́arvi K, Raabe R, Riisager K, Rinta-Antila S, Rogachevskiy A, Saastamoinen A, Sohani M, Tengblad O, Traykov E, Wang Y, Wilhelmsen K, Wilschut H W and Äystö J 2010 *Phys. Rev.* C **81** 024303
[20] Kirsebom O 2013 *Few-Body Systems* **54** 755–759
[21] Laursen K L, Fynbo H O U, Howard A M and Kirsebom O S 2014 *J. Phys: Conf. Series* **569** 012073
[22] Kirsebom O S, Alcorta M, Borge M J G, Cubero M, Diget C A, Dominguez-Reyes R, Fraile L M, Fulton B R, Fynbo H O U, Galaviz D, Garcia G, Hyldegaard S, Jeppesen H B, Jonson B, Joshi P, Madurga M, Maira A, Martin A M, Nilsson T, Nyman G, Obradors D, Perea A, Riisager K, Tengblad O and Turrion M 2009 *Phys. Lett.* B **680** 44–49
[23] Kirsebom O S, Fynbo H O U, Howard A M and Laursen K L 2014 *J. Phys.: Conf. Series* **569** 012013

IOP Conf. Series: Journal of Physics: Conf. Series **863** (2017) 012019 doi:10.1088/1742-6596/863/1/012019

Measurement of the 3-α decay from the Hoyle and the broad 10 MeV states in ^{12}C

M Itoh, S Ando, T Aoki, H Arikawa, S Ezure, K Harada, T Hayamizu, T Inoue, T Ishikawa, K Kato, H Kawamura, Y Sakemi and A Uchiyama

Cyclotron and Radioisotope Center, Tohoku University, Sendai, Miyagi 980-8578, Japan

E-mail: itoh@cyric.tohoku.ac.jp

Abstract. The measurements of the 3-α decay from the Hoyle and the broad 10 MeV state in ^{12}C have been performed using an inverse kinematic method of the ^{12}C(^{12}C,3α)^{12}C reaction at $E_{^{12}C}$ = 110 MeV. In the measurement of the Hoyle state, the upper limit of the direct decay was improved to be 0.2 %. This was inconsistent with the recent reported nonzero value of 0.9 %, but an order larger than the direct decay branch predicted in the 3-α model. The branching ratios of the sequtial decay for the broad 10 MeV state were also obtained. They were higher than those obtained in the RCNP experiment, since they contained the contributions of higer multipole components. Their contributions will be estimated.

1. Introduction

The second 0^+ state at E_x = 7.65 MeV in ^{12}C, which is called the Hoyle state, plays an important role in the creation of the ^{12}C nucleus in stellar nucleosynthesis. However, its structure is not understood well, since the 0^+ state does not appear at such a low excitation energy of 7.65 MeV in the simple shell model. According to the microsopic α cluster model, the Hoyle state is considred to have a dilute 3-α gaslike structure. Tohsaki *et al.* showed it was similar to the Bose-Einstein condensation of α clusters in the nucleus by their model wave function [1]. Recently, some ab initio calculations have beem tried to explain properties of the Hoyle state including the ground state band in ^{12}C [2, 3]. Among them, a lattice approach with chiral effective field theory succeeded in reproducing the excitation energy of the Hoyle state. In their result, the Hoyle state is considered to have a "bent-arm" or obtuse triangular configuration. The algebraic cluster model also succeeded to explain the spectrum of ^{12}C with equilateral triangular configuration [4, 5]. In their model, the Hoyle state is considered to be the A symmetric stretching vibration or breathing mode of the triangular configuration. As described above, the configuration of 3-α clusters for the Hoyle state is still controversial.

It is difficult to determine the structure of the nuclear excited state, particularly the unbound states, experimentally. One possible way is the decay particle measurement. The first experiment was performed by Freer *et al.* [6]. They measured the decay 3-α particles from the Hoyle state via the ^{12}C(^{12}C,3α)^{12}C reaction and found the Hoyle state decay to the 3-α particles through the ground state of ^{8}Be + α channel and the upper limit of the direct decay branch was 4 %. In 2011, Raduta *et al.* reported the direct decay branch of 17 $\pm$ 5 % from the Hoyle state [7]. They measured the correlation function of three α particles emmited from the ^{40}Ca + ^{12}C reaction at the incident energy of 25 MeV per nucleon. In their results, the branching ratio of the direct decay with an equal energy of 3-α particles was 7.5 $\pm$ 4.0 %. They emphasized this was evidence for the α cluster condensation. However, it was inconsistent with the result of the first experiment by Freer *et al.*

This inconsistency has been tested by two experiments at first. One is the analysis of the ^{10}C + C (Be) reaction by Manfredi *et al.* [8]. They obtained the upper limit of the 3.9 % for the direct decay from the Hoyle state, which was consistent with the result of the first experiment by Freer *et al.* Another is the experiment of the ^{11}B(^{3}He, $d3\alpha$) reaction by Kirsebom *et al.* [9]. They highly improved the upper limit of the direct decay to 5×10^{-3}. In this way, the huge branching ratio of 17 % for the direct decay was excluded. Recently, the nonzero value of the direct 3-α decay branch from the Hoyle state was reported by Rana *et al.* [10]. They claimed their obtained nonzero branching ratio of 0.9 % for the direct decay was consistent with the upper limit obtained by Kirsebom *et al.* Therefore, more precise measurement was desired in order to verify this nonzero branching ratio obtained by Rana *et al.*

In this paper, the measurements of the 3-α decay from not only the Hoyle state but also the broad 10 MeV state via the ^{12}C(^{12}C,3α)^{12}C reaction at $E_{^{12}C}$ = 110 MeV are reported. The structure of the broad 10 MeV state in ^{12}C is also still unclear. In our previous studies [11, 12], we pointed out the possibility of consisting of two components of the broad 0^+ state at 10 MeV. In order to investigate the structure of the broad 10 MeV state, we studied the decay mechanism by the measurement of 3-α decay.

2. Experiment

The experiments were performed at the Cyclotron and Radioisotope Center (CYRIC), Tohoku University. The $^{12}C^{4+}$ beams were accelerated up to 110 MeV by AVF cyclotron, and bombarded to the self-supported carbon foil with a thickness of 50 μg/cm^2. The target was tilted to the beam direction in order to minimize the energy loss of the recoil ^{12}C particle in the target. The decay 3-α particles were detected by the double-sided silicon strip detector (DSSD) with a size of 50×50 mm^2 and with a thickness of 1500 μm which has 16×16 strips oriented vertically in the front side and horizontally in the rear side. Recoil ^{12}C particles were caught by a silicon detector with a thickness of 150 μm at 67°. The silicon detectors were kept cooling around 0°C during the experiment. The ^{12}C particle was identified by the time of flight method. From a complete kinematics method, in which all particles participated in the reaction were measured, we reconstructed the excitation energy of ^{12}C and determined the decay mechanism of the Hoyle state.

In the measurement of the broad 10 MeV state, recoil ^{12}C particles were not measured because their energies were too small to be measured at the forward scattering angle of the ^{12}C beam. Two DSSDs were set at extremely forward angles for the detection of the decay α particles. Since the beam was stopped just in front of the DSSD, background events were much larger than the measurement of the Hoyle state. Therefore, the trigger signal was made from the event of the multiplicity 2 and over. Aluminum plates with a thickness of 200 μm were installed in front of the DSSDs to stop ^{12}C particles of the beam halo and elastic scattering, which became main background particles without them. In this measurement, the excitation energy of ^{12}C was determined from the relative kinetic energies between decay 3-α particles. Figure 1 shows the excitation energy spectrum of ^{12}C obtained the measurement of the broad 10 MeV state.

3. The 3-α decay from the Hoyle state

To visualize the energy correlation of the decay 3-α particles, the symmetric Dalitz plot is adopted in the analysi of the Hoyle state. Figure 2 shows the symmetric Dalitz plots obtained in the present measurement and the Monte Carlo simulation. The number of events in each simulation plot in Fig. 2 was same as that of the experimental plot. Three decay mechanisms were

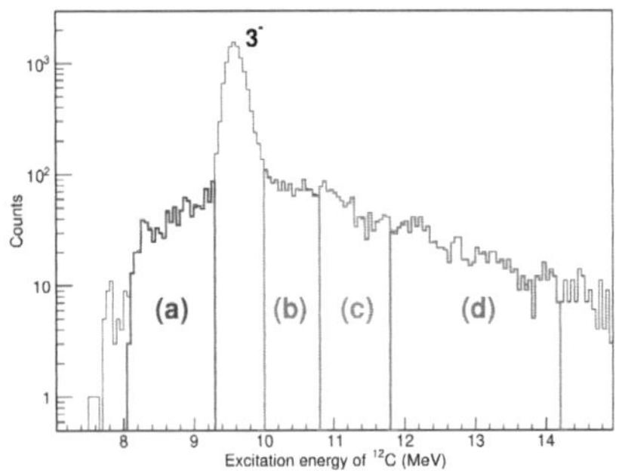

Figure 1. The excitation energy spectrum of ^{12}C obtained in the measurement of the broad 10 MeV state. The broad 10 MeV state were divided into 4 regions (a, b, c, d). See text.

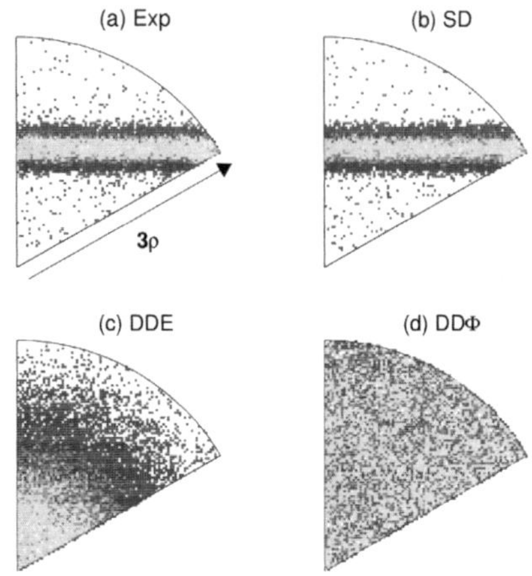

Figure 2. The symmetric dalitz plot of the decay 3-α particles from the Hoyle state. (a) Experimental data, (b) Monte Carlo simulation of the sequential decay (SD), (c) the direct decay with an equal energy of 3-α particles (DDE), (d) the direct decay to the phase space uniformly (DDΦ).

compared with the result of the experiment. The first is the sequential decay (SD) mechanism, in which the Hoyle state decay to three α particles through the ground state of ^{8}Be + α channel. The second is the direct decay with an equal energy of 3-α particles (DDE). The third is the direct decay to the phase space uniformly (DDΦ). The experimental plot seems to be almost same as that of the SD mechanism.

In order to obtain the branching ratio for each decay mechanism, we fitted the energy distribution for the highest energy among the decay 3-α particles with those obtained in the simulation. The decay branch of the SD mechanism was 100%. After statistical treatment, the upper limit of 0.2% for the direct decay mechanisms (DDE + DDΦ) was obtained.

Recently, Ishikawa predicted the branching ratio of the direct decay from the Hoyle state in the framework of the 3-α model [13]. In his result, the direct decay branch was an order magnitude less than our obtained upper limit of 0.2%. We need to improve the sensitivity of the detection of the direct 3-α decay by an order better than this experiment.

4. The 3-α decay from the broad 10 MeV state

Obtained data were analyzed by dividing the excitation energy spectrum into the following 4 regions, (a) 7.8 - 9.3 MeV, (b) 10.0 - 10.8 MeV, (c) 10.8 - 11.8 MeV, (d) 11.8 - 14.2 MeV, as shown in Fig. 1. To investigate the decay mechanism of the broad 10 MeV state, we also carried out the Monte-Carlo simultation for two decay mechanisms, the decay through the ground state of ^{8}Be, ^{8}Be(g.s.), and that through the 2^+ state of ^{8}Be, ^{8}Be(2^+), in each energy region. In the decay from the broad 10 MeV state, the direct 3-α decay could not be distinguished from the ^{8}Be(2^+) channel. Therefore, the energy distributions were fitted with these two decay channels, ^{8}Be(g.s.) and ^{8}Be(2^+). Figure 3 shows the preliminary result for the energy distributions of decay α particles in the c.m. frame. They are reproduced well with the sum of the energy distributions

IOP Conf. Series: Journal of Physics: Conf. Series **863** (2017) 012019 doi:10.1088/1742-6596/863/1/012019

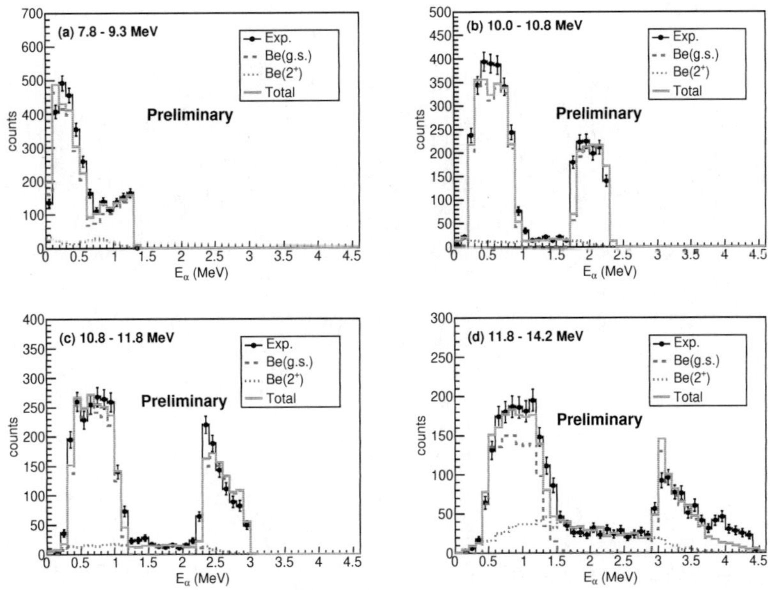

Figure 3. The energy distributions of decay 3-α particles in the energy regions, (a) 7.8 - 9.3 MeV, (b) 10.0 - 10.8 MeV, (c) 10.8 - 11.8 MeV, (d) 11.8 - 14.2 MeV, which correspond to those in Fig. 1.

for the ^{8}Be(g.s.) and ^{8}Be(2^+) channels. The branching ratios of the sequential decay are shown in Fig. 4, where they are compared with those obatined in the RCNP experiment [12]. While the branching ratio of the SD obtained in the RCNP experiment revealed a sharp drop at the excitation energy of 10.8 MeV, that obtained in the CYRIC experiment showed gradually decline. Since momentum transfers of the reaction at CYRIC were larger than those at RCNP, which were $q < 0.8$ fm^{-1} at CYRIC and $q < 0.2$ fm^{-1} at RCNP, respectively, the energy spectrum at CYRIC in Fig. 1 contains higher multipole components, such as the 1^- state at 10.8 MeV, the 2^+ state at 10 MeV and so on. According to Alcorta *et al* [14], the 1^- state at 10.8 MeV decays 100% through the ^{8}Be(g.s.) channel. Therefore, we need to estimate the contributions from higher multipolarity components correctly. The estimations for the higher multipole contributions are in progress.

5. Summary

We performed two measurements of the 3-α decay via the ^{12}C(^{12}C,3α)^{12}C reaction at 110 MeV. The first is the Hoyle state, the second is on the broad 10 MeV state. In the measurement of the Hoyle state, we measured all particles participated in the reaction and obtained the upper limit of the direct decay branch as 0.2 % by comparing with the result of the Monte Carlo simulation. This upper limit was further improved than that obtained by Kirsebom *et al.* [9] and excluded the nonzero direct decay branch of 0.9 % obtained by Rana *et al* [10].

In the second measurement of the broad 10 MeV state, we measured the energy distributions of decay α particles from the broad 10 MeV state. The branching ratios of the sequential decay were acquired in energy regions obtained by dividing the broad 10 MeV state into 4 regions. They were larger than those obtained in the RCNP experiment above 10.8 MeV, although they gradually decreased. The branching ratios of the sequential decay above 10.8 MeV obtained in this experiment included the contributions of higher multipole components. We need to estimate

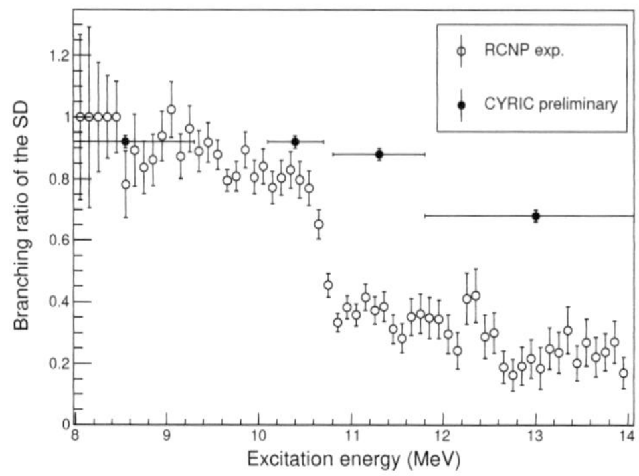

Figure 4. Branching ratio of the sequential decay (SD) from the broad 10 MeV state. Open circles show those obtained from the RCNP experiment [12]. Closed circles show preliminary results at CYRIC.

them correctly. The analysis is in progress.

Acknowledgments
Authors wish to thank S Ishikawa for fruitful discussions and the calculation of the decay branch of the Hoyle and broad 10 MeV states. A part of this work is financially supported by JSPS KAKENHI Grant Number 24740139.

References

[1] Tohsaki A, Horiuchi H, Schuck P and Röpke G 2001 *Phys. Rev. Lett.* **87** 192501
[2] Epelbaum E, Krebs H, Lähde T A, Lee D and Meissner U G 2012 *Phys. Rev. Lett.* **109** 252501
[3] Dreyfuss A C, Laure K D, Dytrych T, Draayer J P and Bahri C 2013 *Phys. Lett.* B **727** 511
[4] Bijker R and Iachello F 2000 *Phys. Rev.* C **61** 067305
[5] Marín-Lámbarri D J *et al* 2014 *Phys. Rev. Lett.* **113** 012502
[6] Freer M *et al* 1994 *Phys. Rev.* C **49** R1751
[7] Raduta Ad R *et al.* 2011 *Phys. Lett.* B **705** 65
[8] Manfredi J *et al.* 2012 *Phys. Rev.* C **85** 037603
[9] Kirsebom O S *et al* 2012 *Phys. Rev. Lett.* **108** 202501
[10] Rana T K *et al* 2013 *Phys. Rev.* C **88** 021601(R)
[11] Itoh M *et al* 2011 *Phys. Rev.* C **84**
[12] Itoh M *et al* 2013 *J. Phys. Conf. Ser.* **436** 012006
[13] Ishikawa S 2014 *Phys. Rev.* C **90** 061604(R)
[14] Alcorta M *et al Phys. Rev.* C **86** 064306

IOP Conf. Series: Journal of Physics: Conf. Series **863** (2017) 012020 doi:10.1088/1742-6596/863/1/012020

α-cluster breaking effects on cluster structures in ^{12}C

Suhara T[1] and Kanada-En'yo Y[2]

[1] Matsue College of Technology, Matsue 690-8518, Japan
[2] Department of Physics, Kyoto University, Kyoto 606-8502, Japan

E-mail: suhara@matsue-ct.ac.jp

Abstract. To clarify the effects of α-cluster breaking on 3α cluster structures in ^{12}C, we investigate ^{12}C using a hybrid model that combines the Brink-Bloch cluster model with the $p_{3/2}$ subshell closure wave function. We have found that α-cluster breaking significantly changes the cluster structures of 0^+ states through orthogonality to lower states. As a result of the structure changes of 0^+ states, the band assignment for the 2_2^+ state is changed. The 0_3^+ state is likely to be the band-head of the 2_2^+ state instead of the 0_2^+ state because the present calculation reproduces experimental data of the low-lying states well with α-cluster breaking. We have found that this band can be considered a linear chain-like band from the analysis of the intrinsic structures.

1. Introduction

In light nuclei, it is known that cluster formation occurs in low-lying states and competes with the shell-model structure. ^{12}C is a typical example of this competition [1, 2, 3, 4, 5, 6]. Because the spin-orbit force usually tends to hinder the cluster formation, the ground state of ^{12}C is not the pure 3α cluster state but is a mixture of 3α cluster structure and jj-coupling shell model structure of the $p_{3/2}$ subshell closure. On the other hand, in excited states of ^{12}C near the threshold energy, developed 3α cluster structures have been discovered. It is important to understand the magnitude of this competition and how much it affects the structures of the ground and excited states of ^{12}C.

In 3α cluster models, the ground and low-lying states have a compact triangle structure, whereas excited states near and above the 3α cluster threshold energy have well-developed cluster structures [7, 8, 9, 10, 11]. For example, the 0_2^+ state, which is well-known as the Hoyle state, is considered to be an α condensate state in which weakly interacting three α clusters occupy an identical lowest orbit of a mean-field potential [9]. Another interesting cluster structure was predicted in the 0_3^+ state, which is considered to be a vibration mode of acute and obtuse triangle configurations [7].

The antisymmetrized molecular dynamics (AMD) and the fermionic molecular dynamics (FMD) models have been successful in describing the nature of ^{12}C from the ground state to the higher excited states [1, 2, 3, 4, 5, 6]. In these models, the ground and low-lying states are a mixture of cluster and jj-coupling shell model configurations. For states near the threshold, these models reproduce the cluster nature, but the detailed cluster structures obtained with these models are somewhat different from those of cluster model results. For example, the 0_3^+ state is considered to be a chain-like structure having an obtuse triangle configuration of three α clusters instead of the vibrational state predicted by the cluster model.

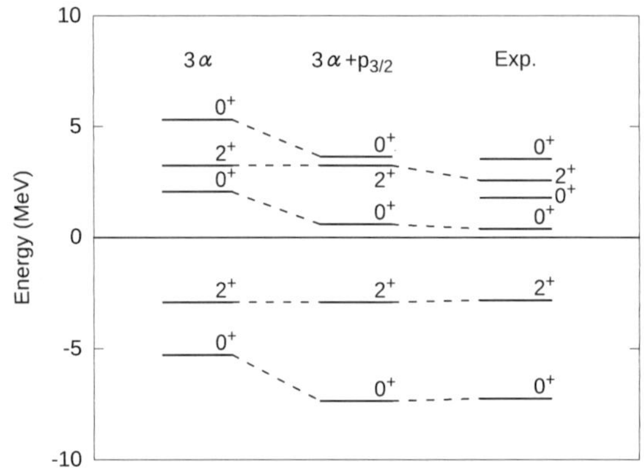

Figure 1. Comparison of calculated energy levels of 0^+ and 2^+ states with experimental ones.

Recently, several interesting experimental data on ^{12}C for excited states near the threshold energy have been reported. A new state, the 2_2^+ state at 9.84 MeV, was discovered [12, 13, 14]. Itoh *et al.* found that the broad 0^+ state at 10 MeV consists of two 0^+ states [15]. A band structure including these new states has been an open problem. In particular, the assignment of the band-head state of the newly measured 2_2^+ state is now controversial.

2. Formalism

In this study, we investigate ^{12}C with a simple model to clarify α-cluster breaking effects on 3α cluster structures. In our model, we superpose the Brink-Bloch (BB) cluster model wave functions and the $p_{3/2}$ subshell closure wave function to incorporate the mixing of cluster breaking components in cluster structures [16]. This model is regarded as an extension of the generator coordinate method of a cluster model. The difference from traditional cluster models is the mixing of the $p_{3/2}$ subshell closure wave function, which is the lowest configuration of the jj-coupling shell model. The results obtained with this model are consistent with the AMD and FMD results even though it is much simpler than those models. We used the same interaction as that used in a previous study of ^{12}C using AMD [17].

3. Results

In Fig. 1, we show the energy levels of 0^+ and 2^+ states calculated with the present model and those obtained with the 3α-cluster model calculation. The former and the latter correspond to calculations with and without α-cluster breaking, i.e., the mixing of the $p_{3/2}$ subshell closure wave function caused by the spin-orbit force, which we call "3α+$p_{3/2}$" and "3α", respectively. Energies are measured from the 3α threshold energy. The theoretical and experimental threshold energies are -82.9 MeV and -84.9 MeV, respectively.

The results for "3α+$p_{3/2}$" agree with experimental results very well except for the absence of a 0^+ state observed around 10 MeV. In particular, the level spacing between 0_1^+ and 2_1^+ states of "3α+$p_{3/2}$" agrees well with the experimental one, which is largely underestimated in "3α". This large spacing comes from the energy gain of the spin-orbit force in the 0_1^+ state with the mixing of the $p_{3/2}$ subshell closure wave function. Any microscopic α cluster model fails to reproduce this large level spacing, and therefore, this is an evidence for α-cluster breaking. The "3α+$p_{3/2}$" result is qualitatively consistent with the AMD and FMD results for ^{12}C [18]. This implies that the major difference between the AMD and FMD results and 3α cluster model calculations is the α-cluster breaking effect, which can be qualitatively simulated by the mixing of only one

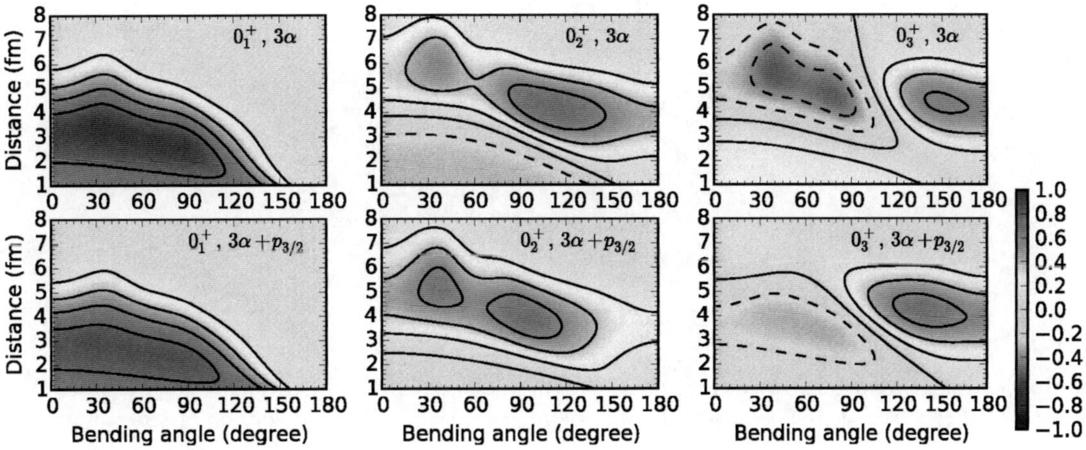

Figure 2. Overlap surfaces between 0^+ states and single BB wave functions.

configuration of the $p_{3/2}$ subshell closure to the 3α cluster model space in the present model.

In Fig. 2, we show overlap surfaces between the 0^+ states and the single BB wave functions. For the BB wave function, the isosceles triangle structure is assumed. These overlap surfaces show the cluster motion projected onto the 3α cluster configuration space for 0^+ states.

First we compare the structure of the 0_1^+ state. For "3α", the wave function has a large overlap around the compact 3α cluster configuration and also contains components of developed cluster configurations with larger inter-cluster distances. For "$3\alpha + p_{3/2}$", the behavior of the overlap surface is similar to that for "3α". However, the absolute amplitude decreases and 3α cluster components become small because of the significant mixing of the $p_{3/2}$ subshell closure component in the 0_1^+ state.

Next we discuss α-cluster breaking effects on the structure of the 0_2^+ state, which reflects the structure change of the 0_1^+ state because of the orthogonal condition to lower states. For "3α", the overlap is distributed over a wide area of well-developed cluster structures, indicating that the 0_2^+ state is described by the superposition of various triangle configurations of three α clusters. For "$3\alpha+p_{3/2}$", the overlap is also distributed over a wide area of well-developed cluster structures and is qualitatively similar to the distribution for "3α". However, quantitatively, the overlap shifts to smaller bending angles and the components for acute triangles are slightly enhanced. Moreover, the maximum peaks also shift toward shorter distances region compared to those for "3α". These results imply that the spatial development of the cluster structure is reduced in the 0_2^+ state with α-cluster breaking.

In the structure of the 0_3^+ state, a qualitative difference is observed between the results with and without α-cluster breaking. For "3α", the phase of the overlap changes as θ increases along the line for $d \sim 5$ fm. This indicates that the 0_3^+ state is the vibration mode of acute and obtuse triangle configurations. For "$3\alpha + p_{3/2}$", the behavior differs from the vibration mode. The amplitude of the negative minimum for the acute triangle configuration is considerably hindered, and as a result, the 0_3^+ state is dominated by the obtuse triangle configuration. Hence, α-cluster breaking produces significant effects not only in the ground state but also in the cluster configurations of excited states. This is surprising because the naive expectation is that α-cluster breaking can affect only ground state properties.

The mixing of the α-cluster breaking in 0^+ states affects the $E2$ transition strengths between 0^+ and 2^+ states. The calculated $E2$ transition strengths from 2^+ states to 0^+ states are shown in Table 1. The $E2$ transition strengths from the 2_2^+ state are important for determining the

Table 1. $E2$ transition strengths from 2^+ states to 0^+ states. The unit is $e^2\text{fm}^4$.

Transition	3α	$3\alpha+p_{3/2}$	Exp.
$2_1^+ \to 0_1^+$	10.8	7.4	7.6 ± 0.4
$2_1^+ \to 0_2^+$	1.4	5.1	2.6 ± 0.4
$2_1^+ \to 0_3^+$	0.4	0.2	
$2_2^+ \to 0_1^+$	4.0	1.1	$1.57^{+0.14}_{-0.11}$
$2_2^+ \to 0_2^+$	183	76.5	
$2_2^+ \to 0_3^+$	64.4	166	

band assignment for this state, which has been attracting significant interest. Its $E2$ transition strengths to the $0_{2,3}^+$ states are remarkably large. Interestingly, the behavior of the magnitude is reversed by α-cluster breaking. Without α-cluster breaking, the $2_2^+ \to 0_2^+$ transition strength is almost three times larger than the $2_2^+ \to 0_3^+$ one. However, with α-cluster breaking, the $2_2^+ \to 0_3^+$ transition strength is larger by a factor of two than the $2_2^+ \to 0_2^+$ one. The strongest $E2$ transition to the 0_3^+ state with α-cluster breaking, in which the reproduction of experimental results is good [18], indicates that the 0_3^+ state is likely to be the band-head of the 2_2^+ instead of the 0_2^+ state. The present results indicate that the band structure of higher excited states is affected by α-cluster breaking caused by the spin-orbit force.

4. Summary

To clarify α-cluster breaking effects on 3α cluster structures in ^{12}C, we investigated ^{12}C with a hybrid model composed of the BB cluster model and the $p_{3/2}$ subshell closure wave function. We have found that α-cluster breaking significantly changes the cluster structures of 0^+ states through orthogonality to lower states. As a result, the band assignment for the 2_2^+ state is changed. As a future work, we will investigate the effects of tensor correlation on cluster structures with a new method to include the tensor correlation in AMD [19].

Acknowledgments

This work was supported by JSPS KAKENHI Grant Nos. 15K17662 and 26400270.

References

[1] Kanada-En'yo Y 1998 *Phys. Rev. Lett.* **81** 5291
[2] Itagaki N, Aoyama S, Okabe S and K. Ikeda 2004 *Phys. Rev.* C **70** 054307
[3] Neff T and Feldmeier H 2004 *Nucl. Phys.* A **738** 357
[4] Kanada-En'yo Y 2007 *Prog. Theor. Phys.* **117** 655
[5] Chernykh M, Feldmeier H, Neff T, von Neumann-Cosel P and Richter A 2007 *Phys. Rev. Lett.* **98** 032501
[6] Kanada-En'yo Y, Kimura M and Ono A 2012 *Prog. Theor. Exp. Phys.* **2012** 01A202
[7] Uegaki E, Okabe S, Abe Y and Tanaka H 1977 *Prog. Theor. Phys.* **57** 1262
[8] Fukushima Y and Kamimura M 1978 *J. Phys. Soc. Jpn.* **44** 225
[9] Tohsaki A, Horiuchi H, Schuck P and Röpke G 2001 *Phys. Rev. Lett.* **87**, 192501
[10] Suhara T, Funaki Y, Zhou B, Horiuchi H and Tohsaki A 2014 *Phys. Rev. Lett.* **112**, 062501
[11] Funaki Y 2015 *Phys. Rev.* C **92** 021302(R)
[12] Freer M *et al* 2009 *Phys. Rev.* C **80** 041303(R)
[13] Itoh M *et al* 2011 *Phys. Rev.* C **84** 054308
[14] Zimmerman W R *et al* 2013 *Phys. Rev. Lett.* **110** 152502
[15] Itoh M *et al* 2013 *J. Phys. Conf. Ser.* **436** 012006
[16] Suhara T, Itagaki N, Cseh J and Płoszajczak M 2013 *Phys. Rev.* C **87** 054334
[17] Suhara T and Kanada-En'yo Y 2010 *Prog. Theor. Phys.* **123** 303
[18] Suhara T and Kanada-En'yo Y 2015 *Phys. Rev.* C **91**, 024315
[19] Myo T, Toki H, Ikeda K, Horiuchi H and Suhara T 2015 *Prog. Theor. Exp. Phys.* **2015**, 073D02

IOP Conf. Series: Journal of Physics: Conf. Series **863** (2017) 012021 doi:10.1088/1742-6596/863/1/012021

Nonlocalized clustering and evolution of cluster structure in nuclei

H Horiuchi

Research Center for Nuclear Physics, Osaka University, Ibaraki 567-0047, Japan

E-mail: horiuchi@rcnp.osaka-u.ac.jp

Abstract. It is shown that the THSR (Tohsaki-Horiuchi-Schuck-Roepke) wave function describe well not only cluster-gas like structures but also ordinary cluster structures with spatial localization of clusters. Based on this fact, the container model has been proposed as a new model of cluster dynamics. For better description of cluster dynamics, extended version of container model has been introduced. The container model of cluster dynamics teaches us how is the evolution of cluster structure which starts from the ground state having shell-model structure to many kinds of cluster states up to the cluster-gas states.

1. Introduction

The THSR wave function was originally devised for exprssing the 3α-condensate gas nature of the Hoyle state [1]. It was found [2] that the 3α THSR wave function has almost 100 % overlap with the 3α RGM/GCM wave function which had been obtained in 1970's. Surprisingly the THSR wave function for the $\alpha+^{16}$O system was shown also to be almost 100 % equivalent to the RGM/GCM wave function for the $\alpha+^{16}$O system [3]. The fact that the THSR wave function which expresses nonlocalized motion of clusters can also describe the localized $\alpha+^{16}$O clustering in the ^{20}Ne inversion-doublet bands forced us to introduce the container model of cluster dynamics. The container model has been extended to multi-container model in order to get better description of various kinds of cluster structures. The extended container model can now be regarded as being a novel cluster model which gives us a new insight into cluster dynamics. On the basis of the container model we can now discuss how is the evolution of cluster structure which starts from the ground state having shell-model structure to many kinds of cluster states up to the cluster-gas states.

2. Hoyle state and 3α condensate-like structure

The 0_2^+ state of ^{12}C (Hoyle state) has been a mysterious state from the shell-model viewpoint. Even for the no-core shell model it has been difficult to reproduce it. The observed reduced α-decay width $\gamma^2_{\rm obs}$ of this state is very large and is comparable to or larger than the single-α value (= Wigner-limit value γ^2_W) [4]; i.e. $\gamma^2_{\rm obs} \geq \gamma^2_W$. This data means that the Hoyle state should have S-wave dominant structure, namely it has dominantly the structure of ^{8}Be(0^+) - α(S-wave). If the Hoyle state has the structure of 3α linear chain as was postulated by H. Morinaga, the reduced α-decay width $\gamma^2_{\rm chain}$ becomes much smaller than the single-α value. It is because the 3α linear-chain 0^+ wave function is composed of not only ^{8}Be(0^+) - α(S-wave) component but also the components of ^{8}Be(2^+) - α(D-wave) and ^{8}Be(4^+) - α(G-wave), resulting in the relation $\gamma^2_{\rm chain} \leq 0.5\gamma^2_W$ [4].

The S-wave dominant structure of the Hoyle state was confirmed by the solution of the 3α problem. I solved 3α equation of OCM (orthogonality condition model) [5] and obtained the result that the Hoyle state has the structure of weakly coupled 3α in relative S waves with large spatial extent giving a large α S-factor, $S^2_\alpha = 1.33$. The results by this 3α OCM was confirmed a few years later by 3α GCM[6] and RGM [7]. Both succeeded to reproduce many observed data in ^{12}C.

About 30 years later, the wave functions of the Hoyle state by 3α GCM and RGM were found to be almost 100 % equivalent to single THSR wave function [2]:

$$|\langle\Phi(\text{single THSR})|\Phi(3\alpha\text{GCM/RGM})\rangle|^2 \approx 100\%. \tag{1}$$

The 3α THSR wave function has a very simple form $\Phi_B(3\alpha) = \mathcal{A}[\prod_{j=1}^3 \exp(-(2/B^2)\tilde{\mathbf{X}}_j^2)\phi(\alpha_j)]$ [1] with $\tilde{\mathbf{X}}_j = \mathbf{X}_j - \mathbf{X}_{\rm C.M.}$ and $\mathbf{X}_j$ standing for the C.M. coordinate of j-th α-cluster. This THSR wave function has a Bose-condensate-like character of α clusters and has clearly an S-wave dominant form.

3. Localized v.s. nonlocalized clustering

The inversion-doublet rotational bands in ^{16}O and ^{20}Ne have been regarded as convincing evidences of spatial localization of clusters [8]. Therefore it was really astonishing that the RGM/GCM wave functions of the inversion-doulet bands of ^{20}Ne were found to be almost 100 % equivalent to single THSR wave functions [3]:

$$|\left\langle\Phi(\text{single THSR})|\Phi(^{16}\text{O}+\alpha\ \text{RGM/GCM})\right\rangle|^2 \approx 100\% \qquad \text{(inversion doublet band states)}. \tag{2}$$

Seeming puzzle why the localized clustering states of ^{20}Ne are well described by THSR wave functions which express nonlocalized clustering can be solved by using the fact that the nucleon density distribution of THSR wave function displays localized clustering of ^{16}O and α clusters [9]. Since the THSR wave function without antisymmetrization of nucleons shows no sign of localized clustering, the nucleon density distribution that shows localized clustering should come from the antisymmetrization operation of nucleons. We can conclude that the spatial localization of ^{16}O and α clusters comes from the inter-cluster Pauli repulsion originating from the antisymmetrization of nucleons [9].

4. Container model of cluster dynamics

We explained in the previous section that the clustering is based on the two fundamental ingredients. One is the nonlocalized cluster dynamics and the other is the inter-cluster Pauli repulsion. The container model of cluster dynamics is the model which explains concretely what are the characteristic features of the nonlocalized cluster dynamics. There are two important points in this model. The first point is that clusters make nonlocalized motion in the container(s). The second point is the recognition that the size parameter(s) of the container(s) are the generator coordinates of the evolution of the cluster structure. In Ref. [1], in the case of ^{12}C, we solved the Hill-Wheeler equation and constructed the energy eigenfunction Ψ_j

$$\sum_B \langle\Phi_{B'}(3\alpha)|(H-E_j)|\Phi_B(3\alpha)\rangle f_j(B) = 0, \quad \Psi_j = \sum_B f_j(B)\Phi_B(3\alpha). \tag{3}$$

The lowest-energy 0^+ solution is the ground state (0^+_1) and the second-lowest-energy 0^+ solution is the Hoyle state (0^+_2). Thus we see that the Hill-Wheeler equation describes the evolution of the cluster structure from the ground state to the Hoyle state.

IOP Conf. Series: Journal of Physics: Conf. Series **863** (2017) 012021 doi:10.1088/1742-6596/863/1/012021

The container model has been extended by introducing two or more containers. In the case of ^{12}C, different values of size parameters B_{kj} were adopted for two Jacobi coordinates $\boldsymbol{\xi}_1 = \boldsymbol{X}_2 - \boldsymbol{X}_1$ and $\boldsymbol{\xi}_2 = \boldsymbol{X}_3 - (1/2)(\boldsymbol{X}_1 + \boldsymbol{X}_2)$; $\Phi_{\boldsymbol{B}_1,\boldsymbol{B}_2}(3\alpha) = \mathcal{A}[\exp(-\sum_j^{x,y,z}\{(1/B_{1j}^2)\xi_{1j}^2 + (3/4B_{2j}^2)\xi_{2j}^2\})\prod_i^3 \phi(\alpha_i)]$ [10, 11]. In Ref. [11] calculated results of the Hoyle-family states by the extended container model were reported. One of the important results is the support to the existence of the 0_3^+ state and its band-like member states in addition to the 0_4^+ state and its band-like member states which have 3α bent-arm structure. The existence of the 0_3^+ state was proposed only by the 3α OCM calculations [12, 13] but has never been supported by fully microscopic calculations including 3α GCM, AMD and FMD calculations.

The extended container model was also successfully applied to the studies of neutron-rich Be isotopes, ^{9}Be and ^{10}Be [14]. In the case of ^{9}Be, it was reported that obtained extended-THSR wave functions for the ground, $5/2^-$, and $7/2^-$ states have large square overlaps with the Brink-GCM wave functions, 0.96, 0.95, and 0.93 %, respectively.

5. Evolution of cluster structure

Clusters make nonlocalized motion in containers under the constraint of inter-cluster Pauli repulsion. The inter-cluster Pauli repulsion originates from the Pauli-forbidden states of the system, $\chi^F(\tilde{\boldsymbol{X}}_1, \cdots, \tilde{\boldsymbol{X}}_N)$ with $\tilde{\boldsymbol{X}}_j = \boldsymbol{X}_j - \boldsymbol{X}_{\text{C.M.}}$, which are defined by

$$\mathcal{A}\{\chi^F(\tilde{\boldsymbol{X}}_1, \cdots, \tilde{\boldsymbol{X}}_N)\phi(C_1)\cdots\phi(C_N)\} = 0. \qquad (4)$$

The physical wave function of cluster motion should be orthogonal to all the Pauli-forbidden states χ^F. It has been shown that this orthogonality requirement is equivalent to the requirement that the physical wave function of cluster motion should be orthogonal to all the two-cluster Pauli-forbidden states between any pair of clusters in the system.

In the case of two-cluster systems, the inter-cluster Pauli repulsion generates the spatial localization of clusters. However in three or more cluster systems, the inter-cluster Pauli repulsion does not necessarily generate the spatial localization of clusters.

In the case of two SU(3)-scalar clusters like α and ^{16}O clusters, the Pauli-allowed states have only $(\lambda, \mu)_{\text{SU(3)}} = (p, 0)$ which is of prolate deformation. However for many clusters, Pauli-allowed states can have variety of deformations.

N	$(\lambda,\mu)/\mu_p^{N(\lambda,\mu)}$								
12	(0, 0) 0.6592								
13	(2, 1) 0.4614								
14	(3, 1) 0.3955	(2, 0) 0.6152	(4, 2) 0.3811	(0, 4) 0.3365					
15	(6, 0) 0.3399	(4, 1) 0.4944	(2, 2) 0.4944	(0, 3) 0.4614	(3, 0) 0.5933	(5, 2) 0.3626	(3, 3) 0.3376	(6, 3) 0.3677	(2, 5) 0.2848
16	(5, 1) 0.5324 0.4237	(3, 2) 0.4635	(1, 3) 0.5191	(4, 0) 0.5982 0.5628	(0, 2) 0.6633	(8, 1) 0.3208	(6, 2) 0.4363 0.3252	(4, 3) 0.4366	(2, 4) 0.4390 0.3856
	(7, 3) 0.3845	(5, 4) 0.3450	(3, 5) 0.2957	(8, 4) 0.4082	(4, 6) 0.2765	(0, 8) 0.2295			
17	(8, 0) 0.3816	(6, 1) 0.5435 0.5192 0.4435	(4, 2) 0.5410 0.4979	(2, 3) 0.5730 0.4944	(5, 0) 0.5933	(3, 1) 0.6406	(1, 2) 0.6691	....	

Figure 1. Pauli-allowed states of 4α system. N is the number of total H.O. quanta. (λ, μ) are the SU(3) labels of the Pauli-allowed states. Numbers below (λ, μ) label are the eigenvalues of the 4α norm kernel of the norm-kernel eigenstates having (λ, μ) label.

In Fig. 1 we show Pauli-allowed states of 4α system which are labeled by the SU(3) label (λ, μ) [15]. N stands for the number of total H.O. quanta of the Pauli-allowed state. The lowest value of N is 12 which is of course equal to the H.O. quantum number of the closed-shell wave function $\Phi_{\rm C.S.} = (1/\sqrt{16!}) \det |(0s)^4(0p)^{12}|$. In more detail, the 4α Pauli-allowed cluster state with lowest quanta $N = 12$ is equivalent to $\Phi_{\rm C.S.}$, which is nothing but the duality property of $\Phi_{\rm C.S.}$. We see clearly for most values of N except $N = 12$ and 13 the Pauli-allowed states have variety of deformations ranging from prolate to oblate.

In Ref. [16] the solution of the Hill-Wheeler equation with 4α THSR wave function is analysed.

$$\sum_B \langle \Phi_{B'}(4\alpha)|(H - E_j)|\Phi_B(4\alpha)\rangle f_j(B) = 0, \quad \Psi_j = \sum_B f_j(B)\Phi_B(4\alpha). \tag{5}$$

The lowest-energy state is almost equal to $\Phi_{\rm C.S.}$, the state around 4α breakup threshold has the property of 4α condensate-like state, and energy eigenstates between these states have ^{12}C + α character. The table of Pauli-allowed states in Fig. 1 teaches us that the formation and evolution of cluster states are largely governed by the inter-cluster Pauli repulsion. In Refs. [16, 17] the eigenstates of this Hill-Wheeler equation were analyzed and it was found that these eigenstates have almost 100 % squared overlaps with single THSR wave functions. Furthermore the size parameters B of these equivalent single THSR wave functions were found to become gradually larger for the larger excitation energies, which means the evolution of cluster structure along the growth of the excitation energy is in parallel with the increase of the system volume.

Acknowledgments
This work is based on the collaboration with the following people whom I would like to thank: B. Zhou, M. Lyu, Z. Ren, C. Xu, Y. Funaki, A. Tohsaki, T. Yamada, T. Suhara, O. Schuck, G. Röpke.

References

[1] Tohsaki A, Horiuchi H, Schuck P and Röpke G 2001 *Phys. Rev. Lett.* **87** 192501
[2] Funaki Y, Tohsaki A, Horiuchi H, Schuck P and Röpke G 2003 *Phys. Rev.* C **67** 051306(R)
[3] Zhou B, Funaki Y, Horiuchi H, Ren Z, Röpke G, Schuck P, Tohsaki A, Xu C and Yamada T 2013 *Phys. Rev. Lett.* **110** 262501
[4] Suzuki Y, Horiuchi H and Ikeda K 1972 *Prog. Theor. Phys.* **47** 1517
[5] Horiuchi H 1974 *Prog. Theor. Phys.* **51** 1266
[6] Uegaki E, Okabe S, Abe Y and Tanaka H 1977 *Prog. Theor. Phys.* **57** 1262
[7] Kamimura M 1981 *Nucl. Phys.* A **351** 456
[8] Horiuchi H and Ikeda K 1968 *Prog. Theor. Phys.* **40** 277
[9] Zhou B, Funaki Y, Horiuchi H, Ren Z, Röpke G, Schuck P, Tohsaki A, Xu C and Yamada T 2014 *Phys. Rev.* C **89** 034319
[10] Zhou B, Funaki Y, Tohsaki A, Horiuchi H and Ren Z 2014 *Prog. Theor. Exp. Phys.* **2014** 101D01
[11] Funaki Y 2015 *Phys. Rev.* C **92** 021302(R)
[12] Kurokawa C and Kato K 2005 *Phys. Rev.* C **71** 021301
[13] Ohtsubo S, Fukushima Y, Kamimura M and Hiyama E 2013 *Prog. Theor. Exp. Phys.* **2013** 073D02
[14] Lyu M, Ren Z, Zhou B, Funaki Y, Horiuchi H, Röpke G, Schuck P, Tohsaki A, Xu C and Yamada T 2015 *Phys. Rev.* C **91** 014313; 2016 *Phys. Rev.* C **93** 054308
[15] Fujiwara Y and Horiuchi H 1981 *Prog. Theor. Phys.* **65** 1901
[16] Funaki Y, Yamada T, Tohsaki A, Horiuchi H, Röpke G and Schuck P 2010 *Phys. Rev.* C **82** 024312
[17] Funaki Y, Horiuchi H and Tohsaki A 2015 *Progr. in Part. and Nucl. Phys.* **82** 78

Clustering effects in fusion evaporation reactions with light even-even N=Z nuclei

L Morelli[1], M D'Agostino[1], M Bruno[1], C Frosin[1], F Gulminelli[2],F Gramegna[3], M Cinausero[3], T Marchi[3], D Fabris[4], M Degerlier[5], G Casini[6], S Barlini[6], M Bini[6], G Pasquali[6], A Olmi[6], S Piantelli[6], S Valdré[6], G Pastore[6], N Gelli[6] and E Vardaci[7]

[1] Dipartimento di Fisica e Astronomia dellUniversità and INFN, Bologna, Italy.
[2] CNRS, UMR6534, LPC, Caen, France and ENSICAEN, UMR6534, LPC, Caen, France.
[3] INFN, Laboratori Nazionali di Legnaro, Legnaro (Padova), Italy.
[4] INFN, Sezione di Padova, Padova, Italy.
[5] University of Nevsehir, Science and Art Faculty, Physics Department, Nevsehir,Turky.
[6] Dipartimento di Fisica e Astronomia dellUniversità and INFN, Firenze, Italy.
[7] Universitá Federico II, Dip. Scienze Fisiche and INFN Sezione di Napoli.

E-mail: luca.morelli@bo.infn.it

Abstract. In the recent years, cluster structures have been evidenced in many ground and excited states of light nuclei [1, 2]. In the currently experimental campaign, the NUCL-EX collaboration has measured the ^{12}C+^{12}C and ^{14}N+^{10}B reactions at 95 MeV and 80 MeV respectively. The experimental data corresponding to complete fusion of target and projectile into an excited ^{24}Mg nucleus was compared to the results of a pure statistical model [3, 4]. In addition, data from ^{12}C+^{12}C have been analyzed to investigate the decay of the Hoyle state of ^{12}C* [12] obtained as an intermediate step in the 6α decay channel of the ^{24}Mg* formed in central events.

1. Introduction

The subject of α-chains as possible building blocks of even-even nuclei has been proposed in the late sixties and α-clustering has become a central issue in nuclear physics ever since. At present, the NUCL-EX collaboration has started a campaign of exclusive measurements of fusion-evaporation reactions employing light nuclei as interacting partners. In this campaign, the first studied system has been the ^{24}Mg compound nucleus. The ^{24}Mg has been populated at the same excitation energy through the ^{12}C+^{12}C and ^{14}N+^{10}B reactions in order to study the influence of the entrance channels. In section 3.1, the results obtained from these first measurements are briefly summarized.

For instance, the Hoyle state [5] of ^{12}C (0^+ at 7.65 MeV) is known to manifest α-clustering effects which play a decisive role in the abundance of this element. Most of the recent results indicate a nearly complete agreement with a sequential decay through an intermediate $^{8}Be_{gs}$ [6, 7], for the Hoyle state . Instead, few other experimental results show also a significant amount of instantaneous 3-α particle decay [8]. In section 3.2, the results from the analysis of central ^{12}C + ^{12}C reactions where the ^{24}Mg* compound nucleus decays in six α-particles, are shown.

2. The experiments

The experiments were performed at the LNL (Laboratori Nazionali di Legnaro), with ^{12}C and ^{14}N beams provided by the XTU TANDEM accelerator. We have studied the two systems ^{12}C + ^{12}C and ^{14}N + ^{10}B, leading to the same compound nucleus ^{24}Mg* at the same excitation energy. The apparatus has been fully described in [9] and hereafter we recall the main characteristics. The experimental apparatus is composed of the GARFIELD detector and the Ring-Counter (RCo) annular detector [10]. The latter apparatus (RCo) covers the forward polar angles ($7^o \div 17^o$) and is divided into three detection stages: an ionization chamber (IC), silicon strips (Si) and CsI(Tl) scintillators. With the current setup, an angular resolution of $\sim \pm\ 0.7^o$ for the polar angle is achieved while $\sim \pm\ 11^o$ for the azimuthal angle is feasible for particles detected in the scintillators. The RCo detector allows reaching energy resolution of percent, charge and mass identification of light particles. The remaning polar angles ($\theta = 30^o \div 170^o$) are covered by the GARFIELD detector composed by a drift ionization chamber followed by CsI(Tl) scintillators.

3. Results

3.1. Central $^{12}C + {}^{12}C$ and $^{14}N + {}^{10}B$ reaction

Since the experimental apparatus allows a complete reconstruction of the events, a Q_{kin}-value distribution estimating the dissipated energy [11] can be built to further investigate all the channels involving alpha particles in both reactions: $Q_{kin} = \sum_{i=1}^{2} E_{\alpha_i} + E_{Z_{Res}} - E_{beam}$, where E_{α_i} and $E_{Z_{Res}}$ are, respectively, the laboratory energy of α particles and residues, and E_{beam} is the energy of the incident projectile. In figure 1 the obtained Q_{kin} distributions for the channels with the maximum α multiplicity associated to the residue of charge Z are displayed for the two reactions.

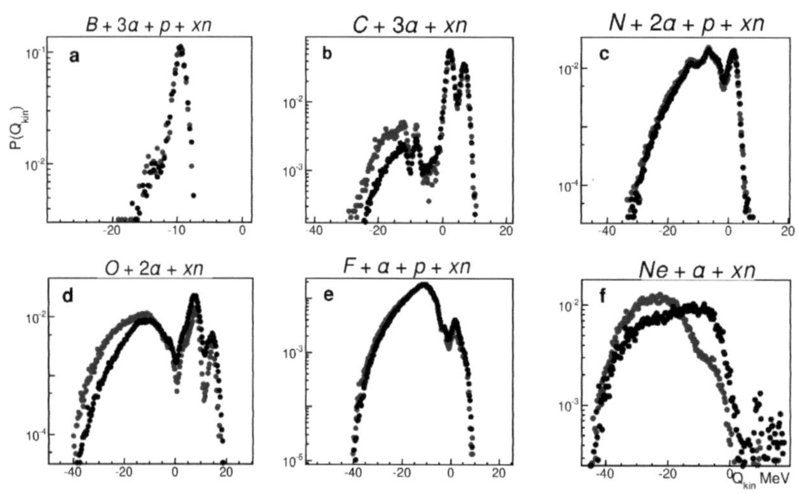

Figure 1. Q$_{kin}$-values distribution for different residues, indicated in each panel, detected in coincidence with the maximum possible number of α-particles. Data from the ^{12}C+^{12}C and ^{14}N+^{10}B reactions are represented by the black dots and by the blu dots, respectively.The Q$_{kin}$ is normalized to the Q-value of ^{14}N+^{10}B data.

In figure 1d), the Q_{kin} distribution for the channel 2α, $^{16}O^{gs/*}$ is shown. One can clearly notice two distinguished peaks corresponding to different α-decay chains, which populate the ^{16}O residue either in its ground state or in one of its excited bound states. A broader distribution is also observed for lower Q_{kin} values due to events associated with the opening of the 4-body channel ^{15}O+$n + \alpha + \alpha$. The two reactions, ^{12}C+^{12}C (black dots) and ^{14}N+^{10}B (blue dots),

Table 1. For the measured evaporation residues, the table shows the most probable experimental channel and its branching ratio alongside with the value predicted by the HFℓ calculations. Every experimental value has an associated error of 5% which takes into account both statistical error and the particle contamination

Z_{res}	Channel	BR_{exp} NB (%)	$BR_{HF\ell}$ NB (%)	BR_{exp} CC (%)	$BR_{HF\ell}$ CC (%)
5	$^{11-xn}$B+3α+p+xn	99	100	99	100
6	$^{12-xn}$C+3α+xn	96	74	98	78
7	$^{15-xn}$N+2α+p+xn	90	95	91	95
8	$^{16-xn}$O+2α+xn	56	15	63	15
9	$^{19-xn}$F+α+p+xn	91	93	92	88
10	$^{22-xn}$Ne+α+nx	47	3	26	3

show an evident difference in the relative population of more and less dissipative events, as can be evinced from figure 1 [1]. In fact, a much higher percentage of (2α,^{16}O) events populates the less dissipative Q-value region in the ^{12}C sample. This larger deviations are common to all the even-Z residues, while a very good agreement for the Q_{kin} distributions for the odd-Z residues is observed. In Table 1, the branching ratios (BR) for the most populated channel in the experimental sample and HFℓ prediction for both reactions, are shown for each residue. One can notice a good agreement between the experimental and statistical model BR of the dominant decay channels for odd Z residues, while discrepancies can be seen for even-Z ones. For the final Carbon, Oxygen and Neon residues, the evaporation chains show a preferential α decay in both reactions. This α excess could be attributed to the presence of residual α correlations in the excited ^{24}Mg or in its daughter nucleus ^{20}Ne, regardless of the entrance channel of the reaction.

3.2. $^{12}C^*$ Hoyle state in $^{24}Mg^*$ decay: the 6-α channel

As anticipated, the 6-α decay channel for the ^{12}C + ^{12}C reaction was investigated to study the decay of the Hoyle state. We have reconstructed the dissipated energy using the Q_{kin} quantity previously defined ($Q_{kin} = \sum_{i=1}^{6} E_i - E_{beam}$). In figure 2 (left panel) the Q_{kin} distribution is shown, together with the cut used for the analysis.

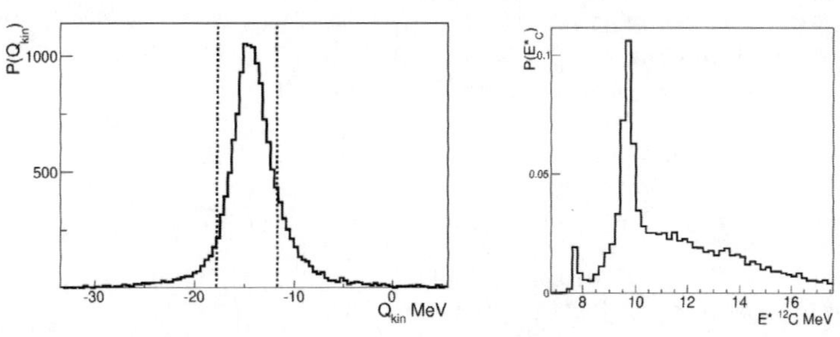

Figure 2. Left panel: Q_{kin} distribution for the six α-particles events. The vertical lines indicate the events retained for the analysis. Right panel: Reconstructed ^{12}C* excitation energy from the three α-particles with the minimum total energy in central collisions.

In order to reconstruct the intermediate ^{12}C* disassembling into 6 alphas of the ^{24}Mg*

[1] with less dissipative we indicate events with Q_{kin} greater than the neutron emission threshold

formed in central collisions we have selected the three (out of six) α-particles corresponding to the minimum $^{12}C^*$ excitation energy. This quantity is shown in right panel of figure 2 where the two peaks correspond to the lowest excited states of $^{12}C^*$ at 7.65 and 9.64 MeV. The former represents the well known Hoyle state. The main debate consists in the interpretation of the decay of this state as sequential, via the $^8Be_{gs}$ formation, or if there is a contribution of instantaneous breakup in alpha-particles. To clarify this point, different observables have been proposed. As an example, we show the energy Dalitz plot where data and HFℓ predictions show the same configuration (see figure 3). The two coordinates of the Dalitz plot are defined as $x_d = \sqrt{3}(e_1 - e_2)$ and $y_d = 2e_3 - e_1 - e_2$, where $e_{i,j,k} = E_{i,j,k}/(E_i + E_j + E_k)$ are the α-particle energies in the ^{12}C frame, normalized to the total energy.

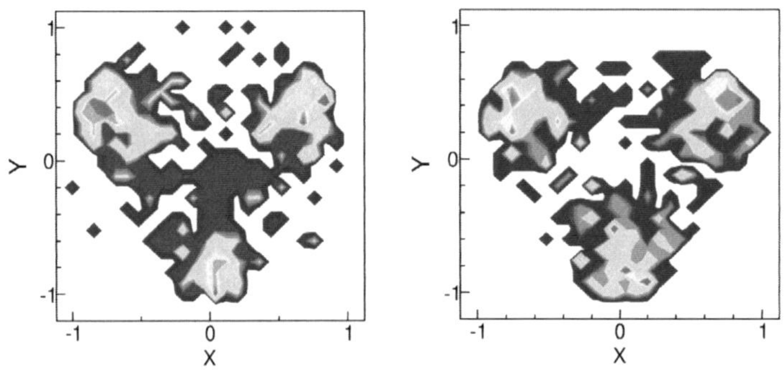

Figure 3. Central collisions. Left panel: experimental energy Dalitz plot. Right panel: HFℓ Dalitz plot.

All the results obtained in inelastic channels in $^{12}C+^{12}C$ reaction [12] and in various reactions at different energies [6, 7] indicate a very low limit of breakup in alpha-particles of the order of permil.

References

[1] Von Oertzen W *2010 Clusters in Nuclei, Lecture Notes in Physics* **818** 109-127
[2] Ikeda K 1968 *Prog. Theor. Phys. (Suppl.) extra number* **464**
[3] Baiocco G *et al* 2013 *Phys. Rev.* C **87** 054614
[4] Morelli L *et al* 2014 *J. Phys.* G: *Nucl. Phys.* **41** 075107
[5] Hoyle F 1954 *The Astrophysical Journal, Supplement Series* **1** 12
[6] Freer M *et al* 1994 *Phys. Rev.* C **107** R1751
[7] Itoh M *et al* 2014 *Phys. Rev. Lett.* **113** 102501
[8] Raduta Ad R *et al* 2011 *Phys. Lett.* B **705** 65
[9] Bruno M *et al* 2013 *Eur. Phys. Journ.* A **49** 128
[10] Pasquali G *et al* 2007 *Nucl. Instr. And Meth.* A **570** 126
[11] Morelli L *et al* 2014 *J. Phys.* G: *Nucl. Phys.* **41** 075108
[12] Morelli L *et al* 2016 *J. Phys.* G: *Nucl. Phys.* **43** 045110

IOP Conf. Series: Journal of Physics: Conf. Series **863** (2017) 012023 doi:10.1088/1742-6596/863/1/012023

Beautiful clusters – boron-9: simulations and decay

Tz Kokalova Wheldon and C Wheldon

School of Physics and Astronomy, University of Birmingham, Edgbaston, Birmingham B15 2TT, UK

E-mail: `t.wheldon@bham.ac.uk`

Abstract. Monte-Carlo simulations for analysis of the ^{9}Be(^{3}He, t)^{9}B* reaction at 33 MeV using coincidences between the high-resolution Munich 3D spectrograph and the large-acceptance Birmingham silicon detector array are presented. These focus on the identification of the proton break-up product and the application to evidencing the first excited $1/2^{+}$ state. Identifying this state has posed a challenge for over 70 years, but would enable the mirror symmetry in this light system to be explored, giving insight into its possible clustered structure. Penetrabilities for the low excitation energy regime are calculated showing the efficacy of the proton-identification method.

1. Introduction

The unstable isotope, ^{9}B has a half-life of approximately 6×10^{-19} s (Ω 0.54$\pm$0.21 ke [1]) and was first discovered in 1940 by Haxby and coworkers [2, 3]. hat is particularly interesting about $^{9}_{5}$B$_{4}$ is the comparison with its mirror partner $^{9}_{4}$Be$_{5}$. The first excited state in ^{9}Be lies at 1.684(7) Me with I^{π} $1/2^{+}$ [1] yet, notably, in ^{9}B, this state has not been firmly established. There have been many experimental studies over the last 70 years, reporting the state anywhere from 0.8-1.9 Me [4] and this activity has been matched by theoretical calculations with a similar energy spread [5, 6, 7]. Solving this persistent mystery is so important due to its relevance to understanding the structure of both the ^{9}B and ^{9}Be states and precision tests of the nuclear force. There are two possible structures for the I^{π} $1/2^{+}$ resonances, either a compact shell-model configuration of core-plus-valence-nucleon, or a more extended nuclear molecular arrangement in which a covalent nucleon binds the two α-particle clusters. Measuring the Coulomb energy di erence (CE) between the two mirror I^{π} $1/2^{+}$ states enables, through its sensitivity to the volume, the structure to be established the clustered state being the larger of the two.

There are several reasons why the state has been di cult to establish in ^{9}B. irstly, in most reactions, the population of any I^{π} $1/2^{+}$ strength is weak. Secondly, the proximity of levels such as the I^{π} $5/2^{-}$ level (Ω 81(5) ke) at 2.361(5) Me , the 2.75(30) Me I^{π} $1/2^{-}$ state (Ω 3.1(2) Me) and the 2.79(3) Me I^{π} $5/2^{+}$ level (Ω 550(40) ke) [1] mean there is significant background obscuring any I^{π} $1/2^{+}$ signal. or some of these the decay mode can be utilised, with coincidence methods, to suppress states. or example the proton channel is open for all states (S_p -0.186) Me , but the α threshold lies at 1.687 Me . or this reason the missing I^{π} $1/2^{+}$ resonance is expected to decay predominantly via proton emission and this can be separated from the 2.36 Me state which decays 99 through the α channel [8, 9]. This is necessary due to the strong population of this and higher-lying levels. The present authors were involved in a recent study that used the Munich Q3 spectrograph [10, 11, 12] in coincidence with the Birmingham large area double-sided silicon strip detector (SS) array to study the ^{9}Be(^{3}He, t)^{9}B* reaction at 33 Me , recording both the e ectile (with high resolution at the Q3 focal plane) and break-up products (in the SS array). That work [8] reported a state at 1.86 Me $\pm$70 ke (stat) $\pm$35 ke (syst) and Ω 650$\pm$160 ke , in close agreement with two other studies using the same reaction without coincidences and at significantly higher beam energies $1.80^{+0.22}_{-0.16}$ Me with Ω 600^{+300}_{-270} ke [13] and 1.850$\pm$0.130 Me with Ω 700^{+270}_{-200} ke [14]. The di erence in Ref. [8] was that, due to the coincidence-filtering, the state observed was isolated and formed a visible feature in the spectrum.

This paper reports details of the Monte-Carlo simulations carried out to aid the analysis of Ref. [8] and also discusses the anomalous decay of the 2.36 Me state. etails of the experimental set-up and other aspects of the analysis are given in Ref. [8].

2. Simulations

The program RESOLUTION8 [15, 16], developed by the Birmingham group, was used to generate events for the following two break-up reactions

$$^9\text{Be} \quad ^3\text{He} \rightarrow t \quad ^9\text{B}^*$$
$$\hookrightarrow p \quad ^8\text{Be}$$
$$\hookrightarrow 2\alpha \qquad (1)$$
$$\hookrightarrow \alpha \quad ^5\text{ i}$$
$$\hookrightarrow p \quad \alpha\,, \qquad (2)$$

each having an overall Q-value of -0.809 Me . In order to speed up the simulation, events were only generated for which the triton laboratory angle lay within the acceptance of the Q3 spectrograph (slits set to $\theta_{\text{in-plane}}$ $\pm1.43°$ and $\theta_{\text{out-of-plane}}$ $\pm2.0°$) before smearing. This facilitated the production of 10^6 events within $\sim$10 min.

The e ects included in the simulation were,

- initial two-body reaction with Q -1.087 Me and a uniform centre-of-mass distribution across the Q3 acceptance;
- entire energy range of the focal-plane; 0.480-2.700 Me generated;
- the break-up of the ^{9}B recoil into either ^{8}Be p (Q 0.186 Me) or ^{5} i α (Q -1.687 Me) for which a uniform distribution in spherical polar angles was generated randomly per event;
- break-up of the heavy fragment into 2α (Q 0.092 Me) or α p (Q 1.965 Me);
- Smearing of the ^{5} i ground-state was added to the second stage of break-up (2) via a random aussian energy distribution centred at E_x(^{5} i) 0 Me with a HM equal to the width of the ^{5} i ground-state, Ω 1.23 Me [17];
- beam divergence and finite beam-spot size;
- energy loss and energy and angular straggling in the target;

These events were analysed with the code used for the experimental data following some preprocessing. Initially, the in- and out-of-plane angles for each detector strip were calculated so that the energy and position information of each particle making up the simulated events could be transformed into strip A C number. This enabled the appropriate channel-by-channel thresholds to be set and dead-strips to be removed. Additionally, the particles incident on the silicon array had their energies smeared to give a front-back energy distribution that matched the experimental data. or the triton events, the centre angle of the Q3 was taken, in this case, 30°. This had one of the largest smearing e ects on the final kinematic reconstruction of the data. All other processing was done with energy-ordered silicon data, ust as for the real data and the reconstructions were done using the same processing code, *i.e.* without knowledge of the particle type for hits in the SS array.

In order to cleanly select events, part of the analysis procedure required a proton gate to be set as, most often, two α particles were detected at forward angles and the proton was not. In order to set the proton gate simulated data for both break-up channels were used and Catania plots constructed. These 2 histograms are based on the rearranged Q-value equation,

$$E_b - E_t - E_1 - E_2 \quad \frac{p_p^2}{2}\frac{1}{m(p)} - Q, \tag{3}$$

where E_b is the beam energy, E_t the triton energy detected at the Q3 focal plane, $E_{1,2}$ the energy of the two detected break-up particles in the SS array and p_p is the momentum of the reconstructed missing particle assumed to be a proton. The proton mass is $m(p)$. The resulting plots for the two decay channels are shown in ig. 1.

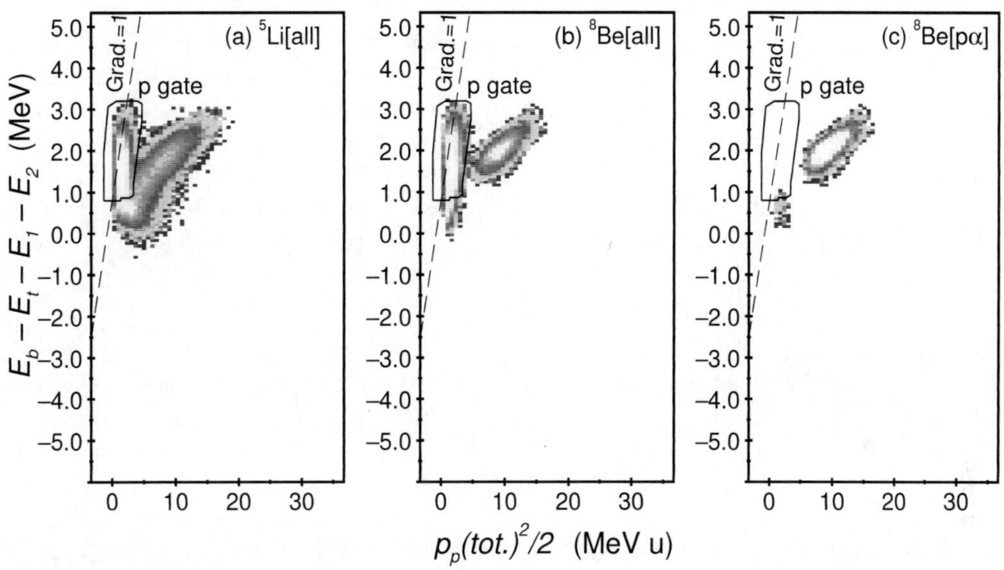

Figure 1. Catania plots for the simulated data for break-up channels (1) and (2) corresponding to the detection of two particles. See text for details. Panel (a) 5 i data with no requirement on which particles were detected. Panel (b) similarly for the ^{8}Be break-up data. Panel (c) as for (b) but with the condition that detected particle (1) is a proton. The dashed line shows the gradient of the loci corresponding to $1/m(p)$ 1. The final proton gate used is shown with the continuous black outline. Note that the y-axis intercept at $p_p/2$ 0 is $-Q$ (see Eqn. 3).

igure 1 panels (a) and (b) show two clear distributions corresponding, largely, to the missing particle being correctly identified as a proton (lying on the dashed lines) and to the missing particle being an α particle (righthand distributions). However, there are some events, visible in ig. 1 (c) for which detected-proton events lie close to the dashed line. The proton gate was therefore set to exclude such events. Though in ig. 1 (c) a small fraction of these lie in the gate, this is the tail of the distribution and only visible due to the large number of simulated events, 1×10^6 for each break-up channel. This is much larger than the number of events in the experimental data.

The results from the analysis are shown in ig. 2(a). Here, the filtered data with all energy and particle cuts are shown, overlaid with a fit, using oigt profiles [18] (continuous blue line).

A clear excess of counts can be seen at 1.86 Me ±70 ke (stat) ±35 ke (syst) with a width of 650(160) Me . The overall $\chi^2/d.o.f.$ 1.13. The broad 2.75 Me and narrow 2.36 Me resonances are also required to fit the data.

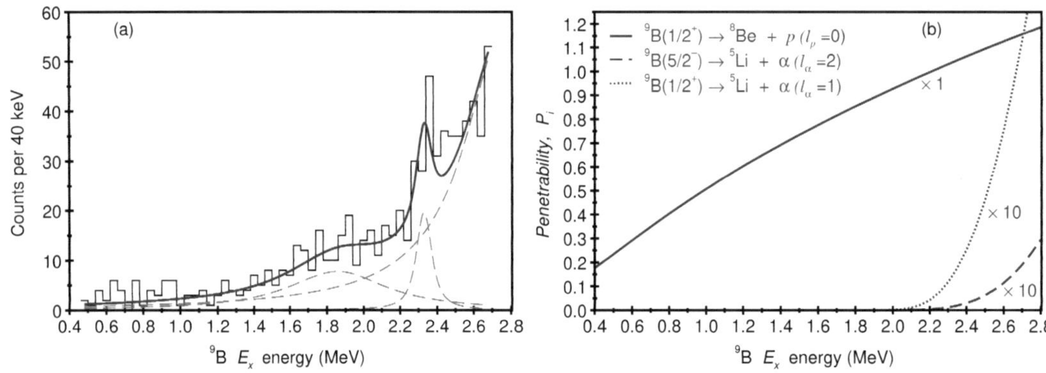

Figure 2. Panel (a) ata (black stepped line) with fit comprising three oigt functions (continuous blue line) with fixed aussian width of 41 ke corresponding to the experimental resolution. The dashed red lines are the individual components the known 2.75 and 2.3 Me states along with a proposed state at 1.86 Me . Modified from ig. 6(a) of Ref. [8]. Panel (b) Penetrabilities (Eqn. 5) for the proton decay of a $1/2^+$ state (continuous blue line), the α decay of a $5/2^-$ state (dashed blue line) such as the known 2.3 Me state and the α decay of a $1/2^+$ state (dotted blue line). Note the di erent vertical scales. See text for details.

3. Decay penetrabilities

In order to compare the decay branches from resonances into di erent channels, i, the reduced width, γ_i^2, is used

$$\gamma_i^2 \quad \frac{\Omega_i}{2P_i}, \tag{4}$$

with the barrier penetrabilities, P_i, removed. These can be written in terms of the wave number, k_i, radius, r_i ($1.4(A_i^{1/3} \quad A_{recoil}^{1/3})$ fm), and the regular, $F_l(k_i r_i)$, and irregular, $G_l(k_i r_i)$, Coulomb wave functions, such that

$$P_i \quad \frac{(k_i r_i)}{[F_l^2(k_i r_i) \quad G_l^2(k_i r_i)]}. \tag{5}$$

The penetrabilities for the two open decay channels below 3 Me (p and α) are plotted in ig. 2(b), calculated using the CKIN code [19] which utilises the CERN ibraries WCLBES code [20, 21]. As can be seen, the proton penetrabilities dominate up to 2.8 Me which makes the decay-channel-filtering described earlier e ective for the $1/2^+$ state. Above the α-decay threshold the α penetrabilities increase rapidly, but even at 2.0 Me , are orders of magnitude smaller than the corresponding proton values. However, beyond this, by 2.3 Me , the $5/2^-$ state overcomes the penetrabilities to decay 99 via α emission [8, 9] suggesting that this level has some α-cluster structure the α preformation giving α emission a preference over proton emission.

4. Summary

The coincidence method for e ectile-tagged break-up data for ^{9}B has been elucidated for the case where two of the three break-up particles have been detected. Monte-Carlo simulations have been used to obtain a selection window for events with an undetected proton, *i.e.* events with two detected α particles. ollowing use of the proton gate, the ^{8}Be ground-state has a significantly suppressed background from mis-identification of the particles and, when combined with other filtering criteria, almost background free ^{9}B excitation spectra have been obtained. inally, the penetrabilities for proton and α decay up to 2.8 Me in ^{9}B have been calculated and plotted to demonstrate the applicability of the decay-channel selection technique.

References

[1] Tilley D R, Kelley J H, Godwin J L, Millener D J, Purcell J E, Sheua C G and Weller H R 2004 *Nucl. Phys. A* **745** 155
[2] Thoennessen M. 2012 *At. Data Nucl. Data Tables* **98** 43
[3] Haxby R O, Shoupp W E, Stephens W E, Wells W H 1940 *Phys. Rev.* **58** 1035
[4] Baldwin T D, *et al.* 2012 *Phys. Rev. C* **86** 034330
[5] Barker F C 2009 *Phys. Rev. C* **79** 017302
[6] Fortune H T and Sherr R 2006 *Phys. Rev. C* **73** 064302
[7] Arai K, Descouvemont P, Baye D and Catford W N 2003 *Phys. Rev. C* **68** 014310
[8] Wheldon C, Kokalova Tz, *et al.* 2015 *Phys. Rev. C* **91** 024308
[9] Gete E, *et al.* 2000 *Phys. Rev. C* **61** 064310
[10] Löffler M, Scheerer H J and Vonach H 1973 *Nucl. Instr. and Meth.* **111** 1
[11] Wirth H-F, Angerer H, von Egidy T, Eisermann Y, Graw G and Hertenberger R 2000 *Beschleunigerlaboratorium München, Annual Report* p71
[12] Wirth H-F 2001 *Ph.D Thesis, Technical University, München*, see http://tumb1.biblio.tu-muenchen.de /publ/diss/ph/2001/wirth.html
[13] Akimune H, Fujimura M, Fujiwara M, Hara K, Ishikawa T, Kawabata T, Utsunomiya H, Yamagata T, Yamasaki K and Yosoi M 2001 *Phys. Rev. C* **64** 041305(R)
[14] Scholl C, *et al.* 2011 *Phys. Rev. C* **84** 014308
[15] Curtis N, *et al.* 1995 *Phys. Rev. C* **51** 1554
[16] Curtis N, *et al.* 1996 *Phys. Rev. C* **53** 1804
[17] Tilley D R, Cheves C M, Godwin J L, Hale G M, Hofmann H M, Kelley J H, Sheu C G and Weller H R 2002 *Nucl. Phys. A* **708** 3
[18] Gubner John A 1994 *J. Phys. A: Math Gen.* **27** L745
[19] Wheldon C 2016 CKIN *Two-body kinematics code*, http://www.np.ph.bham.ac.uk/research_resources/programs/, accessed 01/10/2016
[20] McLaren I 2006 *Wclbes.f subroutine*, http://cernlib.sourcearchive.com/documentation/2006.dfsg.2/ wclbes_8F_source.html, accessed 08/03/2016
[21] Thompson I and Barnett A 1985 *Comput. Phys. Commun.* **36** 363

Probing asymmetric clusters by using isoscalar monopole and dipole transitions

M Kimura[1], Y Chiba[1] and Y Taniguchi[2]

[1]Department of Physics, Hokkaido University, 060-0810 Sapporo, Japan.
[2]Department of Medical and General Sciences, Nihon Institute of Medical Science, Moroyama, Saitama 350-0435, Japan.

E-mail: `masaaki@nucl.sci.hokudai.ac.jp`

Abstract. The isoscalar monopole and dipole transitions and their relationship to the clustering are discussed. By exploiting the Bayman-Bohr theorem, analytical formulae for the transition matrices are derived to show that the cluster states are very strongly populated by the IS monopole and dipole transitions. For the quantitative discussion, the AMD calculations for ^{20}Ne, ^{44}Ti and ^{24}Mg are presented. It is demonstrated that IS monopole and dipole transitions are excellent probe for the clustering.

1. Introduction

At the time when K. Wildermuth, the founder of this conference series, started his study of nuclear clustering [1, 2], the concept of the nuclear clustering and the cluster models were criticized [3]. It was shown that an $SU(3)$ shell model [4] and a cluster model wave functions are mathematically equivalent to each other at the zero inter-cluster distance [3, 5], which is known as the Bayman-Bohr theorem. The theorem was misinterpreted to argue that the cluster model was not describing new aspects of atomic nuclei but just describing a known shell model state.

Today, almost 60 years later, it is interesting to know that the theorem is commonly accepted in an opposite meaning; it is one of the essential and useful theorem for the understanding of nuclear clustering. An interesting and important application of the theorem is the discussion of the isoscalar (IS) monopole and dipole transitions and their relationship to the nuclear clustering [6, 7, 8, 9, 10], which we shall discuss in this contribution.

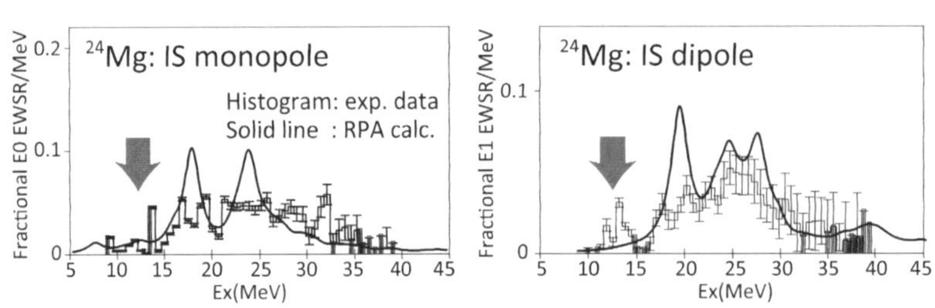

Figure 1. IS monopole and dipole strengths of ^{24}Mg. Histogram shows the experimental data, while the solid line shows the results of RPA calculation.

Figure. 1 shows the IS monopole and dipole strengths of ^{24}Mg [14]. We see that several resonances are observed in the low energy region of $E_x = 10 \sim 15$ MeV (histograms) which are not explained by the RPA calculation (solid lines). We can also find that similar resonances are systematically observed in stable nuclei from ^{12}C to ^{40}Ca [11, 12, 13, 14, 15, 16, 17, 18]. The Bayman-Bohr theorem plays a central role to show that *they are cluster states.* With the help of the theorem, we can prove that the IS monopole and dipole transitions strongly populate cluster states, and hence, they are excellent probe for nuclear clustering. To illustrate it, we first discuss the duality of shell and cluster, and derive analytic formula for transition matrices. For the quantitative discussion, we also present the antisymmetrized molecular dynamics (AMD) calculations for ^{20}Ne, ^{44}Ti and ^{24}Mg [10, 19, 20, 21].

2. Duality of shell and cluster, and IS monopole/dipole transitions

As mentioned above, the Bayman-Bohr theorem states the equivalence of an $SU(3)$ shell model and a cluster model at the zero inter-cluster distance. For example, the shell model wave function for the ^{20}Ne ground state is equivalent to an $\alpha + {}^{16}$O cluster model wave function,

$$\Phi(0_1^+) = \mathcal{A}\{\,(0s)^4(0p)^8(1s0d)^4\,\}_{(8,0)}^{J^\pi=0^+} = n_{N_0}\mathcal{A}\{\,R_{N_0 0}(r)Y_{00}(\hat{r})\phi_\alpha\phi_{\rm O}\,\},\quad N_0 = 8 \tag{1}$$

where the inter-cluster motion between α and ^{16}O clusters is described by a harmonic oscillator wave function $R_{N_0 0}(r)Y_{00}(\hat{r})$. In this way, the theorem tells us that the ground states of atomic nuclei have duality of shell and cluster; they are two-faced like Janus. The correct interpretation of this duality is this: The degrees-of-freedom of cluster excitation is embedded even in an ideal shell model state. Hence, the pronounced cluster states can be populated from the shell model state. For example, by increasing the nodal quantum number of the inter-cluster motion, we get a wave function for an excited 0^+ state,

$$\Phi(0_{ex}^+) = \sum_{N=N_0+2}^{\infty} f_N n_N \mathcal{A}\{\,R_{N0}(r)Y_{00}(\hat{r})\phi_\alpha\phi_{\rm O}\,\}, \tag{2}$$

which has pronounced clustering owing to the increased inter-cluster distance. Another excitation mode is the increase of the orbital angular momentum of inter-cluster motion, that yields a 1^- state with pronounced clustering,

$$\Phi(1_{ex}^-) = \sum_{N=N_0+1}^{\infty} g_N n_N \mathcal{A}\{\,R_{N0}(r)Y_{10}(\hat{r})\phi_\alpha\phi_{\rm O}\,\}. \tag{3}$$

We call these cluster excitations "nodal" and "angular" excitations as shown in Fig. 2. In ^{20}Ne, the 0_4^+ and 1_1^- states at 8.7 MeV and 5.8 MeV are identified as these excited states [22]. Thus, the Bayman-Bohr theorem guarantees that the pronounced cluster states can be populated from the shell model state by activating the degree-of-freedom of cluster excitation.

Now, it must be emphasized that the IS monopole/dipole transitions do realize it. They induce nodal and angular excitations to populate cluster states strongly. It is proved as follows. Using the cluster coordinate defined in Fig. 3, the IS monopole operator is rewritten as[9],

$$\mathcal{M}^{IS0} = \sum_{i=1}^{A}(\boldsymbol{r}_i - \boldsymbol{r}_{\rm cm})^2 = \sum_{i\in C_1}\xi_i^2 + \sum_{i\in C_2}\xi_i^2 + \frac{C_1C_2}{A}r^2. \tag{4}$$

Note that this expression makes it clear that $\mathcal{M}^{\rm IS0}$ will generate nodal excited states, because the last term proportional to r^2 will induce the nodal excitation of the inter-cluster motion. By a similar calculation, one finds that the isoscalar dipole operator is rewritten as [10],

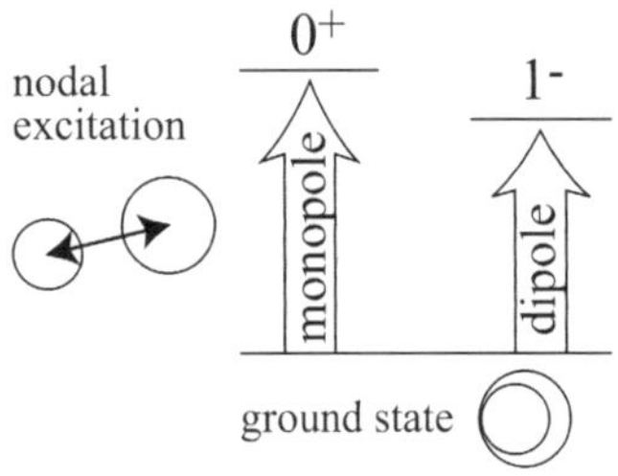

Figure 2. IS monopole and dipole transitions populate nodal and angular excited states.

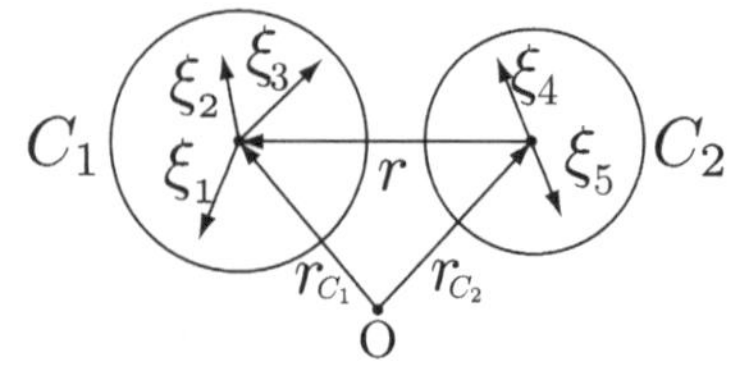

Figure 3. $\boldsymbol{\xi}_i$ denote the internal coordinates, while $\boldsymbol{r}$ is the inter-cluster coordinate. C_1 and C_2 are the masses of clusters.

$$\mathcal{M}_\mu^{IS1} = \frac{5}{3}\left(\frac{C_2}{A}\sum_{i\in C_1}\xi_i^2 - \frac{C_1}{A}\sum_{i\in C_1}\xi_i^2\right)\mathcal{Y}_{1\mu}(\boldsymbol{r}) + \frac{C_1C_2(C_1-C_2)}{A^2}r^2\mathcal{Y}_{1\mu}(\boldsymbol{r}) + ..., \tag{5}$$

where the terms depending on $\mathcal{Y}_{1\mu}(\boldsymbol{r})$ will increase the angular momentum of inter-cluster motion to generate angular excited cluster states. We also see that $\mathcal{M}_\mu^{IS1}$ is amplified for asymmetric cluster systems [23], because those terms depend on the mass asymmetry $(C_1 - C_2)/A$.

Thanks to the Bayman-Bohr theorem, it is possible to derive analytic formulae for these transition matrix by using Eqs. (1)-(5) [9, 10]. In the case of ^{20}Ne, they read

$$M^{IS0} = \langle\Phi(0^+_{ex})|\mathcal{M}^{IS0}|\Phi(0^+_{gs})\rangle = f_{N_0+2}\sqrt{\frac{\mu_{N_0}}{\mu_{N_0+2}}}\langle R_{N_00}|r^2|R_{N_0+20}\rangle \tag{6}$$

$$\begin{aligned} M^{IS1} =& \sqrt{3}\,\langle\Phi(1^-_{ex})|\mathcal{M}^{IS1}|\Phi(0^+_{gs})\rangle = \sqrt{\frac{3}{4\pi}}\left[\frac{3}{5}g_{N_0+3}\sqrt{\frac{\mu_{N_0}}{\mu_{N_0+3}}}\langle R_{N_00}|r^3|R_{N_0+31}\rangle\right. \\ &\left. + g_{N_0+1}\sqrt{\frac{\mu_{N_0}}{\mu_{N_0+1}}}\left\{\frac{48}{25}\langle R_{N_00}|r^3|R_{N_0+11}\rangle + \frac{16}{3}\left(\langle r^2\rangle_\alpha - \langle r^2\rangle_{\mathrm{O}}\right)\langle R_{N_00}|r|R_{N_0+11}\rangle\right\}\right] \end{aligned} \tag{7}$$

where $\langle r^2\rangle_\alpha$ and $\langle r^2\rangle_{\mathrm{O}}$ are the mean-square radius of the clusters. μ_N is the so-called eigenvalue of the RGM norm kernel. By using the amplitudes f_N and g_N calculated by AMD [19, 20], the formulae can be easily estimated as

$$M^{IS0} = 7.67f_{N_0+2} = 5.48\ \mathrm{fm}^2, \quad M^{IS1} = 3.08g_{N_0+1} - 7.36g_{N_0+3} = 5.82\ \mathrm{fm}^3, \tag{8}$$

which are as large as the Weisskopf estimates, $M^{IS0}_{\mathrm{WU}} = 6.37$ fm^2 and $M^{IS1}_{\mathrm{WU}} = 8.44$ fm^3. Thus, the IS monopole and dipole transitions to the cluster states are very strong. These strong transitions to the cluster states should be observed at relatively small excitation energies which are usually below the giant resonances, because the energies of the cluster states are governed by Ikeda threshold rule [24]. This explains why we have many narrow resonances well below the giant resonances. Therefore, the low-lying IS monopole and dipole strengths are the good signature of asymmetric clustering.

3. AMD results for ^{20}Ne, ^{44}Ti and ^{24}Mg

For the quantitative discussion, we have performed AMD calculations for several nuclei. Figure. 4 (a) shows the observed and calculated $\alpha + {}^{16}$O cluster bands in ^{20}Ne. The 0^+_4 and 1^-_1 states

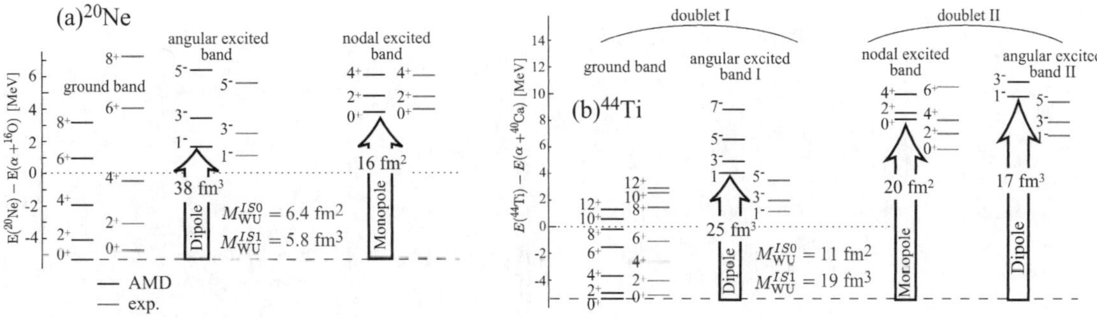

Figure 4. Observed and calculated (a) $\alpha + {}^{16}$O cluster bands in ^{20}Ne, and (b) $\alpha + {}^{40}$Ca cluster bands in ^{44}Ti. Arrows in the figure show the calculated transition matrices.

at 8.7 MeV and 5.8 MeV are identified as nodal and angular excited states [22], and AMD [19, 10] reasonably describes them. The calculated transition matrices are also shown in the figure. One sees that both of the IS monopole and dipole transitions are very strong and much more enhanced than the Weisskopf estimates and the cluster estimates given in Eq. (8). This enhancement owes to the $\alpha + {}^{16}$O clustering in the ground state.

Figure. 4 (b) shows the observed and calculated $\alpha + {}^{40}$Ca cluster bands in ^{44}Ti. The experimental assignment of the cluster bands is summarized as follows [25]. First, the ground state and the angular excited 1^- state at 6.2 MeV constitute a parity doublet denoted by "doublet I". Second, the nodal excited 0^+ state around 11 MeV and another angular excited 1^- around 12 MeV constitute another parity doublet denoted by "doublet II". AMD calculation [20, 10] reasonably reproduces them, although the excitation energies are slightly overestimated. As clearly seen in the figure, the transition strengths to these cluster states are as strong as the Weisskopf estimates as expected. However, they are not amplified as strong as in the case of ^{20}Ne. This is due to the reduction of the α clustering in the ground state of ^{44}Ti which is under the strong influence of the spin-orbit interaction. It is also noted that AMD calculation yields many other excited 0^+ and 1^- states below $E_x = 20$ MeV (not shown in Fig. 4 (b)), which do not have cluster structure. However, it was found that none of them have sizable IS monopole and dipole transition strengths. This suggests that these transitions are very sensitive and selective probe for the clustering. We also performed a similar analysis for ^{28}Si for which readers are directed to the contribution by Y. Chiba in this volume.

Figure. 5 shows the final example. Very high resolution data for IS monopole response of ^{24}Mg [17, 26] is available as shown by the histogram in the figure. The data shows the existence of the many narrow resonances below the giant monopole resonance. Solid lines shows the result of AMD calculation [21] which reasonably describes the distributions of the narrow resonances at small energy region as well as the giant resonance. From the analysis of the spectroscopic factor of AMD wave function, we can say that the resonance around 11 MeV is the $\alpha +^{20}$ Ne state, that at 13 MeV is the ^{12}C+^{12}C state and so on. Thus, the AMD calculations demonstrate that IS monopole/dipole transitions are excellent probe for the clustering.

4. Summary

In summary, we have discussed the relationship between the clustering and IS monopole/dipole transitions to demonstrate how and why they are excellent probe for clustering.

The Bayman-Bohr theorem tells us that the shell model states have duality of shell and cluster, and guarantees that the pronounced cluster states are populated by activating the degree-of-freedom of cluster excitation. This is a reason why the cluster states exist universally

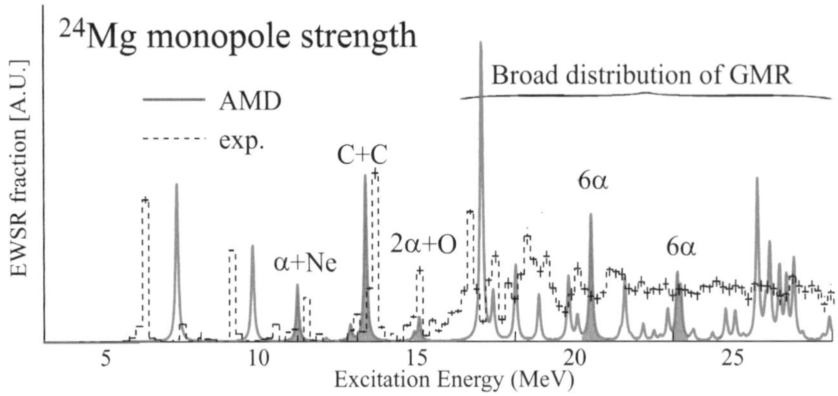

Figure 5. Observed and calculated IS monopole strength distributions of ^{24}Mg.

in atomic nuclei. The theorem is also utilized for deriving analytic formulae of the IS monopole/dipole transitions which show that these transitions do populate the cluster states very strongly.

The AMD calculations for ^{20}Ne, ^{44}Ti and ^{24}Mg are also presented. The results reasonably reproduce the known cluster states in ^{20}Ne and ^{44}Ti, and predict the transitions to the cluster states are very strong and selective. The results for ^{24}Mg is compared with the high resolution data. It is shown that by the analysis of the wave function, we can identify a variety of cluster states in ^{24}Mg such as $^{12}\mathrm{C}+{}^{12}\mathrm{C}$, $\alpha+{}^{20}\mathrm{Ne}$ and $2\alpha+{}^{16}\mathrm{O}$. Thus, the study of cluster states using IS monopole/dipole transitions is really fascinating and look promising.

References

[1] Wildermuth K and Kanellopoulos Th 1958 *Nucl. Phys.* **7** 150
[2] Wildermuth K and Kanellopoulos Th 1958/59 *Nucl. Phys.* **9** 449
[3] Bayman B F and Bohr A 1958/59 *Nucl. Phys.* **9** 596
[4] Elliott J P 1958 *Proc. Roy. Soc.* A **245** 128 and 562
[5] Perring J K and Skyrme T H R 1956 *Proc. Phys. Soc.* **69** 600
[6] Suzuki Y 1987 *Nucl. Phys.* A **470** 119
[7] Kanada-En'yo Y 2007 *Phys. Rev.* C **75** 024302
[8] Kawabata T *et al* 2007 *Phys. Lett.* C **646** 6
[9] Yamada T *et al* 2008 *Prog. Theor. Phys.* **120** 1139
[10] Chiba Y, Kimura M and Taniguchi Y 2016 *Phys. Rev.* C **93** 034319
[11] Youngblood D H, Lui Y W and Clark H L 1997 *Phys. Rev.* C **55** 2811
[12] Youngblood D H, Lui Y W and Clark H L 1998 *Phys. Rev.* C **57** 2748
[13] Lui Y W, Clark H L and Youngblood D H 2001 *Phys. Rev.* C **64** 064308
[14] Chen X *et al* 2009 *Phys. Rev.* C**80** 014312
[15] Itoh M *et al* 2011 *Phys. Rev.* C **84** 054308
[16] Itoh M *et al* 2013 *Phys. Rev.* C **88** 064313
[17] Gupta Y K *et al* 2016 *Phys. Rev.* C **93** 044324
[18] Peach T *et al* 2016 *Phys. Rev.* C **93** 064325
[19] Kimura M 2004 *Phys. Rev.* C **69** 044319
[20] Kimura M and Horiuchi H 2006 *Nucl. Phys.* A **767** 58
[21] Chiba Y and Kimura M 2015 *Phys. Rev.* C **91** 061302
[22] Tilley D R *et al* 1998 *Nucl. Phys.* A **636** 249
[23] Horiuchi H and Ikeda K 1968 *Prog. Theor. Phys.* **40** 277
[24] Ikeda K, Takigawa N and Horiuchi H 1968 *Prog. Theor. Phys. Supplement* Extra Number 464
[25] Yamaya T, Katori K, Fujiwara S, Kato S and Ohkubo S 1998 *Prog. Theor. Phys. Supplement* **132** 73, and references therein
[26] Kawabata T *et al* 2013 *Jour. Phys. Conf. Series* **436** 012009

Study on α-cluster levels in non-4n nuclei using low-energy RI beams

H Yamaguchi[1], D Kahl[1,2], S Hayakawa[1], Y Sakaguchi[1], K Abe[1], T Nakao[1,3], T Suhara[4], N Iwasa[5], A Kim[6,7], D H Kim[7], S M Cha[6], M S Kwag[6], J H Lee[6], E J Lee[6], K Y Chae[6], Y Wakabayashi[8], N Imai[1], N Kitamura[1], P Lee[9], J Y Moon[10,11], K B Lee[10], C Akers[10], H S Jung[11,9], N N Duy[12,13], L H Khiem[12] and C S Lee[9]

[1] Center for Nuclear Study (CNS), University of Tokyo, Wako, Saitama, Japan.
[2] School of Physics and Astronomy, the University of Edinburgh, Edinburgh, UK
[3] Advanced Science Research Center, Japan Atomic Energy Agency, Tokai, Ibaraki, Japan.
[4] Matsue College of Technology, Matsue, Shimane, Japan.
[5] Department of Physics, Tohoku University, Sendai, Miyagi, Japan
[6] Department of Physics, Sungkyunkwan University, Suwon, Republic of Korea
[7] Department of Physics, Ewha Womans University, Republic of Korea
[8] The Institute of Physical and Chemical Research (RIKEN), Wako, Saitama, Japan
[9] Department of Physics, Chung-Ang University, Seoul, Republic of Korea [10] Institute for Basic Science, Daejeon, Korea
[11] Wako Nuclear Science Center (WNSC), KEK, Wako, Saitama, Japan
[12] Institute of Physics, Vietnam Academy of Science and Technology, Hanoi, Vietnam
[13] Dong Nai University, Dong Nai, Vietnam

E-mail: `yamag@cns.s.u-tokyo.ac.jp`

Abstract. Alpha resonant scattering is a simple and promising method to study α-cluster structure in nuclei. It has several good features which enable us to perform measurements with short-lived and relatively low-intense RI beams. Several measurements on alpha resonant scattering have been carried out at CRIB (CNS Radioactive Ion Beam separator), which is a low-energy RI beam separator at Center for Nuclear Study (CNS) of the University of Tokyo. Recent α resonant scattering studies at CRIB, using ^{7}Li, ^{7}Be and ^{10}Be beams with a helium gas target, are discussed.

1. Introduction

CRIB [1,2] is a radio-isotope (RI) beam separator operated by Center for Nuclear Study (CNS), the University of Tokyo, installed in the RIBF facility of RIKEN Nishina Center. CRIB can produce low-energy (< 10 MeV/u) RI beams in flight, using primary heavy-ion beams from the AVF cyclotron of RIKEN (K=70) [3]. The secondary beam is purified with an magnetic analysis using dipole magnets, and with a Wien filter, which can separate the beams according to their velocities. For relatively light RI beams such as ^{7}Be, we can obtain a high purity of almost 100% after the Wien filter.

An experimental method extensively used is the thick-target method in inverse kinematics (TTIK) [4]. In that method, the beam energy is degraded in a thick reaction target, and reactions occur at various center-of-mass energies. Many experiments on proton resonant scatterings have been successfully performed at CRIB with the TTIK method [5–11], most of which are related to the astrophysical (p, γ) reactions. Measurements on the α elastic resonant scatterings with a helium gas target and heavy-ion/RI beams, ^{14}O, ^{21}Na, ^{30}S, ^{7}Li [12] and ^{7}Be [13], have been performed at CRIB. These measurements are to study astrophysical (α, γ) reaction rates, and also α-cluster structures of the compound nuclei. Direct measurement of (α, p) reactions, such as ^{14}O(α, p) [14], ^{11}C(α, p) [15], ^{21}Na(α, p), ^{18}Ne(α, p), ^{30}S(α, p), and ^{22}Mg(α, p), have also been performed with the TTIK method using RI beams at CRIB. Recent results on α resonant scattering measurements are discussed below.

2. ^{7}Li/^{7}Be+α elastic resonant scattering and clustering in ^{11}B and ^{11}C

The exotic cluster structures in the ^{11}B nucleus and its mirror, ^{11}C, are attracting much attention in recent years [16]. The $3/2^-_3$ state in ^{11}B at the excitation energy $E_{\rm ex}$=8.56 MeV is regarded as a dilute cluster state [17], where two α particles and ^{3}He are weakly interacting. In particular, the alpha cluster structure in ^{11}B was studied by measuring its isoscalar monopole and quadrupole strengths in the ^{11}B(d,d') reaction, and the 8.56-MeV state was suggested to have a dilute cluster structure [18, 19].

Using the TTIK method at CRIB, the excitation function of ^{7}Li+α elastic scattering around 180° in the center-of-mass system was measured for the first time, and we observed strong resonant structure in the spectrum [12]. Rotational bands in ^{11}B and ^{11}C, which might be related to the cluster structure, had been discussed in [20, 21]. Based on the experimental result, we proposed a negative-parity cluster band in ^{11}B, which was later found to be consistent with a theoretical calculation [22]. The theoretical interpretation of these cluster states in the band was that they have a 3-body (2α+t) cluster configuration.

We also performed the measurement of ^{7}Be+α resonant elastic and inelastic scatterings at CRIB [13], to study the symmetry in the mirror system. An R-matrix analysis was performed to deduce the parameters of the observed resonance structure, and we proposed a negative parity band in ^{11}C as well. A similar measurement was independently carried out at other facilities [23], but our measurement was with γ-ray detection to identify inelastic scatterings, and some differences in the energy and cross section were found in the obtained spectra [13].

3. Linear-chain cluster levels in ^{14}C

In 1956, Morinaga [24] came up with the novel idea of a particular cluster state: the linear-chain cluster state (LCCS). Despite the pursuit by many scientists for more than half a century, up until now the LCCS has been only hypothetical. Now the LCCS is commonly considered as extreme and exotic, due to its presumed propensity to exhibit bending configurations. A theoretical prediction of LCCS in ^{14}C was made by Suhara and En'yo [25] with an antisymmetrized molecular dynamics (AMD) calculation, yielding a band (J^π = 0^+, 2^+, 4^+) that has a configuration of an LCCS at a few MeV or more above the ^{10}Be+α threshold. A further investigation [26] showed that the AMD wave function has a configuration in which two α particles and two neutrons are located close to each other, while the remaining α particle is relatively further away. This implied that such an LCCS could be experimentally accessible from the ^{10}Be+α channel in a single step.

We applied the ^{10}Be+α resonant scattering method in inverse kinematics [4] to identify the predicted LCCS band in ^{14}C. Our experimental setup was similar to the previous one in the ^{7}Be+α experiment [13], but we placed an extra silicon detector telescope to cover a broader angular range, instead of the NaI detectors. The new setup enabled us to perform a reliable analysis on the angular distribution. The ^{10}Be beam was produced at CRIB with a typical intensity of 2×10^4 particles per second, and the beam purity was better than 95%. The beam was counted with two parallel-plate avalanche counters (PPACs), separated by 30 cm. The ^{10}Be beam at 25 MeV impinged on the gas target, which was a chamber filled with helium gas at

700 Torr and covered with a 20-μm-thick Mylar film as the beam entrance window. α particles recoiling to the forward angles were detected by ΔE-E detector telescopes. We used two sets of detector telescopes in the gas-filled chamber, where each telescope consisted of two layers of silicon detectors with the thicknesses of 20 μm and 480 μm. The main measurement using the helium-gas target was performed for 2 days, injecting 2.2 $\times 10^9$ ^{10}Be particles into the gas target as valid events.

We selected events in which the ^{10}Be beam particle was injected into the target and an α particle was detected at the telescope in coincidence. The scattering position, or equivalently the center-of-mass energy $E_{\mathrm{c.m.}}$, was determined by a kinematic reconstruction on an event-by-event basis. The number of events for each small energy division was converted to the differential cross section $(d\sigma/d\Omega)_{\mathrm{c.m.}}$, using the solid angle of the detector, the number of beam particles, and the effective target thickness. Finally we obtained the excitation function of the ^{10}Be+α resonant elastic scattering for 13.8–19.1 MeV, where events with θ_{lab} = 0–8° ($\theta_{\mathrm{c.m.}}$= 164–180°) were selected, as in Fig. 1. At energies above 15.7 MeV, the excitation function shows a reasonable agreement in the spectral shape with one of the recent measurements [27]. A structure with resonances which may correspond to the theoretical prediction of the LCCS was observed in the spectrum. The resonant information including J^π will be obtained by an analysis with an R-matrix calculation.

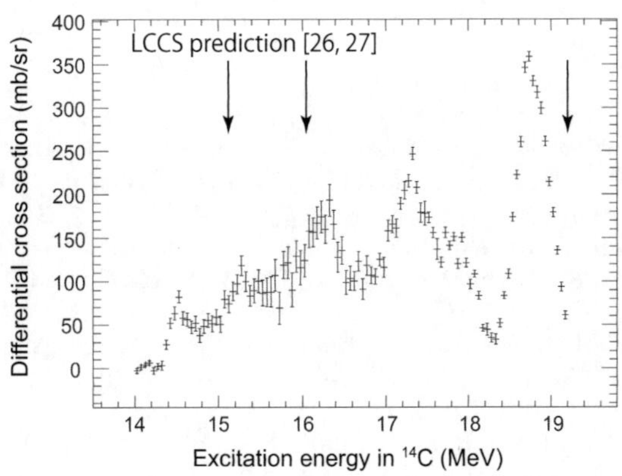

Figure 1. Excitation function of the ^{10}Be+α resonant scattering for θ_{lab}=0–8°.

Acknowledgments

We appreciate Prof. Y. Kanada-Enyo for the discussion and suggestion on the theoretical implication. The experiment was performed at RI Beam Factory operated by RIKEN Nishina Center and CNS, the University of Tokyo. We are grateful to the RIKEN and CNS accelerator staff for their beam production and acceleration. This work was partly supported by JSPS KAKENHI (No. 25800125, 15K17662, and 16K05369) in Japan, and the National Research Foundation Grant funded by Korea Government (Grants Nos. NRF-2009-0093817, NRF-2013R1A1A2063017, NRF-2014S1A2A2028636, and NRF-2016K1A3A7A09005579).

References

[1] Kubono S, Yanagisawa Y, Teranishi T, Kato S, Kishida T, Michimasa S, Ohshiro Y, Shimoura S, Ue K, Watanabe S and Yamazaki N 2002 *Eur. Phys. J.* A **13** 217

[2] Yanagisawa Y, Kubono S, Teranishi T, Ue K, Michimasa S, Notani M, He J J, Ohshiro Y, Shimoura S, Watanabe S, Yamazaki N, Iwasaki H, Kato S, Kishida T, Morikawa T and Mizoi Y 2005 *Nucl. Instr. Meth. Phys. Res.* A **539** 74

[3] Yamaguchi H, Wakabayashi Y, Amadio G, Hayakawa S, Fujikawa H, Kubono S, He J, Kim A and Binh D 2008 *Nucl. Instr. Meth. Phys. Res.* A **589** 150
[4] Artemov K P, Belyanin O P, Vetoshkin A L, Wolskj R, Golovkov M S, Gol'dberg V Z, Madeja M, Pankratov V V, Serikov I N, Timofeev V A, Shadrin V N and Szmider J 1990 *Sov. J. Nucl. Phys.* **52** 408
[5] Teranishi T, Kubono S, Yamaguchi H, He J J, Saito A, Fujikawa H, Amadio G, Niikura M, Shimoura S, Wakabayashi Y, Nishimura S, Nishimura M, Moon J Y, Lee C S, Odahara A, Sohler D, Khiem L H, Li Z H, Lian G and Liu W P 2007 *Phys. Lett.* B **650** 129
[6] Yamaguchi H, Wakabayashi Y, Kubono S, Amadio G, Fujikawa H, Teranishi T, Saito A, He J, Nishimura S, Togano Y, Kwon Y, Niikura M, Iwasa N, Inafuku K and Khiem L 2009 *Phys. Lett.* B **672** 230
[7] He J J, Kubono S, Teranishi T, Hu J, Notani M, Baba H, Nishimura S, Moon J Y, Nishimura M, Iwasaki H, Yanagisawa Y, Hokoiwa N, Kibe M, Lee J H, Kato S, Gono Y and Lee C S 2009 *Phys. Rev.* C **80** 015801
[8] Jung H S, Lee C S, Kwon Y K, Moon J Y, Lee J H, Yun C C, Kubono S, Yamaguchi H, Hashimoto T, Kahl D, Hayakawa S, Choi S, Kim M J, Kim Y H, Kim Y K, Park J S, Kim E J, Moon C B, Teranishi T, Wakabayashi Y, Iwasa N, Yamada T, Togano Y, Kato S, Cherubini S and Rapisarda G G 2012 *Phys. Rev.* C **85** 045802
[9] He J J, Zhang L Y, Parikh A, Xu S W, Yamaguchi H, Kahl D, Kubono S, Hu J, Ma P, Chen S Z, Wakabayashi Y, Sun B H, Wang H W, Tian W D, Chen R F, Guo B, Hashimoto T, Togano Y, Hayakawa S, Teranishi T, Iwasa N, Yamada T and Komatsubara T 2013 *Phys. Rev.* C **88**012801
[10] Jin S J, Wang Y B, Su J, Yan S Q, Li Y J, Guo B, Li Z H, Zeng S, Lian G, Bai X X, Liu W P, Yamaguchi H, Kubono S, Hu J, Kahl D, Jung H S, Moon J Y, Lee C S, Teranishi T, Wang H W, Ishiyama H, Iwasa N, Komatsubara T and Brown B A 2013 *Phys. Rev.* C **88** 035801
[11] Hu J, He J J, Parikh A, Xu S W, Yamaguchi H, Kahl D, Ma P, Su J, Wang H W, Nakao T, Wakabayashi Y, Teranishi T, Hahn K I, Moon J Y, Jung H S, Hashimoto T, Chen A A, Irvine D, Lee C S and Kubono S 2014 *Phys. Rev.* C **90** 025803
[12] Yamaguchi H, Hashimoto T, Hayakawa S, Binh D N, Kahl D, Kubono S, Wakabayashi Y, Kawabata T and Teranishi T 2011 *Phys. Rev.* C **83** 034306
[13] Yamaguchi H, Kahl D, Wakabayashi Y, Kubono S, Hashimoto T, Hayakawa S, Kawabata T, Iwasa N, Teranishi T, Kwon Y, Binh D N, Khiem L and Duy N 2013 *Phys. Rev.* C **87** 034303
[14] Kim A, Lee N H, Han M H, Yoo J S, Hahn K I, Yamaguchi H, Binh D N, Hashimoto T, Hayakawa S, Kahl D, Kawabata T, Kurihara Y, Wakabayashi Y, Kubono S, Choi S, Kwon Y K, Moon J Y, Jung H S, Lee C S, Teranishi T, Kato S, Komatsubara T, Guo B, Liu W P, Wang B and Wang Y 2015 *Phys. Rev.* C **92** 035801
[15] Hayakawa S, Kubono S, Kahl D, Yamaguchi H, Binh D N, Hashimoto T, Wakabayashi Y, He J J, Iwasa N, Kato S, Komatsubara T, Kwon Y K and Teranishi T 2016 *Phys. Rev.* C **93** 065802
[16] Kanada-En'yo Y 2007 *Phys. Rev.* C **75** 024302
[17] Tohsaki A, Horiuchi H, Schuck P and Röpke G 2001 *Phys. Rev. Lett.* **87** 192501
[18] Kawabata T, Akimune H, Fujimura H, Fujita H, Fujita Y, Fujiwara M, Hara K, Hara K Y, Hatanaka K, Ishikawa T, Itoh M, Kamiya J, Kishi S, Nakamura M, Nakanishi K, Noro T, Sakaguchi H, Shimbara Y, Takeda H, Tamii A, Terashima S, Toyokawa H, Uchida M, Ueno H, Wakasa T, Yasuda Y, Yoshida H P and Yosoi M 2004 *Phys. Rev.* C **70** 034318
[19] Kawabata T, Akimune H, Fujita H, Fujita Y, Fujiwara M, Hara K, Hatanaka K, Itoh M, Kanada-En'yo Y, Kishi S, Nakanishi K, amd Y Shimbara H S, Tamii A, Terashima S, Uchida M, Wakasa T, Yasuda Y, Yoshida H P and Yosoi M 2007 *Phys. Lett.* B **646** 6
[20] Ragnarsson I, Åberg S, Håkansson H B and Sheline R 1981 *Nucl. Phys.* A **361** 1
[21] Soić N, Freer M, Donadille L, Clarke N, Leask P, Catford W, Jones K, Mahboub D, Fulton B, Greenhalgh B, Watson D and Weisser D 2004 *Nucl. Phys.* A **742** 271
[22] T Suhara and Y Kanada-En'yo 2012 *Phys. Rev.* C **85** 054320
[23] Freer M, Achouri N L, Angulo C, Ashwood N I, Bardayan D W, Brown S, Catford W N, Chipps K A, Curtis N, Demaret P, Harlin C, Laurent B, Malcolm J D, Milin M, Munoz-Britton T, Orr N A, Pain S D, Price D, Raabe R, Soić N, Thomas J S, Wheldon C, Wilson G and Ziman V A 2012 *Phys. Rev.* C **85** 014304
[24] Morinaga H 1956 *Phys. Rev.* **101** 254
[25] Suhara T and Kanada-En'yo Y 2010 *Phys. Rev.* C **82** 044301
[26] Suhara T and Kanada-En'yo Y 2011 *Phys. Rev.* C **84** 024328
[27] Freer M, Malcolm J D, Achouri N L, Ashwood N I, Bardayan D W, Brown S M, Catford W N, Chipps K A, Cizewski J, Curtis N, Jones K L, Munoz-Britton T, Pain S D, Soić N, Wheldon C, Wilson G L and Ziman V A 2014 *Phys. Rev.* C **90** 054324

^{15}O+α resonant elastic scattering to study cluster states in ^{19}Ne

D Torresi[1], C Wheldon[1], Tz Kokalova[1], S Bailey[1], A Boiano[2], C Boiano[3], M Cavallaro[4], S Cherubini[4], A Di Pietro[4], J P Fernandez Garcia[4], M Fisichella[4], T R Glodariu[5], J Grebosz[6], M La Cognata[4], M La Commara[2,7], M Lattuada[4,8], M Mazzocco[9,10], D Mengoni[9,10], C Parascandolo[2,7], D Pierroutsakou[4], G Pizzone[4,8], C Signorini[10], C Stefanini[9], L Stroe[5], C Spitaleri[4,8], E Strano[9,10] and M Zadro[12]

[1]School of Physics, University of Birmingham, Birmingham, B15 2TT, UK
[2]INFN-Sezione di Napoli, Via Cinthia, I-80126 Napoli, Italy
[3]INFN-Sezione di Milano, Via Celoria 16, I-20133 Milano, Italy
[4]INFN-Laboratori Nazionali del Sud, Via S. Sofia 62 Catania, Italy
[5]TNIPNE, 407 Atomistilor Street, 077125 Magurele, Romania
[6]Department of Nuclear Reactions, Institute for Nuclear Studies, ul. Hoza 69, 00-681 Warsaw, Poland
[7]Dipartimento di Fisica, Universitá di Napoli, Via Cinthia, I-80126 Napoli, Italy
[8]Dipartimento di Fisica, Universitá di Catania, Via A. Doria 64, I-95123 Catania, Italy
[9]INFN-Sezione di Padova, Via F. Marzolo 8, I-35131 Padova, Italy
[10]INFN-Laboratori Nazionali di Legnaro, Viale dellUniversitá 2, I-35020 Legnaro, Italy

E-mail: domenico.torresi@lns.infn.it

Abstract. Clustering phenomena are well known in nuclear physics for stable nuclei, both α-conjugate (N=Z, A=2N), like ^{8}Be, ^{16}O, ^{20}Ne, and non-α-conjugate, like ^{6}Li and ^{7}Li. In general, it is expected that light exotic nuclei may also exhibit cluster behavior. Moving out of the valley of stability configurations can be found where at least one of the clusters is unbound or weakly bound, thus not satisfying the strong internal correlation requirement of classical clusters. This is so-called exotic clustering. The study of such systems presents many difficulties, due, mainly, to the low intensities typical of radioactive ion beams. Therefore, few significant experimental studies have been performed so far. In this work we searched for α-cluster states in ^{19}Ne above its α-decay threshold measuring, for the first time, the ^{15}O(^{4}He,^{4}He) elastic scattering excitation function. Moreover, this study classified low-energy states in Ne in the astrophysically important region in this HCNO-break-out nucleus.

1. Introduction

In recent years the isotope ^{19}Ne has attracted much interest. This is mainly due to the astrophysical importance of the ^{15}O(α,γ)^{19}Ne reaction rate that was identified as a break out from the hot-CNO cycles and a trigger for the following rapid proton capture that populate masses higher then neon [1]. Therefore, the structure of the low-energy spectrum of ^{19}Ne close to the ^{15}O+α threshold (S_α=3.528 MeV) is very important. Moreover, the states in ^{19}Ne near the proton threshold (S_p=6.411 MeV), are again important, as the destruction of the key radio-nuclide ^{18}F [2], occurs mainly via the ^{18}F(p,α)^{15}O reaction.

Neon isotopes exhibit strong α-clustering phenomena, even in the ground states. Neon-20 [3, 4, 5] and ^{21}Ne [6, 7] are two of the best examples of nuclear clustering known to date. Therefore, it is important to extend this cluster study to the neutron poor side of the neon isotopes that is, at the moment, almost totally unknown concerning the clustering point of view. This is needed in order to understand if such clustering phenomena persist in ^{19}Ne and, if so, to what extent. The present work reports the direct measurement of ^{15}O+α elastic scattering using the technique of Thick Target Inverse Kinematics scattering [8] to investigate the structure of the low-lying-energy levels of ^{19}Ne. The method as implemented here allows the elastic scattering excitation function to be measured and enables to measure the α reduced α-width *directly* across a large excitation-energy range using a single beam energy. The nature of the technique makes it ideally suited to use with low-intensity, large-energy-profile radioactive ion beams.

2. The Experimental Technique

The TTIK method [9, 8, 10] with a gas target consists of delivering the beam into a reaction chamber filled with gas (typically ^{4}He) at such a pressure as to fully stop the beam. At the same time the gas acts as both target and degrader of the beam energy. The interaction between projectile and target nuclei can occur at different depths in the chamber along the beam direction, and thus at different energies. Thanks to the different stopping power of the gas for heavy projectiles and light recoils, the light recoiling particles can reach the telescopes placed at the downstream end of the chamber with respect to the entrance window. From the measurement of the recoil energy one can deduce the interaction position along the beam direction and thus the energy of the scattering event. This relationship is unequivocal if the detected light recoils originate from elastic scattering. The α-particles are detected at small angles covering the 0° region in the laboratory system, corresponding to the region around 180° in the center-of-mass. In this condition it is possible to extract the elastic scattering angular distribution in the region where the Coulomb scattering is minimized over the resonant elastic scattering. This technique is particularly suitable for measuring elastic scattering excitation functions with radioactive beams since it strongly reduces the beam time request with respect to the thin target technique.

2.1. The Experimental Apparatus

The ^{15}O radioactive ion beam was delivered by the EXOTIC facility [11, 12] at the Laboratori Nazionali di Legnaro (LNL, Italy) by means of the in-flight technique. A primary 80 MeV $^{15}N^{5+}$ beam was delivered by the LNL-XTU tandem Van de Graaff accelerator and impinged on a H_2 cryogenic gas target at a pressure of 1 bar and a temperature of 24°, corresponding to an equivalent target thickness of about 1.35 mg/cm^2 of H_2. The ^{15}O beam [13], produced by means of the two-body reaction p(^{15}N,^{15}O)n (Q_{val}=-3.54 MeV), was selected and purified from the scattered primary ^{15}N beam and other contaminant nuclei by mean of several optical elements: a 30°-bending magnet, a Wien-filter, two quadrupole triplets and several collimators. Two parallel plate avalanche counters (PPACs) [14] were placed downstream prior to the beam entering the scattering chamber. The PPACs were utilized for on-line monitoring of the secondary ^{15}O beam profile, to provide an event-by-event reconstruction of the trajectory of the beam and for timing purposes. Finally, ^{15}O beam entered the the scattering chamber filled with ^{4}He gas at a pressure of 467 mbar, that was separated from the high vacuum beam line by a 2.2 μm thick Havar window. The ^{15}O beam intensity was in the range 1-2$\times 10^4$ particles/s with $\approx$99% purity. The ^{4}He gas pressure and the temperature were continuously monitored by two capacitance manometers with an accuracy of 0.3% and a thermocouple device with 0.1 K sensitivity. The detection apparatus, illustrated in Fig. 1, consisted of two Double-Sided Silicon Strip Detectors (DSSSDs) from the EXPADES detector array [14]. The detectors were placed, the first at 0° and the second off 0° as shown in figure 1.

IOP Conf. Series: Journal of Physics: Conf. Series **863** (2017) 012026 doi:10.1088/1742-6596/863/1/012026

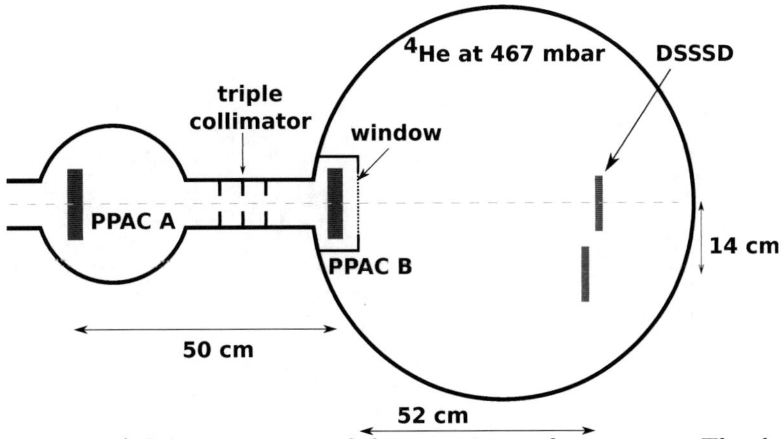

Figure 1. (color online) Schematic view of the experimental apparatus. The detection system consisted of two DSSSDs and two PPACs required for the time-of-flight measurement and beam tracking.

3. Data Analysis

The goal of the analysis is the reconstruction of the elastic scattering excitation function from the energy of the detected α-particles and their impact point on the detectors. The data coming from the two detectors were divided into three angular ranges. The first one corresponding to the whole detector placed at 0° while the dector off 0° was divided in two angular ranges.

The first step is the identification of α-particles produced in elastic scattering events. This was performed through measurements of the time-of-flight between the entering of the particle in the scattering chamber and the following recoiling particle on a DSSSD.

The final results are summarized in Fig. 2 showing the three excitation functions obtained from the two detectors, all three reveal the presence of resonances. Unfortunately the elastic scattering excitation functions corresponding to the detector off 0° have low statistics and very low resolution. This is mainly due to the fact that the effects of the energy and angular spread of the beam profile are larger for particles that are scattered at larger angles. Moreover due to the low statistics obtained for the detector off 0° the angular range is relatively high and this can be a further cause of resolution worsening. Due to the low statistics and low resolution the data at larger angle were not used in the R-matrix calculation (see section 4) for the identification of the spins and parities of the resonances observed in the elastic scattering excitation function.

4. Results and Conclusions

In order to determine the main properties of the observed resonances: widths, partial decay branches, energies and spins, an R-matrix calculation was performed. A comprehensive description of the R-matrix theory can be found in Ref. [15]. A full R-Matrix fit was performed using AZURE2 [16] which uses the Brune transformation. This was done for 0° (θ_{cm} = 180°). The statistics at the other angles were too poor to provide any constraint to the fit. Instead, the spins and energies of states from the mirror nucleus, ^{19}F [17], were used as a starting point for the fit; a powerful technique, with a good agreement between the energy levels in each nucleus. A two channel calculation were performed to fit the data: $^{15}\mathrm{O}(\frac{1}{2}^-) + \alpha(0^+)$ and $^{18}\mathrm{F}(1^+) + p(\frac{1}{2}^+)$, with channel radii for the ith decay channel, $r_i = r_0(A_{recoil}^{\frac{1}{3}} + A_i^{\frac{1}{3}})$. and r_0=1.4 fm. Levels in the 5.3 – 8.9 MeV excitation-energy region have been investigated, with a number of newly observed states in this work. Furthermore, partial widths have been extracted for the observed levels, for many of them it is the first time measure-width data has been available. Reliable cross sections have been measured across the full range of levels populated representing important input to reaction-rate models.

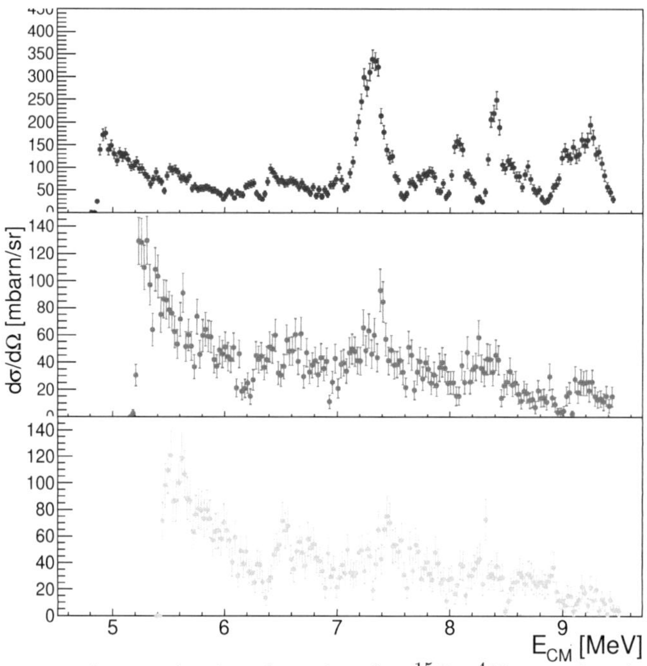

Figure 2. Elastic scattering excitation function for $^{15}O+^{4}He$ at the three different angular values: from top to down ~180°, ~160°, ~155°

Acknowledgments

The present work has received funding from the European Unions Horizon 2020 research and innovation programme under the Marie Sklodowska-Curie grant agreement No 659744. The UK STFC is acknowledged for providing funding under grant ST/L005751/1. M.Z. was partially supported by the Croatian Science Foundation under Project No. 7194.

References

[1] Wiescher M *et al* 2007 *Prog. Part. and Nucl. Phys.* **59** 51
[2] Bardayan D W *et al* 2015 *Phys. Lett.* B **751** 311
[3] von Oertzen W 2001 *Eur. Phys. J.* A **11** 403
[4] Freer M *Rep. Prog. Phys.* **70** 2149
[5] von Oertzen W, Freer M and Kanada-En'yo Y 2005 *Phys. Rep.* **432** 43
[6] Thummerer S *et al* 2003 *J. Phys.* G **29** 509
[7] Wheldon C *et al Eur. Phys. J.* A **26** 321
[8] Artemov K P *et al* 1990 *Sov. J. Nucl. Phys.* **52** 634
[9] Kallman K *et al* 1994 *Nucl. Instr. and Meth. in Phys. Res.* A **338** 413
[10] Zadro M *et al* 2007 *Nucl. Instr. and Meth. in Phys. Res.* B **259** 836
[11] Maidikov V Z *et al.* 2004 *Nucl. Phys.* A **746** 389c
[12] Mazzocco M *et al* 2013 *Nucl. Instr. and Meth. in Phys. Res.* B **317** 223
[13] Torresi D, Wheldon C *et al* 2014 *LNL Annual Report* 19, see *http://www.lnl.infn.it/~annrep/read_ar/2014/contributions/pdfs/019_A_61_A056.pdf*
[14] Pierroutsakou D *et al*, submitted to *Nucl. Instr. and Meth. in Phys. Res.* A
[15] Lane A M and Thomas R G 1958 *Reviews of Modern Physics* **30** 257
[16] Azuma R *et al* 2010 *Phys. Rev.* C **81** 045805
[17] Tilley D R, Weller H R, Cheves C M and Chasteler R M 1995 *Nucl. Phys.* A **595** 1

Some Evidence of the Cluster Struture Inside of ^{9}Be

S M Lukyanov[1], A S Denikin[1], M A Naumenko[1], V Burjan[2], V Kroha[2], J Mrazek[2], K Mendibayev[1], I Sivaček[2], V Glagolev[2], Š Piskoř[2], Yu E Penionzhkevich[1], N K Skobelev[1], Yu G Sobolev[1], W H Trzaska[3] and K Kuterbekov[4]

[1] Flerov Laboratory of Nuclear Reactions, Dubna, Russian Federation
[2] Nuclear Physics Institute, Řež, Czech Republic
[3] Department of Physics, University of Jyväskylä, Jyväskylä, Finland
[4] Eurasian Gumilev University, Astana, Kazakhstan

E-mail: lukyan@jinr.ru

Abstract. Angular distributions of protons, deuterons, tritons and alpha-particles emitted from the reactions in the ^{2}H+^{9}Be-system at E_{lab}=19.5 MeV were measured with an aim to shed light on the internal cluster structure of ^{9}Be and the study of possible cluster transfer of ^{5}He. The analyses suggest a significant contribution of five-nucleon transfer in the reaction channel ^{9}Be(d,^{4}He)^{7}Li.

1. Introduction

Due to its Borromean structure, a special attention has been focused on the ^{9}Be nucleus, the breakup of which can occur directly to two particles and a neutron or via one of two unstable intermediate nuclei such as ^{8}Be or ^{5}He [1, 2]. Scattering of a projectile, such as ^{1,2}H or ^{3,4}He, on a target is a standard tool to study the structure of nuclei. This method involves an angular distribution measurement of elastic and inelastic scattering of projectile-like products. Energy distribution of these products is baring information about internal structure of the interacting and forthcoming nuclei. The angular distributions of the ^{9}Be(^{3}He,^{3}He)^{9}Be, ^{9}Be(^{3}He,^{5}He)^{7}Be, ^{9}Be(^{3}He,^{5}Li)^{7}Li, ^{9}Be(^{3}He,^{6}Be)^{6}He and ^{9}Be(^{3}He,^{6}Li)^{6}Li reaction channels were measured [3, 4] and described within the framework of the optical model, the coupled-channel approach and the distorted-wave Born approximation. The performed analysis of the experimental data shows that the potential parameters are quite sensitive to the exit channel and hence to the cluster structure of the populated states, which allows to make general observations and conclusions regarding the internal structure of the target and residual nuclei. The experiment [3, 4] was designed to study the breakup of ^{9}Be in an attempt to determine the contribution of the ^{8}Be+n and ^{5}He+α channels in the inclusive measurements. We found that the ratio about 2.7:1 may be assigned for the contributions of these two channels, respectively. The determined value ratifies that the ^{5}He+α breakup channel plays an important role.

An other aspect is an attempt to find not only the cluster structure (for instance, ^{5}He) but also clarify how the cluster structure transfer is involved into nuclear reaction mechanism. Indeed, starting from Detraz [5, 6] multi particle-multi hole structures were expected to occur at rather low excitation energies in nuclei. Four-nucleon transfer reactions continue being extensively studied. One may hope that their major features, in spite of the a priori complexity of such a transfer, can be understood assuming that the nucleons are transferred as a whole, strongly correlated in a cluster which has the internal quantum numbers of a free particle.

2. Experimental Method

The experiment was performed using beam energy of a ^{2}H ions at 19.5 MeV using the cyclotron of the INP (Rez, Czech Republic). The average beam current during the experiment was maintained at 10 nA. The self-supporting Be target is prepared from a thin beryllium foil of 99 % purity. To measure (in)elastically scattered ions, a set of four telescopes each consisting of ΔE_0, ΔE, E_r detectors with thicknesses of 12, 100 μm and 3 mm, respectively, were used. The telescopes were mounted at a distance of about 19 cm from the target in a reaction chamber. Particle identification is performed based on the energy-loss measurements of ΔE and residual energy E_r, i.e. by the so-called ΔE-E method. An example of two-dimensional plots (yield versus energy loss ΔE and residual energy E_r) are shown in Fig.1.

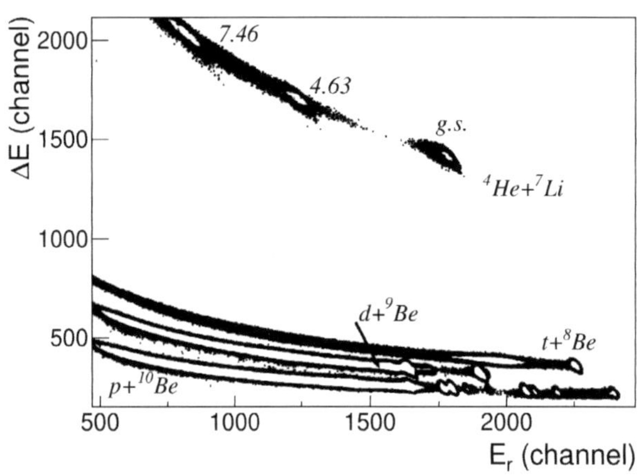

Figure 1. Particle identification plots for the products of the ^{2}H+^{9}Be reaction: p, d, t and ^{4}He. ΔE is the energy loss and E_r is the residual energy. Excited states for ^{7}Li the reaction channel ^{7}Li+α are indicated.

This experimental technique allows for identifying the particles p, d, t, and ^{4}He and determine their total deposited energies. The spectra of total deposited energy are shown in Fig. 2. All peaks, which can be observed in the histograms in Fig. 2, were identified and found to belong to the ground and excited states of ^{10}Be, ^{9}Be, ^{8}Be and ^{7}Li, as the complementary products to detected particles p, d, t and ^{4}He respectively.

3. Results and Data Analysis

The differential cross section for the elastic, inelastic and transfer reaction channels are presented in Fig.3. All calculations and fitting of the experimental data have been performed using the FRESCO[7] code in the framework of the CC method.

The calculation for (in)elastic scattering (Fig.3a) was performed using the method of strong channel coupling in the adiabatic rotational model. It is assumed that the state 2.43 MeV is the first excited state in the rotational band of ^{9}Be. The quadrupole deformation parameter of the target nucleus found from the performed analysis of data is β_2= 0.64, which perfectly coincides with the previous analysis of inelastic scattering of ^{4}He nuclei on the same target [3]. The optical potential parameters used for the calculation are similar to ones obtained in Ref.[8].

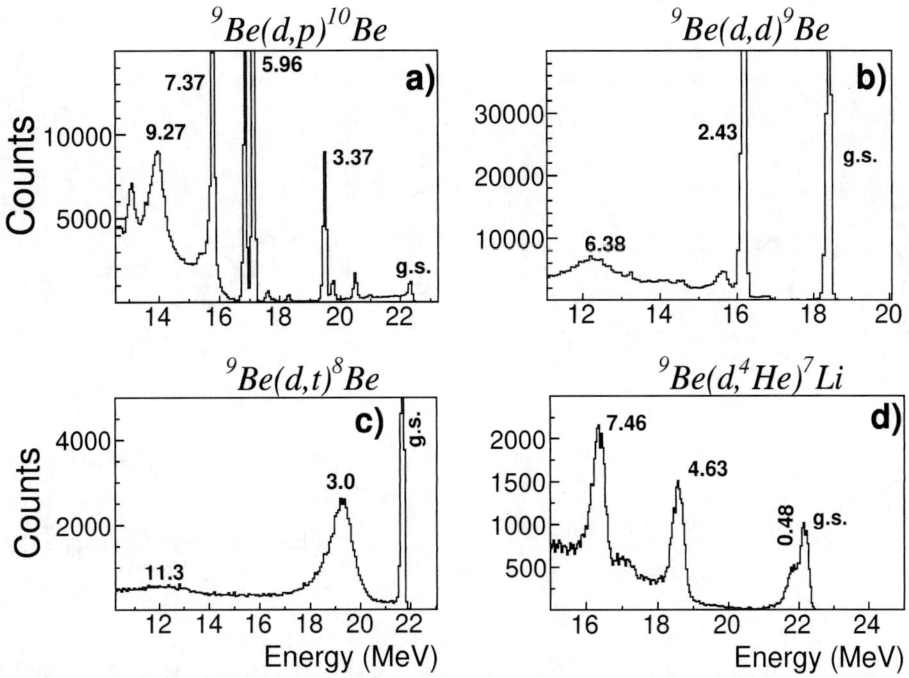

Figure 2. Total deposited energy spectra measured at θ_{lab}=32° for the detected p (panel a), d (panel b), t(panel c) and ^{4}He(panel d). The ground and excited states of ^{7}Li for the case of detected complementary product ^{4}He as well as ground states and excited states for ^{8}Be, ^{9}Be and ^{10}Be in the case of detected td and p, as complementary products, respectively, were unambiguously identified.

Good agreement between experimental data and experimental fitting is observed, as it's shown in Fig. 3a.

A special attention was paied to the ^{4}He+^{7}Li reaction channel, which can proceed through two different channels: either with transfer of d (dashed line in Fig.3b) or transfer of ^{5} (solid line in Fig.3b) from the target to the projectile. In the exit channel both these reactions are indistinguishable from each other. However, in the former case the ^{4}He nuclei are expected to fly "forward" in the center of mass system, whereas in the latter case the alpha particles should preferably fly at the "backward" angles in the center of mass system. The calculations carried out within the DWBA method for these two channels are shown in Fig. 3b as the dashed and solid curves in Fig.3b. Their coherent sum is shown as the black curve which is in good agreement with the experimental points. It is interesting to note that for the correct description of the data amplitude it is necessary to set the sufficiently large values of the spectroscopic factors for the systems ^{9}Be =α+ ^{5}He (S=1.2) and ^{7}Li = d + ^{5}He (S=1.0). In addition, to describe the structure of the angular distributions it is also necessary to assume a 30% admixture of the d-state in the structure of ^{7}Li.

In summary, angular distributions of the differential cross sections for the ^{9}Be(d,d)^{9}Be*, ^{9}Be(d,p)^{10}Be, ^{9}Be(d,t)^{8}Be and ^{9}Be(d,^{4}He)^{7}Li reactions were measured. Experimental angular distributions were described within the optical model, coupled channel approach and distorted wave Born approximation. The optical potential provides good fit of the elastic scattering in the entrance and exit channels. The DWBA calculations are well agreed with the transfer reaction

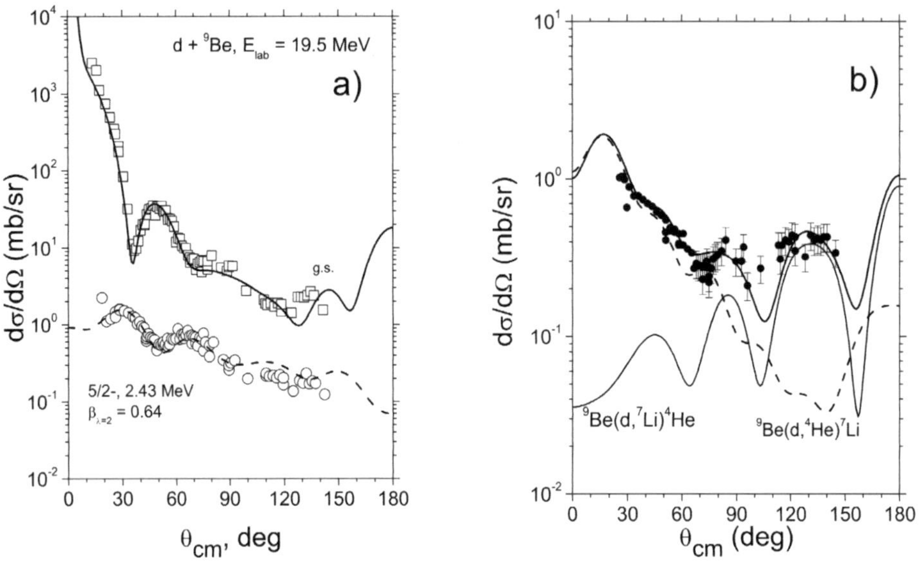

Figure 3. a) Experimental angular distribution for elastic(□) and inelastic scattering(○) of 19.5 MeV deuterons from ^{9}Be target compared with the calculations. The curves are the results of the optical-model and the coupled-channel calculations. b) Experimental angular distribution (•) for ^{4}He+^{7}Li reaction channel (see details in text).

data. The spectroscopic factors of both reactions are close to unit that confirm significant contribution of the considered cluster configurations into the structure of ground states. The analysis shows that the contribution of the compound nucleus mechanism is neglectable. In the (d, ^{4}He) channels, see Fig. 3, the deuteron transfer gives only a small contribution, while a relatively large contribution of ^{5}He transfer was found in agreement to result [8]. This demonstrates that the specific structure of the ^{9}Be nucleus as a weekly bound system of two alpha particles and a neutron strongly favors the five-nucleon transfer compared to deuteron transfer.

The authors are apprciate to the CANAM project (http://users.canam.ujf.cas.cz) for delivering beam time for this experiment.

References

[1] Brown T A D, Papka P and Fulton B R *et al* 2007 *Phys. Rev.* C **76** 054605
[2] Papka P, Brown T A D and Fulton B R *et al* 2007 *Phys. Rev.* C **75** 045803
[3] Lukyanov S M, Denikin A S and Voskoboynik E I *et al* 2014 *J. Phys.* G **41** 035103
[4] Lukyanov S, Harakeh M and Naumenko M *et al* 2015 *World Journal of Nuclear Science and Technology* **5** 265
[5] Détraz C, Duhm H H and Hafner H 1970 *Nucl. Phys.* A **147** 488
[6] Détraz C, Pougheon F and Bernas M *et al* 1974 *Nucl. Phys.* A **228** 39
[7] http://www.ianthompson.org/surrey and http://nrv.jinr.ru
[8] Szczurek A, Bodek K and Krug J *et al* 1989 *Z. Phys.* A **333** 271

IOP Conf. Series: Journal of Physics: Conf. Series **863** (2017) 012028 doi:10.1088/1742-6596/863/1/012028

Evidence of enhanced radius of Hoyle rotational state in ^{12}C inelastic scattering

M Ito

Department of Pure and Applied Physics, Kansai University, 3-3-35 Yamate-cho, suita, Japan

E-mail: itomkl@kansai-u.ac.jp

Abstract. Nuclear radius of three α rotational state in ^{12}C with a life time of 10^{-21} second, which has been expected to have much more extended radius than the ground ^{12}C nucleus, is speculated from systematic analysis of the differential cross section of the α + ^{12}C inelastic scattering. Present analysis predicts about 0.6 $\sim$ 1.0 fm enhancement in the matter radius of the three α rotational state in comparison to the normal radius of the ground state. The spatial extension of the three α rotational state is comparable to the extended radius observed in the neutron halo phenomena.

1. Introduction

Cluster structures, in which a nucleus is decomposed into several sub-units, are realized in the low excited states of light mass nuclei [1]. In the cluster structure, the sub-units are weakly coupled each other, and their matter radius is prominently extended in comparison to the radius of the ground state, which obeys the law of $\propto A^{1/3}$. In particular, such the enhancement of the matter radius is extensive in the 3α structure of the ^{12}C nucleus [2, 3]. The ground state of ^{12}C has a radius of 2.40 fm [2] with the spin-parity of 0^+ (0_1^+), while the excited 0_2^+ state at E_x = 7.65 MeV, which is called the Hoyle state, is considered to have a radius of 3.47 fm with the well developed 3α cluster structure [2]. The extension of the radius in 0_2^+ is predicted to be about 1 fm, which is comparable to the extension of the neutron halos. In this article, we propose a new method to obtain a signature of the enhanced matter radius of the 3α cluster state.

Unfortunately, a direct measurement of the radius of the Hoyle 0_2^+ states with the developed 3α cluster structure is impossible due to its short life time, but there are several attempts to get an evidence of the enhanced matter radius of the 3α Hoyle states from the inelastic scattering of ^{12}C, which excites the Hoyle 0_2^+ state as a final state. In the inelastic scattering of light ions by ^{12}C, an oscillating pattern in the differential cross section of the scattered ion is discussed in connection to the enhanced radius of the final 0_2^+ state with the 3α structure [4, 5, 6]. However, the relation of the extended radius and the oscillating pattern of the cross section of the ^{12}C inelastic scattering still remains unclear [7, 8].

In this article, we focus on the 2_2^+ state, which corresponds to the 3α rotational state of the Hoyle 0_2^+ state [2, 3] with a life of 10^{-21} sec. (width of $\Gamma \sim 1$ MeV), and demonstrate that a direct evidence of the enhanced matter radius in the 2_2^+ state clearly appears in the α + ^{12}C inelastic scattering. In a modern theory, the Hoyle rotational 2_2^+ state, which has been recently identified in experiment [9], is interpreted in terms of an α halo state with a dilute 3α structure [3]. The differential cross section of α + ^{12}C(0_1^+) $\rightarrow$ α + ^{12}C(2_2^+) is compared with the reference

IOP Conf. Series: Journal of Physics: Conf. Series **863** (2017) 012028 doi:10.1088/1742-6596/863/1/012028

reaction of $\alpha + {}^{12}\mathrm{C}(0_1^+) \to \alpha + {}^{12}\mathrm{C}(2_1^+)$, in which the 2_1^+ state is a rotational state of the ground 0_1^+ state with a spatially compact structure. In the inelastic scattering to the $2_{1,2}^+$ states, only the matter radii are prominently different, say 2.4 fm for 2_1^+ and 4.0 fm for 2_2^+ [2]. Thus, the comparison of these two inelastic scattering is expected to give the evidence of the extended matter radius of the 2_2^+ state.

2. Theoretical Framework

We calculate the differential cross sections of an α particle scattered by ^{12}C in the formulation of the microscopic coupled-channels (MCC) calculation [6, 10]. In the MCC calculation, the nuclear potential of α and ^{12}C is calculated from the double folding model, which is symbolically written as a function of the $\alpha - {}^{12}$C relative coordinate of $\mathbf{R}$,

$$V_{f,i}(\mathbf{R}) = N_R \int\int \rho_{f,i}^{(^{12}C)}(\mathbf{r}_1)\rho^{(\alpha)}(\mathbf{r}_2)v_{NN}(\mathbf{s})\, d\mathbf{r}_1 d\mathbf{r}_2 \qquad (1)$$

with $\mathbf{s} = \mathbf{r}_2 - \mathbf{r}_1 - \mathbf{R}$. Here $\mathbf{r}_1$ ($\mathbf{r}_2$) denotes a coordinate measured from the center of mass in the ^{12}C (α) nucleus. $\rho_{f,i}(\mathbf{r})$ represents the diagonal ($f = i$) or transition ($f \neq i$) densities of ^{12}C, which are calculated by the microscopic 3α cluster model, resonating group method (RGM) [2], while $\rho^{(\alpha)}(\mathbf{r})$ denotes the density of α particle, which reproduce the charge form factor of the electron scattering. In Eq. (1), v_{NN} represents the effective nucleon-nucleon (NN) interaction which acts between a pair of nucleon contained in the ^{12}C nucleus and the α particle. In the present calculation, we adopt the DDM3Y (Density Dependent Michigan 3-range Yukawa) interaction [11].

The normalization factor N_R is introduced because the folding potential contains a theoretical ambiguity in its strength. Here this factor is set to $N_R = 1.42$ over all the α incident energy, which is consistent to the MCC calculation in Ref. [6]. In addition to the folding potential, we introduce the absorptive potential with the Saxon-Woods form factor in the diagonal ($f = i$) transition in order to simulate other reaction process [6], and the parameter set of Saxon-Woods is tuned so as to reproduce all of the observed differential cross sections as much as possible. As for the internal excitation of ^{12}C, we include the low-lying collective states (2_1^+ and 3_1^-) and the 3α cluster states (0_2^+, 0_3^+, 2_2^+) in addition to the ground 0_1^+ state.

3. Results

We have solved the coupled-channels equation for the $\alpha + {}^{12}$C scattering at $E_\alpha = 386$ MeV [9] with the nuclear interaction in Eq. (1). The MCC calculation reproduce the differential cross sections of the scattering to 0_1^+, 2_1^+, 3_1^- states. Furthermore, the theoretical calculation nicely reproduce the inelastic scattering to the excited state observed at $E_x \sim 10$ MeV, which is considered as the incoherent mixture of 0^+ and 2^+ strength, by taking the summation of the differential cross sections of the 0_3^+ and 2_2^+ states.

A comparison of the 2_1^+ cross section (dotted curve) and the 2_2^+ one (solid curve) is shown in Fig. 1 (a). The peak position of the 2_2^+ cross section (solid curve) shifts to the forward angular region, and its angular distribution has the shrinkage and rapid fall-down structure in comparison to the 2_1^+ cross section (dotted curve). The shift and shrinkage features are completely consistent to the result of the multi-pole decomposition analysis (MDA) of the experimental cross section, which is performed in the range of $\theta_{c.m.} < 15°$ [9].

In order to analyze the difference of the differential cross sections shown in Fig. 1 (a) more deeply, we have performed the partial wave analysis, in which the angle-integrated cross sections of Fig. 1 (a) are decomposed into the individual components of the incident orbital spin L. The results of the decomposition are shown in Fig. 1 (b). The dotted curve with the crosses shows the partial wave distribution (L-distribution) of the angle-integrated 2_1^+ cross section, while the L-distribution of 2_2^+ is plotted by the solid curve with the open circles.

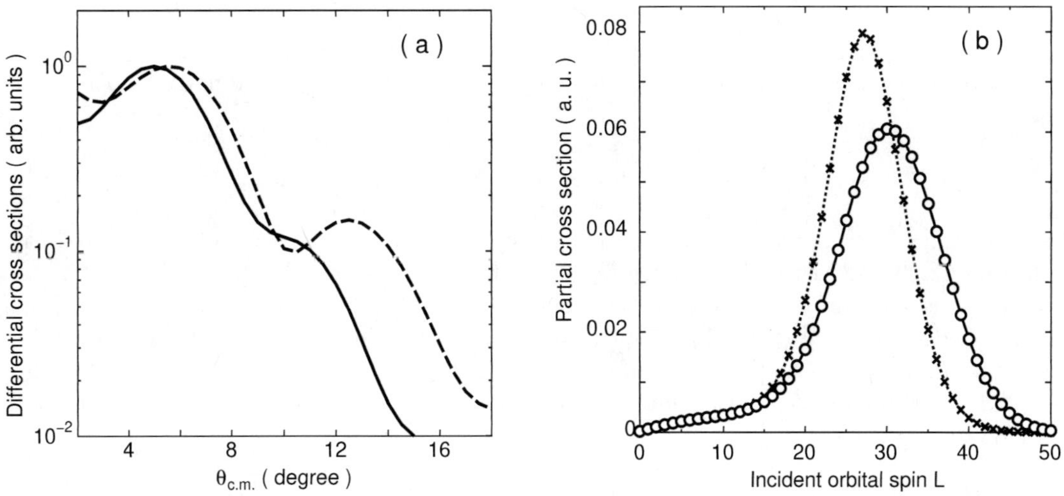

Figure 1. Panel (a): Theoretical differential cross sections of the final state of 2_1^+ (dotted curve) and 2_2^+ (solid curve) in the α + ^{12}C inelastic scattering. The magnitudes of both the cross sections are normalized by the maximum value around $\theta_{c.m.} \sim 5°$. Panel (b): Partial wave decomposition of the angle integrated cross sections calculated as the function of the incident orbital spin L (L-distribution). The open circles with the dotted curve and the solid circles with the solid curve show the partial cross sections of the 2_1^+ and 2_2^+ final states, respectively. The magnitudes of all the partical cross section is normalized by the total cross section, which is the summations of the partial cross sections.

In the comparison of 2_1^+ with 2_2^+, we can clearly see the extended L-distribution of 2_2^+ to the higher L region. This extension of 2_2^+ in the L-space (Fig. 1 (b)) is just opposite to the shrinkage in the θ-space, the differential cross section (Fig. 1 (a)). This is nothing but the uncertainty relation of $L \cdot \theta \sim 1$, which can be hold in the diffraction scattering. According to the classical relation of $L = kb$, where k and b denote the incident wave number and the impact parameter, respectively, the extension of the L-distribution in Fig. 1 (b) clearly means that the production area of the 2_2^+ state is more extended than the area of the 2_1^+ production.

The distribution in the L-space can be transformed into a size of the production area of the final state by applying the method of the scattering radius [12]. In this method, the effective orbital spin $\bar{L}$ for the transition of $0^+ \to 2^+$ is derived according to the following expression:

$$\bar{L}_{2^+\leftarrow 0^+} = \sqrt{\frac{\sum_L \hat{L}^4 \sigma_{2^+\leftarrow 0^+}(L)}{\sum_L \hat{L}^2 \sigma_{2^+\leftarrow 0^+}(L)}} \qquad (2)$$

with a definition of $\hat{L} = \sqrt{L(L+1)}$. Here $\sigma_{2^+\leftarrow 0^+}(L)$ represents the partial cross section for the $0^+ \to 2^+$ transition at the incident orbital spin L. The radius of the final-state production, which is called the scattering radius (R_{SC}), is simply obtained according to the classical relation of $\bar{L} = kR_{SC}$. R_{SC} naturally goes to the matter radius of a target nucleus in the high energy limit of the proton elastic scattering [5, 12, 13]. The resultant scattering radius for 2_1^+ is $R_{SC}(2_1^+) =$ 4.58 fm, while the radius for 2_2^+ is $R_{SC}(2_2^+) = 5.20$ fm. The difference of the scattering radius is $\Delta R_{SC} = R_{SC}(2_2^+) - R_{SC}(2_1^+) = 0.62$ fm.

We have extended the MCC calculation with the method of the scattering radius to the lower α incident energy of $E_\alpha \leq 240$ MeV. In all the energy region considered, ΔR_{SC} ranges from 0.6

fm to 1 fm, which means that $R_{SC}(2_2^+)$ is larger than $R_{SC}(2_1^+)$ over the entire incident energy. From the values of ΔR_{SC}, we can speculate the lower bound of the matter radius of the 2_2^+ state. In a naive consideration, we can image the relation of ΔR_{SC} and the matter radius R_m, such as

$$R_m(2_2^+) \geq R_m(2_1^+) + \Delta R_{SC} . \quad (3)$$

In the present analysis, $\Delta R_{SC} \sim 1$ fm is predicted in the lower energy region of $E_\alpha \leq 240$ MeV. The matter radius of 2_1^+ is still unknown but we can safely assume $R_m(2_1^+) \sim R_m(0_1^+)$. This assumption is because the 2_1^+ state is the first excited state of the ground 0_1^+ state and hence, we can expect that the spatial size of 2_1^+ is not drastically changed from the size of the ground 0_1^+. This assumption is also supported by the 3α RGM calculation [2]. Therefore, the matter radius of the Hoyle rotational state is speculated to $R_m(2_2^+) \geq 3.7$ fm if we employ the experimental value of $R_m(0_1^+) = 2.7$ fm [14].

4. Summary and discussions

In summary, we have shown that the enhanced radius of the 2_2^+ state can be probed via the differential cross section by comparing with the respective cross section of the 2_1^+ state. The inelastic differential cross section of the 2_2^+ state is more shrunk than that of the 2_1^+ state, and this shrinkage is a first evidence of the extended matter radius of the final 2_2^+ state in the inelastic scattering of $\alpha + {}^{12}\mathrm{C}(0_1^+) \to \alpha + {}^{12}\mathrm{C}(2_2^+)$. According to the prescription of the partial wave decomposition, the shrunk structure of the differential cross section corresponds to about 0.6 $\sim$ 1 fm enhancement in the size of the production area of the final 2_2^+ state. The comparison of the yrast 2_1^+ state and the Hoyle rotational 2_2^+ state is a new insight in the discussion of the inelastic scattering of ^{12}C, which was absent in the previous studies about the inelastic scattering to the Hoyle 0_2^+ state [4, 5, 6, 7, 8].

About 1 fm enhancement in the matter radius of the Hoyle rotational state is comparable to the extended radius in ^{11}Li [15], which largely deviates from the systematics of $\propto A^{1/3}$. ^{11}Li corresponds to the excited state from ^{11}B, in which the isospin degrees of freedom is excited. The excitation energy of ^{11}Li, (B.E.(^{11}Li) − B.E.(^{11}B)) is about 34 MeV, while the excitation energy of ${}^{12}\mathrm{C}(2_2^+)$ is just about 10 MeV. Thus, the 1 fm enhancement of the matter radius in ${}^{12}\mathrm{C}(2_2^+)$ is exotic phenomena, which occurs in much lower excitation energy than the neutron-excess nucleus.

References

[1] Ikeda K *et al* 1980 *Prog. Theor. Phys.* **68** 1
[2] Fukushima Y and Kamimura M 1977 *Proc. Int. Conf.Nuclear Structure, J. Phys. Soc. Jpn.* **44** 225; Kamimura M 1981 *Nucl. Phys.* A **351**456
[3] Funaki Y, Thosaki A, Horiuchi H, Schuck P and Reöpke G 2005 *Eur. Phys. J.* A **24** 321 and references therein
[4] Danilov A N, Belyaeva T L, Demyanova A S, Ganchalov S A and Ogloblin A A 2009 *Phys. Rev.* C **80** 054603
[5] Iida K, Koide S, Kohama A and Oyamatsu K 2012 *Mod. Phys. Lett.* A **27** 1250020
[6] Ohkubo S and Hirabayashi Y 2004 *Phys. Rev.* C **70** 041602(R)
[7] Takashina M and Sakuragi Y 2006 *Phys. Rev.* C **74** 054606
[8] Takashina M 2008 *Phys. Rev.* C **78** 014602
[9] Itoh *et al* 2011 *Phys. Rev.* C **84** 054308 (2011)
[10] Ito M, Hirabayashi Y and Sakuragi Y 2002 *Phys. Rev.* C **66** 034307 and references therein
[11] Kobos A M, Brown B A, Hodgson P E, Satchler G R and Budzanowski A 1982 *Nucl. Phys.* A **384** 65
[12] Tomita M, Iwasaki M, Otani R and Ito M 2014 *Phys. Rev.* C **89** 034619
[13] Kohama A, Iida K and Oyamatsu K 2016 *J. Phys. Soc. Jpn* **85** 094201 and references therein
[14] Kim J C, Hicks R S, Yen R, Auer I P, Caplan H S and Bergstrom J C 1978 *Nucl. Phys.* A **297** 301
[15] Tanihata I, Savajolos H and Kanungo R 2013 *Prog. Part. Nucl. Phys.* **68** 215

Study of the cluster structure of ^{10}Be and ^{16}C neutron-rich isotopes by means of intermediate energies breakup reactions

D Dell'Aquila[4], L Acosta[8], L Auditore[5], G Cardella[1], E De Filippo[1], S De Luca[5], L Francalanza[4], B Gnoffo[1], G Lanzalone[2,6], I Lombardo[4], N S Martorana[3,2], A Pagano[1], E V Pagano[2,3], M Papa[1], S Pirrone[1], G Politi[1,3], L Quattrocchi[2], F Rizzo[2,3], E Rosato[4,†], P Russotto[1], A Trifirò[5], M Trimarchi[5], G Verde[1,7] and M Vigilante[4]

[1] INFN-Sezione di Catania, Catania, Italy
[2] INFN-Laboratori Nazionali del Sud, Catania, Italy
[3] Dipartimento di Fisica e Astronomia, Università di Catania, Catania, Italy
[4] Dipartimento di Fisica, Università di Napoli Federico II and INFN Sezione di Napoli, Italy
[5] INFN Gruppo Collegato di Messina and Dipartimento di Fisica e Scienze della Terra, Università di Messina, Italy
[6] Facoltà di Ingegneria e Architettura, Università Kore, Enna, Italy
[7] Institute de Physique Nucléaire d'Orsay, Orsay, France
[8] National Autonomous University of Mexico, Mexico City, Mexico
[†] deceased;

E-mail: `dellaquila@na.infn.it`

Abstract. We describe the results of a new experiment aimed to investigate the possible existence of cluster structures in ^{10}Be and ^{16}C isotopes. They have been investigated at the FRIBs facility of INFN-LNS by means of an invariant mass analysis on correlated projectile breakup fragments carried out with the CHIMERA 4π detector. From the analysis of the ^{6}He+^{4}He channel we found evidence of a new state in ^{10}Be at 13.5MeV excitation energy. Concerning ^{16}C, we investigated ^{6}He+^{10}Be correlated fragments and we found a non-vanishing yield at about 20.5MeV in the corresponding excitation energy spectrum. Finally, we describe few details of a new experiment performed at the FRIBs facility where the CHIMERA detector was coupled to the FARCOS hodoscope, with the aim to improve the presently obtained results.

1. Introduction

The clusterization of light nuclei into α particles usually characterizes the structure of self-conjugated nuclei [1]. It can be imagined as the tendency of nucleons to create bounded sub-units made of quartet of nucleons, due to the stability of α particles. These phenomena are governed by the residual interaction inside nuclei and, for this reason, they are important to understand the properties of nuclear forces [2]. α clustering in nuclei can gain an important role also in Nuclear Astrophysics; for example the existence of cluster states in light nuclei can determine different scenarios in the nucleosynthesis of elements in stars (see for example the case of Hoyle state in ^{12}C [3] or the ^{20}Ne [4, 5, 6]).

IOP Conf. Series: Journal of Physics: Conf. Series **863** (2017) 012029 doi:10.1088/1742-6596/863/1/012029

Cluster structures can also occur in neutron-rich nuclei. These states are generally characterized by large deformation and therefore they can give rise to rotational bands. The extra neutrons can act as sort of covalent particles between α-cluster centres, increasing the stability of the structure and, for this reason, they are usually called *valence neutrons* [7]. Important examples are Be and C isotopes, giving rise respectively to possible dimeric or chain structures [7].

A very interesting beryllium isotope is the ^{10}Be. For this nucleus, theoretical calculations based on the Antisymmetrized Molecular Dynamics model (AMD) predicted the existence of a state close to the helium emission threshold at 6.18MeV (0^+) with a pronounced dimeric structure [8]. The existence of this state has been confirmed by several experiments and a rotational excitation of this highly deformed state has been pointed out at 7.54MeV (2^+), indicating its molecular nature. Recently the existence of a state at about 10.2MeV (4+) has been reported [9], indicating the continuation of this band but, to date, no further members of this band have been suggested in the literature. Ambiguities in the determination of the ground state band are also present in the literature [10].

The carbon isotopic chain has been the subject of several experiments aimed to the study of possible cluster configuration. A study of cluster states in the proton-rich ^{11}C nucleus has been recently published [11]. Resonant elastic experiments have been reported on ^{13}C [12, 13, 14, 15] and ^{14}C [16] nuclei. Anyway, despite the recent theoretical predictions of ref. [17] on the existence of molecular states, very few experimental information are reported for the ^{16}C isotope, for which data at low statistics are published [18, 19].

2. Experimental results

To investigate ^{10}Be and ^{16}C structure we have carried out an experiment at the FRIBs facility of the INFN-Laboratori Nazionali del Sud. A fragmentation cocktail beam (mainly constituted by 10^5 pps ^{16}C at 49.5MeV/u and 4×10^4 pps ^{10}Be at 56MeV/u), characterized by a particle-by-particle tagging system [20], was used to induce sequential breakup reactions on polyethylene

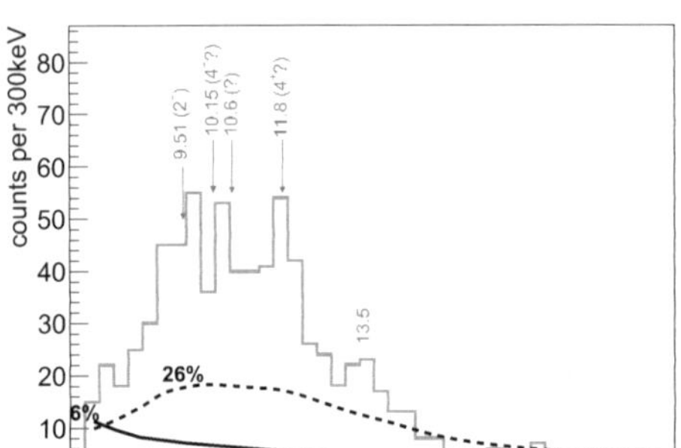

Figure 1. ^{10}Be invariant mass spectrum ($E_{rel} + E_{thr}$) for the ^{4}He+^{6}He break-up channel. The dashed and solid lines represent, respectively, the simulated detection efficiency for inelastic scattering on proton, peaking at 26%, or carbon, peaking at 6%. Arrows indicate the position of known states in the literature.

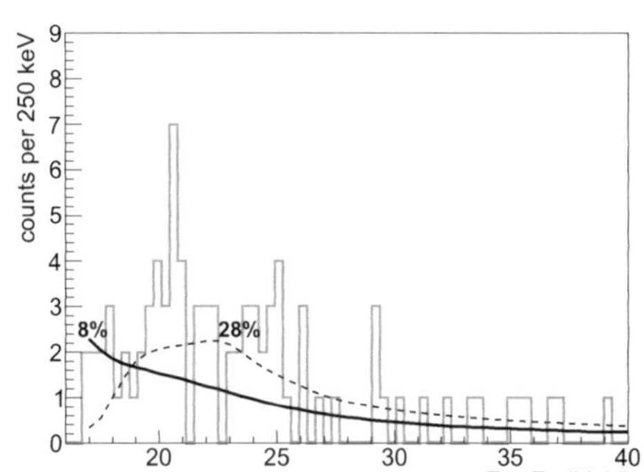

Figure 2. ^{16}C invariant mass spectrum ($E_{rel}+E_{thr}$) for the ^{10}Be+^{6}He break-up channel. The dashed and solid lines represent, respectively, the simulated detection efficiency for inelastic scattering on proton, peaking at 28%, or carbon, peaking at 8%.

IOP Conf. Series: Journal of Physics: Conf. Series **863** (2017) 012029 doi:10.1088/1742-6596/863/1/012029

target (CH_2). Correlation between couples of breakup fragments are used to provide information on the excitation energy of the projectile prior to decay by means of the invariant mass technique [21]. To identify and track particles and fragments we used the CHIMERA 4π multi-detector; detailed information about the functioning and the capabilities of this device can be found in Refs. [22, 23, 24, 25, 26, 27, 28].

The structure of ^{10}Be has been studied by inspecting the ^{4}He+^{6}He invariant mass spectrum ($E_{rel} + E_{thr}$), as described in [29]. The corresponding excitation energy spectrum is reported in the Fig. 1, where arrows are used to indicate the positions of known states in the literature. Peaks, even with poor statistics, are visible in figure, compatible with the presence of states already established in the literature. A further peak is present at about 13.5MeV. This peak is not compatible with detection efficiency effects, as confirmed by the solid and dashed lines (which represent the efficiency curves for excitation induced, respectively, by carbon or hydrogen inside the target), and therefore could be attributed to the existence of a new state in ^{10}Be, as discussed in more details in [30, 31].

A non-vanishing yield is also reported in the ^{6}He+^{10}Be excitation energy spectrum, as shown on Fig. 2, at about 20.5MeV. In analogy to the previous case we could attribute this peak to the population of a new state in ^{16}C. Anyway, in this case the statistics is very limited and this enhancement could be also compatible with a non resonant phase space component.

3. Preliminary results from the CLIR experiment and perspectives

Following the suggestions previously reported, we have recently carried out a new experiment at the FRIBs facility. In this case we used the FARCOS correlator [32, 33] covering the most forward part of the CHIMERA device. FARCOS is based on 3-stages high granularity telescopes made of a 300μm DSSSD, a 1500μm DSSSD and a CsI scintillator based layer. In order to obtain the particle tracking we have firstly developed specific algorithms for the track recognition in the DSSSD detectors. To benchmark the developed algorithms and the data reduction procedure we have studied α-α correlations for two different nuclear systems: ^{16}O+^{11}B and ^{16}O+^{12}C at 55MeV/u. The corresponding correlation functions are reported in the Fig. 2, respectively with the blue and red lines. The result is very nice, indicating the presence of a narrow peak at $\approx$ 92keV relative energy related to the ground state of ^{8}Be, and a broad bump at about 3MeV, evidence of the first excited state in ^{8}Be. Finally, a further small peak is present at $\approx$ 600keV, compatible with the ghost peak due to the ^{9}Be decays, as described for example in [34].

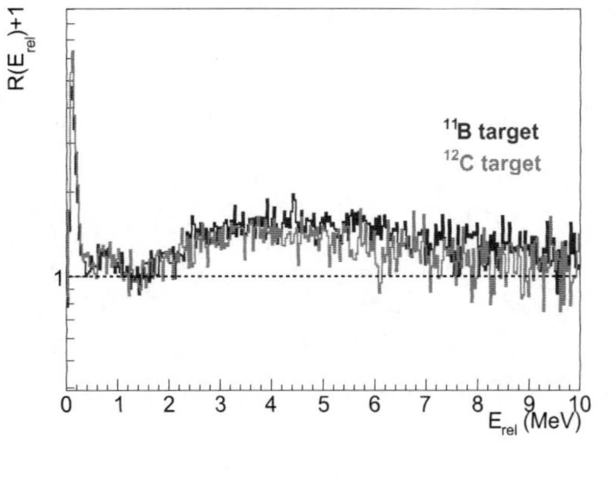

Figure 3. α-α correlation function ($R(E_{rel} + 1)$) in logarithmic scale for two different nuclear collisions ^{16}O+^{11}B (blue line) and ^{16}O+^{12}C (red line) at 55MeV/u. The dashed line represents the line $R(E_{rel}) = 1$.

In the near future we plan to extend the data analysis to the part of the experiment carried out with in-flight beams. We will benefit of the 4π coverage of CHIMERA and the very high granularity given by FARCOS, with the aim of improving the present spectroscopic information especially for the case of ^{16}C, for which no definitive conclusions can currently be drawn.

References

[1] Beck C 2013 *Clusters in Nuclei* vol 1,2,3 (Heidelberg: Springer)
[2] von Oertzen W 1997 *Zeit. Phys.* A **357** 355
[3] Kirsebom O *et al* 2012 *Phys. Rev. Lett.* **108** 202501
[4] Cognata M L *et al* 2015 *Astr. Jour.* **805** 128
[5] Lombardo I *et al* 2013 *J. Phys.* G **40** 125102
[6] Lombardo I *et al* 2015 *Phys. Lett.* B **748** 178
[7] von Oertzen W, Freer M and Kanada-En'yo Y 2006 *Phys. Rep.* **432** 43
[8] Kanada-En'yo Y 1998 *J. Phys.* G **24** 1499
[9] Freer M *et al* 1999 *Phys. Rev. Lett.* **82** 1383
[10] Suzuki D *et al* 2013 *Phys. Rev.* C **87** 054301
[11] Lombardo I *et al* 2016 *J. Phys.* G **43** 45109
[12] Milin M and von Oertzen W 2002 *Eur. Phys. J.* A **14** 295
[13] Freer M *et al* 2011 *Phys. Rev.* C **84** 034317
[14] Lombardo I *et al* 2013 *Nucl. Instrum. Meth. Phys. Res.* B **302** 19
[15] Lombardo I *et al* 2014 *J. Phys. Conf. Series* **569** 012068
[16] Freer M *et al* 2014 *Phys. Rev.* C **90** 054324
[17] Baba T, Chiba Y and Kimura M 2014 *Phys. Rev.* C **90** 064319
[18] Leask P 2001 *Jour. Phys. G: Nucl. Part. Phys.* **27** B9–B14
[19] Ashwood N I *et al* 2004 *Phys. Rev.* C **70** 0644607
[20] Lombardo I *et al* 2011 *Nucl. Phys. B Proc. Suppl.* **215** 272
[21] Grenier F *et al* 2006 *Nucl. Phys.* A **764** 233
[22] Pagano A 2012 *Nucl. Phys. News* **22** 25
[23] DeFilippo E and Pagano A 2014 *Eur. Phys. J.* A **50** 32
[24] Russotto P *et al* 2015 *Phys. Rev.* C **91** 014610
[25] DeFilippo E *et al* 2009 *Acta. Phys. Pol.* B **40** 06011
[26] Lombardo I *et al* 2010 *Nucl. Phys.* A **834** 458c
[27] Cardella G *et al* 2012 *Phys. Rev.* C **85** 064609
[28] Cardella G *et al* 2015 *Nucl. Instr. and Meth.* A **799** 64
[29] Dell'Aquila D *et al* 2016 *EPJ Web of Conf.* **117** 06011
[30] Dell'Aquila D *et al* 2016 *Phys. Rev.* C **93** 024611
[31] Dell'Aquila D 2016 *Nuov. Cim.* C **39** 272
[32] Verde G *et al* 2013 *J. Phys. Conf. Series* **420** 0112158
[33] Pagano E V *et al* 2016 *EPJ Web of Conf.* **117** 10008
[34] Ahmed S *et al* 2004 *Phys. Rev.* C **69** 024303

Cluster structure effects of projectile in ^{7}Li+^{208}Pb reaction

P K Rath[1], Md Ashaduzzaman[1], E Vardaci[1], A Di Nitto[1], G La Rana[1], F Davide[1], A Pulcini[1], D Quero[1], M Mazzocco[2], D Pierroutsakou[3], M Cinausero[4], G Prete[4], M La Commara[1], C Parascandolo[3], M Romoli[3] and N Gelli[5]

[1]Dipartimento di Fisica "E. Pancini", University of Naples "Federico II" and INFN, I-80126 Napoli, Italy.
[2]Dipartimento di Fisica e Astronomia, Università di Padova and INFN - Padova, Padova, Italy.
[3]INFN - Napoli, Napoli, Italy.
[4]INFN - Laboratori Nazionali di Legnaro, Legnaro (Padova), Italy.
[5]INFN - Firenze, Firenze, Italy.

E-mail: vardaci@na.infn.it, rath@na.infn.it

Abstract. Inclusive and exclusive cross sections have been measured for different projectile breakup channels. It is observed that the inclusive cross sections are larger than the exclusive ones. It is found that other processes, besides the direct breakup channel, significantly contribute to the inclusive channels. A clear α+x cluster structure has been found from coincidence data.

1. Introduction

The study of the reaction mechanisms induced by weakly bound stable nuclei of well known cluster structure like ^{6,7}Li and ^{9}Be is of current great interest. The Coulomb break-up of accelerated projectiles (^{7}Li in the present case) into (α+x) particles in the field of a heavy target nucleus is considered as an important phenomenon observed experimentally. Clustering is one of the most important ingredient of the nuclear many body problem [1]. Owing to the recent theoretical and experimental developments, the domain of the cluster studies is rapidly expanding toward highly excited and neutron rich nuclei [2]. Clustering is a general phenomenon [3] not only observed in light neutron-rich nuclei [4], but also in halo nuclei such as ^{11}Li [5] or ^{14}Be, for instance [6].

In the case of weakly bound nuclear systems (nuclei that have a larger breakup probability due to the low breakup thresholds), correlation among nucleons and pairing are manifested among others as an emergence of strong clustering and exotic shapes. In light α-like nuclei, clustering is observed as a general phenomenon at high excitation energy close to the α decay thresholds [7]. This has renewed interest in understanding the role of cluster states in nuclear synthesis [8]. The search for resonant structures in the excitation functions for various combinations of light α cluster (N = Z) nuclei in the energy regime from the Coulomb barrier up to regions with excitation energies of Ex =20-50 MeV remains a subject of contemporary debate [9]. Studying reactions induced by weakly bound nuclei has important outcomes for the cases of radioactive ion beams [10, 11, 12].

In this contribution, we report the data on the production of α particles in ^{7}Li induced reaction on the heavy target ^{208}Pb at near barrier energies. The main objective was to identify the reaction mechanisms responsible for the α particles production in the case of the use of a weakly bound projectile.

2. Experimental Details

The experiment was performed using a ^{7}Li beam delivered by the Laboratori Nazionali di Legnaro Tandem Van de Graaff accelerator having the following beam eneries 31,33,35 and 39 MeV. Beam currents ranged between 5 to 10 nA. A self-supporting ^{208}Pb target of thickness 200 μg/cm^2 was used. Light charged particles were detected with the 4π array (8πLP) set up described detail in [13, 14, 15, 16, 17, 18, 19, 20, 21]. The array is essentially made of two parts: a WALL in forward direction, covering the polar angles 2.5^0 to 34^0 and a BALL part covering the angles up to 163^0. The BALL consists of 126 telescopes. Each telescope consists of a Silicon surface barrier detector as ΔE, (300 μm thick), and CsI(TI) scintillator as E (5 mm thick). The angles from 34^0 to 163^0 are divided into 7 rings A, B, C, D, E, F & G. The WALL is a matrix of 11$\times$11 telescope (the four at the corners and the central one for the exit hole of the beam are missing). This system allows for a very good identification of light charged particles: α, t, d and proton. ^{7}Li was completely stopped in the ΔE in our energy range. The data acquisition [22] has arranged to record for each telescope the Time Vs ΔE and ΔE Vs E_{res} matrices for particle identification. A typical 2D spectrum measured by a telescope at E_{lab} = 33 MeV is shown in Fig.1. One can observe different types of particles (p, d, t and αparticle) detected in ΔE vs. E_{tot} spectra (Fig.1b) are very clearly separated. Fig.1a Shows ΔE-T spectrum for the same energy. One can see from Fig.1a that the elastically scattered ^{7}Li is not observed in the ΔE-E_{tot} spectra as it is stopped in the ΔE (silicon detector).

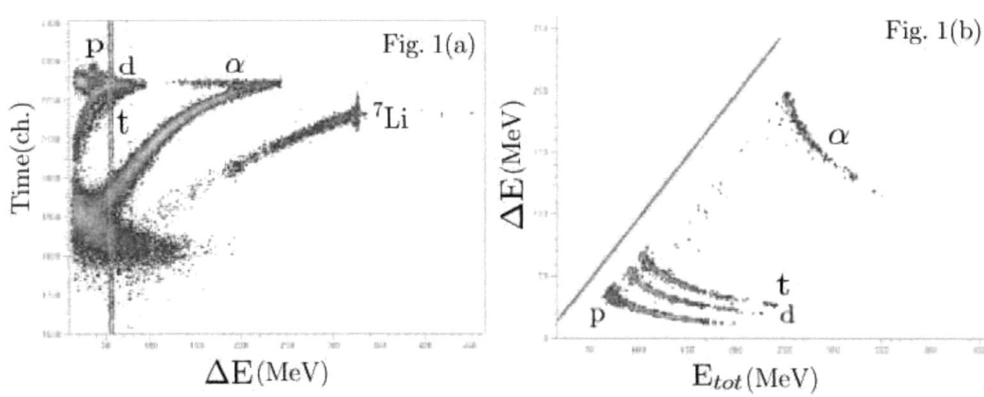

Figure 1. Experimental 2D particle spectra for ^{7}Li+^{208}Pb reaction.(a)ΔE Vs Time spectra at 33 MeV. The telescope was at forward angle (θ =20.6^0 and ϕ =346.53^0). All the particles are clearly observed including the elastic ^{7}Li. (b) same as (a) spectra of ΔE Vs E_{tot} . One can observe that the yields are less and also there is no elastic peak in Fig.1(b) as many particles (α,d,t...) were stopped in the ΔE detector.

3. Results and Discussion

We detected different types of particles emitted during the reaction process (both in singles and coincidence mode) around the Coulomb barrier energies same as in [23, 24]. Experimental 2D spectra are shown in Fig.1. Particles were identified by two methods: 1)using energy loss information from ΔE and Eres for each telescope, 2) ΔE and Time. The time is a mixture of time of flight (TOF) and pulse shape discrimination rise time. From the above figure one can observe that different types of particles (p, d, t and α) have been detected including elastically scattered ^{7}Li. The different particles give indication of the presence of different types of processes involved in the reaction. The case of ^{7}Li is however more interesting as ^{7}Li has one bound excited state below the breakup threshold where as ^{6}Li has none. Coincidences were measured to gain insight on this point. The different channels were identified by making a coincidence between two observed fragments. The particles have different origins: 1) as the direct breakup of ^{7}Li from its resonance state at 4.63MeV into α and triton (t), 2) pickup of a neutron (very unlikely) to become ^{8}Be which breaks into (α+α) or 3) stripping of a neutron which produces ^{6}Li which afterward will break into α + deuteron(d). There was no neutron detector in the whole setup so no measurement was done using neutron in coincidence.

Deuterons and tritons of different energies are clearly separable and are expected to belong to the different types of sources as described above. The inclusive and exclusive angular distributions for the different coincidence measurements are shown in Fig.2. The exclusive differential cross sections were obtained for the following cluster channels (α+α),(α+d) and (α+t).

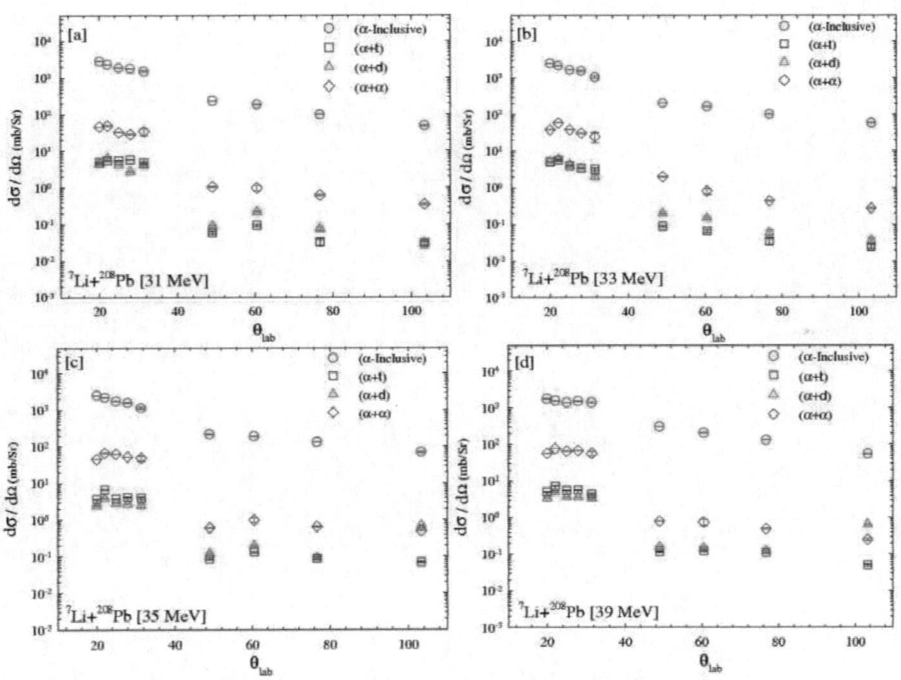

Figure 2. Inclusive and exclusive differential cross sections extracted from the coincidence data for different energies. (a) The differential angular distribution for 31 MeV. (b),(c) and (d) same as (a) only at different energy 33, 35, 39 MeV (see the text for details).

From the above figure one can clearly observe that the inclusive cross section is much larger than the exclusive cross section. This implies not only that the direct breakup channel is important but that processes besides breakup contribute significantly. The same type of conclusion was also reported in [24]. Quantitatively, about two orders of magnitude are missing. We can conclude that other reaction mechanisms which provides significant amount of α to the inclusive cross section need further study.

References

[1] Beck C *et al* 2011 *Acta Phys. Pol.* B **42** 34
[2] Yamada T *et al* 2010 Clusters in Light Nuclei Vol.2 (Springer-Verlag, Berlin)
[3] von Oerzten W and Milin M 2013 Clusters in Nuclei - Vol.3, ed. Beck C, Lecture Notes in Physics
[4] Kanada-Enyo Y and Kimura M 2010 Clusters in Nuclei - Vol.1, ed. Beck C, Lecture Notes in Physics 818 129
[5] Ikeda K et al 2010 Clusters in Nuclei - Vol.1, ed. C. Beck, Lecture Notes in Physics 818 165
[6] Nakamura N and Kondo Y 2012 Clusters in Nuclei - Vol.2, ed. Beck C, Lecture Notes in Physics 848 67
[7] Beck C 2013 J of Phys.: Conf. Series 436 012014
[8] Horiuchi H, Ikeda K and Kato K 2012 *Prog. Theor. Phys. Suppl.* 192 1
[9] Beck C *et al* 1994 *Phys. Rev.* C **49** 2618
[10] Mazzocco M *et al* 2007 *Eur. Phys. J. Special Topics* **150** 363
[11] Signorini C *et al* 2010 *Eur. Phys. J.* A **44** 63
[12] Mazzocco M *et al* 2008 *Nucl. Instr. Meth.* B **266** 4665
[13] Prete G *et al* 1999 *Nucl.Inst. Meth.* A **422** 263
[14] Saxena A *et al* 2002 *Phys. Rev.* C **65** 064601
[15] Vardaci E *et al* 2003 *Phys.At. Nucl.* **66** 1182
[16] La Rana G *et al* 2003 *Eur. Phys. J* A **16** 199
[17] Vardaci E *et al* 2010 *Eur. Phys. J.* A **43** 127
[18] Nadtochy P *et al* 2010 *Phys. Lett.* B **685** 258
[19] Di Nitto A *et al* 2011 *Eur. Phys. J.* A **47** 83
[20] Vardaci E *et al* 2015 *Phys. Rev.* C **92** 034610
[21] Di Nitto A *et al* 2016 *Phys. Rev.* C **93** 044602
[22] Ordine A *et al* 1998 *IEEE Trans. Nucl. Sci.* **45** 873
[23] Signorini C *et al* 2003 *Phys. Rev.* C **67** 044607
[24] D.H. Luong *et al* 2011 *Phys. Lett.* B 695 105

IOP Conf. Series: Journal of Physics: Conf. Series **863** (2017) 012031 doi:10.1088/1742-6596/863/1/012031

Coulomb breakup of the Borromean nucleus ^{22}C

E C Pinilla[1] **and P Descouvemont**[2]

[1] Universidad Nacional de Colombia, Sede Bogotá, Facultad de Ciencias, Departamento de Física, Grupo de Física Nuclear, Carrera 45 N° 26-85, Bogotá D. C., Colombia.

[2] Physique Nucléaire Théorique et Physique Mathématique, Université Libre de Bruxelles (ULB), Brussels, Belgium.

E-mail: ecpinillab@unal.edu.co

Abstract. We study the Coulomb breakup cross section of the ^{22}C nucleus and its relation with three-body resonances and ground state properties. Cross sections are computed with a Coulomb corrected four-body eikonal model and they are constructed by fixing the continuum state and varying the ground state of ^{22}C. These states are calculated by using ^{20}C$-n$ potentials that provide different scattering lengths and resonance energies, of a presumed virtual state and of a d resonance in ^{21}C. We find that the breakup cross section is strongly dependent on the ground state properties and we predict the existence of a 2^+ narrow resonance in ^{22}C.

1. Introduction

The ^{22}C nucleus is the heaviest Borromean nucleus known so far. Its deduced rms radius, r_{rms}, and two neutron separation energy, S_{2n}, present large uncertainties ($r_{rms} = 5.4 \pm 0.9$ fm [1] and, $S_{2n} = 0.42 \pm 0.94$ MeV [1] and $S_{2n} < 300$ keV [2]). As Borromean nuclei can been seen as a core surrounded by two neutrons, three-body models are used to study such as nuclei.

In three-body models, two-body potentials are needed. Those are obtained from experimental information of elastic scattering or by reproducing spectroscopic data. ^{21}C is unbound and little experimental spectroscopic information is known. Recently, Mosby et al. [3] claim that if a ^{21}C virtual state exists, its scattering length should be limited to $|a_0| < 2.8$ fm. In Ref. [4] the recently measured ^{21}C excitation spectrum shows a presumed d resonance.

An important property of halo nuclei is the enhancement of their Coulomb breakup cross sections at low excitation energies. Some theoretical calculations relate this enhancement with a 1^- three-body resonance [5, 6] and others with the weak binding of the ground state [7, 8] of those nuclei.

The goal of the present paper is to study if a possible enhancement in the Coulomb breakup cross section of ^{22}C, for some suitable conditions, is related with a resonance behavior or with its ground state properties. With this aim, we choose three sets of ^{21}C potentials that reproduce different scattering lengths and resonance energies of ^{21}C. In combination with a fixed $n-n$ interaction, these sets provide different ground and continuum states. We fix the continuum state and vary the ground state to observe its effect on the breakup cross sections.

2. Three-body phase shifts and breakup cross sections

Breakup cross sections are computed with the four-body eikonal Coulomb corrected model of Ref. [9] (see Ref. [9] and Ref. [6] for details). This model uses three-body wave functions in hyperspherical coordinates. Bound states are computed variationally and continuum states are computed with the correct asymptotic behavior. We follow Ref. [10] and Ref. [11], respectively.

We need core$-n$ and $n-n$ potentials to compute ground and continuum states of ^{22}C. As $n-n$ potential, we use a Minnesota central interaction with exchange parameter $u=1$ [12]. The ^{20}C$-n$ potentials have the form

$$V_{^{20}\mathrm{C}+n}(r) = -V_0^l f(r) + V_{ls}\boldsymbol{l}\cdot\boldsymbol{s}\frac{1}{r}\frac{d}{dr}f(r), \tag{1}$$

with $f(r)=1/\left[1+\exp(\frac{r-R_c}{a})\right]$, $a=0.65$ fm, $R_c=3.393$ fm, $V_{ls}=35$ MeV and $V_0^{l\neq 2}=42$ MeV. The depths $V_0^{l=0}$ and $V_0^{l=2}$ are given in Table 1. These depths are chosen to provide different scattering lengths of a $1s_{1/2}$ virtual state and $0d_{3/2}$ resonance energies of ^{21}C. The three potentials provide a single bound state in each shell model orbit, which is removed by a supersymmetric transformation to approximately take into account the Pauli principle.

Table 1. Potential depths $V_0^{l=0}$ and $V_0^{l=2}$ of ^{20}C $-\ n$ potentials providing different scattering lengths, a_0, and resonance energies, E_R, of ^{21}C. E_0 notates the ^{20}C$+n+n$ ground state energy.

	$V_0^{l=0}$ (MeV)	$V_0^{l=2}$ (MeV)	E_0 (MeV)	a_0 (fm)	E_R (MeV)
set 1	29.8	47.8	−0.10	− 2.8	0.83
set 2	29.8	48.4	−0.47	− 2.8	0.59
set 3	33.0	47.5	−0.46	−47.6	0.92

Three ground states with each set and a single continuum wave function are computed. We use set 1 to get the continuum state. We calculate the 1^-, 2^+ and 0^+ continuum partial waves, which provide the prominent contributions to the breakup cross sections. Important convergence parameters for computing the wave functions are the maximum hypermomentum, K_{max}, and the number of basis functions, N. As a basis we use the Lagrange-Legendre [13] with $a=90$ fm and $N=100$. We use $K_{\mathrm{max}}=40$ for the ground state and $K_{\mathrm{max}}=30,25,20$ for the $J^\pi=0^+,1^-,2^+$ continuum partial waves.

The breakup of ^{22}C on a ^{208}Pb target is studied at 240 MeV/nucleon. For the breakup calculations, we need core-target and $n-$target potentials. The ^{20}C$-^{208}$Pb potential is taken from Ref. [14] by interpolating the α_{NN} and σ_{NN} parameters. The $n-^{208}$Pb potential is given in Ref. [15].

Figure 1 shows breakup cross sections associated with the three sets and their respective 1^- partial contribution. In general, the cross sections are dominated by their dipole part. However, the cross section calculated with set 2 differs from its dipole contribution showing a bump at low excitation energies. This is related with the non orthogonality of the ground state and the 0^+ continuum partial wave.

The cross section for the weakest bound state exhibits the larger enhancement at low excitation energies. Breakup cross sections computed with set 2 and set 3 are very different, although their corresponding ^{22}C ground states have similar energies. Therefore, we can observe clearly a large effect of the two-body properties, a_0 and E_R, on the ^{22}C excitation spectrum.

In Fig. 2 we present ^{20}C$+n+n+n$ phase shifts computed by following Ref. [11] for the 1^-, 2^+ and 0^+ continuum partial waves. We observe a 2^+ narrow resonance around 0.8 MeV. This is in agreement with the peak structure that shows up in the breakup cross sections, independently of the ground state used. In contrast, we cannot relate a 1^- three-body resonance with the enhancement at low excitation energies of the breakup cross sections.

IOP Conf. Series: Journal of Physics: Conf. Series **863** (2017) 012031 doi:10.1088/1742-6596/863/1/012031

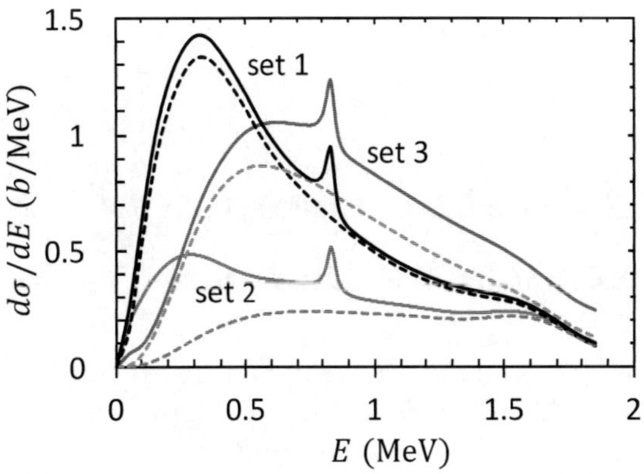

Figure 1. (Color online) Total (solid lines) and dipole (dashed lines) breakup cross sections of ^{22}C with different core$-n$ potentials.

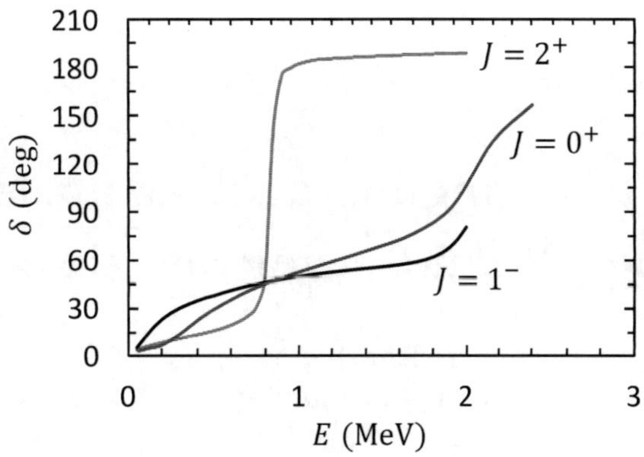

Figure 2. (Color online) Three-body ^{20}C + $n + n$, $J = 0^+$, $J = 1^-$ and $J = 2^+$ phase shifts computed with set 1.

3. Conclusions

We observe a large enhancement of the Coulomb breakup cross section of ^{22}C that is due to the very weakly bound character of the ground state and that is no related with a 1^- three-body resonance. We also find a 2^+ resonance that shows up in the breakup cross section consistent with the three-body phase shift calculation. Of course these results depend on the ^{21}C and ^{22}C properties assumed. However, the existence of a ^{22}C 2^+ resonance seems well established.

Acknowledgments

E.C.P. is financed by the Fondo Nacional de Financiamiento para la Ciencia, la Tecnologia y la Innovación Francisco José de Caldas and the Universidad Nacional de Colombia, Colombia. P.D. acknowledges the support of F.R.S.-FNRS, Belgium.

References

[1] Tanaka K et al 2010 *Phys. Rev. Lett.* **104** 062701
[2] Gaudefroy L et al 2012 *Phys. Rev. Lett.* **109** 202503
[3] Mosby S et al 2013 *Nucl. Phys.* A **909** 69
[4] Enyo H 2015 *Nucl. Phys. News* **25** N° 3 10
[5] Descouvemont P, Pinilla E C and Baye D 2012 *Prog. Theor. Phys. Suppl.* **196** 1
[6] Pinilla E C, Descouvemont P and Baye D 2012 *Phys. Rev.* C **85** 054610
[7] Suzuki Y, Lovas R G, Kabana K and Varga K 2003 *Structure and Reactions of Light Exotic Nuclei* (London: Taylor and Francis).
[8] Nagarajan M, Lenzi S and Vitturi A 2003 *Eur. Phys. J.* A **24** 63
[9] Baye D, Capel C, Descouvemont P and Suzuki Y 2009 *Phy. Rev.* C **79** 024607
[10] Descouvemont P, Daniel C and Baye D 2003 *Phys. Rev.* C **67** 044309
[11] Descouvemont P, Tursunov E M and Baye D 2006 *Nucl. Phys.* A **765** 370
[12] Thompson D R, LeMere M and Tang Y C 1977 *Nucl. Phys.* A **286** 53
[13] Baye D 2015 *Phys. Rep.* **565** 1
[14] Bertulani C A, Hussein M S and Münzenberg G 2001 *Physics of Radioactive Beams* (New York: Nova Science)
[15] Koning A J and Delaroche J P 2003 *Nucl. Phys.* A **713** 231

Disentangling unclear nuclear breakup channels of beryllium-9 using the three-axis Dalitz plot

R Smith[1], M Freer[1], C Wheldon[1], N Curtis[1], S Almaraz-Calderon[2], A Aprahamian[2], N I Ashwood[1], M Barr[1], B Bucher[2], P Copp[3], M Couder[2], X Fang[2], G Goldring[4], F Jung[2], Tz Kokalova[1], S R Lesher[3], W Lu[2], J D Malcolm[1], A Roberts[2], W P Tan[2] and V A Ziman[1]

[1]School of Physics and Astronomy, University of Birmingham, Edgbaston, B15 2TT, UK.
[2]Institute for Structure and Nuclear Astrophysics, Department of Physics, University of Notre Dame, Notre Dame, IN 46556, USA.
[3]Department of Physics, University of Wisconsin - La Crosse, La Crosse, WI 54601, USA.
[4]Department of Particle Physics, Weizmann Institute, 76100 Rehovot, Israel.

E-mail: `r.smith.3@pgr.bham.ac.uk`

Abstract. The three-axis Dalitz plot has been applied to the breakup of a nucleus into unequal mass fragments for the first time. The Dalitz plot allows clear identification of the various breakup channels of ^{9}Be $\to 2\alpha + n$ process. The method has allowed the branching ratio for the 6.38 MeV level in ^{9}Be to be provisionally calculated when examining the ^{9}Be(^{4}He,α)$\alpha\alpha n$ reaction. The effects of non-uniform angular distributions on the Dalitz plot must still be properly investigated along with the effects of contaminant reaction channels. It is proposed that this method could be used to determine the breakup branching ratio of a newly-measured level in this nucleus.

1. Introduction

A *ternary plot* is a barycentric plot of three quantities, which sum to a constant value [1]. Using the ternary plot representation, the three quantities are visualised simultaneously by a two-dimensional position on an equilateral triangle. While working at the University of Birmingham in 1953, R. H. Dalitz was the first person to apply the ternary plot technique to visualising the three-body decay of sub-atomic particles [2]. Therefore, the ternary plot will henceforth be referred to as a *Dalitz plot* in this context. In the centre of mass (CoM) of a decay $X \to a+b+c$, the energies of a, b and c must sum to the decay Q value (in the non-relativistic limit). The Dalitz plot technique is widely used in high energy physics [3].

Only in recent years has the technique been applied in nuclear physics. Such Dalitz plots are used to determine if the Hoyle state in ^{12}C decays through a ^{8}Be$_{g.s.}$ intermediate level or through a direct three-body process [4]. Further, interference patterns in these Dalitz plots are used to determine the angular momentum and parity of ^{12}C resonances [5]. There is no reason why it cannot also be applied to breakups into unequal mass fragments such as ^{9}Be $\to \alpha+\alpha+n$.

Measuring the breakup branching ratios for states in ^{9}Be is important, since the breakup through particular channels is correlated with the strength of the corresponding cluster

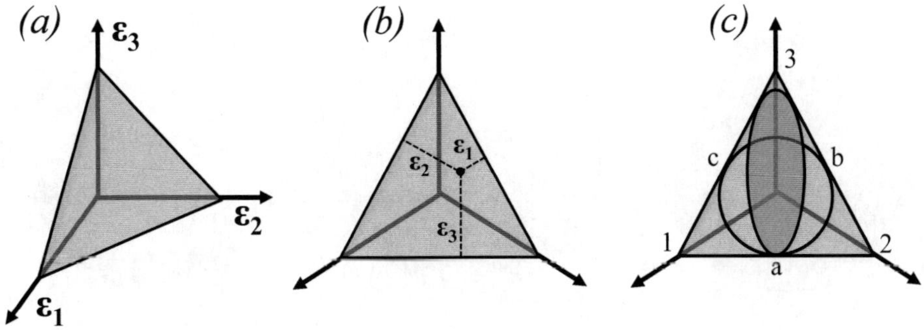

Figure 1. Construction of the Dalitz plot for a decay of a nucleus into three particles. ϵ_i corresponds to the fractional energy of particle i in the CoM of the decaying system. (a) illustrates that if $\sum \epsilon_i$ = constant then points must lie on the shaded triangle. (b) shows the same plot from a different perspective and (c) illustrates how the allowed area changes from a circle to an ellipse for decays into unequal mass fragments (see text for details).

configurations in the wave functions, which can have implications for astrophysical reaction rates [6,7]. The n + ^{8}Be$_{g.s}$, n + ^{8}Be$_{2^+}$, α + ^{5}He$_{g.s}$ and $\alpha + \alpha + \alpha$ channels are expected to dominate for the low-lying levels. This paper demonstrates, through the use of Monte-Carlo simulations and experimental data, that the three-axis Dalitz plot can be applied to analyse the breakup systematics of the ^{9}Be $\to \alpha + \alpha + n$ system.

2. The Dalitz plot

Figure 1 illustrates how to plot three quantities (which sum to a constant) on a single 2D plot. In this case, ϵ_i corresponds to the fractional energy of particle i in the CoM of the decaying system ($\epsilon_i = E_i / \sum_{j=1}^{3} E_j$). Figure 1 ($a$) shows that by plotting three quantities in 3D cartesian coordinates, the resulting point must lie on the shaded triangle. By shifting the viewing perspective along a line normal to the plane of the triangle and passing through the origin, figure 1 (b) is extracted. A point on the triangle is uniquely defined by ϵ_1, ϵ_2 and ϵ_3 which are the distances of each point from each edge of the triangle. If particles 1, 2 and 3 are of equal mass, conservation of linear momentum ensures that points can only lie within the circle of figure 1 (c). For unequal mass fragments, the circle becomes an inscribed ellipse. The points where the circle or ellipse touch the edge of the triangle are denoted by a, b and c and are fixed such that 2b:b3 equals m_2:m_3 (and cyclical for sides 12 and 31). The x, y coordinates of a point on this triangle are related to ϵ_i by $x = \sqrt{3}(\epsilon_2\text{-}\epsilon_1)/2$ and $y = (2\epsilon_3\text{-}\epsilon_2\text{-}\epsilon_1)/2$.

3. Application to beryllium-9 breakup

An experiment was performed to measure the ^{9}Be(^{4}He,α)$\alpha\alpha n$ reaction, at 22 and 26 MeV beam energies (detailed in Ref. [8]). A new resonance was recorded at 3.8 MeV, and the necessity for reliable branching ratio measurements in order to determine the angular momentum of the state, was highlighted. The Dalitz plot method was applied to this experimental data in order to test if it was a valid method of calculating reliable branching ratios. The $7/2^-$ state at 6.38 MeV was chosen as a test subject because branching ratios are already well constrained [6]. The experimental analysis is similar to that of Ref. [8], however, the specific condition that decays proceeded via the ^{8}Be ground state was not required. Hence, ^{9}Be excitation spectra were reconstructed for events which break up via any of the open channels. Events in the excitation region between 6.1 and 6.7 MeV were selected to consider breakups of the 6.38 MeV level.

It was possible to select which final state α particles arise from the breakup of ^{9}Be. The

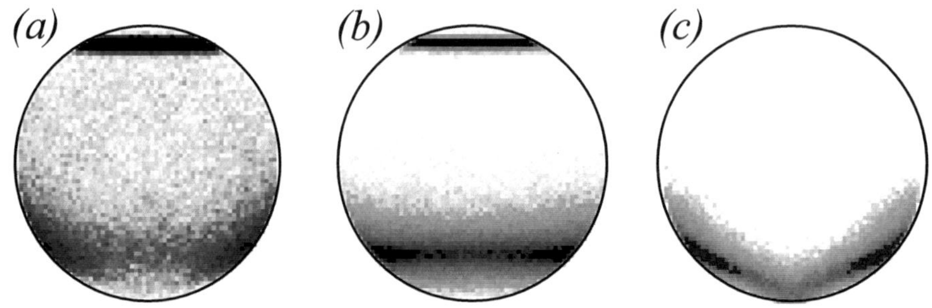

Figure 2. Dalitz plots for the breakup of the 6.38 MeV level in ^{9}Be. Darker regions indicate more counts. (a) depicts the experimental data within an excitation cut of $6.1 < E_x < 6.7$, (b) depicts a simulation of the $^{8}\mathrm{Be}_{g.s}$ (upper line) and $^{8}\mathrm{Be}_{2+}$ (lower line) channels and (c) shows the simulation of decays through the $^{5}\mathrm{He}_{g.s}$ channel. In the case of decays through $^{8}\mathrm{Be}_{g.s}/^{5}\mathrm{He}_{g.s}$, the first emitted n/α takes the majority of the decay energy, hence the event concentrations in these regions (top centre/ bottom left and right, respectively). Sequential decays through the broad $^{8}\mathrm{Be}_{2+}$ state lead to a more even distribution of energies.

energy and momentum of the undetected neutron was calculated from momentum conservation. The momenta of the breakup particles were converted into the CoM frame of the decaying ^{9}Be. Their fractional energies in this frame (ϵ_i) were used as the axes of a Dalitz plot. To force the Dalitz plot into a circular, rather than elliptical, shape, the axes were chosen to be ($4\epsilon_{\alpha_1}$, $4\epsilon_{\alpha_2}$, ϵ_n). The resulting Dalitz plot is shown in figure 2 (a). Simulations of the three decay processes with proper threshold dependencies, experimental resolution, but assuming isotropic particle emission, are shown in figures 2 (b) and 2 (c).

4. Conclusions

Figure 2 (a) shows contributions from all three sequential breakup channels as simulated in figures 2 (b) and 2 (c). Each decay contribution is well separated when compared with similar studies using other types of Dalitz plots [6]. One-dimensional projections of the Dalitz plot permitted a quantitative comparison between simulation and experiment. Provisional branching ratios were extracted which are comparable with previous measurements [6]. The effects of non-uniform angular correlations on each breakup stage and contaminant reaction channels need to be considered in order to be confident of these values. Lower levels in ^{9}Be such as the new 3.8 MeV state are much closer to the decay thresholds and it is predicted that the signatures from each channel will strongly overlap on the Dalitz plot. Their exact manifestations will depend on the angular distributions and it remains to be seen how separable each channel will be.

[1] Brazzle B and Tapp A 2016 *The Physics Teacher* **54(4)** 216
[2] Dalitz R H 1953 *Philos. Mag.* **44** 1068
[3] Alff C, Berley D, Colley D, Gelfand N, Nauenberg U, Miller D, Schultz J, Steinberger J, Tan T H, Brugger H, Kramer P and Plano R 1962 *Phys. Rev. Lett.* **9** 325
[4] Itoh M, Ando S, Aoki T, Arikawa H, Ezure S, Harada K, Hayamizu T, Inoue T, Ishikawa T, Kato K, Kawamura H, Sakemi Y and Uchiyama A 2014 *Phys. Rev. Lett* **113** 102501
[5] Kirsebom O S, Alcorta M, Borge M J G, Cubero M, Diget C A, Dominguez-Reyes R, Fraile L M, Fulton B R, Fynbo H O U, Hyldegaard S *et al* 2010 *Phys. Rev.* C **81** 064313
[6] Brown T A D, Papka P, Fulton B R, Watson D L, Fox S P, Groombridge D, Freer M, Clarke N M, Ashwood N I, Curtis N *et al* 2007 *Phys. Rev.* C **76** 054605
[7] Papka P, Brown T A D, Fulton B R, Watson D L, Fox S P, Groombridge D, Freer M, Clarke N M, Ashwood N I, Curtis N *et al* 2007 *Phys. Rev.* C **75** 045803
[8] Smith R, Wheldon C, Freer M, Curtis N, Kokalova Tz, Almaraz-Calderon S, Aprahamian A, Ashwood N I, Barr M, Bucher B, *et al* 2016 *Phys. Rev.* C **94** 014320

Structure and decay pattern of linear-chain states in neutron-rich Carbon isotopes

T Baba and M Kimura
Department of Physics, Hokkaido University, 060-0810 Sapporo, Japan
E-mail: baba@nucl.sci.hokudai.ac.jp

Abstract. The linear-chain states of ^{14}C are studied by using the antisymmetrized molecular dynamics. The calculated properties of the linear-chain states were compared with the observed data. The results for the π-bond linear-chain states reasonably agree with the observation. It is also shown that the linear-chain states decay to the excited states of ^{10}Be. Hence, we regard that this unique decay pattern is a strong evidence of the linear-chain formation.

1. Introduction

The alpha clustering strongly influences the structure of light nuclei and many different types of α cluster structure have been studied for a long time. The linear-chain (LC) configuration of 3α clusters [1] in which α particles are linearly aligned is a famous and exotic cluster structure. The interest in the linear-chain configuration is further reinforced by the physics of unstable nuclei, because the valence neutrons may stabilize it by their glue-like role [2, 3, 4, 5, 6, 7, 8, 9, 10].

Recently, rather promising candidates of LC configuration in ^{14}C were independently reported by several groups [11, 12]. However, there are several discrepancies between the theory [6] and experiment which must be resolved to establish the LC formation in ^{14}C. For this purpose, we performed AMD calculation. From the obtained wave functions, we estimated the α decay widths of the LC states as well as those of other cluster and non-cluster states.

2. Theoretical Framework

In this study, we uses the Gogny D1S interaction as an effective interaction. We employ the parity-projected wave function Φ^π as the variational wave function,

$$\Phi^\pi = \frac{1+\pi\hat{P}_x}{2}\Phi_{int}, \quad \pi = \pm, \quad \Phi_{int} = \mathcal{A}\{\varphi_1, \varphi_2, ..., \varphi_A\}. \tag{1}$$

Here φ_i is the single particle wave packet which is a direct product of the deformed Gaussian spatial part [13], spin (χ_i) and isospin (ξ_i) parts,

$$\varphi_i(\mathbf{r}) = \exp\left\{-\sum_{\sigma=x,y,z} \nu_\sigma\left(r_\sigma - \frac{Z_{i\sigma}}{\sqrt{\nu_\sigma}}\right)^2\right\} \otimes (a_i\chi_\uparrow + b_i\chi_\downarrow) \otimes (\text{proton or neutron}). \tag{2}$$

The variational parameters $\mathbf{Z_i}$, a_i, b_i and ν_σ are determined by the constrained energy variation. After the energy variation, we perform the angular momentum projection and generator coordinate method (GCM) to obtain the excitation spectrum of ^{14}C. To estimate the α reduced widths, we used an approximate method suggested in Ref. [14].

IOP Conf. Series: Journal of Physics: Conf. Series **863** (2017) 012033 doi:10.1088/1742-6596/863/1/012033

3. Results

3.1. Intrinsic wave functions obtained by the constrained energy variation.

Fig. 1 shows the intrinsic density distributions of ^{14}C obtained by the quadrupole constrained energy variation. The panel (a) shows the the ground state without any clustering. At deformed region, we found several cluster states with different valence neutron configurations. In the oblate deformed region, we found the triangular configuration of 3α particles with valence neutrons in *sd*-shell. In the prolate deformed region, we found two different LC configurations. The panel (c) shows the configuration which we call π-bond LC, while the panel (d) shows the σ-bond LC. As clearly seen in there valence neutron distributions, they have two valence neutrons in the π and σ molecular orbits [2].

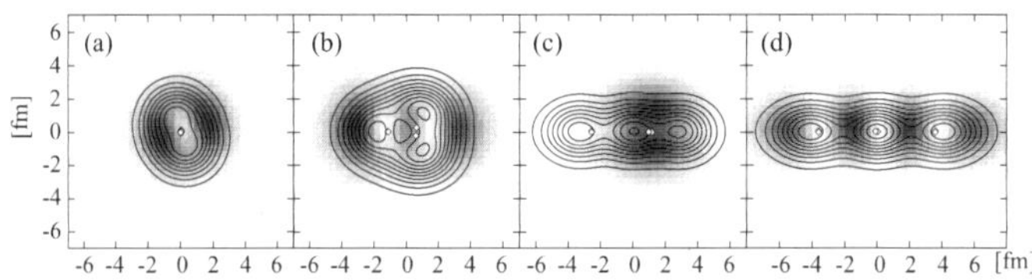

Figure 1. The density distribution of the positive states. The contour show the proton densities, while the color plots show the single particle orbits occupied by the most weakly bound neutron.

By superposing those wave functions and performing GCM calculation, we obtained the excitation spectrum of the positive-parity states shown in Fig. 2. In addition to the ground band, the cluster configurations shown in Fig. 1 (b)-(d) generate three rotational bands. The triangular band is built on the 0^+ state at 7.5 MeV and dominantly composed of the cluster configuration shown in Fig. 1 (b). The linear chain configurations (Fig. 1 (c) and (d)) generate

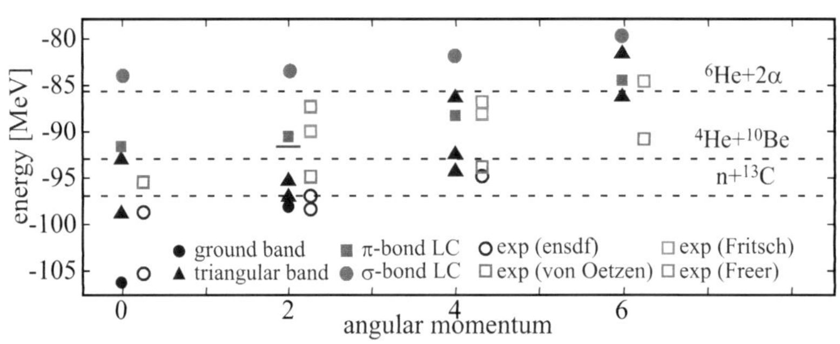

Figure 2. The calculated and observed LC candidates. Open boxes show the observed LC reported by Refs. [4, 11, 12]. Filled symbols show the present result [].

two rotational bands named π-bond and σ-bond LC which are respectively built on the 0^+ states at 14.6 and 22.2 MeV. We see that the energy of the calculated π-bond LC is very close to the candidates observed by the resonant scattering [11, 12], while the states reported by von Oertzen [4] are close to the triangular band that is composed of the configuration shown in Fig. 1 (b).

Figure 3 (a) shows the α reduced widths (RW) for the decays to the ground band of ^{10}Be (0_1^+, 2_1^+ and 4_1^+ states). It is clear that only the π-bond LC has large RW and reasonably agrees with the observed decay width to the ^{10}Be(0_1^+). Since both of the observed and calculated excitation energies and decay widths agree to each other, we consider that the formation of the π-bond LC in ^{14}C is rather convincing. We also note that the present calculation predicts that π-bond LC

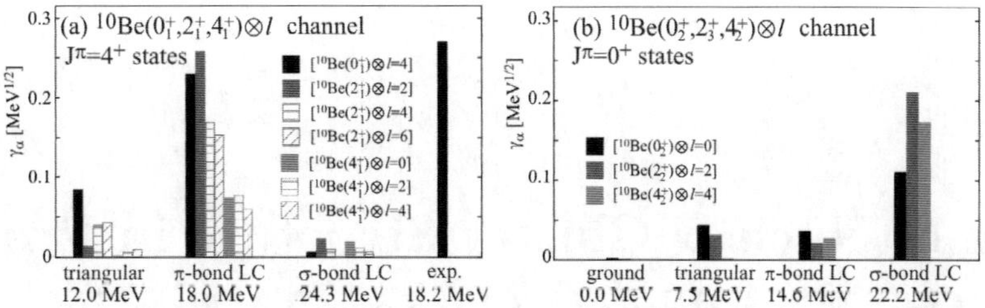

Figure 3. The calculated α-decay reduced widths compared with the observation Ref. [11]. The panel (a) shows the decays of the 4^+ states to the ground band of ^{10}Be. Panel (b) shows the decay of the 0^+ states to the excited band of ^{10}Be.

has larger RW to the ^{10}Be(2_1^+) state than the ^{10}Be(0_1^+) state, which owes to the strong coupling nature of the LC. Therefore, we suggest that the observation of the decay to the ^{10}Be(2_1^+) state further firms the assignment of LC band. It is also interesting to note the decay pattern of the σ-bond LC (Fig. 3 (b)). We predict that the σ-bond LC primary decays to the excited states of ^{10}Be (0_2^+, 2_3^+ and 4_2^+ states) which also have the valence neutrons in σ orbit. Very recently the candidates of the σ-bond LC were reported by an experiment [16] which observed the resonances at 22 and 24 MeV which primary decay to the ^{10}Be(0_2^+). We expect that further experiments will establish the formation of LC with π and σ bonding.

4. Summary

In summary, we have performed the AMD calculations to identify the LC bands in ^{14}C. We found three different cluster configurations which we call triangular and π- and σ-bond LC. We have shown that π-bond LC band is built on the 0^+ state at 14.6 MeV, while the σ-bond LC band is built on the 0^+ state at 22.2 MeV. The excitation energy and decay width of the π-bond LC reasonably agree with the observation, and hence, we conclude that the formation of the π-bond LC is rather convincing. We predict that the π-bond LC also decays to the ^{10}Be(2_1^+), while the σ-bond LC will decay to the ^{10}Be(0_2^+, 2_3^+). The observation of these decay pattern will establish the assignment of the LC bands in ^{14}C.

References

[1] Morinaga H 1956 *Phys. Rev.* **101** 254
[2] von Oertzen W, Freer M and Kanada-En'yo Y 2006 *Phys. Rep.* **432** 43
[3] Itagaki N, Okabe S, Ikeda K and Tanihata I 2001 *Phys. Rev.* C **64** 014301
[4] von Oertzen W and Bohlen H. G 2003 *C. R. Physique* **4** 465
[5] Ashwood N. I *et al* 2004 *Phys. Rev.* C **70** 064607
[6] Suhara T and Kanada-En'yo Y 2010 *Phys. Rev.* C **82** 044301
[7] Maruhn J. A, Loebl N, Itagaki N and Kimura M 2010 *Nucl. Phys.* A **883** 1
[8] Furutachi N and Kimura M 2011 *Phys. Rev.* C **83** 021303(R)
[9] Baba T, Chiba Y and Kimura M 2014 *Phys. Rev.* C **90** 064319
[10] Dell'Aquila D *et al* 2016 *Phys. Rev.* C **93** 024611
[11] Freer M *et al* 2014 *Phys. Rev.* C **90** 054324
[12] Fritsch A *et al* 2016 *Phys. Rev.* C **93** 014321
[13] Kimura M 2004 *Phys. Rev.* C **69** 044319
[14] Kanada-En'yo, Suhara T and Taniguchi Y 2014 *Prog. Theor. Exp. Phys.* **2014** 073D02
[15] Baba T and Kimura M 2016 *arXiv:1605.05567 [nucl-th]*
[16] Z. Y. Tian *et al.*, CPC in print; arXiv:1607.00157

IOP Conf. Series: Journal of Physics: Conf. Series **863** (2017) 012034 doi:10.1088/1742-6596/863/1/012034

Influence of Nuclear Cluster Structure in Proton-Induced Pre-Equilibrium Composite Particle Emission

A A Cowley[1,2]

[1] Department of Physics, Stellenbosch University, Private Bag X1, Matieland 7602, South Africa

[2] iThemba Laboratory for Accelerator Based Sciences, P O Box 722, Somerset West 7129, South Africa

E-mail: aac@sun.ac.za

Abstract. Recent statistical multistep analyses of the pre-equilibrium reaction ^{93}Nb(p, α) at incident energies of 65 and 100 MeV are reviewed. It is shown that the α-cluster structure of the target nucleus is prominently displayed at the lower incident energy, whereas at the higher value its presence is obscured by the dominance of three-nucleon pickup. This result appears to be simply associated with a difference in the details of the effective momentum mismatch experienced by a cluster knockout mechanism, as opposed to a pickup process.

1. Introduction

It has been known for many years that the mechanism of pre-equilibrium (p, α) reactions reveals itself as closely related to a basic statistical multistep process. Cross section and analyzing power angular distributions over the complete range of possible emission energies on various target nuclei are described successfully, for example, by the quantum mechanical theory of Feshbach, Kerman and Koonin (FKK)[1]. The multistep character of the pre-equilibrium reaction is reflected in large excursions in the analyzing power values at high emission energy. Towards lower emission energies, however, the analyzing power angular distributions become featureless as the successive multisteps effectively wash out the polarization transferred by the projectile to the nuclear system. In addition, the shapes of analyzing power angular distributions are especially sensitive to competition between cluster knockout and three-nucleon pickup as part of the statistical multistep chain. In this way Bonetti *et al.* [2] found that the ^{58}Ni(p, α) reaction at an incident energy of 72 MeV displays the features associated with cluster knockout exclusively. This result was recently confirmed for ^{93}Nb(p, α) at a close incident energy of 65 MeV [3, 4]. Thus preformed clusters in ^{58}Ni and ^{93}Nb are implicated as clearly influencing the shapes of the analyzing power angular distributions at incident energies around 70 MeV. Very surprisingly, on ^{93}Nb, but at an incident energy of 100 MeV, three-nucleon pickup clearly dominates [4]. An explanation of this unusual incident-energy dependence is offered in Ref. [4] as resulting from the basic momentum mismatch between the incident and exit channels in a (p, α) reaction. The purpose of this paper is to review the evidence presented in Ref. [4, 5], and to draw conclusions regarding the possible trend expected for the target dependence of the observed phenomenon.

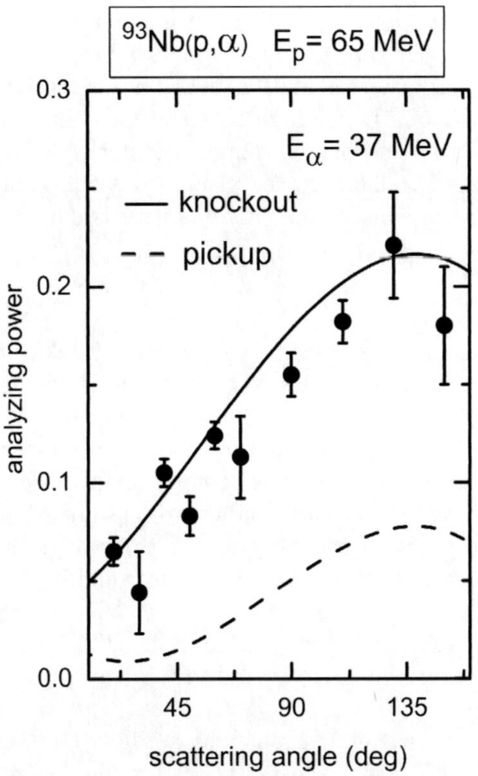

Figure 1. Analyzing power angular distribution for the ^{93}Nb(p,α) reaction at an incident energy of 65 MeV and an α-particle emission energy of 37 MeV. The solid curve is a calculation based on α-cluster knockout as the terminating reaction in the statistical multistep chain, whereas the dashed curve is for three-nucleon pickup leading to α-particle emission. The figure is adapted from one published in Ref. [4].

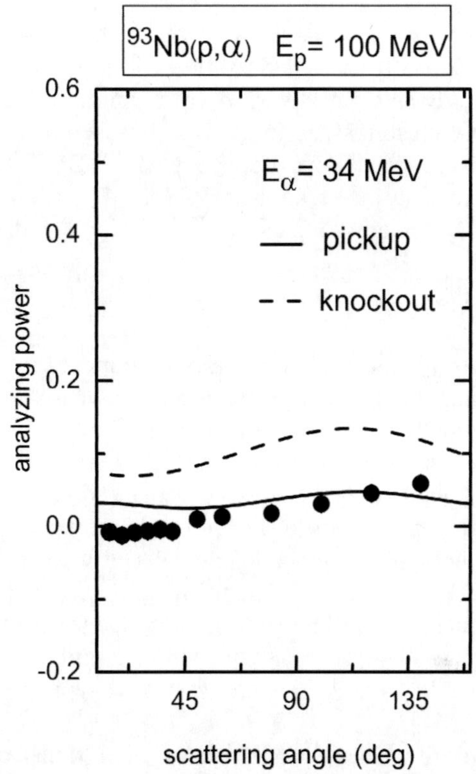

Figure 2. Analyzing power angular distribution for the ^{93}Nb(p,α) reaction at an incident energy of 100 MeV and an α-particle emission energy of 34 MeV. The meaning of the theoretical curves are switched from those in Fig. 1 to emphasize that in this figure the calculation in which pickup is the terminating step agrees with the experimental values. The figure is adapted from one published in Ref. [4].

2. Incident-energy trend of competition between cluster knockout and pickup in the reaction ^{93}Nb(p,α)

Theoretical FKK predictions [4] for either cluster knockout or three-nucleon pickup as the final step in the statistical multistep chain are compared in Fig. 1 with measured [6] analyzing power angular distributions for the reaction ^{93}Nb(p,α) at an incident energy of 65 MeV. Although the shapes of the theoretical angular distributions are qualitatively similar, the absolute magnitudes differ by more than a factor of two, which is a significant difference. This demonstrates the dominance of an α-cluster mechanism in the reaction process. Similar calculations for ^{93}Nb(p,α) are displayed in Fig. 2 for an incident energy of 100 MeV. However, at this higher incident energy we find that the comparison between the theoretical results and the experimental analyzing power distribution favours pickup. Note that in both figures preferred theoretical results are consistently indicated by solid curves, which means that the convention for Fig. 1 is different from that of Fig. 2, as indicated.

Clearly the identity of the dominant reaction mechanism for ^{93}Nb(p,α) at an incident energy of 65 MeV is consistent with that found previously for ^{58}Ni(p,α) reaction at an incident energy of 72 MeV. It now needs to be understood why at 70 MeV the (p,α) reaction displays properties normally associated with α clustering in the target nucleus, whereas at 100 MeV it is insensitive to the cluster structure. This is discussed in Refs. [4, 5]. Use is made of the fact that in distorted-wave Born approximation (DWBA), which is an important ingredient of the FKK formulation (see for example Ref. [7]), the cross section is related to radial integrals which in turn are roughly a function [5] of the overall momentum mismatch q. This quantity is defined as [4]:

$$q = q_{\text{out}} - \frac{\Delta L}{R} - q_{\text{in}} \ , \qquad (1)$$

where q_{in} and q_{out} are the linear momenta in the incident- and exit channels, respectively, ΔL is the angular momentum transfer and R is the nuclear radius associated with either an α-cluster knockout or a three-nucleon pickup system relative to the core part of the target nucleus. As formulated Eq. 1 applies to small scattering angles. As discussed in Refs. [4, 5] α-cluster knockout is expected to be associated with a low (or even zero) angular momentum transfer, whereas in pickup fairly large transfers may occur. Roughly gaussian shapes are assumed for the momentum dependencies of the wave functions of both an α-cluster and a three-nucleon system. The large momentum difference in knockout restricts the cross section to roughly the exponentially-falling part of the wave function. The cross section for pickup, on the other hand, is favoured by the contribution of angular momentum transfer which maximizes it. This gives an incident-energy dependence of the pickup cross section which is centred around $q = 0$ for $\Delta L = 8$.

Results of explicit DWBA calculations are presented in Ref. [4], of which Fig. 3 is an adapted version. Cross sections associated with either pickup or cluster knockout have been extracted from the experimental results as described in Ref. [4], and theoretical calculations are normalized to the extracted cross section values. The dominant reaction mechanism for ^{93}Nb(p,α) as a function of incident energy, as then reflected in the corresponding analyzing power angular distribution, is simply determined by the process which has the largest cross section at a given incident-energy value. In Fig. 3 dashed arrows compare relative cross section magnitudes at incident energies of 65 and 100 MeV for the two reaction mechanisms. For knockout the cross section is substantially larger at an incident energy of 65 MeV than the corresponding contribution from pickup. At an incident energy of 100 MeV the situation is reversed.

Apart from the kinematic influence of momentum mismatch as discussed above, the absolute cross sections associated with knockout or pickup processes are also determined by either cluster spectroscopic factors or three-nucleon amplitudes, respectively. This fact implies that the target structure also influences the extent of the observed competition between the two mechanisms.

3. Summary and conclusion

Results for the pre-equilibrium ^{93}Nb(p,α) reaction at incident energies of 65 and 100 MeV have been discussed in terms of the signature of an α-cluster structure of the target nucleus. It was explained how cluster knockout and three-nucleon pickup compete to determine the characteristic shape of the analyzing power angular distribution by means of which the dominant reaction is identified. The proposed explanation of the apparent change in the mechanism with a relatively small increase in the incident energy from 65 MeV to 100 MeV appears to be confirmed with explicit DWBA calculations. Clearly the target dependence of the phenomenon should be explored in future work.

IOP Conf. Series: Journal of Physics: Conf. Series **863** (2017) 012034 doi:10.1088/1742-6596/863/1/012034

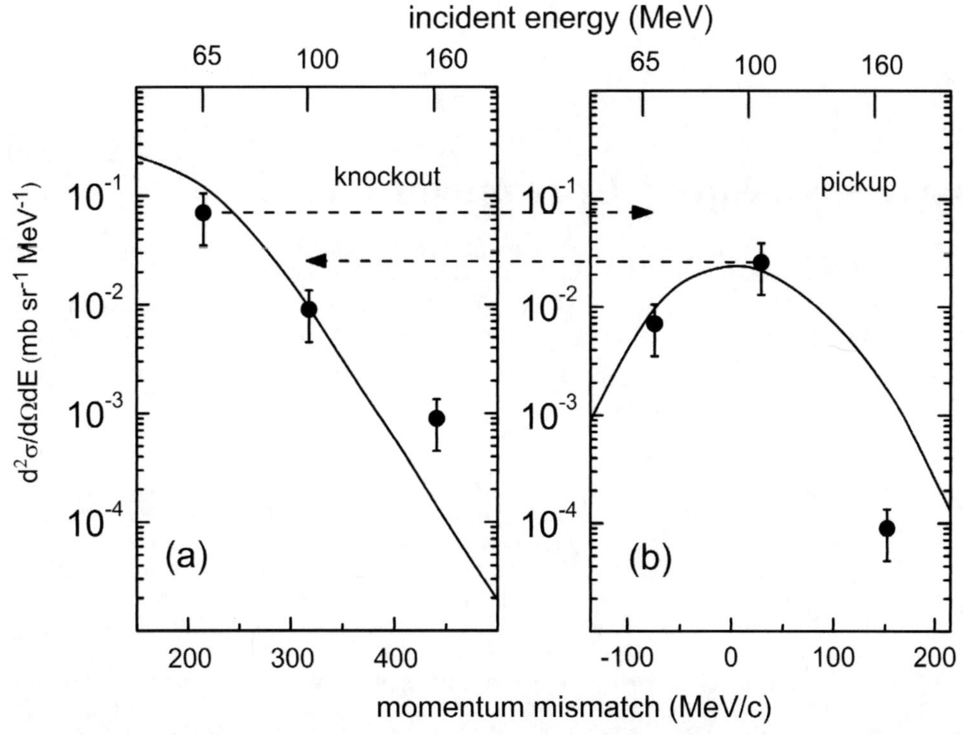

Figure 3. Double-differential cross section for the ^{93}Nb(p,α) reaction corresponding to yields from a knockout mechanism (a) and pickup (b). Results at emission energies 20 MeV lower than the incident energies are shown as a function of overall momentum difference (scales on the lower horizontal axes) and the incident energy is indicated on the top horizontal axes. The curves are DWBA predictions. The figure is adapted from one shown in Ref. [4].

Acknowledgement
This work was performed with funding from the South African National Research Foundation (NRF) under Grant No. 80833. The financial support is gratefully acknowledged.

References

[1] Feshbach H, Kerman A and Koonin S 1980 *Ann. Phys. (N. Y.)* **125** 429
[2] Bonetti R, Crespi F and Kubo K-I 1989 *Nucl. Phys.* A **499** 381
[3] Dimitrova S S, Cowley A A, Zemlyanaya E V and Lukyanov K V 2014 *Phys. Rev.* C **90** 054604
[4] Cowley A A, Dimitrova S S, Zemlyanaya E V, Lukyanov K V and Van Zyl J J 2016 *Phys. Rev.* C **93** 034624
[5] Cowley A A, Dimitrova S S, Zemlyanaya E V, Lukyanov K V and Van Zyl J J 2016 *EPJ Web of Conferences* **107** 08004
[6] Sakai H, Hosono K, Matsuoka N, Nagamachi S, Okada K, K. Maeda K and Shimizu H 1980 *Nucl. Phys.* A **344** 41
[7] Dimitrova S S, Cowley A A, Van Zyl J J, Zemlyanaya E V and Lukyanov K V 2014 *Phys. Rev.* C **89** 034616

Inclusive breakup of Borromean nuclei

M S Hussein[1,2,3], **B V Carlson**[1] **and T Frederico**[1]
[1]Instituto Tecnológico de Aeronáutica, DCTA,12.228-900 São José dos Campos, SP, Brazil
[2]Instituto de Estudos Avançados, Universidade de São Paulo C. P. 72012, 05508-970 São Paulo-SP, Brazil
[3]Instituto de Física, Universidade de São Paulo, C. P. 66318, 05314-970 São Paulo,-SP, Brazil

E-mail: hussein@if.usp.br

Abstract. We derive the inclusive breakup cross section of a three-fragment projectile nuclei, $a = b + x_1 + x_2$, in the spectator model. The resulting four-body cross section for observing b, is composed of the elastic breakup cross section which contains information about the correlation between the two participant fragments, and the inclusive non-elastic breakup cross section. This latter cross section is found to be a non-trivial four-body generalization of the Austern formula [1], which is proportional to a matrix element of the form, $\langle \hat{\rho}_{x_1,x_2} |[W_{x_1} + W_{x_2} + W_{3B}]| \hat{\rho}_{x_1,x_2} \rangle$. The new feature here is the three-body absorption, represented by the imaginary potential, W_{3B}. We analyze this type of absorption and supply ideas of how to calculate its contribution.

1. Introduction

The theory of inclusive breakup was developed in the 80's [1, 2, 3, 4] for a reaction induced by a two-fragment projectile, $a = b + x$ with b taken as the detected spectator fragment, and x being the interacting participant fragment. The HM theory was applied to the alpha spectra (b = α) using the eikonal approximation for the distorted wave in [5]. The IAV and the HM theory were based on the post form where the interaction is V_{xb}, while the UT theory is based on the prior form with the interaction given by $U_b + U_x - U_a$. The HM expression for the inclusive non-elastic breakup cross section was further greatly analyzed in [6]. In a recent publication [7], Potel et al. tested the Surrogate Method [8] in the case of (d, p) reaction on the actinide nuclei to be used in projected breeder reactors. For this purpose, they employed the theory of inclusive non-elastic breakup reactions, where the proton is treated as a spectator, merely scattering off the target, and the neutron is captured by the target. Other papers on the (d, p) reaction were also published in 2015 dealing with the same issue [9, 10, 11, 12]. Ref. [9] also discussed the application of this hybrid picture (direct breakup followed by compound nucleus formation of the subsystem) to the reaction ^{6}Li $+^{209}$Bi $\rightarrow \alpha + X$, at $E_{Lab} = 24$ MeV and 32 MeV.

At near-barrier energies the three-body Austern formula alluded to above [1] can in principle calculate the incomplete fusion part of the total fusion cross section. The extension of the Austern formula to the case of three fragment and Borromean projectiles, such as the weakly bound stable nucleus ^{9}Be = $\alpha + \alpha$ + n, and Borromean, two-neutron halo, nuclei like ^{6}He = α + n + n, ^{11}Li = ^{9}Li + n + n, ^{14}Be = ^{12}Be + n + n, and ^{22}C = ^{20}C + n + n, is certainly important as more data on these reactions have become available. Data on complete fusion and total fusion around the Coulomb barrier are currently being obtained and analyzed using the effective two-body "four-body" Continuum Discretized Coupled Channels model, which is basically unable

IOP Conf. Series: Journal of Physics: Conf. Series **863** (2017) 012035 doi:10.1088/1742-6596/863/1/012035

to calculate the incomplete fusion part of the total fusion cross section [16], requiring urgent derivation and developments of the the four-body inclusive breakup cross section. This is the purpose of this Contribution.

2. The inclusive breakup of two-fragments projectile cross section within the spectator model

The formula for the inclusive breakup of a two-fragment projectile has been derived by [1] using a three body model. The form of the cross section is,

$$\frac{d^2\sigma_b}{dE_b d\Omega_b} = \frac{d^2\sigma_b^{(EB)}}{dE_b d\Omega_b} + \frac{d^2\sigma_b^{(INEB)}}{dE_b d\Omega_b} \tag{1}$$

The above formula was derived using a sum rule obtained by [13, 14]. The elastic breakup is given by,

$$\frac{d^2\sigma_b^{(EB)}}{dE_b d\Omega_b} = \frac{2\pi}{\hbar v}\rho_b(E_b)\sum_{\mathbf{k}_x}\left|\langle\chi_x^{(-)}\chi_b^{(-)}|V_{bx}|\Psi_0^{(+)}\rangle\right|^2\delta(E-E_b-E_{\mathbf{k}_x}) \tag{2}$$

The inclusive non-elastic breakup part of the cross section has come to be known as the Austern formula which involves the reaction cross section of the "captured" fragment, calculated with a full three-body scattering wave function [1].

$$\frac{\sigma^2_{b,INBU}}{dE_b d\Omega_b} = -\frac{2}{\hbar v}\rho_b(E_b)\left\langle\hat{\rho}_x\left|W_x\right|\hat{\rho}_x\right\rangle \tag{3}$$

where $\rho_b(E_b) = \mu_b k_b/[(2\pi)^3\hbar^2]$, is the density of states of the observed fragment, b, W_x is the imaginary part of the xA optical potential, and $\hat{\rho}_x(\mathbf{r}_x) = (\chi_b^{(-)}|\Psi_{3B}^{(+)}\rangle$, is the source function of the participant fragment, x. At much higher deuteron energies or other breaking up projectiles, researchers relied on the very simple but physically transparent Serber model [15], which is a natural limiting approximation of the Austern formula.

3. Extension to three-fragments projectiles

As in the derivation of the inclusive breakup cross section for two-fragment projectile, we use the spectator model to calculate the inclusive breakup cross section of projectiles of the type $a = b+x_1+x_2$, with b representing the spectator, observed, fragment, while x_1 and x_2 correspond to the two participating, interacting, fragments. The result of the calculation, given in great details in [17], is,

$$\frac{d^2\sigma_b}{dE_b d\Omega_b} = \frac{d^2\sigma_b^{EB}}{dE_b d\Omega_b} + \frac{d^2\sigma_b^{INBU}}{dE_b d\Omega_b} \tag{4}$$

where the four-body elastic breakup cross section is,

$$\frac{d^2\sigma_b^{EB}}{dE_b d\Omega_b} = \frac{2\pi}{\hbar v}\rho_b(E_b)\sum_{\mathbf{k}_X}\left|\langle\chi_X^{3B(-)}\chi_b^{(-)}|[V_{bx_1}+V_{bx_2}]|\Psi_0^{4B(+)}\rangle\right|^2\delta(E-E_b-E_{\mathbf{k}_X}) \tag{5}$$

and

$$\frac{d^2\sigma_b^{INEB}}{dE_b d\Omega_b} = -\frac{2}{\hbar v}\rho_b(E_b)\langle\hat{\rho}_X|(W_{x_1}+W_{x_2}+W_{3B})|\hat{\rho}_X\rangle \tag{6}$$

with the source function $\hat{\rho}_X \equiv \hat{\rho}_{x_1,x_2}$ given by,

$$\hat{\rho}_X(\mathbf{r}_{x_1},\mathbf{r}_{x_1}) = (\chi_b^{(-)}|\Psi_0^{4B(+)}\rangle = \int d\mathbf{r}_b(\chi_b^{(-)}(\mathbf{r}_b))^\dagger\Psi_0^{4B(+)}(\mathbf{r}_b,\mathbf{r}_{x_1},\mathbf{r}_{x_1}) \tag{7}$$

Eq. (6) is referred to here as the Carlson-Frederico-Hussein (CFH) cross section. The elastic breakup component of the cross section, Eq. (5), contains in the final state the distorted wave of the observed fragment, b, and the full scattering wave function of the $x_1 + x_2$ system in the potential $U_{x_1} + U_{x_2} + V_{x_1,x_2}$. Thus the four-body EBU cross section contains invaluable information about the $X \equiv x_1 + x_2$ correlation both in the full incident 4B scattering wave function and the final $x_1 + x_2$ wave function, $\chi_X^{3B(-)}$. The Inclusive non-elastic breakup piece (INBU), the CFH cross section, $\frac{d^2\sigma^{INBU}}{dE_b d\Omega_b}$, Eq.(6), contains the four-body source function $\hat{\rho}_X$, the density of states of the observed b fragment, and the imaginary potential of the 3B, $x_1 + x_2 + A$, system, $W_{x_1} + W_{x_2} + W_{3B}$. with the genuine 3B imaginary potential representing the absorption of the two interacting fragments. The CFH cross section can be expressed in terms of a 4B reaction cross section as,

$$\frac{d^2\sigma_b^{INEB}}{dE_b d\Omega_b} = \rho_b(E_b)\sigma_{R,4B} \tag{8}$$

$$\sigma_{R,4B} = \frac{2}{\hbar v}\langle\hat{\rho}_X|(W_{x_1} + W_{x_2} + W_{3B})|\hat{\rho}_X\rangle = \sigma_{R,x_1} + \sigma_{R,x_2} + \sigma_{R,3B} \tag{9}$$

where $\sigma_{R,3B} = (2/\hbar v_a)\langle\hat{\rho}_X|W_{3B}|\hat{\rho}_X\rangle$. This 3B absorptive potential, W_{3B}, is an hitherto not well studied one [18] and corresponds to the principal result of this Contribution. The detailed derivation of this potential, as well as the interpretation of its structure are given in [17]. Its leading contribution corresponds to the excitation of the target by one of the fragments and its de-excitation by the other fragment as represented schematically in Fig. 1.

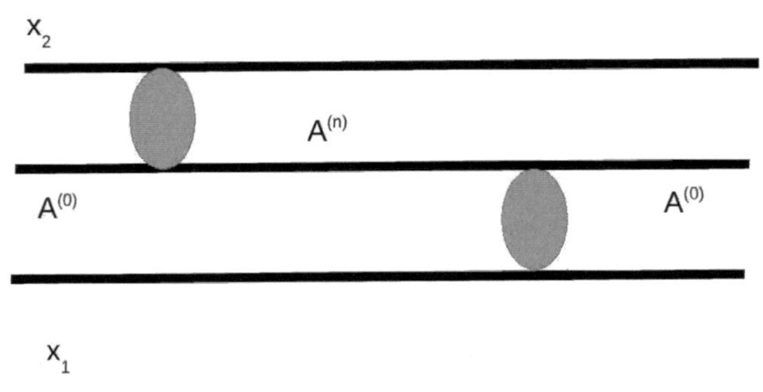

Figure 1. Representation of the leading term in the 3-body absorptive component W_{3B}.

4. DWBA analysis of the 4-body inclusive breakup cross section

In the two-fragments projectiles case, the 3-body wave function was expanded in terms of distorted waves and the post interaction. The leading term came to be known as the Ichimura-Austern-Vincent (IAV) cross section [2]. Using the prior form for the interaction, the leading DWBA breakup cross section was derived by Udagawa and Tamura (UT) [3]. Finally using the post form for the interaction but using the DWBA wave function to replace the 3-body wave function $\Psi_{3B}^{(+)}$ in Eq. (3), Hussein and McVoy (HM) derived their HM cross section [4]. It was later shown through use of the Faddeev equations and other techniques [19, 20], that the source function $\hat{\rho}_x^{(+)} \approx \hat{\rho}_{x,IAV}^{(+)} = G_x^{(+)}(\chi_b^{(-)}|V_{xb}|\chi_a^{(+)}\phi_a\rangle$, can be written as,

$\hat{\rho}^{(+)}_{x,IAV}(\mathbf{r_x}) = \hat{\rho}^{(+)}_{\mathbf{x,UT}}(\mathbf{r_x}) + \hat{\rho}^{(+)}_{\mathbf{x,HM}}(\mathbf{r_x})$, where $\hat{\rho}^{(+)}_{x,UT}(\mathbf{r_x}) = \mathbf{G}^{(+)}_{\mathbf{x}}(\chi^{(-)}_{\mathbf{b}}|[\mathbf{U_b} + \mathbf{U_x} - \mathbf{U_a}]|\chi^{(+)}_{\mathbf{x}}\phi_{\mathbf{a}}\rangle$, and $\hat{\rho}^{(+)}_{x,HM} = (\chi^{(-)}_{b}|\chi^{(+)}_{a}\phi_{a}\rangle$. Accordingly the INBU cross section acquires a form composed of three terms; the UT cross section plus the HM cross section plus the interference term. The relative importance of these components was discussed in [10].

A similar DWBA analysis of the four-body cross section, the CFH equation, Eq. (6), would require the use of the Faddeev-Yakubosky equations [21] for the 4B wave function $\Psi^{4B(+)}_{0}$. The leading term in the dominant component of the FY equations would result in a the four-body equivalent of the IAV, the UT and the HM cross sections. This analysis is in progress and will be reported in a future publication. The four-body HM cross section is, on the other hand, easy to derive. It hinges on approximating the 4-body scattering wave function, $\Psi^{4B(+)}_{0}$ by the DWBA distorted wave times the ground state wave function of the projectile. The 4B inclusive non-elastic breakup theory discussed above, Eqs. (6, 7), namely the CFH cross section, is potentially important for reactions involving Borromean nuclei as well in reactions involving the transfer of two neutrons and the population of Giant Pairing Vibration states [22]. The capture or transfer of the two neutrons is accounted for by the 3B absorptive potential, W_{3B}, and the strength of this transition is measured by the value of $\sigma_{R,3B}$.

Acknowledgments
We thank the Brazilian Agencies, FAPESP, CNPq and CAPES for the partial support. MSH acknowledges a Senior Visiting Professorship at ITA under the auspices of the CAPES/ITA PVS Program.

References

[1] Austern N, Iseri N Y, Kamimura M, Kawai M, Rawitscher G and Yahiro M 1987 *Phys. Rep.* **154** 125
[2] Ichimura M, Austern N and Vincent C M 1985 *Phys. Rev.* C **32** 431
[3] Udagawa T and Tamura T 1981 *Phys. Rev.* C **24** 1348
[4] Hussein M S and McVoy K W 1985 *Nucl. Phys.* A **445** 124
[5] Hussein M S and Mastroleo R C 1989 *Nucl. Phys.* A **491** (1989) 468
[6] Canto L F, Donangelo R, Mattos L, Hussein M S and Lotti P 1998 *Phys. Rev.* C **58** 1107
[7] Potel G, Nunes F M and Thompson I J 2015 *Phys. Rev.* C **92** 034611
[8] Escher J E, Burke J T, Dietrich F S, Scielzo N D Thompson I J and Younes W 2012 *Rev. Mod. Phys.* **84** 353
[9] Lei J and Moro A M 2015 *Phys. Rev.* C **92** 044616
[10] Lei J and Moro A M 2015 *Phys. Rev.* C **92** 061602(R)
[11] Carlson B V, Capote R and Sin M 2016 *Few-Body Syst.* **57** 307
[12] Ducasse Q. et al., arXiv:1512.06334 [nucl-ex]
[13] Kasano A and Ichimura M 1982 *Phys. Lett.* B **115** 81
[14] Hussein M S 1984 *Phys. Rev.* C **30** 1962
[15] Serber R 1950 *Phys. Rev.* **80** 1098
[16] Diaz-Torres A and Thompson I J 2002 *Phys. Rev.* C **65** 024606
[17] Carlson B V, Frederico T and Hussein M S, arXiv:1607.05783
[18] Hussein M S, Carlson B V, Frederico T and Tarutina T 2004 *Nucl. Phys.* A **738** 367
[19] Hussein M S, Frederico T and Mastroleo R C 1990 *Nucl. Phys.* A **511** 269
[20] Ichimura M 1990 *Phys. Rev.* C **41** 834
[21] Yakubovsky O A 1967 *Yad. Fiz.* **5** 1312 [*Sov. J. Nucl. Phys.* **5** 1312 (1967)].
[22] Cappuzzello F, Carbone D, Cavallaro M, Bond' M, Agodi C, Azaiez F, Bonaccorso A, Cunsolo A, Fortunato L, Foti A, Franchoo S, Khan E, Linares R, Lubian J, Scarpaci J A and Vitturi A 2015 *Nature Communications* **6** 6743

IOP Conf. Series: Journal of Physics: Conf. Series **863** (2017) 012036 doi:10.1088/1742-6596/863/1/012036

Probing surface distributions of α clusters in ^{20}Ne via α-transfer reaction

T Fukui,[1] Y Taniguchi,[2] T Suhara,[3]
Y Kanada-En'yo[4] and K Ogata[5]
[1]Nuclear Data Center, Japan Atomic Energy Agency, Tokai, Ibaraki 319-1195, Japan
[2]Department of Medical and General Sciences, Nihon Institute of Medical Science, Moroyama, Saitama 350-0435, Japan
[3]Matsue College of Technology, Matsue, Shimane 690-8518, Japan
[4]Department of Physics, Kyoto University, Kyoto 606-8502, Japan
[5]Research Center for Nuclear Physics, Osaka University, Ibaraki, Osaka 567-0047, Japan

E-mail: `fukui.tokuro@jaea.go.jp`

Abstract. Although several analyses of the α-transfer reaction ^{16}O(^{6}Li, d)^{20}Ne using the distorted wave Born approximation have been performed with the aim to clarify the α-cluster structure of ^{20}Ne, they have resulted in poor information on the clustering due to insufficiency of models for the analyses. In this work we propose a precise model employing the coupled-channels Born approximation with the microscopically calculated cluster wave function to elucidate the surface manifestation of the α-particle in the ground state of ^{20}Ne. Our calculation has shown improvement of the theoretical calculation compared with that in the previous analyses. We have confirmed numerically that the angular distributed cross section at the forward angles probes only the surface region of the α-cluster wave function. The breakup e ect of ^{6}Li has been investigated and concluded that only the back coupling is important for the (^{6}Li, d) reaction.

1. Introduction

So far a large number of theoretical and experimental studies have been carried out to investigate the α-cluster state in nuclei. Nevertheless no direct evidence of the clustering in bound states of nuclei is obtained. In particular for the α-cluster structure of ^{20}Ne in its ground state, several experimental works using the α-transfer reaction ^{16} (^{6} i, d)^{20}Ne [1 3] and its inverse reaction ^{20}Ne(d,^{6} i)^{16} [4] were reported. However in these papers the discussions on the clustering in the ground state are not enough. urthermore due to ambiguities in reaction models with the distorted-wave Born approximation (BA) used for their analyses, unphysical values of the spectroscopic factor were reported, i.e., some of them exceed unity. In this work we reanalyze the α-transfer reaction reported previously by adopting a precise theoretical framework described below. e aim to extract the information of the surface manifestation of the α-cluster structure in the ground state of ^{20}Ne.

2. Theoretical framework

The transfer reaction is described with the coupled-channels Born approximation (CCBA), in which the three-body (α d ^{16}) model is adopted to explicitly take into account the excitation of ^{6} i into the α d continuum. The excitation to the continuum, i.e., the virtual breakup, is

IOP Conf. Series: Journal of Physics: Conf. Series **863** (2017) 012036 doi:10.1088/1742-6596/863/1/012036

handled by means of the continuum-discretized coupled-channels (C CC) method [5 7]. The relative wave function of the α-^{16} system in the ground state of ^{20}Ne is calculated by the microscopic cluster model with generator coordinate method (CM) [8 10].

The input interactions of our model are the optical model potential (MP) of the α (d)-^{16} system, the α d binding potential, and the e ective nucleon-nucleon interaction in the CM calculation. or the MP we adopt a phenomenological potential [11, 12] for each subsystem, whereas the binding potential is assumed to be a multi-range aussian interaction [13]. The olkov No. 2 e ective interaction [14] is used in the CM calculation. The detail of the calculation is expressed in our full paper [15].

3. Results and discussion

In ig. 1(a) we compare the calculated transfer cross sections at 20.0 (solid line) and 42.1 Me (dashed line) as a function of the deuteron emitting angle θ with the experimental data [1, 2]. The cross section at 20.0 Me is multiplied by 10. Each lines are normalized to the data with a factor of 0.261 (0.769) at 20.0 (42.1) Me . ne of our main achievements is that description of the di raction pattern, in particular the first and second peak of the cross section at the forward angles, is significantly improved at each incident energy compared to the previous BA analyses [1, 2]. urthermore we obtain the reasonable values of the normalization factor compared to them reported in the previous BA as 2.70 [2] and 2.59 [1] at 20.0 and 42.1 Me , respectively. Note that the normalization factor exceeding unity is unphysical because it expresses the probability of the cluster configuration in ^{20}Ne. It is found that the ambiguities of both the ^{6} i- MP and the α-^{16} wave function cause the unphysical normalization. In our CCBA model based on the three-body model, the ambiguity of ^{6} i- MP is reduced because we use only the MPs of the subsystems. In addition, our microscopic cluster model of ^{20}Ne can be regarded to be precise because, as shown in Ref. [16], the calculated spectra of the ground-state band as well as the root-mean-square radius of ^{16} are consistent with the measured ones.

Another important finding is that the transfer cross section probes only the tail of the α-^{16} wave function. This is numerically confirmed by comparison of the calculated cross

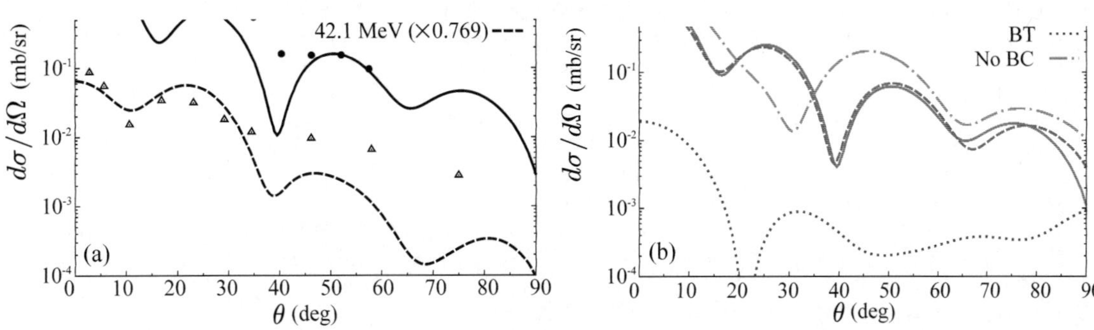

Figure 1. (a) Comparison of the calculated cross section of ^{16} (^{6} i, d)^{20}Ne at 20.0 (solid line) and 42.1 Me (dashed line) with the experimental data [1, 2]. Each line is normalized by multiplying by a factor indicated in the legend. At 20.0 Me the line and the dots are multiplied by 10. (b) E ect of each transfer process on the cross section at 20.0 Me . The results of the ET and BT are shown by the dashed and dotted lines, respectively. The solid line is the result involving both the ET and BT, whereas the dash-dotted line is the ET one without the BC.

section employing the microscopic cluster wave function with that employing a phenomenological oods-Saxon potential for the α-^{16} system. rom a detailed analysis [15], we conclude that at 20.0 (42.1) Me the transfer cross section at the forward angles $\theta \lesssim 40^\circ$ probes the wave function at surface region $r \gtrsim 5$ (4) fm, where r is the relative distance between α and ^{16} .

Next we investigate the reaction mechanism by decomposing the transition process into two paths. ne is the transition from the ground state of ^{6} i, which is called the elastic transfer (ET). The other is the breakup transfer (BT), which describes the transition from the continuum state of ^{6} i. Note that the ET involves the breakup e ect as the back coupling (BC), which is the channel coupling between the ground and continuum states. In ig. 1(b) the calculated cross sections of each transition process at 20.0 Me are shown. e see that the full result (solid line) involving both the ET and the BT coincides well with that only with the ET, whereas the calculation of the ET without the BC (dash-dotted line) results in the significantly di erent shape. Moreover the result of the BT (dotted line) is negligibly small. These facts indicate that only the BC plays an essential role. e have similar conclusion for the 42.1 Me case. It should be noted that this does not mean failure of conventional BA analyses employing a ^{6} i- MP because such an MP should contain the BC implicitly in its imaginary part if it is constructed properly. Same conclusion as in the present study was mentioned in Ref. [17] for another α-transfer reaction ^{13}C(^{6} i$, d)^{17}$. In Ref. [18] although the importance of the breakup e ect of ^{6} i in ^{12}C(^{6} i$, d)^{16}$ was reported, whether only the BC is essential was not investigated. To clarify the universality of the breakup e ect in (^{6} i$, d$) reactions, further analyses with the CCBA model are necessary.

4. Summary

e have analyzed the α-transfer reaction ^{16} (^{6} i$, d)^{20}$Ne by means of the CCBA model with the microscopic cluster wave function. e have obtained the improvement in terms of the shape of the angular distribution as well as the normalization factor compared to the previous BA analyses. It has been found that the cross section at the forward angles probes only the tail region of the α-^{16} relative wave function. As for the breakup e ect of ^{6} i, it has been pointed out that only the BC plays an important role.

References

[1] Becchetti F D, Jänecke J and Thorn C E 1978 *Nucl. Phys.* 313
[2] Anantaraman N, Gove H, Lindgren R, Tōke J, Trentelman J, Draayer J, Jundt F and Guillaume G 1979 *Nucl. Phys.* A **313** 445
[3] Tanabe T, Yasue M, Sato K, Ogino K and Kadota Y 1981 *Phys. Rev.* C **24** 2556
[4] Oelert W, Chung W, Betigeri M, Djaloeis A, Mayer-Böricke C and Turek P 1979 *Phys. Rev.* C **20** 459
[5] Kamimura M, Yahiro M, Iseri Y, Sakuragi Y, Kameyama H and Kawai M 1986 *Prog. Theor. Phys. Suppl.* **89** 1
[6] Austern N, Iseri Y, Kamimura M, Kawai M, Rawitscher G and Yahiro M 1987 *Phys. Rep.* **154** 125
[7] Yahiro M, Ogata K, Matsumoto T and Minomo K *Prog. Theor. Exp. Phys.* **2012** 01
[8] Hill D L and Wheeler J A 1953 *Phys. Rev.* **89** 1102
[9] Griffin J J and Wheeler J A 1957 *Phys. Rev.* **108** 311
[10] Brink D M 1966 *International School of Physics "Enrico Fermi", XXXVI* (New York: Academic Press)
[11] Michel F, Albinski J, Belery P, Delbar T, Grégorie G, Tasiaux B and Reidemeister G 1983 *Phys. Rev.* C **28** 1904
[12] Dave J H and Gould C R 1983 *Phys. Rev.* C **28** 2212
[13] Sakuragi Y, Yahiro M and Kamimura M 1986 *Prog. Theor. Phys. Suppl.* **89** 136
[14] Volkov A B 1965 *Nucl. Phys.* **74** 33
[15] Fukui T, Taniguchi Y, Suhara T, Kanada-En'yo Y and Ogata K 2016 *Phys. Rev.* C **93** 034606
[16] Kanada-En'yo Y, Suhara T and Taniguchi Y 2014 *Prog. Theor. Exp. Phys.* **2014** 073D02
[17] Fukui T, Ogata K and Yahiro M 2011 *Prog. Theor. Phys.* **125** 1193
[18] Adhikari S, Basu C, Thompson I J, Sugathan P, Jhinghan A, Golda K S, Babu A, Singh D, Ray S and Mitra A K 2014 *Phys. Rev.* C **89** 044618

IOP Conf. Series: Journal of Physics: Conf. Series **863** (2017) 012037 doi:10.1088/1742-6596/863/1/012037

Photodisintegration cross section of ^{9}Be up to $E_\gamma = 16$ MeV in the $\alpha + \alpha + n$ three-body model

Y Kikuchi[1], **M Odsuren**[2], **T Myo**[3,4] **and K Katō**[5]

[1] RIKEN Nishina Center, Wako 351-0198, Japan
[2] Nuclear Research Center, National University of Mongolia, Ulaanbaatar 210646, Mongolia
[3] General Education, Faculty of Engineering, Osaka Institute of Technology, Osaka 535-8585, Japan
[4] Research Center for Nuclear Physics (RCNP), Osaka University, Ibaraki 567-0047, Japan
[5] Nuclear Reaction Data Centre, Faculty of Science, Hokkaido University, Sapporo 060-0810, Japan

E-mail: yuma.kikuchi@riken.jp

Abstract. We investigate the photodisintegration cross section (PDXS) of ^{9}Be in low and higher energy regions by using the $\alpha + \alpha + n$ three-body model and the complex scaling method (CSM). The calculated photodisinegration cross section of well reproduces the recent experimental data, and shows that the transitions into the ^{8}Be(2^+) + n states dominate the cross section in the energy region of $6 \leq E_\gamma \leq 16$ MeV.

1. Introduction

The photodisintegration cross section (PDXS) of ^{9}Be is very important in investigations of not only astrophysical reaction cross sections for fomation of heavior elements but also nuclear structures of an $\alpha+\alpha+n$ cluster system. As for the former, a very sharp peak corresponding to the $1/2^+$ state has been observed at the energy just above the ^{8}Be+n threshold and our recent investigation is also reported in this meeting by Odsuren[1]. About the second problem, it was reported that the observed dipole strength in ^{9}Be at the exciation energy of ~ 8 MeV exhausts 10% of Thomas-Reiche-Kuhn sum rule and almost all the cluster dipole sum rule [2].

In this talk, we discuss the PDXS due to the electromagnetic transitions into resonant and continuum states of ^{9}Be applying the complex scaling method (CSM) [3, 4] to an $\alpha + \alpha + n$ three-body model. The observed cross section below $E_\gamma \sim 6$ MeV are reproduced with several resonant states corresponding to the experimetal data on excited states of ^{9}Be. The broad bump with the peak position at 8 MeV is also well expained in our calculation, which shows the electric-dipole transition from the ground state of ^{9}Be to the ^{8}Be(2^+) + n continuum state.[5]

2. Complex-scaled $\alpha + \alpha + n$ three-body model

We solve the Schrödinger equation for the $\alpha + \alpha + n$ three-body model using the orthogonality condition model [6]:

$$\hat{H}\Psi^\nu_{J^\pi} = E_\nu \Psi^\nu_{J^\pi}, \qquad \langle \Phi_{PF} | \Psi^\nu_{J^\pi} \rangle = 0, \tag{1}$$

where the second equation means the orthogonality condition for the solution $\Psi^{\nu}_{J^{\pi}}$ to be orthogonal to the Pauli forbidden states Φ_{PF} in the relative motion of α-α and α-n. The Hamiltonian for the relative motion of the $\alpha + \alpha + n$ three-body system is given as

$$\hat{H} = \sum_{i=1}^{3} t_i - T_{c.m.} + \sum_{i=1}^{2} V_{\alpha n}(\xi_i) + V_{\alpha\alpha} + V_{\alpha\alpha n}. \tag{2}$$

We employ the KKNN potential [7] for $V_{\alpha n}$, and the folding-type nuclear and Coulomb potentials for $V_{\alpha\alpha}$. Furthermore, we introduces the $\alpha+\alpha+n$ three-body potential $V_{\alpha\alpha n}$ of a single Gaussian form with the strength v_3 and the range paprameter μ so as to reproduce the binding energy of ^{9}Be and the energy positions of the observed peaks in the PDXS.

We solve the eigenvalue problem for the complex-scaled Schrödinger equation using the coupled rearrangement-channel Gaussian expansion method [8] where the spatial part of the wave function is expanded with the complex-range Gaussian basis functions.

The cross sections of $^9\text{Be}(3/2^-) + \gamma \rightarrow \alpha + \alpha + n$ by the electro-magnetic dipole transitions σ_{EM1} are expressed as the following form:

$$\sigma^{\gamma}(E_\gamma) = \sigma_{E1}(E_\gamma) + \sigma_{M1}(E_\gamma), \quad \sigma_{EM1}(E_\gamma) = \frac{16\pi^3}{9}\left(\frac{E_\gamma}{\hbar c}\right)\frac{dB(EM1, E_\gamma)}{dE_\gamma}. \tag{3}$$

Using the energy eigenvalues and eigenstates of the complex-scaled Hamiltonian $\hat{H}^{\theta}$ and the extended completeness relation [9] in the CSM, the electro-magnetic dipole transition strength is calculated as

$$\frac{dB(EM1, E_\gamma)}{dE_\gamma} = -\frac{1}{\pi}\frac{1}{2J_{gs}+1} Im\left[\sum_{\nu}\langle\tilde{\Psi}_{gs}(\theta)||\hat{O}^{\theta}_{EM1}||\Psi^{\nu}(\theta)\rangle\frac{1}{E-E^{\theta}_{\nu}}\langle\tilde{\Psi}^{\nu}(\theta)||\hat{O}^{\theta}_{EM1}||\Psi_{gs}(\theta)\rangle\right], \tag{4}$$

where J_{gs} and $\Psi_{gs}(\theta)$ are the total spin and the wave function of the ground state, respectively, and $\hat{O}^{\theta}_{EM1}$ is an electro-magnetic dipole transition operator.

3. Cluster dipole resonance

We determine the strength $v_3 = 1.10$ MeV and the range parameter $\mu = 0.02$ fm^{-2} of the $\alpha+\alpha+n$ three-body potential so as to fit the binding energy, charge and matter radii of the ^{9}Be ground state. In Fig. 1, calculated PDXS is shown in comparison with experimantal data. Although the peak positions are fitted by adjusting the strength v_3 for each spin-parity state, the magnitudes and widths of the observed cross section are well reprodced simultaneously. We see that the lowest peak just above the $^8\text{Be}(0^+)+n$ threshold energy ($E_\gamma = 1.6654$ MeV) comes from the $E1$ transition into the $1/2^+$ state, and the calculated cross section shows that the strength below the $^8\text{Be}(0^+) + n$ threshold is negligibly small. We have no resonance solutions corresponding to the peak, and discussed that the origin of the peak is a virtual state [1]. The second and third peaks at $E_\gamma = 2.5$ and 3.0 MeV in the cross section come from the transitions into the resonances of $5/2^-$ and $5/2^+$ states, respectively. The $M1$ transition into the $1/2^-$ resonance has a sizable contribution to the cross section at around $E_\gamma = 2.7$ MeV. The transitions into the $3/2^-$ and $3/2^+$ states play minor roles in the PDXS below $E_\gamma = 6$ MeV. The energy level diagram is also shown in Fig. 2, where we show the levels which are obtained using $v_3 = 1.10$ MeV commonly for all the spin-parity states for reference.

From Fig. 1, we see that the calculated PDXS has a broad peak at $E_\gamma \sim 8$ MeV, similar to the experimental data. The calculation indicates that the peak is dominated by the $E1$ transitions into $3/2^+$ and $5/2^+$. In the recent measurement, it is reported that the energy integrated cross section for the enhanced dipole strength measured for $4 \leq E_\gamma \leq 16$ MeV is estimated to be

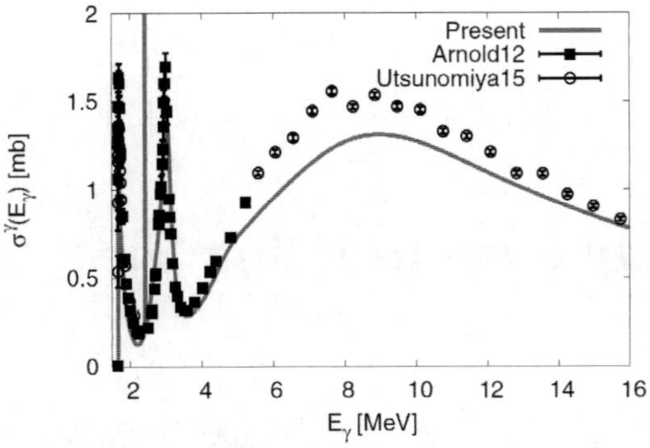

Figure 1. Calculated PDXS up to E_γ =16 MeV. The solid squares and open circles represent experimental data taken from Refs. [11, 2], respectively.

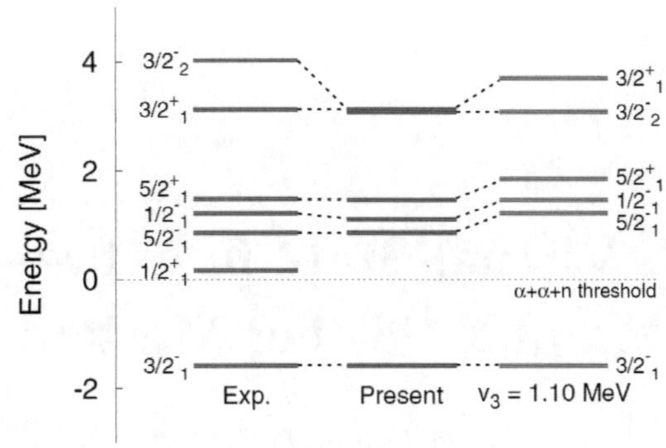

Figure 2. Comparison of the energy diagrams from the present calculation, experimetal data [10] and calculations using v_3 = 1.10 MeV for all the spin-parity states.

11.3 mb·MeV as a lower limit [2]. It is also suggested that this energy-integrated cross section exhausts 10% of the Thomas-Reiche-Kuhn (TRK) sum rule and almost all of the energy-weighted cluster dipole sum rule. We calculated the energy-integrated cross section by integrating σ_{E1} over the energy interval of $4 \leq E_\gamma \leq 16$ MeV. The calculated energy-integrated cross section is 12.1 mb·MeV for this energy interval, which is consistent with the experimental value (11.3 mb·MeV). On the other hand, we obtain the energy-integrated cross sections for $E_\gamma < 4$ MeV and $E_\gamma > 16$ MeV as 0.954 and 8.40 mb·MeV, respectively. It is seen that the energy-integrated cross section for $4 \leq E_\gamma \leq 16$ MeV is 56.5 % of the total amount in the present calculation.

We calculated the decomposition of the PDXS into different families of the nonresonant continuum states. The result shows that the cross sections at $E_\gamma \sim 8$ MeV are dominated by the contributions from the ^{8}Be(2^+) + n continuum states, while the contribution from the ^{8}Be(0^+) + n continuum solutions is negligible in the cross section [5]. This result is consistent with the previous study [12].

4. Conclusion

The photodisintegration of ^{9}Be in the energy region lower than E_γ =16 MeV was investigated by using the $\alpha + \alpha + n$ three-body model and the complex scaling method. It was shown that the dipole strength at $E_\gamma \sim 8$ MeV is understood to be caused by the single-neutron excitation from the ^{8}Be(2^+) $\otimes \nu p_{3/2}$ configuration in the ground state.

References

[1] Odsuren M, Kikuchi Y, Myo T, Aikawa M and Katō K 2015 *Phys. Rev.* C **92** 014322
[2] Utsunomiya H *et al* 2015 *Phys. Rev.* C **92** 064323
[3] Aoyama S, Myo T, Katō K and K. Ikeda 2006 *Prog. Theor. Phys.* **116** 1
[4] Myo T, Kikuchi Y, Masui H and Katō K 2014 *Prog. Part. Nucl. Phys.* **79** 1
[5] Kikuchi Y, Odsuren M, Myo T and Katō K 2016 *Phys. Rev.* C **93** 054605
[6] Saito S 1969 *Prog. Theor. Phys.* **41** 705; 1977 *Prog. Theor. Phys. Suppl.* **62** 11
[7] Kanada H, Kaneko T, Nagata S and Nomoto M 1979 *Prog. Theor. Phys.* **61** 1327
[8] Hiyama E, Kino Y and Kamimura M 2003 *Prog. Part. Nucl. Phys.* **51** 223
[9] Myo T, Ohnishi A and Katō K 1998 *Prog. Theor. Phys.* **99** 801
[10] Tilley D *et al* 2004 *Nucl. Phys.* A **745** 155
[11] Arnold C W *et al* 2012 *Phys. Rev.* C **85** 044605
[12] Okabe S, Abe Y and Tanaka H 1979 *Prog. Theor. Phys.* **61** 1049

A Virtual State in a Two-Body System Using the Complex Scaling Method

M Odsuren[1], Y Kikuchi[2], T Myo[3,4], M Aikawa[5], G Khuukhenkhuu[1] and K. Katō[6]

[1] School of Engineering and Applied Sciences, Nuclear Research Center, National University of Mongolia, Ulaanbaatar 210646, Mongolia
[2] RIKEN Nishina Center, Wako 351-0198, Japan
[3] General Education, Faculty of Engineering, Osaka Institute of Technology, Osaka 535-8585, Japan
[4] Research Center for Nuclear Physics (RCNP), Osaka University, Ibaraki 567-0047, Japan
[5] Faculty of Science, Hokkaido University, Sapporo 060-0810, Japan
[6] Nuclear Reaction Data Centre, Faculty of Science, Hokkaido University, Sapporo 060-0810, Japan

E-mail: odsurenn@gmail.com

Abstract. In association with the property of the $1/2^+$ state observed just above the ^{8}Be$(0^+)+n$ threshold energy in ^{9}Be, we investigate the photo-disintegration cross section of an s-state in a schematic two-body system using the complex scaling method.

1. Introduction

The ^{9}Be nucleus is one of the most interesting objects in nuclear cluster studies, because this nucleus is considered a typical non-alpha nucleus which has a neutron in addition to the ^{8}Be core consisting of two α-clusters. In particular, it has been a long-standing problem to clarify the structure of the excited $1/2^+$ state in $A = 9$ nuclei (^{9}Be and ^{9}B). Recently, applying the complex scaling method (CSM) [1, 2, 3] to the $\alpha + \alpha + n$ three-body model for ^{9}Be, we have shown that the sharp peak of the photo-disintegration cross section observed just above the ^{8}Be$(0^+) + n$ threshold is well explained as a $1/2^+$ virtual state of the ^{8}Be$(0^+) + n$ two-body configuration [4].

In this report, we discuss the virtual state nature in more detail using a schematic two-body model and calculate the cross section of the virtual state. We investigate the peak in the cross section, which has a different shape from the Breit-Wigner form. Furthermore, we observe that the phase shift behavior of the virtual state can also be described using the eigenvalue solutions of the CSM. In this contribution, we discuss that the CSM is very useful to study the virtual state.

2. $1/2^+$ state in ^{9}Be

The $\alpha + \alpha + n$ system is one of the Borromean systems. There is only one bound state, which is corresponding to the ^{9}Be ground state with $J^\pi = 3/2^-$. The first excited state is a $1/2^+$ unbound state above the $\alpha + \alpha + n$ threshold, and has been observed as a sharp peak in the

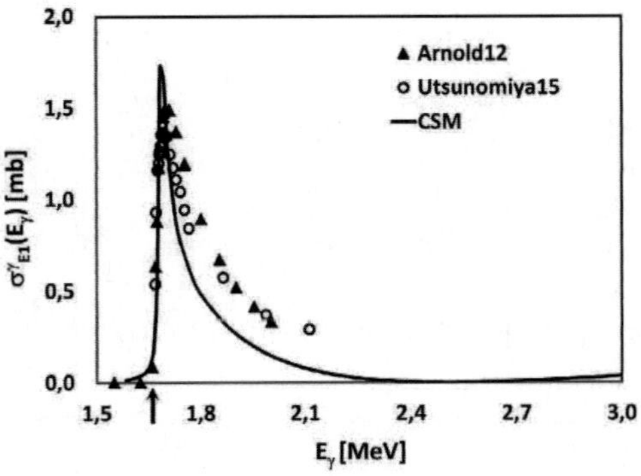

Figure 1. Calculated photo-disintegration cross sections in comparison with the experimental data of the ^{9}Be($1/2^+$) state. The open circles and olid triangles represent the experimental data [5, 6]. The arrow indicates the threshold energy of the ^{8}Be(0^+) + n channel.

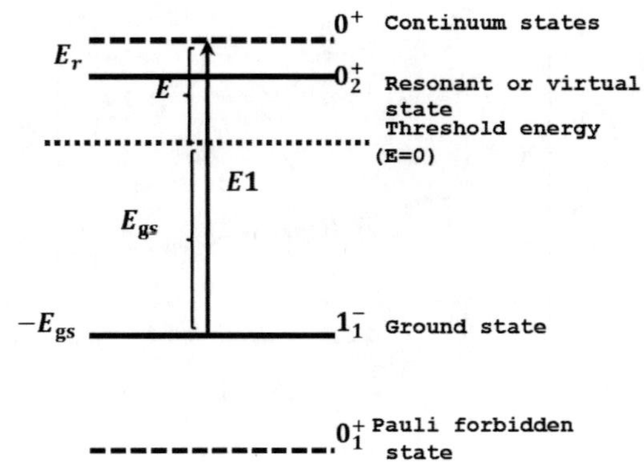

Figure 2. Energy scheme of the schematic potential.

photo-disintegration cross sections [5, 6]. The photo-disintegration cross section is observed above but not below the ^{8}Be(0^+) + n threshold. The behavior of the cross sections suggests a dominant two-body component of ^{8}Be(0^+) plus s-wave neutron, while there is no barrier to keep the neutron inside a potential.

We performed the complex scaling calculation of an $\alpha + \alpha + n$ model to look for a three-body resonance or a ^{8}Be(0^+, 2^+) + n coupled channel resonance for the $1/2^+$ state. In the results, we could not find a sharp resonance corresponding to the peak of the photo-disintegration cross section. As for the experimental sharp peak, we can well reproduce its position and shape as shown in Figure 1 without resonance solutions. The conclusion of our study is that the peak of the photo-disintegration cross section is due to the virtual state of ^{8}Be(0^+) + n.

3. Virtual state in two-body model with the complex scaling method

To understand the photo-disintegration cross section due to the virtual state, we investigate a simple two-body potential model for ^{8}Be+n system in the CSM. The CSM has been believed not possible to describe virtual states corresponding to S-matrix poles on the negative imaginary axis on the complex momentum plane. However, we show that continuum solutions of the CSM describe the virtual state through the phase shift of the s-wave.

The simple two-body potential model is expressed by the Hamiltonian $H = -\frac{1}{2}\nabla^2 + V(r)$, where $V(r) = V_1 \exp(-0.16r^2)$. Using the complex-scaled Hamiltonian $H(\theta)$, we solve the complex-scaled Scrödinger equation $H(\theta)\Psi^{\nu}_{J^\pi}(\theta) = E^{\theta}_{\nu}\Psi^{\nu}_{J^\pi}(\theta)$.

The energy levels of $J^\pi = 0^+$ and 1^- obtained by solving the complex-scaled Scrödinger equation are presented schematically in Figure 2. The potential strength V_1 is taken so as to reproduce one bound 0^+ of s-waves. But, this 0^+ solution is assumed as the Pauli forbidden state, because this model simulates ^{8}Be(0^+)+n. So, in this model, the 1^- solution describes the ground state, and there is no 0^+ bound state. To see the transition of the excited 0^+ solution from unbound to bound states, we change the strength V_1 for 0^+ state.

We can calculate the scattering phase shift using the eigenvalue solutions of $H(\theta)$ and $H_0(\theta)$

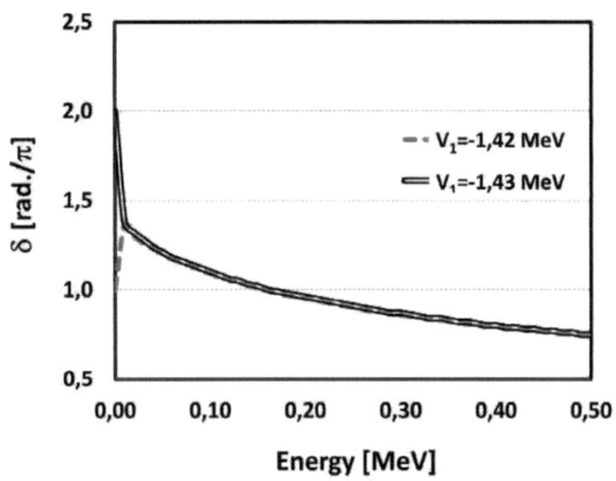

Figure 3. Calculated phase shifts at $V_1 = -1.42$ MeV and -1.43 MeV. The dashed red and open black curves are phase shift calculated at $V_1 = -1.42$ MeV and -1.43 MeV, respectively.

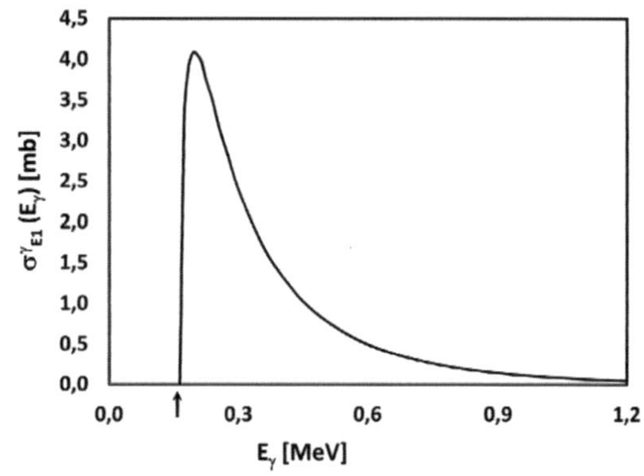

Figure 4. Calculated photo-disintegration cross sections for the Gaussian potential with the strength $V_1 = -1.42$ MeV. The arrow indicates the threshold energy.

in the CSM [7], where $H_0(\theta)$ is the free-Hamiltonian without potentials.

$$\delta(E) = N_b\pi + \sum_{r=1}^{N_r^\theta}\left\{-\cot^{-1}\left(\frac{E-E_r^{res}}{\Gamma_r/2}\right)\right\} + \sum_{c=1}^{N_c^\theta}\left\{-\cot^{-1}\left(\frac{E-\epsilon_c^{Re}}{\epsilon_c^{Im}}\right)\right\} - \sum_{k=1}^{N}\left\{-\cot^{-1}\left(\frac{E-\epsilon_{0k}^{Re}}{\epsilon_{0k}^{Im}}\right)\right\}, \tag{1}$$

where the eigenvalues are classified into the numbers of N_b bound states, N_r^θ resonant states and N_c^θ continuum states for a given θ; $N = N_b + N_r^\theta + N_c^\theta$ for total number of eigenstates. The energy eigenvalues of $H_0(\theta)$ are given as $\epsilon_{0k}^{Re} - i\epsilon_{0k}^{Im}$ with $k = 1, \cdots, N$. In Figure 3, the calculated phase shifts are shown for $V_1 = -1.42$ and -1.43 MeV, where an additional bound state appears suddenly at $V_1 = -1.43$ MeV. The shape of phase shifts of $V_1 = -1.42$ MeV indicates a virtual state behavior.

Using the virtual state solution with $V_1 = -1.42$ MeV, we also calculate the $E1$ strength and the photo-disintegration cross section. The result is shown in Figure 4, and the peak is obtained at a lower energy region in the similar way as was observed in ^{9}Be.

4. Conclusion

To understand the structure and the origin of the peak of the photo-disintegration cross section for the $1/2^+$ state in ^{9}Be, we investigated the simple two-body potential model in detail in the CSM. It was shown that the virtual state can be described by the CSM through calculations of the phase shift. The $E1$ strength is also calculated to show the similar peak as observed for the $1/2^+$ state in ^{9}Be.

References

[1] Ho Y K 1983 *Phys. Rep.* **99** 1
[2] Aoyama S, Myo T, Katō K and Ikeda K 2006 *Prog. Theor. Phys.* **116** 1
[3] Myo T, Kikuchi Y, Masui H and Katō K 2014 *Prog. Part. Nucl. Phys.* **79** 1
[4] Odsuren M, Kikuchi Y, Myo T, Aikawa M and Katō K 2015 *Phys. Rev.* C **92** 014322
[5] Utsunomiya H *et al* 2015 *Phys. Rev.* C **92** 064323
[6] Arnold C W *et al* 2012 *Phys. Rev.* C **85** 044605
[7] Odsuren M, Katō K, Aikawa M and Myo T 2014 *Phys. Rev.* C **89** 034322

IOP Conf. Series: Journal of Physics: Conf. Series **863** (2017) 012039 doi:10.1088/1742-6596/863/1/012039

A Theoretical Study of Deuteron-induced Surrogate Reactions

B V Carlson[1], R Capote[2] and M Sin[3]

[1]Instituto Tecnológico de Aeronáutica, DCTA,12.228-900 São José dos Campos, SP, Brazil
[2]NAPC-Nuclear Data Section, International Atomic Energy Agency, 1400 Vienna, Austria
[3]Department of Nuclear Physics, University of Bucharest, P.O. Box MG-11, 70709 Bucharest-Magurele, Romania

E-mail: brett@ita.br

Abstract. We use the zero-range post-form DWBA approximation to calculate deuteron elastic and nonelastic breakup cross sections and estimate the breakup-fusion cross section that could serve as a surrogate for a neutron-induced reaction cross section. We compare the angular momentum dependence of the breakup-fusion compound nucleus formation cross section with that of the corresponding neutron-induced cross section.

1. Introduction

Deuteron-induced reactions have long been studied as substitutes - surrogates - for neutron-induced reactions [1, 2]. Competition between elastic and inelastic breakup, absorption of only a neutron or a proton and absorption of the deuteron must be taken into account to determine the formation or not of a compound nucleus. The breakup-fusion reactions, in which either only a neutron or a proton is absorbed, form compound nuclei with a wide range of excitation energies and angular momenta, but are the cross sections of interest as surrogates. Such measurements of the ^{238}U(d,pf) cross section have been performed recently[3, 4, 5]. A theoretical estimate of this cross section requires the cross section for deuteron breakup - neutron fusion with the target. We estimate this by the nonelastic breakup cross section and compare its angular momentum dependence with the corresponding neutron-induced CN formation cross sections.

2. The nonelastic and breakup-fusion cross sections

The inclusive proton emission cross section from breakup can be separated into an elastic and nonelastic part[6, 7], denoted here by bu for elastic breakup and bf for nonelastic breakup, as

$$\frac{d^3\sigma}{dk_p^3} = \frac{d^3\sigma^{bu}}{dk_p^3} + \frac{d^3\sigma^{bf}}{dk_p^3} \,. \tag{1}$$

The contribution due to elastic breakup is the double differential elastic breakup cross section [8, 9] integrated over the neutron momentum. The nonelastic breakup cross section takes the form of an expectation value of the imaginary part of the optical potential,

$$\frac{d^3\sigma^{bf}}{dk_p^3} = -\frac{2}{\hbar v}\frac{1}{(2\pi)^3}\left\langle \Psi_n(\vec{k}_p,\vec{r}_n;\vec{k}_d)\right| W_n\left(\vec{r}_n\right)\left|\Psi_n(\vec{k}_p,\vec{r}_n;\vec{k}_d)\right\rangle \,, \tag{2}$$

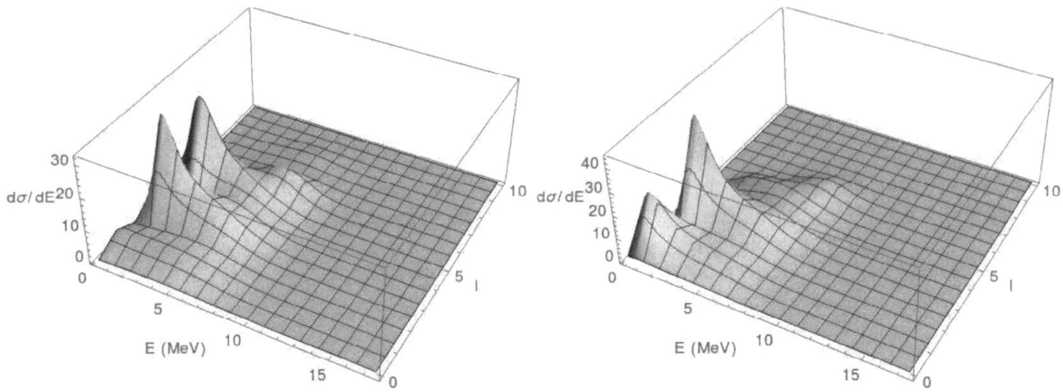

Figure 1. (a) Energy - angular momentum distribution of (d,p) differential CN formation cross section for 20 MeV deuterons incident on ^{238}U. (b) Renormalized partial wave cross section for neutrons incident on ^{238}U.

where the effective neutron wave function is given by

$$\left|\Psi_n(\vec{k}_p,\vec{r}_n;\vec{k}_d)\right\rangle = \left(\tilde{\psi}_p^{(-)}(\vec{k}_p,\vec{r}_p)G_n^{(+)}(\vec{r}_n,\vec{r}_n')\left|v_{pn}(\vec{r})\right|\psi_d^{(+)}(\vec{k}_d,\vec{R})\phi_d(\vec{r})\right\rangle . \quad (3)$$

This wave function can be well-approximated in the zero-range approximation by including the finite range correction of Ref. [10].

To perform numerical calculations as well as to obtain CN formation cross sections, the wave functions and matrix elements are expanded in partial waves. The deuteron breakup - neutron fusion cross sections at a given value of the neutron orbital angular momentum and kinetic energy are estimated by summing the nonelastic breakup partial wave cross sections over the possible values of the deuteron and proton angular momenta.

The Koning-Delaroche global optical potentials [11] are used in the proton and neutron channels while the potential of Ref. [12] is used to describe the deuteron scattering. The elastic breakup matrix elements are only conditionally convergent. The most efficient means of performing the integrals is their extension to the complex plane[13], which usually limits the numerical integration to at most several hundreds of fm.

Inclusive elastic and nonelastic (d,p) and (d,n) breakup cross sections have been calculated recently by our group in the post form, as well as by others, using both the prior and post forms of the DWBA and have verified their consistency [14, 15, 16, 17]

3. (d,p) breakup-fusion as a surrogate neutron-induced reaction

To more easily compare the neutron reaction cross section with the nonelastic (d,p) breakup cross section, shown in the left panel of Fig. 1 as a function of neutron energy and orbital angular momentum for a deuteron energy of 20 MeV, we have renormalized the former so as to reproduce the angular momentum summed nonelastic (d,p) cross section as a function of the neutron energy,

$$\frac{d\tilde{\sigma}_n}{dE}(E,l) = \frac{d\sigma_{(d,p)}}{dE}\frac{\sigma_n(E,l)}{\sigma_n(E)} \quad (4)$$

The renormalized neutron reaction cross section is shown in the right-hand panel of Fig. 1. We immediately observe the greater extension in angular momentum of the inclusive nonelastic (d,p) cross section. To better quantify the difference, we compare the average angular momenta of the distributions as a function of the neutron energy E, shown by the red curves in Fig. 2.

IOP Conf. Series: Journal of Physics: Conf. Series **863** (2017) 012039 doi:10.1088/1742-6596/863/1/012039

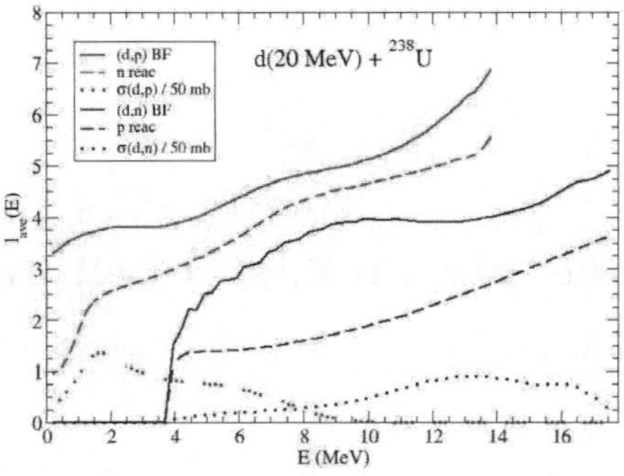

Figure 2. Average angular momentum as a function of energy of the inclusive (d, p) and neutron reaction cross sectuions (in red). The scaled (d, p) differential cross section is also shown for reference. The corresponding (d, n) and proton reaction cross sections are shown in blue.

We observe that the difference in average angular momentum of the (d, p) and neutron reaction cross sections is largest at low neutron energy, but is still about one unit of angular momentum in the region of the energy peak of the (d, p) reaction. A similar conclusion holds for the (d, n) reaction and a proton-induced one, as can be seen by comparing the blue curves in the figure.

We thus find that a (d, p) reaction typically furnishes more angular momentum to the compound nucleus than a neutron-induced reaction at the same neutron energy. This difference limits the extent to which the (d, p) reaction can substitute a neutron-induced one. A more complete study of these effects is in progress.

Acknowledgments

BVC acknowledges partial support from the Brazilian funding agencies FAPESP (Project 2009/00069-5) and the CNPq (Project 306692/2013-9) as well as from the International Atomic Energy Agency (Research Contract 17740).

References

[1] Britt H C and Cramer J D 1970 *Phys. Rev.* C **2** 1758
[2] Escher J E, Burke J T, Dietrich F S, Scielzo N D, Thompson I J and Younes W 2012 *Rev. Mod. Phys.* **84** 353
[3] Ducasse Q *et al* 2013 *EPJ Web of Conferences* **62** 08002
[4] Marini P *et al* 2015 CERN Proceedings-2015-001 195
[5] Ducasse Q *et al* 2015 arXiv:1512.0633
[6] Kasano A and Ichimura M 1982 *Phys. Lett.* **115** 81
[7] Austern N, Iseri Y, Kamimura M, Kawai M, Rawitscher G and Yahiro M 1987 *Phys. Rep.* **154** 125
[8] Baur G and Trautmann D 1976 *Phys. Rep.* **25** 293
[9] Baur G, Rösel F, Trautmann D and Shyam R 1984 *Phys. Rep.* **111** 333
[10] Buttle P J A and Goldfarb L J B 1964 *Proc. Phys. Soc.* **83** 701
[11] Koning A J and Delaroche J P 2003 *Nucl. Phys.* A **713** 231
[12] Han Y, Shi Y and Shen Q 2006 *Phys. Rev.* C **74** 044615
[13] Vincent C M and Fortune H T 1970 *Phys. Rev.* C **2** 782
[14] Lei J and Moro A M 2015 *Phys. Rev.* C **92** 044616; 2015 *Phys. Rev.* C **92** 061602(R)
[15] Potel G, Nunes F M and Thompson I J 2015 *Phys. Rev.* C **92** 034611
[16] Carlson B V, Capote R and Sin M 2015 CERN Proceedings-2015-001 163
[17] Carlson B V, Capote R and Sin M 2016 *Few-Body Syst.* **57** 307

IOP Conf. Series: Journal of Physics: Conf. Series **863** (2017) 012040 doi:10.1088/1742-6596/863/1/012040

Isomeric cross section ratios for the reactions $^{86(87)}Sr(\alpha,n(2n))^{89mg}Zr$

T V Chuvilskaya

Skobeltsyn Institute of Nuclear Physics, Lomonosov Moscow State University, Moscow II9991, Russia

E-mail: tatchuv@anna19.sinp.msu.ru

Abstract. Calculations of the excitation functions of high-spin and low-spin isomeric states in one and the same nucleus as well as the respective isomeric cross section ratios (ICSR) are carried out for the reactions $^{86(87)}Sr(\alpha,n(2n))^{89mg}Zr$.The reliability of advanced data-containing codes EMPIRE and TALYS are analyzed. The results of the measurements of isomeric cross section ratios themselves and comparison of these results with the ones calculated by the discussed codes are proved to be a promising test of the codes and their capability to reproduce angular momentum dynamics of the discussed and related reactions.

1. Introduction. ICSR as a tool for nuclear reaction studies

Excitation functions of various isotopes and energy dependences of ICSR obtained in typical α-induced nuclear reactions such as (α,n), (α,2n), etc. are presented in many papers, see, for example Refs. [1–11]. The calculations of the total excitation functions performed by use of modern computer codes demonstrate the similarity of the resulting curves and a reasonable agreement of the computed absolute values with the measured data. Various experimental and theoretical results reliably indicate that at least for near-barrier energies of the incident α-particle the compound nucleus mechanisms are dominating. In contrast to that behaviour of the ICSR curves is very different in accordance with spin characteristics of the initial, the intermediate and the final states. In many cases it turns out to be impossible to reproduce the ICSR in calculations if even a broad variety of starting assumptions is exploited. Thus some properties of the angular momentum dynamics in these reactions and spin dependence of nuclear level density are not disclosed yet.

Off-beam measurement of reaction product radiation is a popular method in studies of excitation functions and ICSR of various nuclear reactions. Distinctive features of this method are high efficiency and reliable recognition of radioactive residues. Experiments of such a type are rather convenient and labour-saving. At the same time analysis of the excitation functions of the isotope production and isomeric cross section ratios allows one to reveal reaction mechanisms of a residue production and properties of the intermediate states involved in the cascade. Indeed, usually the yield of the products strongly depends on the density of these states, particle capture and emission strength functions, discrete spectra of a nucleus under study, etc.

The isomeric cross section ratio which characterizes the relative yield of high- and low-spin isomers of one and the same isotope is determined to a large measure on spin distributions of the intermediate products of the reaction as well as on the properties of gamma-cascade in a final residue. Thus the results of the measurements of the ICSR indicate the dynamics of distribution of angular momentum over the reaction products. In particular, ICSR observed in alpha-particle-induced reactions reveal the spin dependence of the alpha-cluster strength in the energy area under study.

Recently the analysis of the excitation functions and the ICSR is based on large-scale data-containing codes, for example, EMPIRE [12] and TALYS [13]. Codes of such a type allow one to take into account various reaction mechanisms as well as a large number of characteristics of nuclear reactions and nuclei involved in them. Thus the codes turn out to be beneficial in search for various associations of these characteristics with the results of the discussed measurements.

2. Results and conclusions

Two reactions producing one and the same isomeric pair: $^{86(87)}Sr(\alpha,n(2n))^{89mg}Zr$ at the energy ranges 13-29(27) MeV are analyzed. Experimental studies of ICSR produced by these reactions were carried out by us earlier [4] using off-beam measurements of induced activity of members of the isomeric pair.

The results of the measurements and the calculations of ICSR in the framework of, EMPIRE and TALYS codes are presented in Figs. 1 and 2.

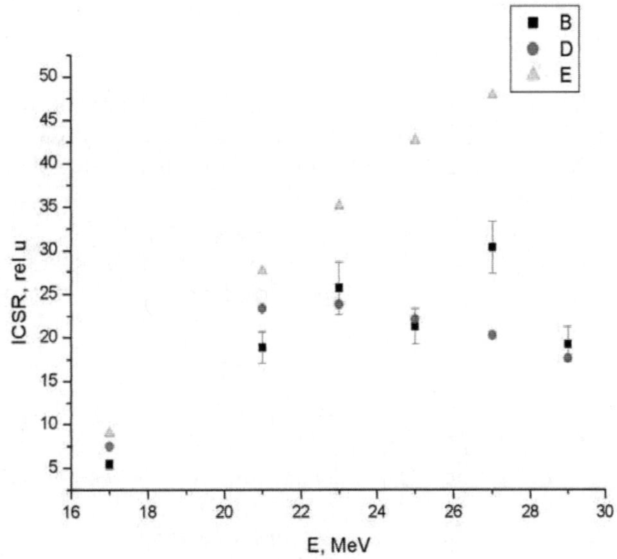

Figure 1. Values of the isomeric cross section ratios σ_g/σ_m ($9/2^+/1/2^-$) for the reaction $^{86}Sr(\alpha,n)$ ^{89mg}Zr: D – calculated by TALYS, E – calculated by EMPIRE-3.1, B – experiment.

These figures demonstrate both similarity and distinction in behaviour of ICSR of the respective reactions. On the one hand the values of ICSR of both reactions are very large and in this sense aresimilar. On the other hand the (α,n) reaction reveals unusual maximum of ICSR at an energy point of 27 MeV, while the (α,2n) reaction demonstrates monotonic increasing of this value. TALYS code turns out to be preferable for description of the first reaction, at the same time both it and EMPIRE code fall short of the ideal in description of the second reaction. The analysis of some other ICSR data by these codes may result sometimes in a more drastic disagreement between the experimental and results (see, for example [11]). Such examples provide reasons to perform the more wide analysis of related experiments with the intent to upgrade the codes.

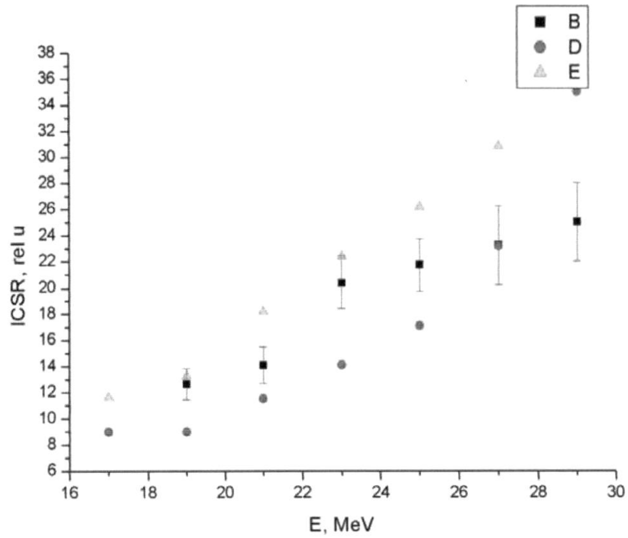

Figure 2. The came as in Fig. for the reaction ^{87}Sr(α,2n) ^{89mg}Zr: D – calculations by TALYS, E – calculated by EMPIRE–3.1, B – experiment.

In conclusion it should be emphasize that codes under discussion are constructed as a synthesis of more or less standard theoretical models and a broad spectrum of nuclear data. Nevertheless delicate properties of nuclear structure including, probably, properties of high-spin alpha-cluster states turn out to be lost in the input of these codes

The measurements of the isomeric cross-section ratios and comparison of the results with the ones obtained in the calculations seem to be a good test of the capability of nuclear reaction codes and a tool to demonstrate a manifestation on unusual properties of highly-excited states including their clustering.

References

[1] Brenquindj C L et al. 1979 *J. Inorg. Nucl. Chem.* **41** 617
[2] Grent I S, M. Rathle M 1979 *J. Phys. G; Nucl. Phys.* **5** 1741
[3] Newton G W A , Robinson V J, Shaw E M 1981 *J. Inorg. Nucl. Chem.* **43** 2227
[4] Avchukhov V D et al. 1980 *Izv. AN SSSR; ser. fiz.* **44** 155
[5] Baskova K A et al. 1983 *Ukrainskii Fizicheskii Zhurnal* **28** 1552
[6] Baskova K A et al. 1984 *Sov. J. Nucl. Phys. USSR* **39** 340
[7] Chuvilskaya T V et al. 1989 *Izv. AN SSSR; ser. fiz.* **53** 2262
[8] Tulinov A F et al. 1993 *Izv. RAN; ser. fiz.* **57** 135
[9] Nagame Y et al. 1990 *Phys. Rev. C* **41** 889
[10] Sudar S, Qaim S M 2006 *Phys. Rev. C* **73** 034613
[11] Chuvilskaya T V 2015 *JPS of Japan, Conference Proceedings* **6** 030083
[12] Herman M, Marcinkowski A, Stankewitz K 1984 *Comp. Phys. Comm.* **33** 373; https://nddx4.bnl.gov/gf/project/empire/
[13] Koning A J, Hilaire S and Duijvestijn M 2009 *Talys-1.2. A nuclear reaction program, User manual. December 22*

Microscopic time-dependent description of alpha-cluster transfer and incomplete fusion in reactions with ^{12}C and ^{16}O near Coulomb barrier

V V Samarin

Flerov Laboratory of Nuclear Reactions, Joint Institute for Nuclear Research, Dubna, Moscow region, 141980, Russia

samarin@jinr.ru

Abstract. The time-dependent Schrödinger equation has been numerically solved for the α-cluster wave function the in reaction ^{12}C+^{51}V at the energies near and above the Coulomb barrier. The α-target potential included an imaginary part as in the optical model. The α-cluster mean field of the shell model for ^{12}C and ^{16}O nuclei based on the results of Feynman's integrals method is proposed. The evolution of the α-cluster probability density was calculated for the complete and incomplete fusion as well as transfer channels. The calculated cross sections show satisfactory agreement with the experimental data for (^{12}C,α) and (^{12}C,2α) transfer channels.

1. Potentials of cluster model for reactions with ^{12}C and ^{16}O nuclei

The nuclei ^{12}C and ^{16}O have an α-cluster structure [1], which may influence the processes of nucleus-nucleus collisions. The ground states of the three- and four-body systems ^{12}C (3α) and ^{16}O (4α) were studied by Feynman's continual integrals method in Euclidean time $\tau = it$ (similar to Ref. [2]). Information on the cluster-cluster interaction in the external region (before the Coulomb barrier) may be obtained from elastic scattering data [3, 4]. In the optical model the well-known Akyüz-Winther potential [5] is widely used. Some function with the maximum in the center may be used as the potential in the internal region. This maximum is a result of the Coulomb repulsion and the Pauli Exclusion Principle for nucleons. The nuclear part of the α-α-potential has the form of the sum of two terms of the Woods-Saxon type

$$V_{\alpha-\alpha}(r) = -V_0\left[1+\exp\left(\frac{r-R_V}{a_V}\right)\right]^{-1} + U_0\left[1+\exp\left(\frac{r-R_U}{a_U}\right)\right]^{-1}. \qquad (1)$$

For ^{12}C and ^{16}O nuclei the same parameter values $U_0 = 38$ MeV, $R_V = 3.73$ fm, $R_U = 2.21$ fm, $a_V = a_U = 0.512$ fm were used, whereas the values of V_0 were slightly different, *viz.* 29.75 MeV and 29.00 MeV, respectively. These values give the theoretical binding energies 7.2±0.2 MeV for ^{12}C and 14.5±0.2 MeV for ^{16}O which are close to the experimental values (taken from [6]) 7.37 MeV and 14.53 MeV, respectively. These values also provide absence of the bound state for the ^{8}Be nucleus. The form (1) was used for the mean field of the α-cluster shell model for ^{16}O, ^{12}C and ^{8}Be nuclei. This potential is similar to the potentials used in Refs. [7, 8]. The values $V_0 = 53.8$ MeV, $R_V = 2.05$ fm, $U_0 = 58$ MeV, $R_U = 1.4$ fm, $a_V = a_U = 0.25$ fm for the α-^{12}C potential give the separation energy of the α-particle and the charge distribution similar to the experimental data [6]. The potential *V*(*r*) of interaction of the α-cluster with heavy target nuclei has the imaginary part (as in the optical model) for description of absorption of the α-cluster (its dissolving in nucleons) in the central part of the nucleus. The real part of *V*(*r*) has the form (1) with the Akyüz-Winther parametrization [5] of the attractive part. The parameters of the repulsive part of the potential are chosen to provide the equality of the energy of the α-cluster ground state in the potential well near the surface region [9, 10] to $-E_{sep,\alpha}$, where $E_{sep,\alpha}$ is the separation energy of the α-particle.

2. Time-dependent approach to study of reactions with cluster nuclei

The time-dependent Schrödinger equation is numerically solved (similar to Ref. [11]) for the α-cluster wave function in the collision ^{12}C+^{51}V at energies near the Coulomb barrier. The dynamic rearrangement of α-clusters is taken into account by changing the time-dependent potential. After "crossing of the surfaces" the ^{12}C(core+α)+^{51}V nuclear system transforms into ^{8}Be(α+α)+^{55}Mn, and later into ^{4}He+^{59}Co. This approximation may be considered as a generalization of the approach of orbital momentum bins for different reaction channels: fusion, (^{12}C,α) and (^{12}C,2α) transfer as well as (^{12}C,3α) break-up [12].

The probability density for one of the three independent α-clusters of ^{12}C is shown in figure 1. In the grazing collision (figure 1*a*) α-cluster transfer takes place with the probability $p(b)$, where b is the impact parameter of the collision. At close contact of nuclei (figure 1*b*) the emission of the α-cluster by the break-up of the projectile nucleus and by the decay of excited states of the transferred α-cluster in the target nucleus take place and decrease the transfer probability $p(b)$.

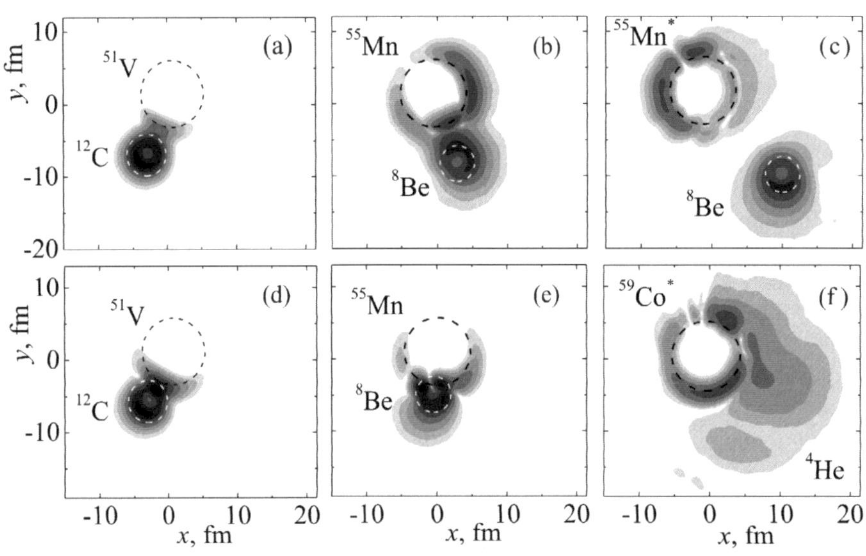

Figure 1. The probability density for the α-cluster of ^{12}C in the collision with ^{51}V at $E_{lab} = 51$ MeV: (a – c) α-cluster transfer at impact parameter of the collision $b = 6.5$ fm; (d – f) α-cluster transfer with emission of the ^{4}He by the break-up of the projectile nucleus and by the decay of excited states of the transferred α-cluster in the target nucleus at $b = 5.5$ fm. Radii of circumferences equal the radii of nuclei $R_i = r_0 A_i^{1/3}$, $r_0 = 1.25$ fm.

The α-cluster transfer probability $p(b)$ is shown in figure 2*a*. The probabilities of complete fusion $P_0(b)$, incomplete fusion ("^{8}Be" capture") for (^{12}C,α) channel $P_1(b)$, and α-transfer ("α-capture") for (^{12}C,2α) channel $P_2(b)$ are

$$P_0(b) = p^3(b),\ P_1(b) = 3p^2(b)[1 - p(b)],\ P_2(b) = 3p(b)[1 - p(b)]^2. \quad (2)$$

The corresponding cross sections σ_i, $i = 1,2,3$ are calculated by integration over impact parameter b

$$\sigma_i = 2\pi\int_0^\infty P_i(b)bdb. \quad (3)$$

The theoretical cross sections calculated by the formulas (2), (3) show satisfactory agreement with the experimental data [13] for (^{12}C,α) and (^{12}C,2α) channels (figure 2*b*). The approximation of independent transfer of α-clusters is not quite correct for calculation of the complete fusion cross section because the result at energies above the Coulomb barrier is less than the experimental values and the fusion cross section calculated by tunneling of the partial waves of ^{12}C nucleus through the barrier [6]. Nevertheless, the proposed model may be used for the analysis of experimental data for reactions with ^{12}C, ^{16}O and other α-cluster nuclei.

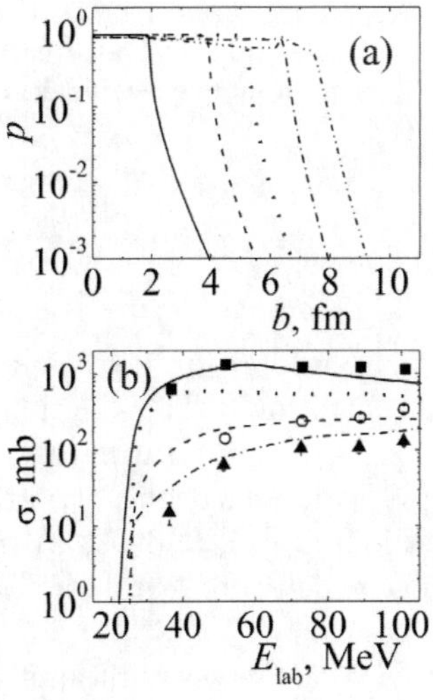

Figure 2. (a) α-cluster transfer probability for E_{lab} = 26, 30, 36, 51 and 100 MeV (solid, dashed, dotted, dash-dotted and dash-dot-dotted curves, respectively). (b) The experimental cross sections for complete fusion (squares), incomplete fusion for (^{12}C,α) channel (circles) and α-transfer for (^{12}C,2α) channel (triangles) from Ref. [13]. The dotted, dashed and dash-dotted curves show the corresponding theoretical cross sections calculated by the formulas (2), (3); the solid curve is the fusion cross section calculated by tunneling of the partial waves of ^{12}C nucleus through the Coulomb barrier [6].

References

[1] Freer M 2007 *Rep. Prog. Phys.* **70** 2149
[2] Samarin V V and Naumenko M A 2016 *Bull. Rus. Acad. Sci. Phys.* **80** 283
[3] Afzal S A, Ahmad A A Z and Ali S 1969 *Rev. Mod. Phys.* **41** 247
[4] Ali S and Bodmer A R 1969 *Nucl. Phys.* **80** 99
[5] Winther A 1994 *Nucl. Phys.* A **572** 191
[6] http://nrv.jinr.ru/nrv/
[7] Langanke K and Stademann R 1984 *Phys. Rev.* **C 29** 40
[8] Langanke K 1982 *Nucl. Phys.* **A 373** 493
[9] Winslow G 1954 *Phys. Rev.* **54** 1032
[10] Samarin V V 2014 *Bull. Rus. Acad. Sci. Phys.* **78** 1124
[11] Zagrebaev V I, Samarin V V and Greiner W 2007 *Phys. Rev.* **C 75** 035809
[12] Siwek-Wilczynska K, Du Marchie van Voorthuysen E H, van Porta J, Siemssen R H and Wilczynski J 1979 *Nucl. Phys.* **A 330** 150
[13] Parker D J, Asher J, Conlon T W, and Naqib I 1984 *Phys. Rev.* **C 30** 143

Four- and three-body breakup mechanism of ^{6}Li elastic scattering

S Watanabe[1], T Matsumoto[2], K Ogata[3] and M Yahiro[2]

[1]RIKEN Nishina Center, Wako, Saitama 351-0198, Japan
[2]Department of Physics, Kyushu University, Fukuoka 812-8581, Japan
[3]Research Center for Nuclear Physics (RCNP), Osaka University, Ibaraki, Osaka 567-0047, Japan

E-mail: shin.watanabe.vf@riken.jp

Abstract. We investigate a breakup mechanism of ^{6}Li elastic scattering on heavy targets (T = ^{209}Bi or ^{208}Pb) with the four-body version of the continuum-discretized coupled-channels method (four-body CDCC). Four-body CDCC successfully reproduces measured elastic cross sections with no adjustable parameter, and we can then clearly discuss the four-body dynamics. Our analysis shows that $d\alpha$ breakup (^{6}Li + T $\to d + \alpha$ + T) is much more essential than $np\alpha$ breakup (^{6}Li + T $\to n + p + \alpha$ + T) in ^{6}Li scattering.

1. Introduction

In reactions of weakly-bound nuclei, projectile breakup plays an important role and the treatment of projectile-breakup effects is essential to describe scattering. The continuum-discretized coupled-channels method (CDCC) was proposed as a method for treating breakup effects [1, 2, 3]. Nowadays, CDCC is applied to not only three-body scattering (two-body projectile + T) but also four-body scattering (three-body projectile + T) [4, 5, 6, 7], where T denotes a target. CDCC for three- and four-body scattering are now called three- and four-body CDCC, respectively.

^{6}Li + ^{209}Bi scattering near the Coulomb barrier energy ($E_{\rm b}^{\rm Coul} \approx 30$ MeV) was first analyzed with three-body CDCC based on the d+α + ^{209}Bi three-body model [8]. However, the calculation could not reproduce the measured elastic cross section without introducing a normalization factor for the optical potentials. This problem was solved by four-body CDCC based on the $n + p + \alpha$ + ^{209}Bi four-body model [6]. The four-body calculation describes the experimental data without introducing any adjustable parameter. As an interesting finding, we showed that three-body CDCC can reproduce the cross section if the phenomenological d-optical potential is replaced by the single-folding potential that does not include d-breakup effects [6]. This suggests that d (*i.e.* the n-p subsystem of ^{6}Li) hardly breaks up during ^{6}Li scattering. In this work, we investigate the breakup mechanism within the four-body CDCC framework and validate the evidence.

2. Decomposition of the CDCC model space

We only recapitulate the treatment of model space in four-body CDCC; see Ref. [3, 7] for the detail. In four-body CDCC, the Schrödinger equation is solved in the model space P spanned by

IOP Conf. Series: Journal of Physics: Conf. Series **863** (2017) 012042 doi:10.1088/1742-6596/863/1/012042

the ground and discretized-continuum states of ^{6}Li: $P = \sum_{\gamma=0}^{N} |\Phi_\gamma\rangle \langle\Phi_\gamma|$, where Φ_γ represents the γ-th eigenstate, and the $\gamma = 0$ and $\gamma = 1$–N correspond to the ground and discretized-continuum states, respectively. The Φ_γ are obtained as pseudostates by using the Gaussian expansion method [9].

In this paper, we investigate the breakup mechanism by restricting the model space P. For this purpose, we first specify whether the breakup state Φ_γ ($\gamma = 1$–N) is the $d\alpha$-dominant or $np\alpha$-dominant breakup state by calculating the squared overlap between Φ_γ and the d ground state $\psi^{(d)}$: $\Gamma_\gamma^{(d\alpha)} = \left|\langle\phi^{(d)}|\Phi_\gamma\rangle\right|^2$. If $\Gamma_\gamma^{(d\alpha)}$ is larger (smaller) than 0.5, the state is defined as a $d\alpha$-dominant state $\Phi_\gamma^{(d\alpha)}$ ($np\alpha$-dominant state $\Phi_\gamma^{(np\alpha)}$). With the $d\alpha$- and $np\alpha$-dominant state above, the CDCC model space P can be decomposed into the three parts $P = P_0 + P_{d\alpha} + P_{np\alpha}$, where

$$P_0 = |\Phi_0\rangle \langle\Phi_0|, \; P_{d\alpha} = \sum_\beta |\Phi_\beta^{(d\alpha)}\rangle \langle\Phi_\beta^{(d\alpha)}|, \; P_{np\alpha} = \sum_\delta |\Phi_\delta^{(np\alpha)}\rangle \langle\Phi_\delta^{(np\alpha)}|. \quad (1)$$

In the following discussion, we calculate cross sections by switching on and off to clarify the reaction dynamics.

3. Results

First, we show the validity of four-body CDCC. Figure 1 shows the angular distribution of elastic cross section for ^{6}Li + ^{209}Bi scattering at 32.8 MeV. The three-body CDCC calculation (dashed line) underestimates the experimental data as reported in Ref. [8]. We then apply four-body CDCC in order to explain this discrepancy. Four-body CDCC (solid line) perfectly reproduces the experimental data without introducing any adjustable parameter. ^{6}Li + ^{209}Bi scattering near the Coulomb barrier energy is thus described by four-body CDCC. Therefore, we can clearly discuss the breakup mechanism below.

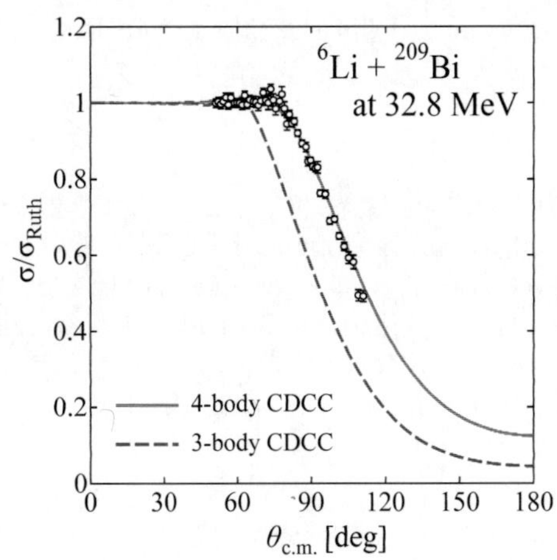

Figure 1. (Color online) Elastic cross sections (divided by the Rutherford cross section) for ^{6}Li + ^{209}Bi scattering at 32.8 MeV. The solid (dashed) line represents the result of four-body CDCC (three-body CDCC) calculation. The experimental data is taken from Refs.[10, 11].

Next, we show ^{6}Li + ^{208}Pb scattering at 39 MeV in Fig 2, which is almost the same as ^{6}Li + ^{209}Bi scattering at 32.8 MeV. The solid and dotted lines correspond to the full and 1ch calculations, respectively. These are nothing but the calculations in P and P_0, respectively. The difference comes from breakup effects and the full calculation reproduces the experimental data well by virtue of breakup effects.

Now, we switch on only the subspace $P_{np\alpha}$ or $P_{d\alpha}$ from P_0 in order to investigate the breakup mechanism. The dot-dashed line (a) represents the calculation of $P_0 + P_{np\alpha}$ and it is close to 1ch calculation (dotted line). On the other hand, the dashed line (b) corresponds to the calculation of $P_0 + P_{d\alpha}$ and it simulates the full calculation (solid line) almost perfectly. It should be noted that the number of $d\alpha$-dominant states is much less than that of $np\alpha$-dominant states in the present model space P. As seen above, $d\alpha$ breakup is favored in ^{6}Li scattering. This property is now called *$d\alpha$-dominance*, and we have found that the *$d\alpha$-dominance* is realized in a wide energy range [7]. It has been thus confirmed that d (*i.e.* the n-p subsystem of ^{6}Li) hardly breaks up during ^{6}Li scattering.

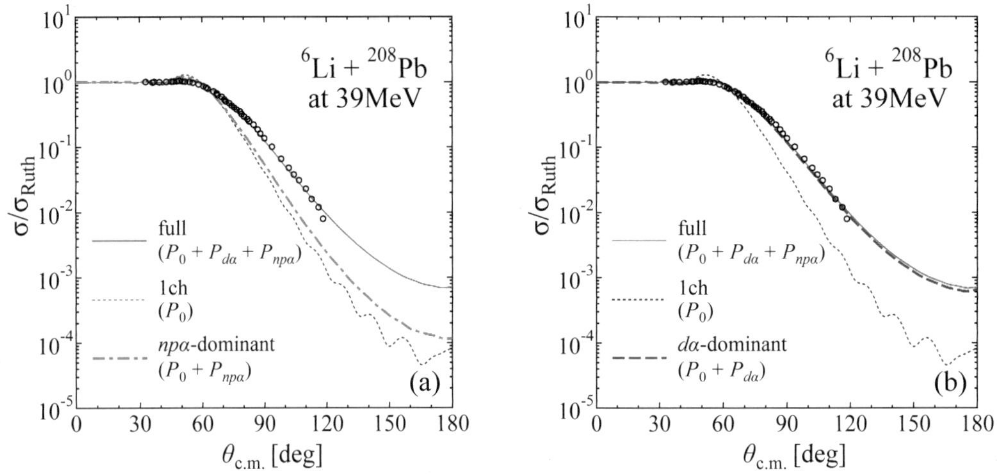

(a) Effects of $P_{np\alpha}$ (b) Effects of $P_{d\alpha}$

Figure 2. (Color online) Elastic cross sections for ^{6}Li + ^{208}Pb scattering at 39 MeV. The solid and dotted lines correspond to the results of full and 1ch calculations, respectively. The dot-dashed line (a) represents the calculation with the model space $P_0 + P_{np\alpha}$, whereas the dashed line (b) shows the calculation with $P_0 + P_{d\alpha}$. The experimental data is taken from Ref. [12].

4. Summary

We have investigated four-body dynamics of ^{6}Li elastic scattering ($n + p + \alpha + \mathrm{T}$, T = ^{209}Bi or ^{208}Pb). The elastic scattering are successfully described in the four-body CDCC framework without introducing any adjustable parameter. We can then clearly analyze the breakup mechanism. Our analysis shows that the $d\alpha$-dominant breakup is much more essential compared with the $np\alpha$-dominant breakup for describing the scattering (*$d\alpha$-dominance*). This justifies the fact that d (*i.e.* the n-p subsystem of ^{6}Li) hardly breaks up during ^{6}Li scattering.

References

[1] Kamimura M *et al* 1986 *Prog. Theor. Phys. Suppl.* **89** 1
[2] Austern N *et al* 1987 *Phys. Rep.* **154** 125
[3] Yahiro M *et al* 2012 *Prog. Theor. Exp. Phys.* 01A206
[4] Matsumoto T *et al* 2004 *Phys. Rev.* C **70** 061601(R)
[5] Matsumoto T *et al* 2006 *Phys. Rev.* C **73** 051602(R)
[6] Watanabe S *et al* 2012 *Phys. Rev.* C **86** 031601
[7] Watanabe S *et al* 2015 *Phys. Rev.* C **92** 044611
[8] Keeley N *et al* 2003 *Phys. Rev.* C **68** 054601
[9] Hiyama E *et al* 2003 *Prog. Part. Nucl. Phys.* **51** 223
[10] Aguilera E F *et al* 2000 *Phys. Rev. Lett.* **84** 5058
[11] Aguilera E F *et al* 2001 *Phys. Rev.* C **63** 061603
[12] N Keeley *et al* 1994 *Nucl. Phys.* A **571** 326

IOP Conf. Series: Journal of Physics: Conf. Series **863** (2017) 012043 doi:10.1088/1742-6596/863/1/012043

Fusion-fission of superheavy nuclei and clustering phenomena

M G Itkis, I M Itkis, G N Knyazheva and E M Kozulin

Joint Institute for Nuclear Research, 141980 Dubna, Russia.

E-mail: **itkis@jinr.ru**

Abstract. Results of the study of mass-energy distributions of binary fragments for a wide range of nuclei with Z = 82-122 produced in reactions with heavy ions at energies close and below the Coulomb barrier are reported. The role of the shell effects, the influence of the entrance channel asymmetry and the deformations of colliding nuclei on the mechanism of the fusion-fission and quasifission processes are discussed. The observed peculiarities of the mass and energy distributions of reaction fragments are determined by the shell structure of the formed fragments.

1. Introduction

Nowadays more than 30 nuclei around the superheavy elements (SHE) region have been synthesized in fusion reactions of heavy nuclei. The existence of these nuclei is completely determined by proton and neutron closed shells at Z = 114-120 and N = 184 (island of stability). In the collisions of two massive nuclei many different reaction mechanisms such as elastic and inelastic scatterings, deep inelastic reactions, fusion and de-excitation of a compound nucleus, fusion-, quasi- and fast fission may take place. The existence of deep valleys on potential energy surface caused by shell effects is likely to be responsible for the manifestation of clustering phenomena in different reaction channels.

In low-energy collisions of heavy ions the quasifission (QF) process caused by the shell effects dominates hindering formation of compound nucleus [1]. QF is a transitional mechanism between deep-inelastic collisions and complete fusion, in which the composite system separates in two main fragments without forming a compound nucleus (CN). QF happens to be the most important mechanism that prevents the formation of SHE in the fusion of heavy nuclei.

This paper presents the results of the study of mass-energy distributions of binary fragments for a wide range of nuclei with Z = 82-122 produced in reactions heavy ions at energies close and below the Coulomb barrier. Velocity vectors of binary reaction products were measured using the two-arm time-of-flight spectrometer CORSET [2].

2. Shell effects in quasifission

In the last decade a great success in the synthesis of SHE with Z = 112–118 was achieved bombarding actinide targets with ^{48}Ca ions (hot fusion reactions). In these reactions, in contrast to the cold fusion, neutron excess in the composite systems leads to difference in a cluster composition of the compound system (due to the neutron shell with N = 184) as well as in its decay products.

One can see from Figure 1 that mass-energy distributions of binary reaction products, obtained in the ^{48}Ca+^{208}Pb, ^{232}Th, ^{238}U, ^{244}Pu, ^{248}Cm reactions change from triangular shape for the reaction with spherical target ^{208}Pb [3], where CN-fission process dominates, to the wide two-humped shape in the case of reactions with actinide targets [4] determined by the QF process. This onset of quasifission for the latter reactions might be explained by the target deformation and, consequently, the orientation effect that favors the QF process which manly leads to the formation of the clusters in the exit channel of the reaction.

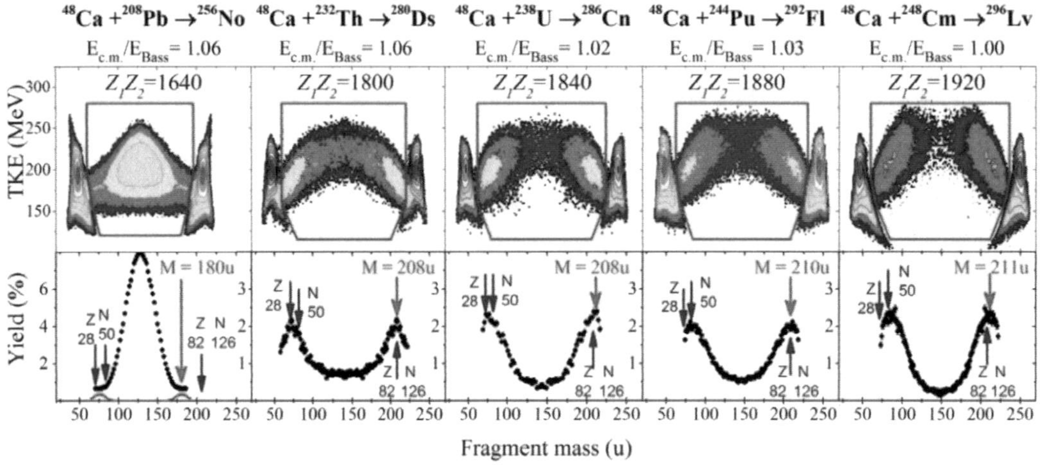

Figure 1. Top: Mass-TKE matrices for the reactions $^{48}Ca + ^{208}Pb, ^{232}Th, ^{238}U, ^{244}Pu, ^{248}Cm$ at energies close to the Coulomb barrier; bottom: open circles are mass distributions for fission-like fragments inside the regions limited by red lines on mass-TKE matrices; arrows indicate the positions of neutron and proton shells.

Generally, in heavy-ion-induced reactions the formation of asymmetric QF fragments is connected with the strong influence of the nuclear shell at $Z = 82$ and $N = 126$ (double magic lead). In fact for the $^{48}Ca + ^{232}Th$ and $^{48}Ca + ^{238}U$ reaction the maximum yield corresponds to fragments with masses around 208 u. However, in reactions with heavier targets the maximum is shifted up to 211 u for the reaction $^{48}Ca + ^{248}Cm$. Notice, for the reaction $^{64}Ni + ^{238}U$, the maximum yield of asymmetric QF fragments corresponds to the heavy mass 215 u. As it was shown in [5] the shells in light fragment at $Z = 28$ and $N = 50$ could be effective, together with the shell $Z = 82$ and $N = 126$, and could lead to the shift of the asymmetric QF peak.

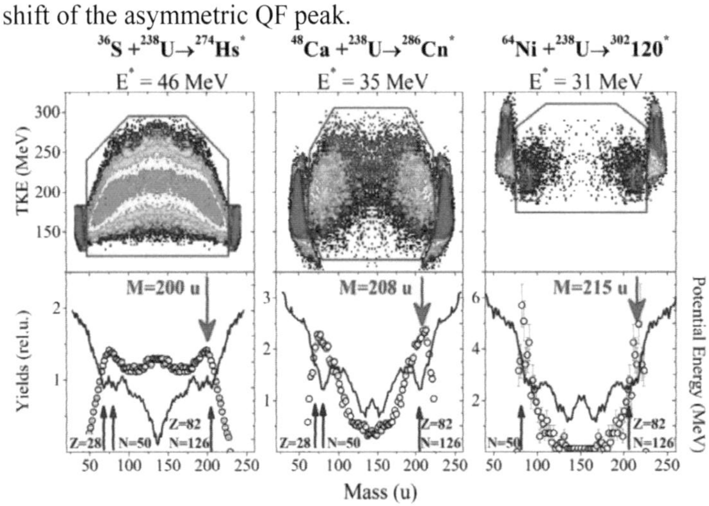

Figure 2. Top: Mass-TKE matrices for the reactions $^{36}S, ^{48}Ca, ^{64}Ni + ^{238}U$ at energies close to the Coulomb barrier; bottom: open circles are mass distributions for fission-like fragments inside the regions limited by red lines on mass-TKE matrices; solid lines are the potential energies as a function of mass.

Figure 2 shows the mass-energy distributions of binary fragments obtained in the reactions of ^{36}S, ^{48}Ca, ^{64}Ni ions with uranium target at energies close to the Coulomb barrier. The Coulomb factors are very different for these systems: $Z_1Z_2 = 1472$, 1840 and 2576 for the reactions $^{16}S + ^{238}U$, $^{48}Ca + ^{238}U$ and $^{64}Ni + ^{238}U$, respectively. In the bottom panel of Fig. 2 the potential energy at scission point calculated with NRV code [6] using the proximity model along with the experimental mass distributions are shown. A strong correlation between the maxima in mass distributions and the minima of the potential energy is likely to be responsible for the shift of the heavier mass asymmetric peaks. These effects on the fragments formation due to the nuclear shells are evidences of clustering phenomena. Another noticeable case of the importance of shell closures in driving the mass flow is reported in [7] for the system ^{88}Sr + ^{176}Yb.

3. Fission of heavy and superheavy nuclei

3.1. Bimodal fission

The question about symmetric and asymmetric modes in low-energy nuclear fission arose immediately after the discovery of nuclear fission. The numerous theoretical works [8, 9, 10] showed that multimodality is caused by the valley structure of the deformation potential energy surface of a fissioning nucleus.

The phenomenon of bimodal fission, discovered in the 1980's for the case of spontaneous and low-energy fission of nuclei in the Fm-Rf region [11], means the co-existence of two fission modes with different total kinetic energies for the same symmetric mass division. Bimodality is connected with a possibility for these nuclei to have in both fragments the number of neutrons and protons close to magic numbers Z = 50 and N = 82 (Super-Short (SS) mode). This possibility will sharply disappear when moving from Fm to more heavy nuclei. However, calculations show that the SS-valley should exist up to $^{270,\,272}$Hs and even $^{278,\,290}$Ds nuclei.

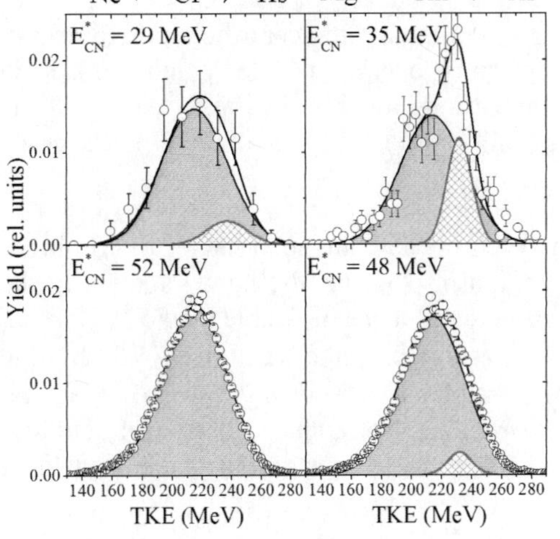

Figure 3. TKE distributions of fragments with masses $A_{CN}/2 \pm 20$ u obtained in the reactions ^{22}Ne + ^{249}Cf (left panels) and ^{26}Mg + ^{248}Cm (right panels). High and low kinetic energy components are given as hatched and filled regions, respectively.

The TKE distributions of symmetric fragments with masses $A_{CN}/2 \pm 20$ u for the reactions ^{22}Ne + ^{249}Cf and ^{26}Mg + ^{248}Cm are presented in Fig. 3. At sub-barrier energies the TKE distributions of symmetric fragments have a complex structure with mean TKEs and its dispersions higher than predicted by the liquid drop model (LDM). At the same time the mass distributions for both nuclei are symmetric. The structural features of the TKE distributions at low excitation may arise due to the fact that in symmetric fission both fragments are close to the spherical neutron shells with N = 82. A similar behavior could be observed when both fission fragments are close to the spherical proton shell Z = 50. Thus the properties of mass and TKE distributions show the characteristic features for bimodal fission.

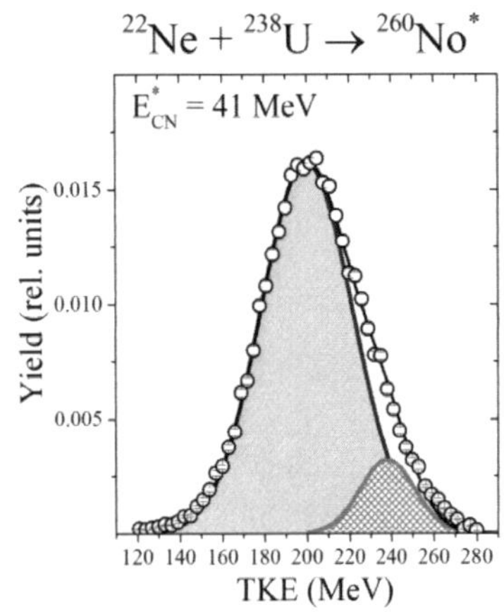

Figure 4. The TKE distribution of fission fragments obtained in the reaction $^{22}Ne + ^{238}U$ and its' decomposition into high and low energy components.

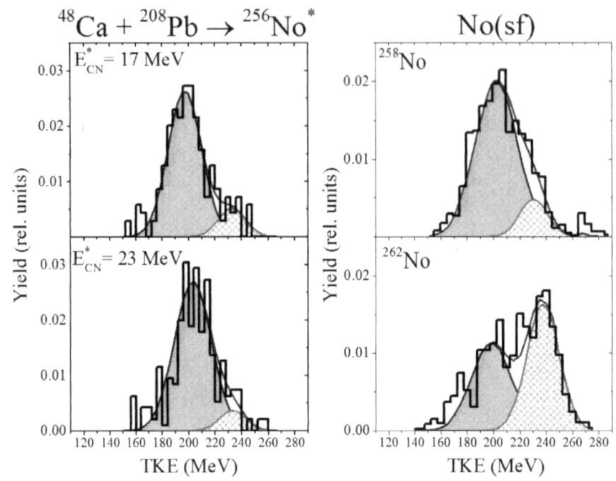

Figure 5. The TKE distributions of fragments The TKE distributions of fission fragments obtained in the reaction $^{48}Ca + ^{208}Pb$ [3] and in spontaneous fission of ^{258}No and ^{262}No taken from [12].

The TKE distribution of symmetric fragments obtained in the $^{22}Ne + ^{238}U$ reaction at an excitation energy of $^{260}No^*$ of 41 MeV is shown in Fig. 4. For this reaction the contribution of high energy component is about 7%. The presence of SS-mode in energy distributions of fragments of the $^{256}No^*$ induced fission at low excitation energies was observed in [3], although the contribution high energy component in this case is rather small: about 2.5% at an excitation energy of 17 MeV and about 1.7% at 23 MeV (see Fig. 5 left panel). In the case of spontaneous fission of ^{258}No and ^{262}No [12] the contributions of high energy component are about 14% and 54%, respectively (see Fig. 5 right panel).

3.2. Superasymmetric fission

For the first time the superasymmetric mode was observed in the compound nuclei fission in Pb region [13]. The enhancement of the mass yield in the region 65-75 u for the light fragment in the fission of ^{213}At and ^{210}Po compound nuclei is connected with the influence of double magic Ni ($Z = 28$, $N = 50$) and double magic Sn ($Z = 50$, $N = 82$). Notice that in this case the ratio $A_H/A_L \approx 2$, and the yield is around 0.01%. The superasymmetric fission mode with mass ratio of $A_H/A_L \approx 2.5$, caused by the closed shells $Z = 28$ and $N = 50$, was also found in thermal-neutron-induced fission of actinides nuclei [14]. In this case only the light fragment is close to the double magic Ni and the yield of superasymmetric mode does not exceed 10^{-4}%.

The mass and energy distributions of fragments obtained in the reaction $^{22}Ne + ^{238}U$ are shown in figures 6 and 7. The mass distribution of the symmetric fragments has a nearly Gaussian shape and the average TKE shows a parabolic dependence on fragment mass typical for fission of excited compound nuclei as established by the LDM, whereas in the mass region around 52/208 u ($A_H/A_L \approx 4.3$), that corresponds to the formation of fissioning pair of two magic nuclei Ca/Pb, an increase of fragment yields was observed. Moreover the total kinetic energy for these fragments is found to be about ~30 MeV higher than predicted by the LDM. The higher TKE indicates that the asymmetric fragments originate from more compact scission configuration as compared to normal fission. In the low panel of figure 6 the experimental mass distribution together with the prediction of W. Greiner for thermal neutrons induced fission of ^{255}Fm is presented. According to this calculation the yield of about 0.01% is expected due to the influence of the closed shells, while for spontaneous fission of ^{255}Fm the yield of 10^{-4}% is expected for this superasymmetric mode [15].

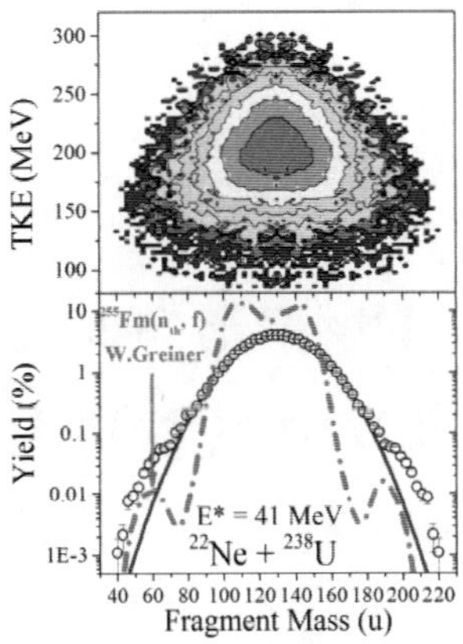

Figure 6. Double differential cross section and mass yield for the $^{22}Ne + {}^{238}U$ reaction. Thick line represents the LDM calculation, dash-dot line is W. Greiner calculations for $^{255}Fm(n_{th},f)$

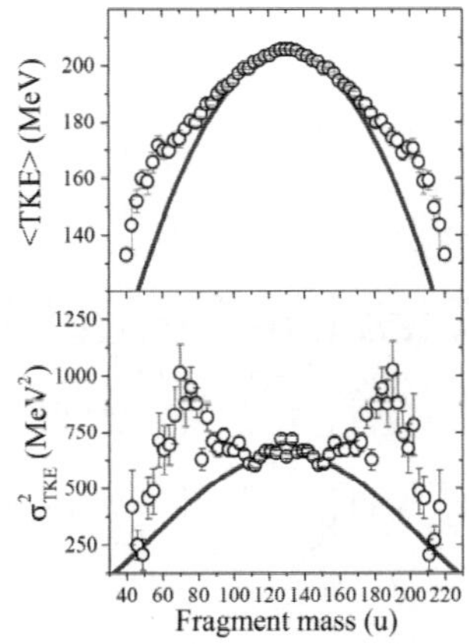

Figure 7. The mean TKE and its' dispersion as a function of fragment mass. The thick lines represent the LDM calculations.

In present work the superasymmetric mode with ratio $A_H/A_L \approx 4.3$ caused by the influence of double magic Ca and double magic Pb has been observed in fission of excited $^{260}No^*$ compound nucleus. At an excitation energy of 41 MeV the yield of these fragments is about 5×10^{-2}%.

4. Conclusion

The clustering phenomena have a great impact on the heavy-ion reaction products. In the case of the quasifission process at least one of the fragment mass is determined mainly by nuclear shells with Z = 28, 82 and N = 50, 126; whereas in the case of the fusion-fission the nuclear shells with Z = 50 and N = 82 play an important role in the fragment formation.

The static deformation of the reaction partners is responsible for the evolution of the composite system and the appearing of the clusters in the exit channel for quasifission process.

The bimodal fission caused by clustering phenomena was observed for fission of superheavy nuclei $^{271,274}Hs^*$ and $^{256}No^*$. For the compound nucleus $^{260}No^*$ formed in the reaction $^{22}Ne + {}^{238}U$ at the initial excitation energy $E^* = 41$ MeV the bimodal fission as well as superasymmetric fission with mass ratio $A_H/A_L \approx 4.3$ were observed.

References

[1] Itkis M G, Vardaci E, Itkis I M, Knyazheva G N and Kozulin E M 2015 *Nucl. Phys.* A **944** 204
[2] Kozulin E M *et al.* 2008 *Instrum. Exp. Tech.* **51** (1) 44
[3] Prokhorova E V *et al.* 2008 *Nucl. Phys.* A **802** 45
[4] Kozulin E M *et al.* 2014 *Phys. Rev.* C **90** 054608
[5] Itkis I M *et al.* 2011 *Phys. Rev.* C **83** 064613
[6] http://nrv.jinr.ru/nrv
[7] Kozulin E M *et al.* 2014 *Phys. Rev.* C **89** 014614
[8] Moller P, Nix J R and Swiatecki W J 1987 *Nucl. Phys.* A **469** 1
[9] Pashkevich V V 1988 *Nucl. Phys.* A **477** 1
[10] Brosa U, Grossmann S and Muller A 1990 *Phys. Rep.* **197** 167
[11] Hulet E K *et al.* 1986 *Phys. Rev. Lett.* **56** 313
[12] Hulet E K *et al.* 1994 *Phys. Atom. Nucl.* **57** 1165
[13] Itkis M G, Okolovich V N, Russanov A Ya and Smirenkin G N 1985 *Z. Phys.* A **320** 433
[14] Rochmann D *et al.* 2004 *Nucl. Phys* A **735** 3
[15] Greiner W 2001 *Proc. Int. Workshop on Fusion Dynamics at the Extremes (Dubna, Russia, 25-27 May 2000)* ed Yu Ts Oganessian and V I Zagrebaev (Singapore: World Scientific) p 1

Pre-Scission Model Calculation of Fission Fragment Mass and Total Kinetic Energy Distributions for Even-Even Fm, No and Rf Isotopes

N Carjan[1,2], F A Ivanyuk[3] and Yu Ts Oganessian[1]
[1] Joint Institute for Nuclear Research, 141980 Dubna, Moscow Region, Russia
[2] National Institute for Physics and Nuclear Engineering "Horia Hulubei", Reactorului 30, RO-077125, POB-MG6, Magurele-Bucharest, Romania
[3] Institute for Nuclear Research, Kiev, Ukraine

Abstract. The main properties of the fission fragments in spontaneous fission of even-even isotopes of Fm (Z=100), No (Z=102) and Rf (Z=104) are estimated using a pre-scission point model. The underlying potential energy surfaces are calculated with Strutinsky's shell correction procedure. The parametrization of the nuclear shapes is based on Cassini ovals generalized by the inclusion of three additional shape parameters: α_1, α_4 and α_6. It represents a natural way to describe scission configurations. The corresponding fragment-mass distributions are estimated supposing they are due to thermal fluctuations in the mass asymmetry degree of freedom. A detailed comparison with all existing data for Fm, No and Rf isotopes is presented. For these three series of isotopes the experimentally observed transition from asymmetric to symmetric fission, that happens with increasing mass number A, is well reproduced. In lighter isotopes (e.g. ^{254}Fm and ^{254}Rf) two mass-asymmetric fission modes are predicted to occur with comparable yields: one having relatively compact and the other relatively elongated scission configurations. On the other hand, in heavier isotopes (e.g. ^{264}Fm and ^{264}Rf) the fragment-mass distributions are predicted to be narrow single-peaked around A/2 corresponding to essentially one compact fission mode. We call this type of fission "super-symmetric". The corresponding distributions of the total kinetic energy of the fragments are also calculated (in the point-charge approximation) and compared with measurements. Despite the fact that the dynamical effects were neglected, we have obtained a quantitative agreement with the experimental data.

1. Introduction

The scission-point model [1] was recently improved and used to calculate fission-fragment mass and total kinetic energy distributions for Fm, No, Rf and Sg isotopes [2].

The scission process (from the beginning of the neck rupture at finite radius ($r_{neck} \approx 2.0$ fm) till the total absorption of the neck stubs by the fragments) is extremely fast [3]. During this transition the mass distribution is frozen and the distance between the centers of mass of the nascent fragments D_{cm} stays practically unchanged. It is therefore "just-before scission" that the above mentioned fission-fragment properties have to be estimated (and not when the fragments are already separated). This is the first improvement brought to the traditional approach.

The second improvement concerns the description of the corresponding pre-scission shapes in the lemniscate coordinate system (R, x) [4]. It is most convenient since the basic lines R=const (defined by a parameter ϵ) represent a sequence of shapes (Cassinian ovals) that are surprisingly close to the sequence of shapes of a fissioning nucleus. Therefore some of these ovals represent two nascent fragments separated by a thin neck. The expansion of such particular ovals in series of Legendre polynomials is used to generate the nuclear shapes along the scission line. Apart from the elongation parameter ϵ, in present work two other relevant shape parameters are included: α_1 (the mass asymmetry) and α_4 (the neck radius). In addition minimization with respect to α_6 is performed.

This improved model was able to describe the observed transition from asymmetric to symmetric fission that occurs, for the above mentioned isotopes, when the mass of the fissioning nucleus increases. A natural continuation of this study would be to make a direct comparison of all existing experimental distributions for the spontaneous fission of Fm, No and Rf isotopes with the model predictions and this is the object of the present work. As a results we will infer how reliable is to apply the model in the unexplored region of even heavier nuclei [5].

2. Computational details

As mentioned above, the variety of the pre-scission shapes involved are expansions of selected Cassini ovals in series of Legendre polynomials $P_n(x)$,

$$R(x) = R_0[1 + \sum_n \alpha_n P_n(x)], \tag{1}$$

where R_0 is the radius of the spherical nucleus and α_n are shape parameters [4].

Instead of elongation ϵ it is convenient to introduce a new parameter α [4] such as a zero neck shape has always α=1.0 irrespective of the α_n values. α slightly smaller than 1.0 corresponds to two fragments connected by a thin neck. In [2] and here we choose α=0.98 [6].

With this shape parametrization [Eq. (1)] we calculate the potential energy of deformation using the microscopic - macroscopic approach [7]:

$$E_{def} = E_{def}^{LD} + E_{shell} \tag{2}$$

where E_{def}^{LD} is the macroscopic liquid-drop energy and E_{shell} contains the microscopic shell and pairing corrections calculated with a Woods-Saxon type potential and parameters from [8].

Keeping $\alpha = 0.98$ we calculate the potential energy of deformation as a function of the chosen shape parameters. First we use α_1 (which introduces a left-right (mass) asymmetry in the scission shape) and α_4 which, acting on the quadrupole deformation of each fragment, makes the scission shapes more compact ($\alpha_4 < 0$) or more elongated ($\alpha_4 > 0$). Next we add α_6 and minimize the potential energy surfaces with respect to it.

Each point $(\alpha_1, \alpha_4, \alpha_6)$ on these surfaces has a certain probability to be realized. Supposing statistical equilibrium for the collective degrees of freedom normal to the fission direction [9], the distribution of these probabilities is

$$P(\alpha_1, \alpha_4, \alpha_6) \propto e^{-E_{def}(\alpha_1,\alpha_4,\alpha_6)/T_{coll}} \,. \tag{3}$$

Projecting on the α_1 axes

$$Y(\alpha_1) = \sum_i P(\alpha_1, \alpha_{4i}, \alpha_{6i}) / \sum_{ij} P(\alpha_{1j}, \alpha_{4i}, \alpha_{6i}) \,, \tag{4}$$

one obtains the fission fragment mass distribution since α_1 determines the mass asymmetry $\eta = (A_F^H - A_F^L)/A$.

T_{coll} is an unknown parameter that controls the overall width of the distribution. It is however E_{def} that determines the main features of the calculated distribution: the most probable masses and the relative intensities of the fission modes. In particular it is the stiffness $\partial^2 E_{def}/\partial\eta^2$ that controls the stability against variation of the number of nucleons in the fragments, i.e., their relative widths. Smaller or larger T_{coll} values can only reveal or washout the E_{def} structures.

Finally for each shape defined by $(\alpha_1, \alpha_4, \alpha_6)$ one calculates the distance between the centers of mass of the nascent fragments D_{cm}. Then one estimates, in the point-charge approximation, the Coulomb interaction of the fragments supposing uniform charge densities $Z_L/Z_H = A_F^L/A_F^H$.

$$E_{coul}^{int} = e^2 Z_L \times Z_H / D_{cm} = TKE \tag{5}$$

if one neglects TKE_0, the kinetic energy in the fission direction at $\alpha = 0.98$.

The TKE distribution is obtained using the formula:

$$Y(TKE) = \sum_j \sum_i P(\alpha_{1j}, \alpha_{4i}, \alpha_{6i}) \times exp[-(TKE_{ij} - TKE)^2/\Delta E^2]/(\Delta E\sqrt{\pi}) \tag{6}$$

that accounts for the finite energy resolution through the parameter ΔE. The $P(\alpha_{1j}, \alpha_{4i}, \alpha_{6i})$ are the Boltzmann factors given by Eq. (3). The value of ΔE was chosen to be equal to 20 MeV. It corresponds to an experimental energy resolution with a FWHM = 16 MeV.

3. Comparison with existing experimental data

Numerous studies of fission fragment properties in the spontaneous fission of trans-einsteinium isotopes have been performed in the 80's. These nuclei were obtained as evaporation residues in fusion reactions of light ion beams with actinide targets. Refs. [10, 11, 12, 13, 14, 15] are only few examples. The measured mass and total kinetic distributions for Fm, No and Rf isotopes

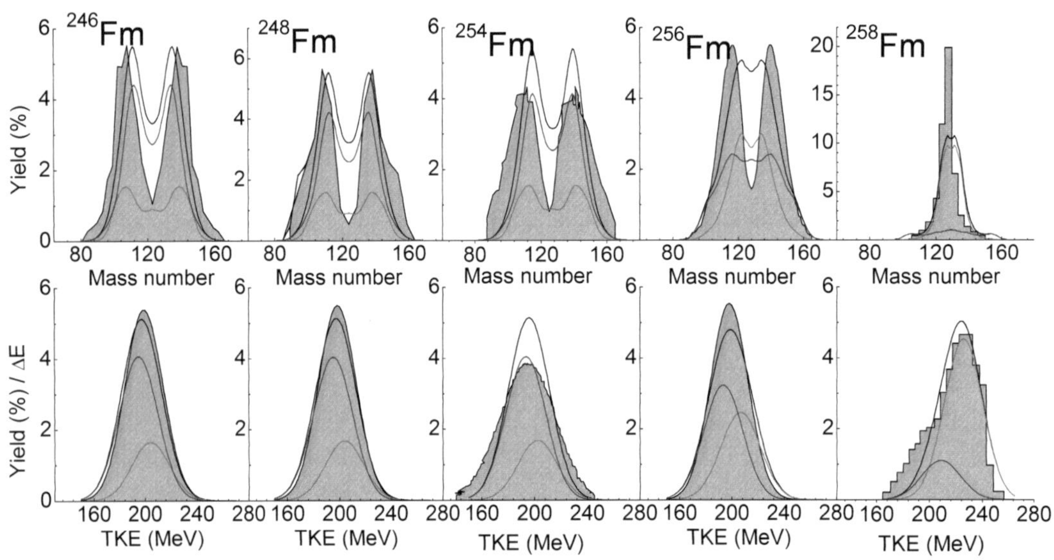

Figure 1. Experimental (histogram) and calculated (black) fragment mass and TKE distributions for even-even Fm isotopes. The red and blue curves represent the compact and the elongated modes respectively. The parameters of the averaging of the energy distributions are: $T_{coll} = 1.0\,\mathrm{MeV}$, $TKE_0 = 15.0\,\mathrm{MeV}$ and $\Delta E = 20.0\,\mathrm{MeV}$ for ^{258}Fm and $T_{coll} = 2.0\,\mathrm{MeV}$, TKE_0=0 and $\Delta E = 20.0\,\mathrm{MeV}$ for the rest of the rest of the isotopes.

are compared with pre-scission model calculations in Figs. 1-3 respectively. It is worth noting that except for the ^{258}Fm the same values of the model parameters ($T_{coll} = 2.0\,\mathrm{MeV}$, $TKE_0 = 0$ and $\Delta E = 20.0\,\mathrm{MeV}$) are used for all nuclei. The agreement is surprisingly good taking into account the simplicity of the theoretical approach. Small differences are however noticeable. For Fm isotopes the transition from asymmetric to symmetric fission is not as sharp as observed. For No and Rf isotopes the TKE's are slightly overestimated.

The compact-symmetric narrow peak observed for ^{258}Fm could not be reproduced using the same model parameters as for the rest of the nuclei. In order to explain the data we need to vary T_{coll} and TKE_0. The nucleus ^{262}Rf is the heaviest even-even nucleus for which the spontaneous fission properties have been measured [16]. As it is seen in Fig. 3, our calculations agree also with these most recent data. The mass distributions for Fm isotopes are most difficult to explain quantitatively (see Fig. 3). From ^{246}Fm to ^{256}Fm the measured distributions have

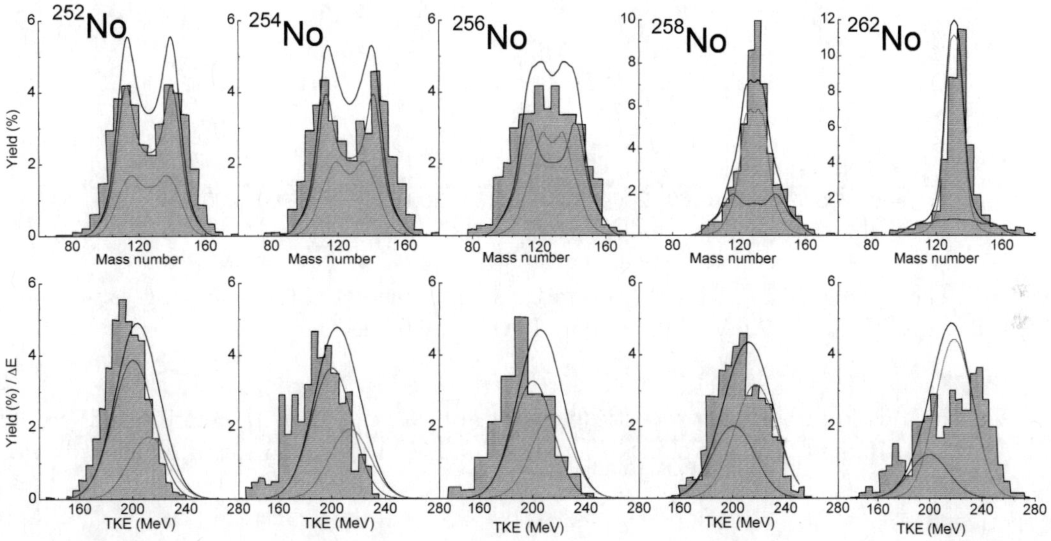

Figure 2. The same as in Fig.1 for No isotopes. The parameters of the averaging of the energy distributions are: $T_{coll} = 2.0\,\text{MeV}$, $TKE_0 = 0$ and $\Delta E = 20.0\,\text{MeV}$.

two equally pronounced peaks while the calculations show a transition from two separated to two overlapping peaks.

Few comments on the agreement with data are appropriate. One should not attempt to reproduce every details of the mass distributions since they are sensitive to relative variations of the potential energy E_{def} of the order of 0.2 MeV and no model can achieve such precision. Concerning the sharp transition from ^{256}Fm to ^{258}Fm, it has not been quantitatively explained so far. As it was pointed out from the beginning [13], all fission models (including the Strutinsky's procedure used here) tend to average over the contributions of several Nilsson levels and therefore are incapable to describe abrupt changes from one nucleus to its neighbor.

4. Summary and conclusions

The main properties of the fission fragments in spontaneous fission of even-even isotopes of Fm, No and Rf are estimated using the pre-scission point model. The underlying potential energy surfaces are calculated with Strutinsky's shell correction procedure in a three dimensional deformation space.

The parametrization of the nuclear shapes chosen here is based on generalized Cassini ovals and represents a natural way to describe scission configurations. We think it is this choice that allows us to explain, with only three shape parameters, the observed features.

Despite the fact that the dynamical effects were neglected, we have obtained quantitative agreement with the experimental data. These nuclei are the heaviest spontaneous fissioning

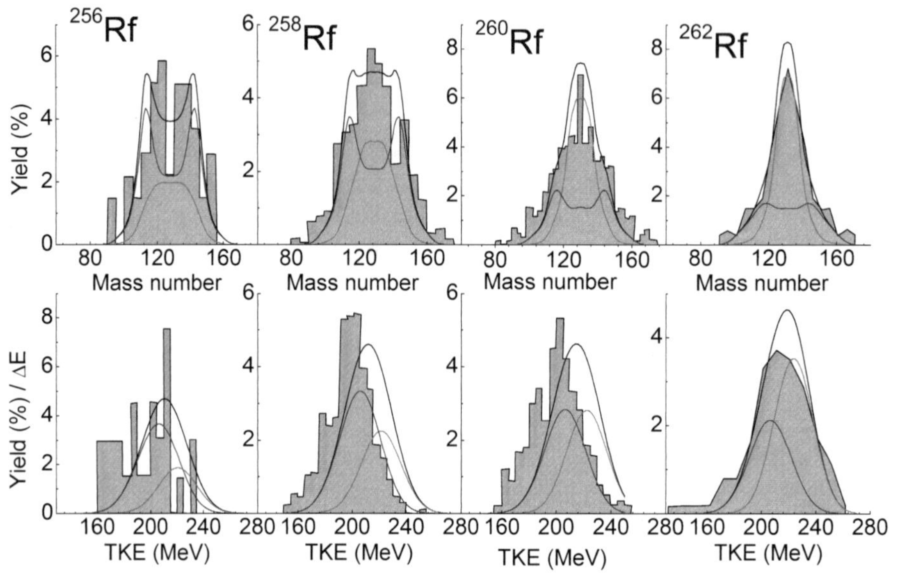

Figure 3. The same as in Fig.1 for Rf isotopes. The parameters of the averaging of the energy distribution are: $T_{coll} = 2.0\,\mathrm{MeV}$, TKE_0=0 and $\Delta E = 20.0\,\mathrm{MeV}$.

nuclei for which fission fragment properties have been measured. It is therefore safe to use the model in the unexplored region of super-heavy nuclei (Z $\geq$ 110) [5] and this is our next project.

References

[1] Wilkins B D, Steinberg E P and Chasman R R 1976 *Phys. Rev.* C **14** 1832
[2] Carjan N, Ivanyuk F A, Oganessian Yu and Ter-Akopian G 2015 *Nucl. Phys.* A **942** 97
[3] Rizea M and Carjan N 2013 *Nucl. Phys.* A **909** 50
[4] Pashkevich V 1971 *Nucl. Phys.* A **169** 275
[5] Oganessian Yu and Utyonkov V 2015 *Rep. Prog. Phys.* **78** 036301
[6] Pashkevich V 1988 *Nucl. Phys.* A **477** 1
[7] Brack M, Damgaard J, Jensen A S, Pauli H C, Strutinsky V M and Wong C Y 1972 *Rev. Mod. Phys.* **44** 320
[8] Ivanova S P, Komov A L, Malov L A and Soloviev V G 1976 *Sov. J. Part. Nucl.* **7** 175
[9] W. Norenberg, Second Symposium on Physics and Chemistry of Fission, Vienna, 1969, IAEA, Vienna, 1969, p. 51.
[10] Bemis C E, Ferguson R L, Plasil F, Silva R J, Pleasonton F and Hahn R L 1977 *Phys. Rev.* C **15** 705
[11] Hoffman D C, Wilhelmy J B, Weber J, Daniels W R, Hulet E K, Lougheed R W, Landrum J H, Wild J F and Dupzyk R J 1980 *Phys. Rev.* C **21** 972
[12] Hoffman D, Lee D, Ghiorso A, Nurmia M and Aleklett K 1980 *Phys. Rev.* C **92** 1581
[13] Hulet E K *et al* 1989 *Phys. Rev.* C **40** 770
[14] Wild J F, Hulet E K, Lougheed R W, Moody K J, Bandong B B, Dougan R J and Veeck A 1994 *J. All. Comp.* **213/214** 86
[15] Hoffman D C and Lane M R 1995 *Radiochim. Acta* **70/71** 135
[16] Lane M R, Gregorich K E, Lee D M, Mohar M F, Hsu M, Kacher C D, Kadkhodayan B, Neu M P, Stoyer N J, Sylwester E R, Yang J C and Hoffman D C 1996 *Phys. Rev.* C **53** 2893

IOP Conf. Series: Journal of Physics: Conf. Series **863** (2017) 012045 doi:10.1088/1742-6596/863/1/012045

Observation of shape isomers states in fission fragments

D V Kamanin[1], Yu V Pyatkov[2,1], A A Alexandrov[1], I A Alexandrova[1], N Mkaza[3], V Malaza[3], E A Kuznetsova[1], A O Strekalovsky[1], O V Strekalovsky[1] and V E Zhuchko[1]

[1]Joint Institute for Nuclear Research, Dubna, Russia
[2]National Nuclear Research University MEPhI (Moscow Engineering Physics Institute), Moscow, Russia
[3]University of Stellenbosch, Faculty of Military Science, Military Academy, Saldanha 7395, South Africa

E-mail: **kamanin@jinr.ru**

Abstract. We discuss the manifestations of a new original effect appeared at crossing of th metal foils by fission fragments. We have observed significant mass deficit in the total mass M of the fission fragments detected in coincidence with ions knocked out from the foil. It was shown that at the large angles of scattering of the knocked–out ions from the foil predominantly conventional elastic Rutherford scattering takes place. As the result M corresponds to the mean mass of the mother system after emission of fission neutrons (no missing mass). In contrast, in near frontal impacts fission fragment misses essential part of its mass. Residual nuclei at least for the fragments from the heavy mass peak show magic nucleons composition.

1. Experiments and results

The layout of one of the experiments (marked below Ex1) dedicated to passing of the fissio fragments (FFs) through the metal foils [1–4] is presented in figure 1. The geometry of the source un is shown in detail in figure 1a. Absorber foil was placed at 1 mm distance from the Cf source. Typical FF overcomes such distance approximately in 0.1 ns. Ternary events were analyzed. It means that three fragments were actually detected in coincidence in three different silicon detectors. For the sake of convenience, the FFs from such events are labeled as M_1, M_2 and M_3 in an order of decreasing masses in the ternary event. Mass distribution for ternary events is shown in figure 2. The total mass o the two heavier fragments M_1+M_2 is plotted vs. the mass of the lighter fragment M_3. As can be inferred from the figure almost in all the events where the lightest fragment corresponds to th knocked out Cu ions (M_3 is around 64 amu, figure 2a) we observe missing mass of FFs. Similar results were also obtained using Ti and Ni foils.

The effect of Rutherford scattering accompanied by the loss of the mass of the scattered fission fragment takes place also in scattering on ^{16}O and ^{27}Al nuclei from the source backing. At the same time at least some part of the FFs scattered on the ^{12}C nuclei from the organic foil undergo scatterin without the mass loss. Corresponding events are marked by the arrow.

Next very interesting result obtained is presented in figure 2b. It is the mass spectrum of the fragments scattered in the foil and detected in coincidence with the knocked out ions. As can be inferred from the figure this spectrum radically differs from the FF mass distribution in conventional binary fission. The masses of known magic nuclei are marked in the figure by the arrows. The most pronounced peak in the experimental mass spectrum corresponds to the magic ^{128}Sn nucleus in contrast with the mass yields in conventional binary fission.

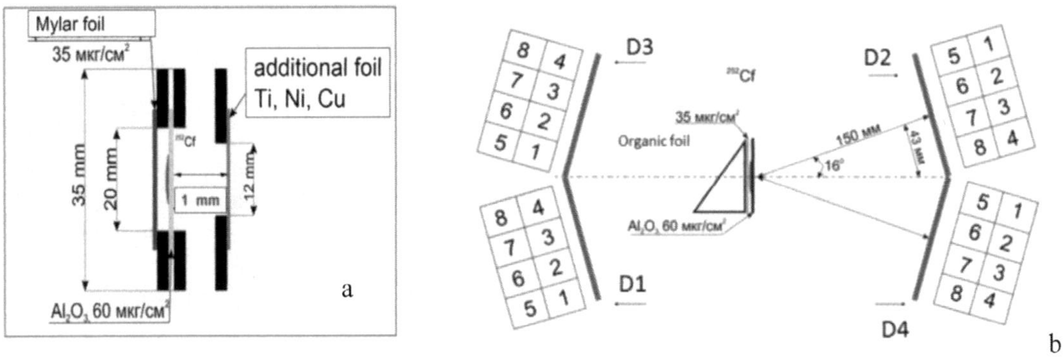

Figure 1. Layout of the experiment Ex1. ^{252}Cf source and additional foils placed around (a) and positions of the detectors used (b). Micro-channel-plate based start timing detector is placed in the center of the setup while four mosaics eight PIN diodes each provide measuring both FF time-of-flight and energy.

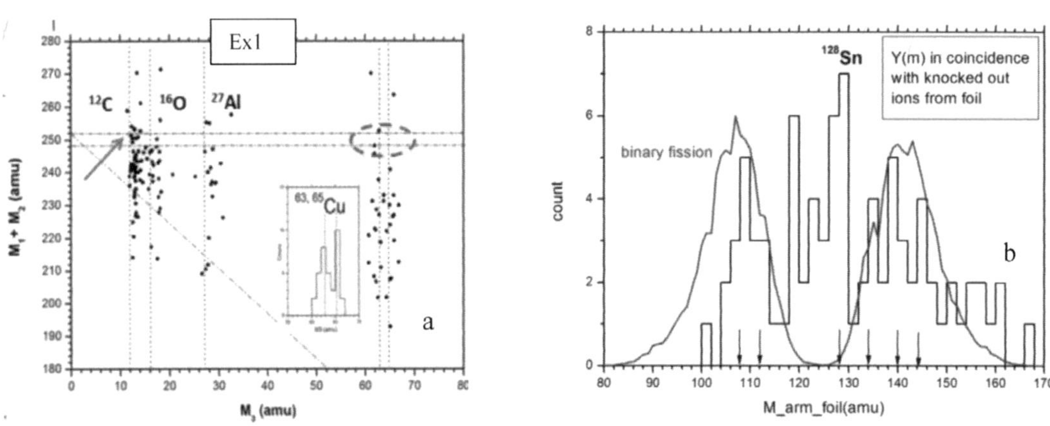

Figure 2. Results for ternary events from Ex1. Cu foil 0.83 μkm thick was used as the additional absorber (a). Mass spectrum of the fragments detected in coincidence with the ions knocked out from the foil and in the same arm with them (b). See text for details.

Additional information concerning the effect under study was obtained in the experiment Ex2. The experiment was performed at the LIS (Light Ions Spectrometer) spectrometer in FLNR of JINR. The layout of the setup is shown in figure 3. LIS is a double-armed time-of-flight spectrometer which includes three micro-channel based timing detectors and two PIN diodes. Each PIN diode provides information for estimating both FF energy and time-of-flight. Metal foils (degraders) of different thicknesses can be placed in the detector TD1. The aperture for fission fragments detected in coincidence in the opposite PIN diodes does not exceed 3^0. The data acquisition system consists of the fast digitizer CAEN DT5742 and a personal computer. The digital images of all the signals were obtained for further off-line processing. Mass reconstruction procedure used is presented in [5]. The

IOP Conf. Series: Journal of Physics: Conf. Series **863** (2017) 012045 doi:10.1088/1742-6596/863/1/012045

construction of the spectrometer allows measuring of the FF mass using "two-velocities" (using time-of-flights at the bases L3 and L4) and "velocity-energy" (using time-of-flight at the base L5 and energy measured by PIN1) approaches simultaneously. Corresponding mass values will be marked below M_{tt} and M_{te}. Thus, we know the mass of each FF before and after it crosses the degrader-foil in TD1.

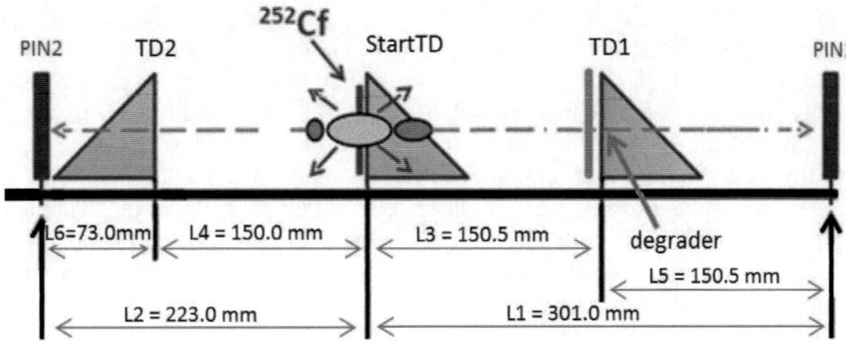

Figure 3. Layout of the LIS spectrometer used in Ex2. It includes three timing detectors (Start TD, TD1, TD2), two PIN diodes and ^{252}Cf (sf) source. Additional metal foil (degrader) can be placed in TD1 detector.

Mass correlation distribution M_{tt}–M_{te} obtained for the heavy FFs mass peak is presented in figure 4. Copper foil 4.11 microns thick was placed in the TD1 detector. Pronounced grid-like structure is vividly seen. Projection of the plot onto M_{te} axis is shown in figure 4b. The spectrum agrees well with this obtained earlier in Ex2 (figure 2b).

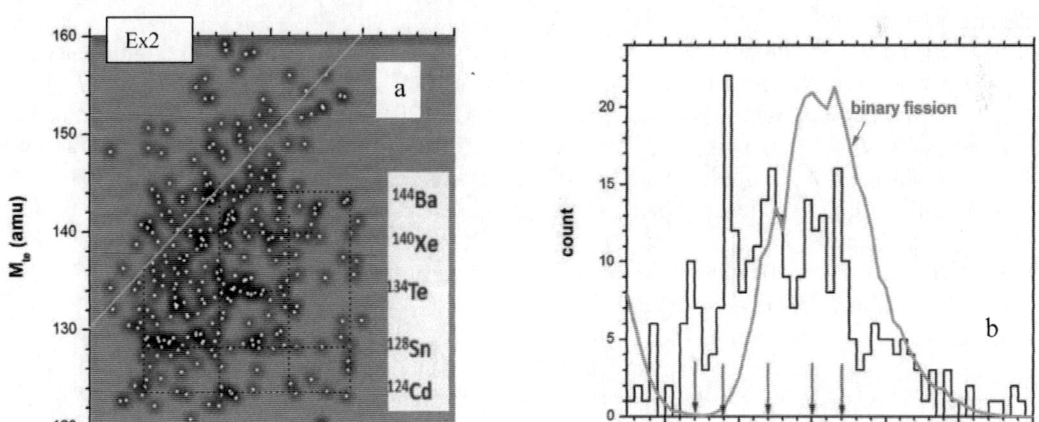

Figure 4. Mass correlation distribution M_{tt}–M_{te} for the heavy FFs mass peak (a) obtained in Ex2. Copper foil 4.11 microns thick was placed in the TD1detector. Dotted lines are drawn to guide the eye. Horizontal lines correspond to known spherical and deformed magic nuclei. Tilted white line corresponds to the equation $M_{tt} = M_{te}$. Projection of the shown mass distribution onto M_{te} axis (b). Masses of known magic nuclei (listed in the panel in figure 4a) are marked by arrows. Mass spectrum of the FFs from conventional binary fission is shown in gray.

IOP Conf. Series: Journal of Physics: Conf. Series **863** (2017) 012045 doi:10.1088/1742-6596/863/1/012045

2. Discussion

The mass distribution in figure 4 vividly shows event by event how changes the initial mass M_{tt} of the heavy fragment after passing the foil. Full distribution consists of two parts, namely FFs crossed the foil without missing mass and those which lost part of their masses. Both parts are almost equal by number of events in specific conditions of this experiment. The result observed corresponds to the hypothesis that initial fragment consists of one of the magic cores (listed in the inset in figure 4a) and additional nucleons adding the mass of the core up to the total mass of the fragment. These additional nucleons are missed due to inelastic scattering of the initial fragment in the foil. Thus, apparently fission fragment is burned in shape isomer state. A life time of the typical FF isomer state is estimated to be more than 15ns (mean FF flight-time at L3 in Ex2). Projection of the plot onto M_{te} axis (figure 4b) differs radically from the mass distribution of heavy fragments in conventional binary fission. The strongest difference is observed for the masses associated with magic isotopes of ^{128}Sn and ^{124}Cd. The spectrum in figure 4b agrees well with this shown in figure 2b obtained earlier in Ex1. The latter spectrum can be called the spectrum of "brake-up residuals" detected in coincidence with a nucleus knocked out from the foil in the same spectrometer arm.

3. Conclusions

1. Experimental evidences of the binary brake-up of the fission fragments when passing of metal foil were obtained for the first time.
2. In the light of the results obtained fission fragment is supposed to be born in the shape isomer state which looks like di-nuclear system consisting of the magic core and lighter cluster.
3. A low limit of a life-time of typical shape-isomer state in heavy fission fragments is estimated to be 15 ns.

Acknowledgments

This work was supported in part by the Department of Science and Technology of the Republic of South Africa (RSA), Bundesministerium für Bildung und Forschung (Germany).

References

[1] Pyatkov Yu V, Kamanin D V, Alexandrov A A, Alexandrova I A, Kondratyev N A, Kuznetsova E A, Strekalovsky A O, Strekalovsky O V, Zhuchko V E 2014 *Proc. 21th Int. Sem. on Interaction of Neutrons with Nuclei*, 20–25 May 2013, Alushta, Ukraine (Dubna) p. 127

[2] Pyatkov Yu V, Kamanin D V, Alexandrov A A, Alexandrova I A, Kondratyev N A, Kuznetsova E A, Strekalovsky A O, Strekalovsky O V, Zhuchko V E 2014 FIAS Interdisciplinary Science Series, Nuclear Physics: Present and Future, ed. W. Greiner (Springer) p. 79

[3] Pyatkov Yu V *et al.* 2015 *Proc. 22th Int. Sem. on Interaction of Neutrons with Nuclei*, 27–30 May 2014, Dubna, Russia (Dubna) p. 83

[4] Pyatkov Yu V, Kamanin D V, Alexandrov A A, Alexandrova I A, Kondratyev N A, Kuznetsova E A, Strekalovsky A O, Strekalovsky O V, Zhuchko V E, Mkaza N 2015 *Int. Sym. on Exotic Nuclei "EXON-2014"*, 08-13 September 2014, Kaliningrad, Russia, ed. Yu E Penionzhkevich and Yu G Sobolev (World Scientific) p. 383

[5] D.V. Kamanin et al., Proceedings of the 21th International Seminar on Interaction of Neutrons with Nuclei, Alushta, Ukraine, 20–25 May 2013. Dubna 2014, p. 107.

Collinear cluster tri-partition – the brightest observations and their treating

Yu V Pyatkov[1,2], D V Kamanin[2], J E Lavrova[1], N Mkaza[3], V Malaza[3] and A O Strekalovsky[1]

[1]National Nuclear Research University MEPhI (Moscow Engineering Physics Institute), Moscow, Russia
[2]Joint Institute for Nuclear Research, Dubna, Russia
[3]University of Stellenbosch, Faculry of Military Science, Military Academy, Saldanha 7395, South Africa

E-mail: **yvp_nov@jinr.ru**

Abstract. Careful studies of the fission fragments mass correlation distributions let us to reveal specific linear structures in the region of a big missing mass. It became possible due to applying of effective cleaning of this region from the background linked with scattered fragments. One of the most pronounced structure looks like a rectangle bounded by the magic nuclei. The fission events aggregated in the rectangle show a very low total kinetic energy. We propose possible scenario of forming and decay of the multi-cluster prescission configuration decisive for the experimental findings. This approach is valid as well for treating of another rare decay modes discovered in the past.

1. Introduction

A lot of our previous publications [1–3] were devoted to experimental findings treated as a manifestation of a new multibody decay channel of low excited actinide nuclei. In the bulk of experiments two fragments were actually detected while deficit of their total mass comparing the mass of the mother system ("missing" mass) served a sign of at least ternary decay. Keeping in mind that the detected fragments fly apart almost collinearly and at least one of them shows magic nucleon composition, we have called this decay channel "collinear cluster tri-partition (CCT)" in order to underline the likeness with known cluster decay or "lead radioactivity" where magic lead cluster plays a key role. To date the results obtained were discussed in several theoretical works [4, 5 and references therein], but they are far from the quantitate description of the phenomenon under study. Here we discuss possible scenario of one of the most pronounced CCT modes.

2. Experiments and results

Manifestations of the analyzing multibody decay mode in the fission fragment (FF) mass correlation distributions are presented in figure 1.

IOP Conf. Series: Journal of Physics: Conf. Series **863** (2017) 012046 doi:10.1088/1742-6596/863/1/012046

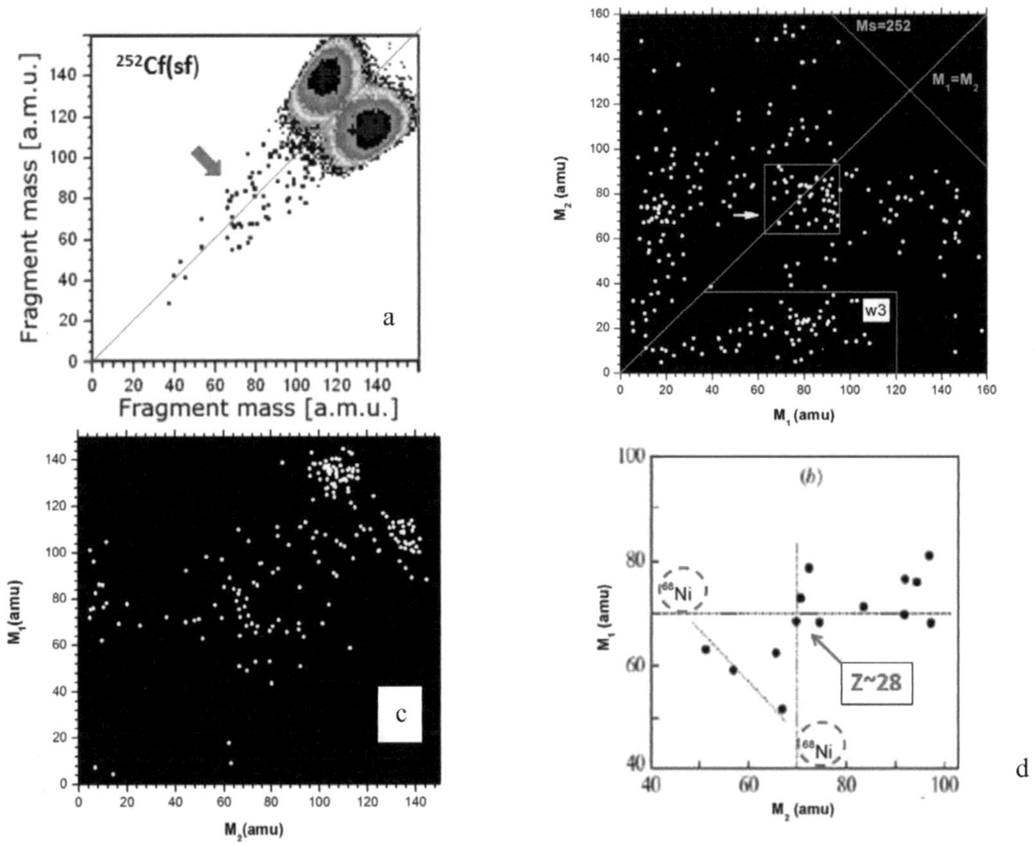

Figure 1. Mass distribution of the complimentary fission fragments (FFs) from ^{252}Cf(sf) selected by requirement of their approximately equal velocities and momenta (a). Mass distribution obtained for the events with the FFs momenta beyond the "tails" of scattered FFs and experimental neutron multiplicity $n \geq 1$ (b). Mass correlation distribution for the FFs from the reaction ^{235}U(n_{th}, f) with approximately equal velocities and momenta (c). FFs from the same reaction with approximately equal velocities, momenta and nuclear charges (d). The vertex of the rectangular structure corresponds to the nuclear charge z ~ 28 (Ni isotopes). See text for details.

The spectrometers used were based on the modules of the FOBOS setup [1, 2]. The TOF-E (time-of-flight vs. energy) method for the measurements of two FF masses in coincidence with two detectors placed at 180 degrees was used in both experiments under discussion. The TOF of the fragment has been measured over a flight path of 50 cm between the "start" detector, placed next to the ^{252}Cf source and the "stop" detectors formed by position-sensitive avalanche counters (PSAC). PSACs provided also the fragment emission angle with a pre-scission of 1^0. The energies of those coincident fragments which passed through the PSACs were measured in the Bragg ionization chambers (BIC).

For searching for the isotropic component of the neutrons emitted in fission the "neutron belt" consisting of 140 ^{3}He filled neutron counters was assembled in a plane perpendicular to the symmetry axis of the spectrometer which serves as the mean fission axis at the same time.

In order to estimate fragment nuclear charge the drift time of a track formed after stopping of a fragment in the gas volume of the BIC was measured in the experiment dedicated to the ^{235}U(n_{th}, f) reaction (see [1, 2] for more details).

Coming back to the figure1 we attract your attention to the rectangular structure marked by the arrows in all the plots. The low vertex of the structure lies in the vicinity of the point (68, 68) amu associated with magic Ni isotopes. The structure is reproduced in main in three independent experiments and at different FFs selection conditions. The yield of the events joined in the rectangular structure in figure 1a does not exceed 10^{-5} per binary fission. The total kinetic energy (TKE) of the events in the structure lies in the range (90–140) MeV. They have almost equal momenta. Comparing a total number of events in plot 1b with similar one for $n \geq 2$ (does not shown here) in the frame of the mathematical model of the neutron registration channel used [2] we came to conclusion that a neutron source which could provide such difference, corresponds to the multiplicity $n = 2$ if neutrons are emitted isotropically or $n = 7$ if they are emitted from the fully accelerated FFs.

3. Discussion

Observation of the Ni isotopes in fission of actinides has a long lived history. In experiment at the gamma-sphere using gamma-gamma coincidence technique [6] two modes were reported for the Ba/Mo partition. The modes essentially differ by the multiplicity of the emitted neutrons. To the second unusual mode neutron multiplicity $n = 7–10$ neutrons and TKE ~ 154 MeV were assigned. In order to reproduce these experimental findings it was supposed [7] the following cluster configuration of the ^{252}Cf nucleus in the scission point (row №1 in Table 1).

Table 1. Clustering leading to binary and multibody decays of ^{252}Cf nucleus

<table>
<tr><th>№</th><th>Prescission cluster configuration</th><th>Comments</th></tr>
<tr><td>1.</td><td>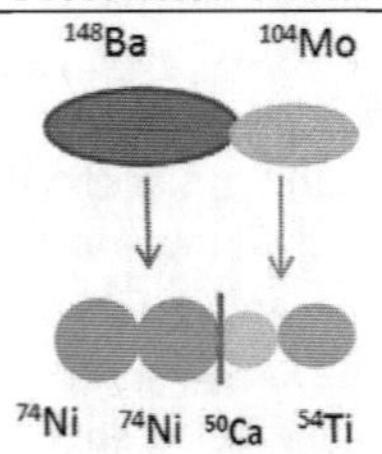
</td><td>Binary fission via 4–cluster prescission configuration [6, 7].</td></tr>
<tr><td>2.</td><td>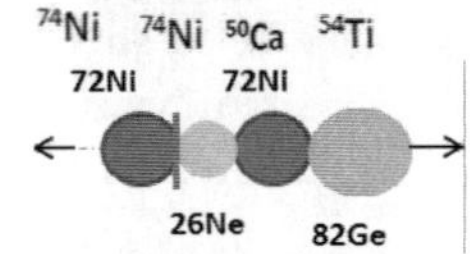
</td><td>Binary fission via 4–cluster prescission configuration [11 and this work].</td></tr>
<tr><td>3.</td><td>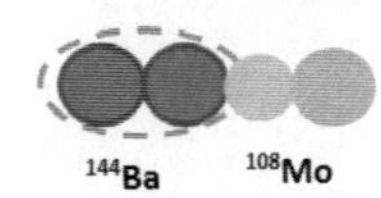
</td><td>Prescission configuration of ^{252}Cf.</td></tr>
<tr><td>4.</td><td>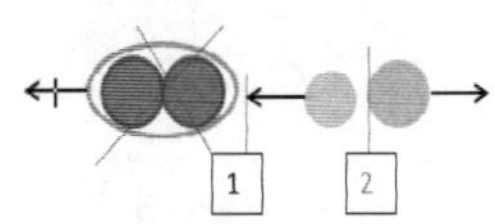
</td><td>The missed fragments are marked by the red oblique cross. The sequence of ruptures is marked by the numbers in the rectangles.</td></tr>
<tr><td>5.</td><td>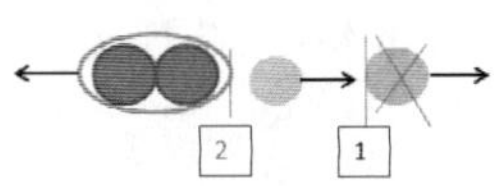
</td><td><u>True ternary fission (CCT) takes place.</u></td></tr>
<tr><td>6.</td><td>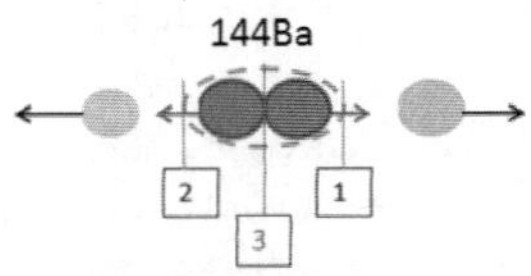
</td><td></td></tr>
<tr><td>7.</td><td>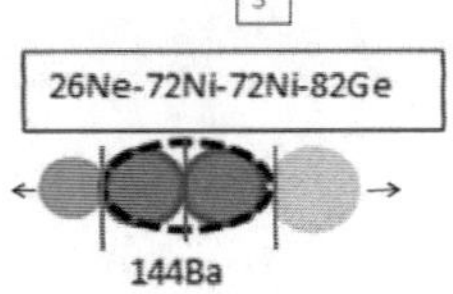
</td><td><u>Quaternary decay</u> via 4–cluster prescission configuration (this work) appears to occur.</td></tr>
</table>

In studies of far asymmetric fission of actinides a shoulder in the FFs mass yields linked with Ni isotopes was revealed [8–10]. Corresponding mass splits show rather low TKE (~ 140MeV for ^{252}Cf(sf) [8]. It can be provided if the following prescission configuration is supposed (row №2 in Table 1).

The data of our experiment performed in JYFL (Jyväskylä, Finland) gives another example of clustering in spontaneous fission of ^{252}Cf. The experiment was performed using a double armed time-of-flight spectrometer based on timing detectors with micro channel plates and silicon surface barrier detectors [11]. TOF-E method was applied to measure the FFs masses. Due to the construction of the timing detectors the spectrometer had relatively low transparency for the detected FFs. In the FFs mass correlated distribution we have found both the events corresponding to the missed ^{144}Ba fragment and the events where one of the fragments is equal to ^{144}Ba. We suppose the prescission cluster configuration in this case is similar to that proposed in [7] (row № 3 in Table 1). The scission scenario for the events with missed ^{144}Ba fragment is shown in the next row № 4 while the opposite case when this fragment was detected is illustrated in row № 5. Thus we suppose true ternary fission takes place. One of two fragments flying in the same direction can be missed at the flight pass due to the limited transparency of the timing detectors.

In order to reproduce very low TKE (~90 MeV) for the events around the vertex of the rectangle (72, 72) amu in our data in figure 1, it should be assumed prescission configuration and sequence of ruptures shown in Table 1, row №6. After two initial ruptures residual ^{144}Ba nucleus clusterised into two ^{72}Ni clusters should stay almost at rest. In this case after scission of ^{144}Ba nucleus TKE of Ni clusters will be equal to their interaction energy at the Coulomb barrier to be also ~90 MeV. Such hypothetical scission scenario is supported by the events from box w3 in figure 2b. The difference between mass of the mother system and total masses of two complimentary fragments from the box reproduces well the total masses of the fragments from the rectangle marked by the arrow in figure 1b. In other words, the box w3 joins the utmost clusters from the configuration №6 in Table 1. Basing on the parameters of these events we have estimated kinetic energy of the central fragment just after utmost clusters fly apart. It does not exceed some MeV to be in line with the hypothesis put forward.

Summing up, in the frame of scenario of sequential ternary scission of four-cluster prescission configuration of mother nucleus (the last row of Table 1) we reproduce quite satisfactory and self consistently experimental data including prediction that two neutrons are emitted isotropically i.e. from the neutron source to be at rest.

References

[1] Pyatkov Yu V *et al* 2010 *Eur. Phys. J.* A **45** 29
[2] Pyatkov Yu V *et al* 2012 *Eur. Phys. J.* A **48** 94
[3] Kamanin D V, Pyatkov Yu V 2013 Clusters in Nuclei *Lecture Notes in Physics* **875** vol. 3 ed. Beck C p. 183.
[4] von Oertzen W, Nasirov A K, Tashkhodjaev R B 2015 *Phys. Lett* B **746** 223
[5] Vijayaraghavan K R, von Oertzen W, Balasubramaniam M 2012 *Eur. Phys. J.* A **48** 27
[6] Wu S-C, Donangelo R, Rasmussen J O, Daniel A V, Hwang J K, Ramayya A V, Hamilton J H 2000 *Phys. Rev.* C **62** 041601
[7] Shneidman T M, Adamian G G, Antonenko N V, Ivanova S P, Jolos R V, Scheid W 2002 *Phys. Rev.* C **65** 064302
[8] Barreau G, Sicre A, Caïtucoli F, Asghar M, Doan T P, Leroux B, Martinez G, Benfoughal T 1985 *Nucl. Phys.* A **432** 411
[9] Goverdovskiy A A, Mitrofanov V F, Khiyachkov V A 1995 *Phys. Atom. Nucl.* **58** 1546 (in Russian)
[10] Rochman D, Tsekhanovich I, Gönnenwein F, Sokolov V, Storrer F, Simpson G, Serot O 2004 *Nucl. Phys.* A **735** 3
[11] Yamaletdinov S, University of Jyväskylä, research report №9/2007

IOP Conf. Series: Journal of Physics: Conf. Series **863** (2017) 012047 doi:10.1088/1742-6596/863/1/012047

Fission dynamics at high excitation energies investigated in complete kinematics measurements

J L Rodríguez-Sánchez[1], J Benlliure[1], J Taïeb[2], H Álvarez-Pol[1], L Audouin[3], Y Ayyad[1], G Bélier[2], G Boutoux[2], E Casarejos[4], A Chatillon[2], D Cortina-Gil[1], T Gorbinet[2], A Heinz[5], A Kelić-Heil[6], B Laurent[2], J F Martin[2], C Paradela[1], E Pellereau[2], B Pietras[1], D Ramos[1], C Rodríguez-Tajes[7], H Simon[6] and J Vargas[1]

[1] Universidad de Santiago de Compostela, E-15782 Santiago de Compostela, Spain
[2] CEA, DAM, DIF, F-91297 Arpajon, France
[3] Institut de Physique Nucléaire d'Orsay, F-91406 Orsay, France
[4] Universidad de Vigo, E-36200 Vigo, Spain
[5] Chalmers University of Technology, SE-412 96 Gothenburg, Sweden
[6] GSI Helmholtzzentrum für Schwerionenforschung GmbH, D-64291 Darmstadt, Germany
[7] Grand Accélérateur National d'Ions Lourds, F-14076 Caen, France

E-mail: `joseluis.rodriguez.sanchez.phy@gmail.com`

Abstract. Light-charged particles emitted in proton-induced fission reactions on ^{208}Pb have been measured at different kinetic energies: 370A, 500A, and 650A MeV. The experiment was performed by the SOFIA collaboration at the GSI facilities in Darmstadt (Germany). The inverse kinematics technique was combined with a setup especially designed to measure light-charged particles in coincidence with fission fragments. The data were compared with different model calculations to assess the ground-to-saddle dynamics. The results confirm that transient and dissipative effects are required for an accurate description of the fission observables.

1. Introduction

Measurements of light particles have been used as a "clock" to estimate the fission time [1, 2] because they are emitted along the fission path up to the scission-point configuration. In particular, Lestone et al. [2] used fusion-fission reactions to investigate the pre- and postsaddle fission times by comparing proton- and α-particle multiplicities with statistical model calculations. Their work demonstrated that this observable is sensitive to dissipation. Similar conclusions were obtained by Fröbrich and Gontchar [3] by comparing the same data with dynamical calculations based on the Langevin equation.

In this work we took advantage of the high excitation energies, and low angular momenta and shape distortion induced by spallation reactions. In this kind of reaction, light-charged particles can be emitted during the first stage of the reaction, the so-called intranuclear cascade, or in the deexcitation of the compound nucleus. In the latter case, we can assume that light-charged particles are mostly emitted before the saddle point because of the shorter saddle-to-scission paths, where the most probable deexcitation channels are the emission of neutrons and γ-rays. The measurement of these particles in coincidence with fission is a difficult task in direct kinematics experiments due to the fact that fission fragments cannot always be detected, and it is even harder to identify them in terms of their atomic or mass number [4]. For this reason, our novel experimental setup together with the inverse kinematics technique represent an ideal scenario because fission fragments and light-charged particles leave the target with high kinetic energies in the forward direction, facilitating their detection with high efficiency and precision.

2. Experiment

The description of the experimental setup used in this work can be found in Refs. [5, 6], and here, we only describe the most important features.

A multisampling ionization chamber and a time projection chamber were placed upstream of a liquid-hydrogen target to provide us the beam identification and the position of the beam on the target, respectively. Then, a double multisampling ionization chamber, placed downstream of the liquid-hydrogen target, was used to detect both fission fragments simultaneously and permitted us an unambiguous identification of fission [6]. In addition, the tracking capabilities of this last detector allowed us to select fission events produced at the target position [7]. Light-charged particles were detected in a Time-of-Flight Wall (ToF Wall) placed behind the target at a distance of 140 cm. This detector consisted of two detection planes made of segmented plastic-scintillators, each 50 cm long, 6 cm wide and 1 cm thick. The first plane was formed by six horizontal paddles and the second one by six vertical paddles, which left a square hole of 12.5×12.5 cm^2 in the center for the transmission of the fission fragments.

3. Results

In Fig 1(a) we display the probability for the emission of particles with $Z = 1$ in fission events as function of the total multiplicity of light-charged particles per event for the reaction ^{208}Pb + p at 370A (open triangles), 500A (solid circles), and 650A (open squares) MeV. The maximum of these distributions appears for small values of the total multiplicity, as expected because most of the fissioning systems are produced in peripheral collisions where only a few number of nucleons are removed. Then, the tails observed towards larger multiplicities can be attributed to the production of lighter fissioning nuclei with higher excitation energies [8]. As can also be seen in the figure, the probability for fission events with large multiplicities of light-charged particles emitted in coincidence, as well as the maximum of the distributions, increase with the bombarding energy. This tendency can be attributed to the increase of the violence of the reaction, which means more particles are emitted during the intranuclear cascade and deexcitation processes.

In Fig. 1(b) we show the average values of the multiplicity distributions of Fig. 1(a) as a function of the projectile bombarding energy. These data are compared with different model calculations performed by using the intranuclear cascade model INCL4.6 [9] coupled to the deexcitation code ABLA07 [10]. As can be observed in the figure, dynamical calculations based on the Kramers approach and considering a constant value of the reduced dissipation parameter of $\beta = 4.5 \times 10^{21}$ s^{-1} deduced in previous works [11, 12] (long-dashed line) overestimate the data. However, ABLA07 calculations considering dissipative and transient time effects for the same value of the reduced dissipation parameter (solid line) provide a good description. Therefore, these calculations reveal that this observable is sensitive to transient time effects.

On the other hand, ABLA07 calculations assuming values of $\beta = 10 \times 10^{21}$ s^{-1} (dotted line) and $\beta = 20 \times 10^{21}$ s^{-1} (dot-dashed line) clearly underestimate the data. This fact indicates that this observable is also sensitive to the value of the reduced dissipation parameter. This comparison also allows us to contrain the dissipation parameter obtaining a value around $\beta = 4.5$ $\times 10^{21}$ s^{-1} that is in excellent agreement with the results found in other works [2, 3, 11, 12, 13, 14].

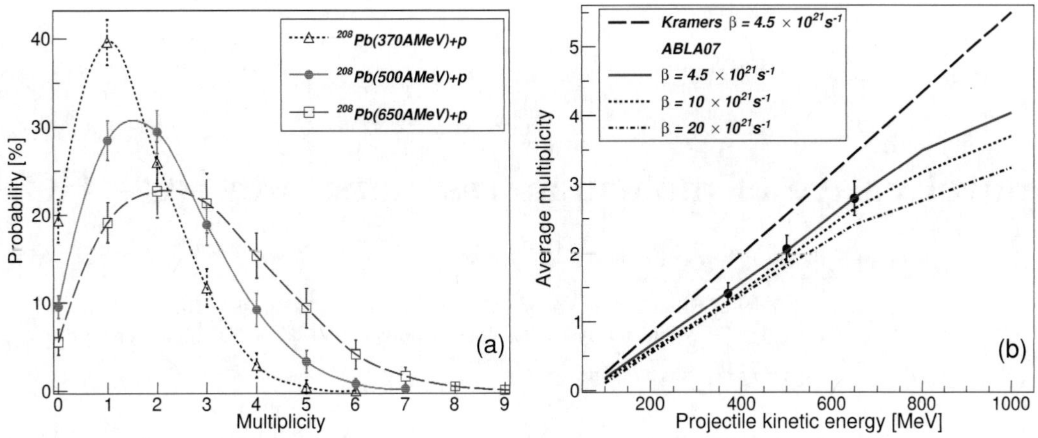

Figure 1. (Color online) (a) Multiplicity distribution of particles with $Z = 1$ for the reaction ^{208}Pb + p at different kinetic energies: $370A$ (open triangles), $500A$ (solid circles), and $650A$ (open squares) MeV. The lines are to guide the eye. (b) The average multiplicity of particles with $Z = 1$ from (a) are compared with different model calculations (lines).

4. Conclusions

We have investigated proton-induced fission of ^{208}Pb in inverse kinematics at different bombarding energies. We have used a highly efficient detection setup that permitted us for the first time to identify the atomic number of the fission fragments in coincidence with the light-charged particles. The data were compared to different model calculations, and the results obtained with respect to the temperature dependence of the dissipation parameter and its magnitude are consistent with the conclusions obtained in most of the works related with the investigation of fission [1, 3, 11, 12, 14, 13, 15, 16]. Moreover, we also conclude that dissipative and transient time effects are needed to explain the measured multiplicities of light-charged particles at high excitation energies above 100 MeV.

References

[1] Hinde D J *et al* 1989 *Nucl. Phys.* A **502** 497
[2] Lestone J P 1993 *Phys. Rev. Lett.* **70** 2245
[3] Fröbrich P *et al* 1993 *Nucl. Phys.* A **556** 281
[4] Lott B *et al* 2001 *Phys. Rev.* C **63** 034616
[5] Martin J F *et al* 2015 *Eur. Phys. J.* A **51** 174
[6] Rodríguez-Sánchez J L *et al* 2015 *Phys. Rev.* C **91** 064616
[7] Rodríguez-Sánchez J L *et al* 2014 *Phys. Rev.* C **90** 064606
[8] Rodríguez-Sánchez J L *et al* 2015 *Phys. Rev.* C **92** 044612
[9] Boudard A *et al* 2013 *Phys. Rev.* C **87** 014606
[10] Kelić A, Ricciardi M V and Schmidt K H, Proceedings of Joint ICTP-IAEA Advanced Workshop on Model Codes for Spallation Reactions, ICTP Trieste, Italy, 48 February 2008, edited by Filges D, Leray S, Yariv Y, Mengoni A, Stanculescu A and G. Mank (IAEA INDC(NDS)-530, Vienna, 2008), pp. 181-221.
[11] Schmitt C *et al* 2010 *Phys. Rev.* C **81** 064602
[12] Ayyad Y *et al* 2015 *Phys. Rev.* C **91** 034601
[13] Nadtochy P N *et al* 2010 *Phys. Lett.* B **685** 258
[14] Ye W and Tian J 2016 *Phys. Rev.* C **93** 044603
[15] Jacquet D and Morjean M 2009 *Prog. Part. Nucl. Phys.* **63** 155
[16] Vardaci E *et al* 2015 *Phys. Rev.* C **92** 034610

IOP Conf. Series: Journal of Physics: Conf. Series **863** (2017) 012048 doi:10.1088/1742-6596/863/1/012048

A novel visage of molecular resonances in ^{12}C + ^{12}C

E Uegaki[1] **and Y Abe**[2]

[1]Graduate School of Engineering Science, Akita University, Akita 010-8502, Japan
[2]Research Center for Nuclear Physics, Osaka University, Ibaraki 567-0047, Japan

E-mail: uegaki@phys.akita-u.ac.jp

Abstract. Resonances observed in the ^{12}C + ^{12}C collisions are studied by the molecular model. Firstly, normal-modes analysis made around the equilibrium equator-equator touching configuration gives a basic insight into dynamics of the system. The quantum numbers of the states also indicate rotational motions of the constituent ^{12}C nuclei, which relate the normal modes to the corresponding inelastic ^{12}C + ^{12}C channels. Next, low-lying 11 states are taken up as bases, and Coriolis coupling has been diagonalized. Many excited states are obtained with intermediate widths, which is just in accord with the experimental observation. For example, the second and third excited states have characteristic feature as low-lying mutual 2^+ excitation with the spin vectors of ^{12}C nuclei along the molecular z'-axis. Those states appear very close to the molecular ground state with the higher spin of $\Delta J = 2$, to form a bunch. Resultant excitation functions are surprisingly in good agreement with the experiments, which lights up a new physical picture of molecular structures of the high-spin ^{12}C + ^{12}C resonances.

1. Introduction

Well above the Coulomb barrier of the ^{12}C + ^{12}C system, series of resonances have been found with high spins over $10\hbar$, which exhibit prominent peaks in the elastic and inelastic 2^+ channels [1]. Band Crossing Model (BCM), based on the double resonance mechanism, has successfully explained resonance mechanism with the aligned configurations of the orbital angular momentum and the spins of the excited states of ^{12}C nuclei [2].

We have revisited the resonances with the new molecular model [3, 4], which has been developed for the analyses of the high-spin resonances observed in ^{24}Mg + ^{24}Mg and ^{28}Si + ^{28}Si. The model has an advantage to describe various geometrical configurations of two constituent deformed nuclei with the method of normal modes around the stable energy minimum under strong nucleus-nucleus interaction, which gives intuitive understanding on nuclear structure of the resonances. For the oblate-oblate systems such as ^{28}Si – ^{28}Si and ^{12}C – ^{12}C, the stable configurations have a small axial asymmetry as a whole and thus exhibit a series of low-lying excited states with $K = 2$ or 4, in which spins of the constituent nuclei incline to the molecular z'-axis. Those states naturally have an interesting property of spin disalignments with the orbital angular momentum, which is contrary to the alignments in BCM. How to resolve an apparent contradiction?

In rotating systems, Coriolis coupling plays an important role in optimization of the centrifugal energy. Actually, a diagonalization of the coupling has resulted in an admixture of the $K = 1$ configuration in the molecular ground state and thus the state has aligned components similar to those of BCM [5]. Furthermore, interesting is that the $K = 2$ and $K = 4$ states are expected to be also observed through a weak coupling to the elastic channel caused by chains

IOP Conf. Series: Journal of Physics: Conf. Series **863** (2017) 012048 doi:10.1088/1742-6596/863/1/012048

of Coriolis coupling. Those nearby or overlapping states appear as a bunch of resonances or a single gross peak, which is apparently different from the interpretation of BCM. We will discuss below how the present study has revealed revolutionarily refined features of the resonances as well as consistent aspects with BCM.

2. Brief reminder of the model and the method of normal modes

Assuming a constant deformation and axial symmetry of the constituent nuclei, we have seven degrees of freedom $(q_i) = (\theta_1, \theta_2, \theta_3, R, \alpha, \beta_1, \beta_2)$ to describe dynamics of the dinuclear system, as illustrated in fig. 1. The relative vector (R, θ_2, θ_1) of two ^{12}C nuclei defines the molecular z'-axis. Orientations of the symmetry axes of ^{12}C are described with Euler angles (α_i, β_i) referring to the molecular axes. α_1 and α_2 of the rotations around z'-axis are combined into normal modes; the total K-rotation defined with $\theta_3 = (\alpha_1 + \alpha_2)/2$ and the twisting degree of freedom with $\alpha = (\alpha_1 - \alpha_2)/2$.

At first, the kinetic energy is given classically in terms of angular velocities, and then the quantization is done to describe the whole rotating system quantum-mechanically. The resultant kinetic energy is classified into three parts, as $\hat{T} = \hat{T}_{\mathrm{rot}} + \hat{T}_{\mathrm{int}} + \hat{T}_{\mathrm{C}}$, where $\hat{T}_{\mathrm{rot}}$ denotes the rotational energy of the whole system described with the total angular momentum operator $\hat{\boldsymbol{J}}'$, $\hat{T}_{\mathrm{int}}$ the energy of internal motions, and finally $\hat{T}_{\mathrm{C}}$ the Coriolis coupling.

The nucleus-nucleus potential is calculated by a folding model with nucleon-nucleon interaction DDM3Y. The potential is a function of the internal degrees of freedom $(R, \alpha, \beta_1, \beta_2)$, which specify geometrical configurations. Consistently with the coordinate system, we introduce a rotation-vibration type wave function specified with K-quantum number as basis one, $\Psi_\lambda \sim D^J_{MK}(\theta_i)\chi_K(R, \alpha, \beta_1, \beta_2)$, where χ_K describes the internal motions.

In order to analyze dynamics of multidimensional internal motions, we calculate the effective potential for given J and K,

$$V_{JK}(R, \alpha, \beta_1, \beta_2) = V_{\mathrm{interaction}}(R, \alpha, \beta_1, \beta_2) + T'_{\mathrm{rot}}(J, K; R, \alpha, \beta_1, \beta_2), \qquad (1)$$

where T'_{rot} denotes a multidimensional centrifugal potential with specified J and K, reduced from the rotational kinetic energy operator $\hat{T}_{\mathrm{rot}}$. In fig. 2, a multidimensional energy surface is displayed for $J = 14$. A local minimum is obtained around $\beta's = 90°$, $\alpha = 0°$ and $R = 5.7$ fm. The equilibrium configuration is displayed in r.h.s. of fig. 2, which is called as *equator-equator* (E-E) configuration. The barrier height from the bottom of the local minimum is 4.8 MeV located at $R = 7.5$ fm. Strong centrifugal potential makes the configuration to be stable for $J =$ 10–18, while for the smaller J, the stable configuration is found in belly-equator or belly-belly one, and for the larger J the potential pocket disappears.

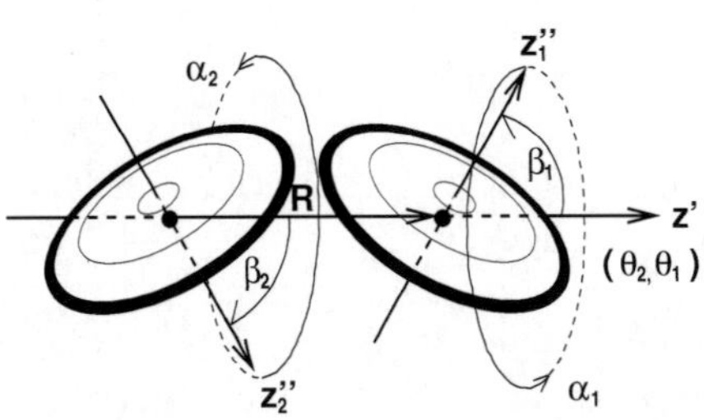

Figure 1. Dinuclear molecular coordinates; seven degrees of freedom of ^{12}C + ^{12}C (oblate-oblate system) are described with $(q_i) = (\theta_1, \theta_2, \theta_3, R, \alpha, \beta_1, \beta_2)$.

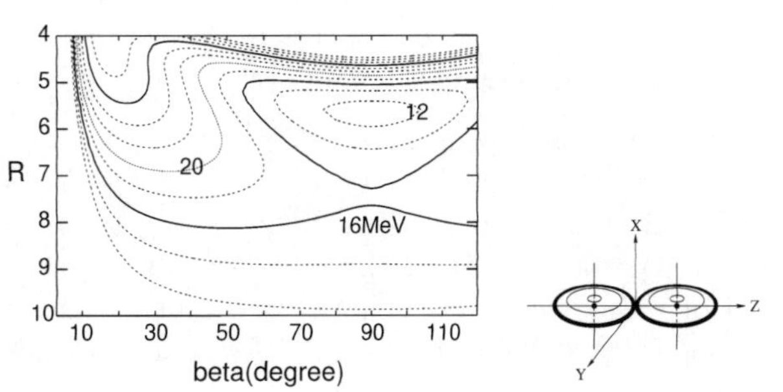

Figure 2. Contour map of the effective potential energy V_{JK} with $J = 14$ and $K = 0$ for the $R - \beta(\beta_1 = \beta_2)$ degrees at $\alpha = \pi/2$. Contours are in 2 MeV.

IOP Conf. Series: Journal of Physics: Conf. Series **863** (2017) 012048 doi:10.1088/1742-6596/863/1/012048

Couplings among various molecular configurations are taken into account by the method of normal modes around the equilibrium at $\beta_i = 90^\circ$, which gives rise to the molecular modes of excitation. We expand the effective potential $V_{JK}(R, \alpha, \beta_1, \beta_2)$ into a quadratic form to describe normal modes. Combining the kinetic energy operator $\hat{T}_{\rm int}$, we obtain hamiltonian H_0, which is almost separated into each internal degree of freedom. This means the internal degrees of freedom $(R, \alpha, \Delta\beta_1, \Delta\beta_2)$, where $\Delta\beta_i$ denote $\beta_i - \pi/2$, are the normal modes themselves. Thus we use basis wave functions, $\Psi_{JMK}(\theta_i; R, \alpha, \beta_1, \beta_2) \sim D^J_{MK}(\theta_i) f_n(R) \phi_\nu(\alpha) \varphi_{n_1}(\Delta\beta_1) \varphi_{n_2}(\Delta\beta_2)$, where f_n, φ_{n_1} and φ_{n_2} are H.O. functions, while ϕ_ν is a rotational function $e^{i\nu\alpha}$ as the confinement is very weak in α-degree due to the shallow potential pocket. The energy of the basis state is given as $E_{JK}(n, \nu, n_1, n_2) = E_0(R_{\rm e}) + \{J(J+1) - K^2 - 1\}\hbar^2/2\mu R_{\rm e}^2 + \{K^2 - 2\}\hbar^2/4I + \nu^2\hbar^2/4I + (n + 1/2)\hbar\omega_R + (n_1 + n_2 + 1)\hbar\omega_\beta$, where $\hbar\omega_\beta \sim 5$ MeV. In fig. 3, the spectrum of the normal modes with $J = 12$ is displayed, classified with quantum numbers $K = 0$–4. The axially asymmetric shape of the whole composite system displayed in r.h.s. of fig. 2 allows states with $K = 2m$, m being integer, in low excitation. The system has states with $K = 2m+1$ for the $1\hbar\omega_\beta$ excitation.

3. Coriolis coupling and characteristic features of solutions with K-mixing

With the normal-mode hamiltonian, the total hamiltonian is written as

$$H = H_0 + \hat{T}_{\rm C} + (higher\ order). \qquad (2)$$

Generally in the rotating system, we have Coriolis coupling $\hat{T}_{\rm C}$, which can be described with a simple product $-\boldsymbol{J'} \cdot (\boldsymbol{S'_1} + \boldsymbol{S'_2})/\mu R^2$ of the total angular momentum operator $\boldsymbol{J'}$ and the spins of ^{12}C $\boldsymbol{S'_i}$. Note that they refer to the molecular axes as designated with prime. In the model, it is given with many terms; $\boldsymbol{J'}$ consists of differential operators of (θ_i), which are combined into lowering and raising operators of K, $J'_\pm = J'_x \pm iJ'_y$, respectively. As for the spins $\boldsymbol{S'_i}$, they are described usually with differential operators of Euler angles β_i, i.e., given with those of the internal variables $(\alpha, \beta_1, \beta_2)$. We consider the Coriolis terms as residual coupling, which causes K-mixing with $\Delta K = 1$. The mixing induces aligned components. Actually, in the molecular ground state, a $(K = 1, \nu = 1)$ mixed component appears to bring strong alignments in the single 2^+ channel [5].

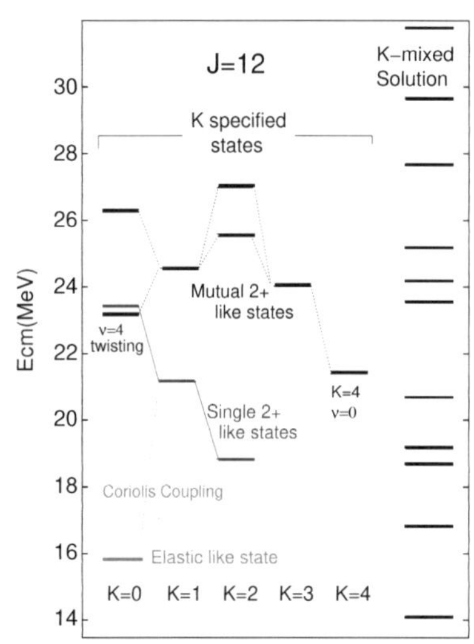

Figure 3. Energy spectra of molecular normal modes of ^{12}C + ^{12}C for $J = 12$, specified with $K = 0$ to 4 from the l.h.s. Diagonalization of the Coriolis coupling gives a new spectrum shown in the rightmost column.

The molecular quantum numbers $(J, K; n, \nu, n_1, n_2)$ indicate rotational motions of two constituent ^{12}C nuclei by $\{n_1, \mu_1 = (K + \nu)/2\}$ and $\{n_2, \mu_2 = (K - \nu)/2\}$, respectively. μ_i indicates the quantum of the rotation along the molecular z'-axis, while n_i corresponds to the rotations perpendicular to the z'-axis. Thus a set of numbers $I_i = n_i + |\mu_i|$ is related to the spins of ^{12}C nuclei, like a channel assignment. This gives us tokens of the basis wave functions corresponding to inelastic channels. We take 11 bases for the single 2^+ and mutual 2^+ channels; states with $n_1 + n_2 = 3$ and 4 are omitted as the energies are high. In fig. 3, single 2^+ and mutual 2^+ assignments are depicted with different colors and connected lines. In the rightmost column of fig. 3, levels of the K-mixed solutions caused by the Coriolis coupling are displayed for $J = 12$. In table 1, for those resultant states, probabilities in each channel are analyzed. Surprisingly, characteristic feature of dominant mutual 2^+ components with $L = J - 2$ is found

Table 1. Properties of resonance states with $J = 12$ indicated by probability distributions.

Resonance states	elastic	single 2^+	mutual 2^+
ground state[1]	36%	52%(52% $L = J - 2$)	10%
1st exc. state	31%	33% (25% $L = J$)	34% (18% $L = J - 4$, 16% $L = J - 2$)
2nd exc. state[2]	7%	15%	74% (62% $I = 4, L = J - 2$)
3rd exc. state[3]	4%	9%	82% (69% $I = 2, L = J - 2$)

[1] $K = \nu = 0$ lowest energy configuration (55%) and $K = \nu = 1$ (26%) mixed state.
[2] $K = 4$ (43%) and $K = 3$ (28%) mixed state.
[3] $K = 0$ twisting state (30%), butterfly state (14%) and ($K = 1, \nu = 3$) state (40%).

in the second and third excited states, which appear just below $E_{\rm cm} = 20$ MeV. Those states are inferred to be originally $K = 4$ state and $K = 0$ twisting state with relatively low excitation energies due to zero-point oscillation of the β-vibration at the E-E configuration, which have the property of spin disalignments $L = J$, because the ^{12}C spins are along the molecular z'-axis. However they both affected with $K = 3$ or $K = 1$ mixing, which brings spin alignments. Such a mechanism is similar to that of BCM for the molecular ground state [5].

4. Molecular spectrum and excitation functions

We have found the stable E-E configuration for high spins from $J = 10$ to $J = 18$, and the corresponding molecular bands are shown in fig. 4. Of course, there are many possible molecular bands for lower spins, to be observed around the Coulomb barrier region. We have obtained many molecular excited states for the same spin, the number of which are comparable with the

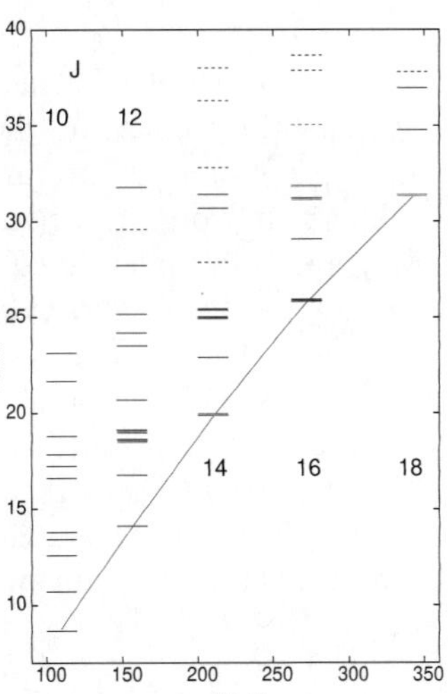

Figure 4. Molecular band structure for $J = 10$–18. The ground band is connected for eye-guide. The levels with widths larger than 4 MeV are displayed by dashed lines.

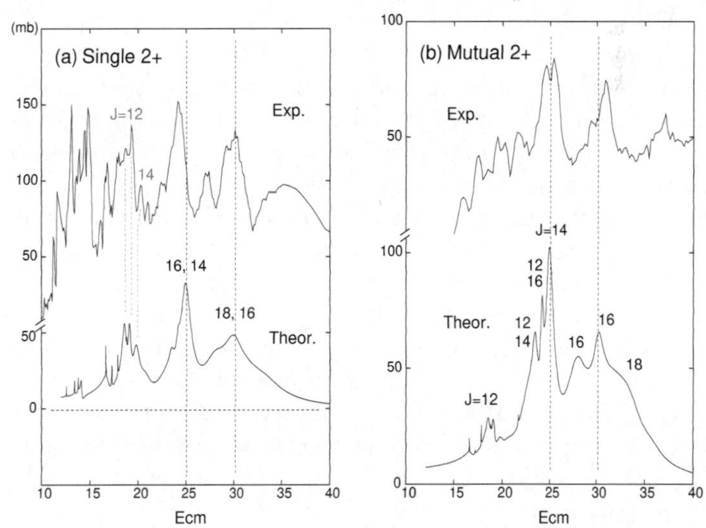

Figure 5. Theoretical excitation functions: (a) for the single 2^+ channel, and (b) for the mutual 2^+ channel. The levels of $J = 10$–18 ($\Gamma \leq 4$MeV)are included in. Spins J are indicated for some resonance peaks. At many peaks, different J contributions are overlapping.

experimental observations. Surprisingly, widths of those states are not large due to very small amounts of the elastic components. That is consistent with that the elastic components are almost exhausted in the molecular ground state and the first excited state. In fig. 5, theoretical excitation functions for the single 2^+ and mutual 2^+ channels are compared with the experiments [1]. Note that the calculated resonance energies are not precisely correct for higher spins such as $J = 16$ and 18, as harmonic approximation is adopted for the radial motions in shallow potential pockets. So the level energies of those spins are shifted by -1 MeV for $J = 16$, and -1.5 MeV for $J = 18$, respectively.

Ledoux et al. assigned spins of three resonance states around $E_{\rm cm} = 20$ MeV as $J = 12$ for lower two states (18.4, 19.3 MeV) and $J = 14$ for the 20.3 MeV state [6]. This spin assignment is very important, because spins of all three resonances are suggested to be $J = 12$ by Cormier [1]. In the present results, the molecular ground state with $J = 14$ appears at 19.9 MeV in good agreement with the Ledoux's data. With $J = 12$, the ground state is obtained at 14 MeV, while the second and third excited states appear around 19 MeV (see fig. 4). Thus three close-by resonances obtained around $E_{\rm cm} = 19$ MeV are in good correspondence with three intermediate peaks on the gross peak observed in the single 2^+ channel. Note that a few highly excited states with $J = 10$ also overlap in the region $E_{\rm cm} = 17$–20 MeV of the gross peak, which is expected to correspond with the observation of $J = 10$ resonances [7]. As for another gross peak just below 15 MeV, three $J = 10$ states and the $J = 12$ molecular ground state appear to overlap, which is similar situation around $E_{\rm cm} = 19$ MeV. However shape of the gross peak is not seen in fig. 5(a), as the resonance peaks are too narrow due to lack of $^{20}\text{Ne} + \alpha$ and $^{16}\text{O} + {}^{8}\text{Be}$ decay contributions.

In the region $E_{\rm cm} \geq 25$ MeV displayed in Fig. 5(b), peaks of the mutual 2^+ excitation are prominent as well as those of the single 2^+. Around $E_{\rm cm} = 25$ and 30 MeV, bunches of peaks are well reproduced. There are contributions from different J in the bunches again. Such a structure seems to be due to the excitation energy of 2_1^+ state of ^{12}C very close to the centrifugal energy difference for $\Delta L = 2$.

5. Conclusions

High-spin resonances of $^{12}\text{C} + {}^{12}\text{C}$ has been studied by the dinuclear molecular model. Normal-modes analysis made around the E-E configuration provides good perspectives both for the dynamics of the system and for the rotational motions of ^{12}C nuclei corresponding to the channels. Intrinsic states favored by the $^{12}\text{C} + {}^{12}\text{C}$ interaction, especially, excitations with non-aligned configurations due to rotations around the molecular z'-axis play an important role. K-mixing by the role of the Coriolis coupling changes those states into spin-aligned characteristics, as well as the molecular ground state, consistent with BCM.

Those *multi-excited molecular states* weakly couple to the elastic channel and thus are to be observed as resonances with intermediate widths. As a result, characteristic features of excitation functions in the single 2^+ and mutual 2^+ channels, i.e., the intermediate structures on the top of the prominent gross peaks have been very well reproduced over a wide energy range. This gives rise to a novel refined view to the known molecular resonances. Experiments to explore nature of those resonance structures are strongly called for.

References

[1] Cormier T M *et al* 1978 *Phys. Rev. Lett.* **40** 924
[2] Abe Y, Kondō Y and Matsuse T 1980 *Prog. Theor. Phys. Suppl.* No. **68** 303
[3] Uegaki E and Abe Y 1993 *Prog. Theor. Phys.* **90** 615
[4] Uegaki E and Abe Y 2012 *Prog. Theor. Phys.* **127** 831; *Prog. Theor. Phys.* **127** 877
[5] Uegaki E and Abe Y 2014 *J. Phys.: Conf. Series* **569** 012091
[6] Ledoux R J *et al* 1983 *Phys. Rev.* C **27** 1103
[7] Fortune H T *et al* 1977 *Phys. Rev.* C **15** 439 and *references therein*

IOP Conf. Series: Journal of Physics: Conf. Series **863** (2017) 012049 doi:10.1088/1742-6596/863/1/012049

Experimental investigations of the sub-Coulomb $^{12}C+^{12}C$ and $^{12}C+^{16}O$ reactions

S Courtin[1,2,3], G Fruet[1,2], D G Jenkins[4], C L Jiang[5], M Heine[1,2], D Montanari[1,2,3], D Santiago-Gonzalez[5], M Avila-Coronado[5], L G Morris[4], A Goasduff[6], E Rehm[5], B Back[5], D Bourgin[1,2] and C Beck[1,2]

[1] *IPHC, Université de Strasbourg, F-67037 Strasbourg, France*
[2] *CNRS, UMR7178, F-67037 Strasbourg, France*
[3] *USIAS, F-67083 Strasbourg, France*
[4] *Department of Physics, University of York, Heslington, York YO10 5DD, UK*
[5] *Physics Division, Argonne National Laboratory, Argonne IL 60439, USA*
[6] *INFN Laboratori Nazionali di Legnaro, I-35020 Legnaro (Padova), Italy*

E-mail: **Sandrine.Courtin@iphc.cnrs.fr**

Abstract. Cluster resonances in light heavy-ion systems like $^{12}C+^{12}C$ and $^{12}C+^{16}O$ may have a major impact on astrophysics stellar scenarios. Resonant radiative capture reactions have been studied for these systems at energies at and slightly below their Coulomb barriers to investigate the possible ^{12}C-^{12}C and ^{12}C-^{16}O molecular origin of the resonances. Spins have been attributed to the resonances and specificities of their γ-decay have been identified. At deep sub-barrier energies, a fusion cross section measurement using the particle-γ coincidence technique is discussed for the $^{12}C+^{12}C$ system. A new project is presented to possibly extend the $^{12}C+^{12}C$ S low-energy S factor study.

1. Introduction

Alpha cluster configurations are features of light nuclei, known to occur close to their α decay thresholds as proposed by H. Horiuchi in the late 1960s [1]. From the theoretical point of view, the cluster nature of a nuclear state could be characterized by the projection of the state wave function onto a cluster subspace, part of the full space used to describe the state, in a spectroscopic factor picture. From the experimental point of view, the cluster nature of a nucleus can be related to: *i.* its fragmentation properties into alpha or multiples of alpha substructures, *ii.* its γ decay, specific of a highly deformed structure [2].

The first part of this contribution concerns the γ decay of cluster resonances in the $^{12}C+^{12}C$ and $^{12}C+^{16}O$ systems in reactions close to the Coulomb barrier (CB).

The second part will focus on measurements of fusion cross sections at deep sub-barrier energies in the $^{12}C+^{12}C$ system. Such cross sections are related to nuclear astrophysics since the ^{12}C burning is an important feature of the evolution of massive stars and the associated nucleosynthesis. The $^{12}C+^{12}C$ reaction is known to be highly resonant from energies at the CB to a few MeV per nucleon. These resonances have been discussed in terms of molecular configurations in ^{24}Mg. Interestingly enough, such configurations have been predicted down to the Gamow window [3] but not yet observed directly.

The last part of this contribution will discuss the STELLA project which aims at measuring sub-barrier fusion cross sections of astrophysical interest in systems for which cluster resonances have been identified.

2. Radiative capture reactions in the $^{12}C+^{12}C$ and $^{12}C+^{16}O$ systems

The $^{12}C+^{12}C$ and $^{12}C+^{16}O$ radiative capture reactions have been performed at Triumf (Vancouver) using the Dragon spectrometer to select the compound nucleus at 0° [4-6], on resonances previously observed by Sandorfi et al. in the 1980s [7]. A study of the γ decay of resonances around the Coulomb barrier, in an entrance region with $E_{c.m.}$ ranging from 6 to 9 MeV has been performed for both systems using the Triumf (Vancouver, Canada) BGO array. Results of this campaign are summarized in Table 1. Spins of the populated resonances have been attributed by measuring γ angular distributions. These are in good agreement with previous fusion data. Measurements have been performed on and off resonance. For both systems, enhanced feeding of deformed states of ^{24}Mg and ^{28}Si has been observed at bombarding energies on resonance. At the lowest bombarding energy, for both systems, an enhanced feeding of T=1 states has been observed. A possible interpretation could be that the giant quadrupole resonance (GQR) is excited, built on the ^{24}Mg prolate ground state for $^{12}C+^{12}C$ and on the prolate excited band in the case of $^{12}C+^{16}O$. In fact, excitation energies in the compound nuclei ^{24}Mg and ^{28}Si, corresponding to the radiative capture at the CB are rather large (20 to 25 MeV). These excitation energies fall into the range of giant quadrupole resonances (GQR). For example, the experimental maximum of the GQR in ^{28}Si is at 18.8 MeV [8], the GQR built on the prolate band of ^{28}Si (starting at 6.7 MeV) would thus lie around 25.5 MeV. Interestingly enough, in ^{28}Si, the B(M1) values extracted from our experimental data for the transition from the resonance to the T = 1 states are around 1.25 W.u., among the strongest measured in this mass region [9].

System	$^{12}C+^{12}C$			
$E_{c.m.}$ (MeV)	8	7.5	6.8	6
Measured J^{π}	4^+-2^+	4^+	0^+-2^+	2^+
Specificities of the decay	Enhanced feeding of deformed states			Enhanced feeding of 1^+ T=1 states
System	$^{12}C+^{16}O$			
$E_{c.m.}$ (MeV)	9	8.8	8.5	6.6
Measured J^{π}	6^+	5^-	5^--6^+	2^+
Specificities of the decay	Enhanced feeding of deformed states			Enhanced feeding of 2^+ and 1^+ T=1 states B(M1) ~1.25 W.u.

Table 1. : Summary of the results of the Triumf $^{12}C+^{12}C$ and $^{12}C+^{16}O$ radiative capture campaign (see text).

At lower energies, the $^{12}C+^{12}C$ fusion reaction has strong synergies with astrophysics; it plays a crucial role in several scenarios like the formation of white drafs, the nucleosynthesis in massive stars and their further evolution, the C ignition in type Ia supernovae, and in suberbusts of neutron stars [10]. Stellar temperatures imply that Gamow energies for light heavy-ion fusion reactions are far below the corresponding Coulomb barriers. The associated cross sections are therefore as small as the sub-nanobarn range, making direct experimental investigations very challenging.

3. Fusion cross sections in the $^{12}C+^{12}C$ system

The $^{12}C+^{12}C$ fusion reaction has been the object of a realm of experimental investigations in the past [11-17]. The lowest-energy measured resonance in this reaction is at $E_{c.m.}$ = 2.1 MeV [16], only partially overlapping with the high-energy part of the Gamow window (E_{Gamow} = 1.5 MeV at T = 10^7 K). Techniques used to measure the $^{12}C+^{12}C$ S factors were based on the detection of particles, i.e. p and α from the $^{12}C(^{12}C,p)^{23}Na$ and $^{12}C(^{12}C,\alpha)^{20}Ne$ channels respectively, or the detection of γ-rays from the ^{23}Na and ^{20}Ne evaporation residues. All these experiments were subject to large backgrounds from hydrogen or deuterium contaminations of the target, proton and deuteron recoil particles, which could explain some deviations in the experimental results.

Recently, C.L. Jiang et al. proposed to overcome the problem by using a particle-γ coincidence technique, allowing almost no-background measurements [18]. This technique was used to revisit the $^{12}C+^{12}C$ low-energy fusion reaction at the Argonne National Laboratory using the Gammasphere large volume germanium array for γ detection in coincidence with particles detected in 3 annular double-sided silicon strip detectors of type S1 and S2 from MICRON, covering ~ 20 % of 4π [19]. The corresponding bombarding energies are E_{Lab} = 9.74, 8.56, 7.01 and 6.06 MeV. The results are reported in Fig. 1 (red circles), along with results from previous campaigns and extrapolations of the S factors

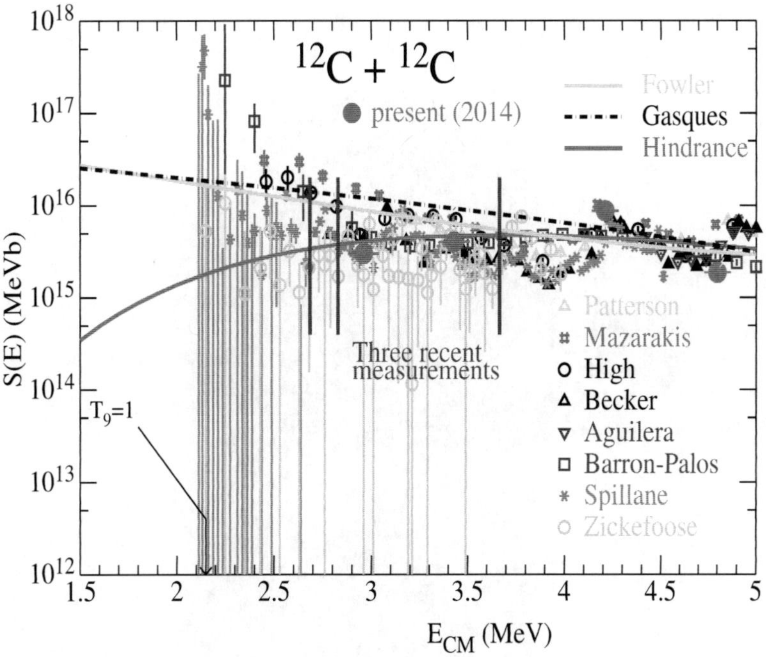

to low energies. A new Gammasphere $^{12}C+^{12}C$ experiment was performed very recently using the same set-up (Feb. 2016). The corresponding energies are reported on the Figure with vertical blue lines (three recent measurements). The analysis of this last run is under progress.

Fig. 1: Experimental S factors for $^{12}C+^{12}C$. The present results are indicated with red circles. Previous data is taken from Ref. [11-17].

It can be noticed that the technique allows to considerably reduce the data error bars. Yet at low energies, all extrapolations are still compatible with existing experimental data. These extrapolations show more that an order of magnitude differences in the astrophysics region (< $E_{c.m.}$ = 2.5 MeV). The extrapolation indicated by the red line accounts for fusion hindrance effects which have been observed in deep sub-barrier fusion reactions in medium mass systems [20], which could also play a role in the light systems of astrophysics interest. Moreover it should be noted that molecular resonances could be present in this region, and could impact the stellar evolution scenarios.

There would be considerable value in persuing experimental investigations of the $^{12}C+^{12}C$ reaction using the particle-γ coincidence technique, possibly with high beam intensities. Such a project will be briefly described in the next part of this contribution.

4. Perspectives and future projects

The STELLA project (STELlar LAboratory) has been developed to investigate fusion cross sections of astrophysics relevance, hindrance effects and eventual molecular resonances in light heavy-ion systems like $^{12}C+^{12}C$, $^{12}C+^{16}O$, $^{16}O+^{16}O$.

The project will use the particle-γ coincidences technique. Alpha and proton channels will be identified in double-sided silicon strip detectors in coincidence with γ-rays in 1.5″x2.0″ (diameterx length) $LaBr_3$ scintillators in a compact configuration. Details on the γ detectors can be found in Ref. [21]. Efficiency simulations for the γ array in this configuration are presented in Ref. [22]. The setup is presented in Fig. 2. The STELLA mobile station will be installed for its 1st Phase in 2016-2017 at the Andromede 4 MV NEC Pelletron® electrostatic single stage accelerator in Orsay (France) and will make use of high intensity beams (I >1 μA) [23,24]. A fast rotating target system has been developed to allow heat dissipation in the ^{12}C target and to sustain high beam intensities.

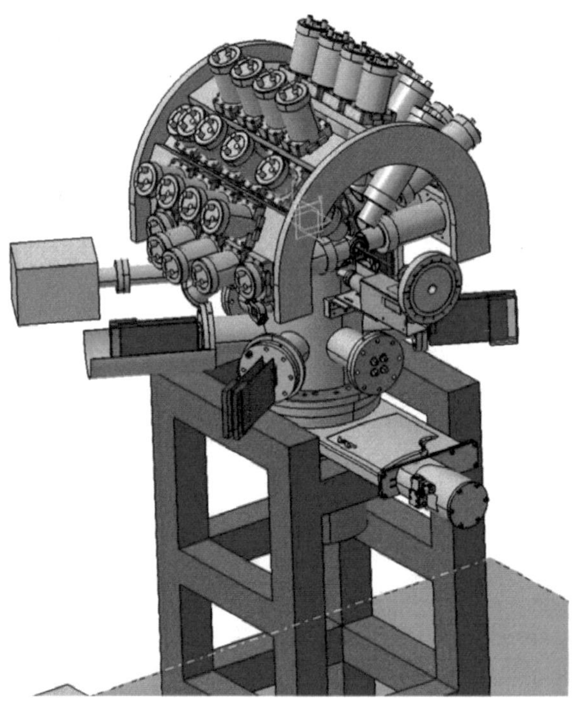

Fig. 2 : Design of the STELLA experimental station showing on top the 36 $LaBr_3$ detectors in a cylindrical configuration [22].

A 96 channels digital data acquisition system based on μTCA cards is developed to be used in the context of the STELLA project.

Commissioning of the new experimental set-up will start in 2016.

Acknowledgments

This work was supported by the IdEx Attractivity grant of the University of Strasbourg and by the University of Strasbourg Institute for Advanced Studies (USIAS).

IOP Conf. Series: Journal of Physics: Conf. Series **863** (2017) 012049 doi:10.1088/1742-6596/863/1/012049

References

[1] Horiuchi H and Ikeda K 1968 *Prog. Theor. Phys. (Jpn.)* **40** 277
[2] Jenkins D G and Courtin S 2015 *J. Phys.* G*: Nucl. Part. Phys.* **42** 034010
[3] Suzuki Y and Hecht K T 1982 *Nucl. Phys.* A **388** 102
[4] Jenkins D G *et al* 2007 *Phys. Rev.* C **76** 044310
[5] Lebhertz D *et al* 2012 *Phys. Rev.* C **85** 034333
[6] Goasduff A *et al* 2014 *Phys. Rev.* C **89** 014305
[7] Sandorfi A M Treatise on Heavy-Ion Science, Vol.2, p.52, Ed. Allan Bromley, Plenum Press, New-York 1984, and references therein
[8] Youngblood D H , Lui Y W and Clark H L 2007 *Phys. Rev.* C **76** 027304
[9] Endt P M 1998 *Nucl. Phys.* A **633**
[10] Rolfs C E and W. S. Rodney, *Cauldrons in the Cosmos*, The Univ. of Chicago Press, 1988
[11] Patterson L J *et al* 1969 *The Astrophys. J.* **157** 367
[12] Mazarakis M *et al* 1973 *Phys. Rev.* C **7** 1280
[13] High M D and Cujec B 1977 *Nucl. Phys.* A **282** 181
[14] Barron-Palos L R *et al* 2006 *Nucl. Phys.* A **779** 218
[15] Aguilera E F *et al* 2006 *Phys. Rev.* C **73** 064
[16] Spillane T *et al* 2007 *Phys. Rev. Lett.* **98** 122
[17] Zickefoose J 2010 PhD Thesis : « ^{12}C + ^{12}C *Fusion : Measurement and Advances Toward the Gamow Energy* » University of Connecticut
[18] Jiang C L *et al* 2012 *Nucl. Instr. and Meth. in Phys.* A **682** 12 ; Jiang C L *et al* 2013 *Journ. of Phys. Conf. Series* **420** 012076
[19] Santiago-Gonzales D *et al* 2016 *EPJ Web Conf.* **117** 09011
[20] Jiang C L *et al* 2002 *Phys. Rev. Lett.* **89** 052701
[21] Regan P 2012 *Appl. Rad. Iso.* **70** 1125
[22] Heine M *et al* 2016 *Journ. of Phys. Conf. Series* in press
[23] Eller M J *et al* 2016 *Nucl. Instr. Meth. B* **365** 367
[24] http://ipnwww.in2p3.fr/ANDROMEDE,384?lang=en

IOP Conf. Series: Journal of Physics: Conf. Series **863** (2017) 012050 doi:10.1088/1742-6596/863/1/012050

^{16}O + ^{16}O molecular structures of superdeformed states in S isotopes

Y Taniguchi[1]

[1]Faculty of Health Sciences, Nihon Institute of Medical Science, 1276 Shimogawara, Moroyama, Saitama, Japan

E-mail: `yasutaka@nims.ac.jp`

Abstract. Structures of excited states in S isotopes are investigated by using the antisymmetrized molecular dynamics and generator coordinate method (GCM). The GCM basis wave functions are calculated via energy variation with a constraint on the quadrupole deformation parameter β. By applying the GCM after parity and angular momentum projections, the coexistence of positive- and negative-parity superdeformed (SD) bands are predicted in $^{33\text{-}36}$S except for negative-parity states in ^{36}S. The SD bands have structures of ^{16}O + ^{16}O + valence neutron(s) in molecular orbitals around the two ^{16}O cores in a cluster picture. The configurations of the valence neutron(s) in the SD states are δ and/or π molecular orbitals.

1. Introduction

Clustering and deformation play important roles in nuclear structures. In S isotopes, ^{16}O + ^{16}O structures are expected to develop because S isotopes are analogous of Be isotopes. Be isotopes are considered to have structures of α + α + valence neutrons in molecular orbitals[1–8]. An α and an ^{16}O are double-closed-shell nuclei, and when α is replaced with ^{16}O in Be isotopes, they become S isotopes. In fact, existence of superdeformed (SD) states that contain a large amount of ^{16}O + ^{16}O cluster structure components has been predicted[9, 10], which is an analogue of α + α structure of ^{8}Be. The dominant components of the SD states in ^{32}S are predicted to be $4\hbar\omega$ configurations in the spherical shell-model picture.

By a γ-spectroscopy experiment, various positive- and negative-parity states have been observed[11]. Low-lying states are considered to be $0\hbar\omega$ or $2\hbar\omega$ neutron excited states. And many additional states are observed. But theoretical study about many-particle-many-hole states is insufficient.

In this paper, structures of positive- and negative-parity SD states in ^{34}S is discussed in details. The SD states have structures of ^{16}O + ^{16}O clusters and valence neutrons in molecular orbitals. Molecular structures developed in $^{33\text{-}36}$S are also discussed. Details of this work is reported in Ref. [12].

2. The GCM and AMD models

In this work, wave functions of ground and excited states are calculated by using the antisymmetrized molecular dynamics (AMD) and the generator coordinate method (GCM), which is superposition of AMD wave functions to diagonalize Hamiltonian. The Gogny D1S force

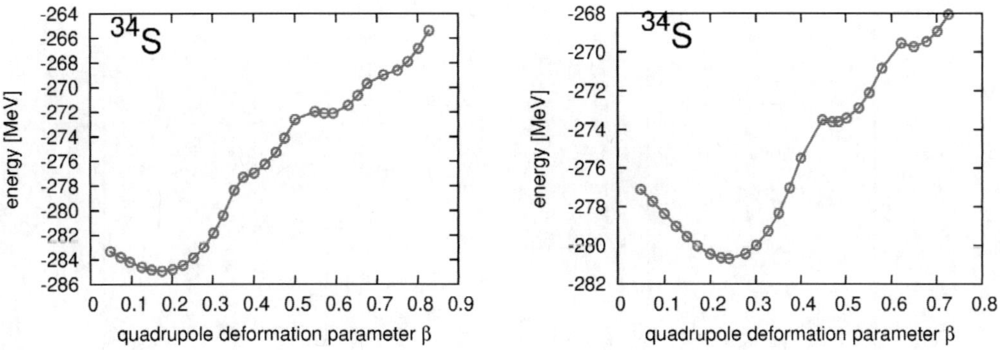

Figure 1. Energy curves of ^{34}S as functions of quadrupole deformation parameter β. Left and right panels are for positive- and negative-parity, respectively.

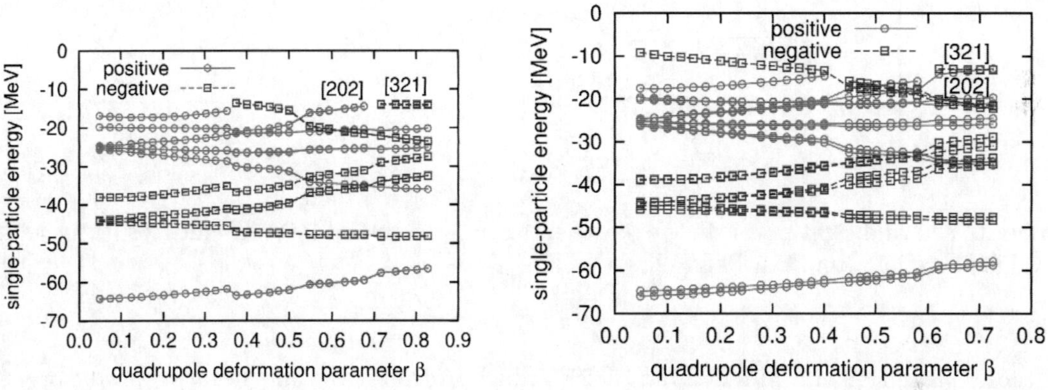

Figure 2. Single-particle energies of as functions of quadrupole deformation parameter β for positive- (left) and negative-parity (right) states. Circles and squares show positive- and negative-parity orbits, respectively. Numbers in brackets show the Nilsson quanta for the two highest orbits of neutrons in SD region. This figure is taken from Ref. [12].

is used as an effective Hamiltonian. An AMD wave function is a Slater determinant of Gaussian wave packets, and parameters in the wave functions such as position of each wave packet and spin directions are optimized by variational calculation with a constraint on quadrupole deformation parameter β. Obtained wave functions are projected onto eigen states of parity and angular momentum and superposed. Final wave functions are obtained by diagonalizing Hamiltonian.

3. Results

Figure 1 shows β-energy surface obtained by energy variational calculations for positive- and negative-parity states, respectively, in ^{34}S. β-energy surfaces have local minima and a shoulder at SD region around $\beta \sim 0.6$, which implies coexistence of positive- and negative-parity SD states.

Figure 2 shows single-particle orbits of neutrons as functions of quadrupole deformation parameter β for positive- and negative-parity states, respectively. Circles and squares show positive- and negative-parity orbits, respectively. In SD region, the two highest orbits are flat

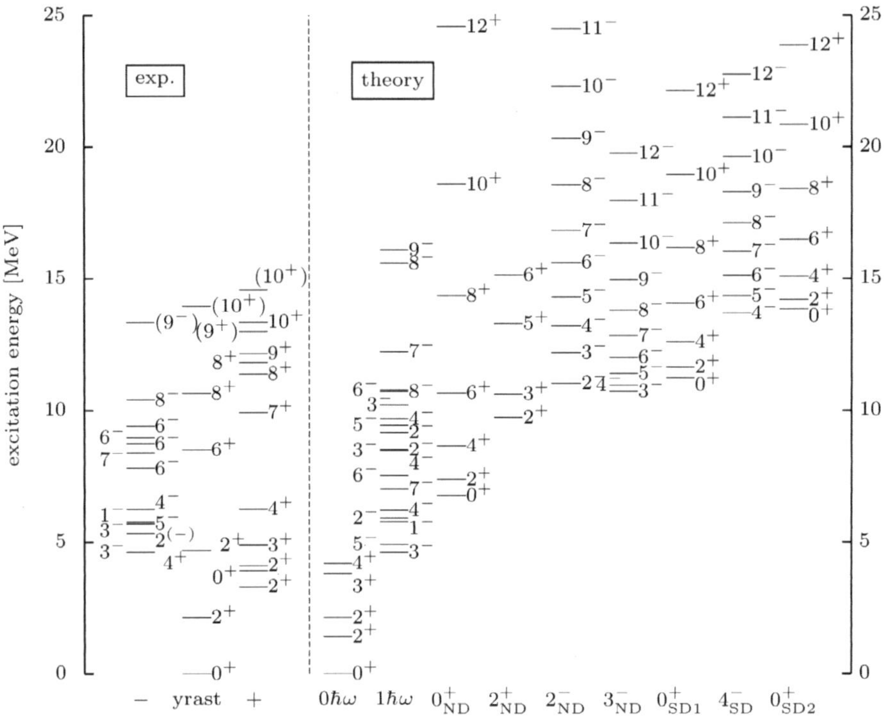

Figure 3. Left and right parts show experimental and theoretical level scheme, respectively, in ^{34}S. This figure is taken from Ref. [12].

and those Nilsson quanta are [202] and [321], which are positive- and negative-parity orbits, respectively. For lower 16 orbits, they have $2\hbar\omega$ excited configurations. Proton orbits have same configurations as lower 16 orbits of neutrons. Totally, the ^{32}S core part of SD region in ^{34}S has $4\hbar\omega$ excited configurations, which are same as configurations of predicted SD states in ^{32}S [^{32}S(SD)][10]. The ^{32}S(SD) core part has neck structure as well as a prediction of SD states. They show that structure of these wave functions are ^{32}S(SD) core + [202] and [321] neutrons.

Superposing those wave functions after parity and angular momentum projection, level scheme is obtained as shown in Fig. 3. Various rotational band are obtained. Three bands labeled $K^\pi = 0^+_{\rm SD1}$, $4^-_{\rm SD}$, and $0^+_{\rm SD2}$, coexist, which are SD bands. Those three SD bands have multi-particle-multi-hole excited configurations for both of proton and neutron parts. Configurations of dominant components of $K^\pi = 0^+_{\rm SD1}$, $4^-_{\rm SD}$, and $0^+_{\rm SD2}$ bands are ^{32}S(SD) core + $[202]^2$, [202][321] and $[321]^2$ valence neutrons, respectively, in the Nilsson picture.

The SD states are interpreted as ^{16}O + ^{16}O molecular structure in a cluster picture. Configuration of ^{32}S(SD) core is same as those of SD states in ^{32}S, which contain a large amount of ^{16}O + ^{16}O cluster structure components[10]. Valence neutrons have [202] and [321] configurations in a Nilsson picture, and the [202] and [321] orbits correspond to δ and π orbitals around two ^{16}O cores, respectively, as shown in Fig. 4. The left part of Fig. 4 shows schematic pictures of phase of $0d$ orbits around two ^{16}O cores (dotted circles) for $|l_z| = 2$, 1, and 0 components. By linear combination of $0d$ around two ^{16}O cores, δ, π, and σ orbitals (right part) are generated from $|l_z| = 2$, 1, and 0 orbits, respectively, around two ^{16}O cores. Figure 4(a) shows a $|l_z| = 2$ orbital, which is a δ orbital. Superposing two $|l_z| = 2$ orbits around ^{16}O cores, a molecular orbital becomes right part. It has no node for z-direction. The $0d$ orbits has no node for radial direction, so Nilsson quanta of a δ orbital is [202]. Figure 4(b) shows a $|l_z| = 1$

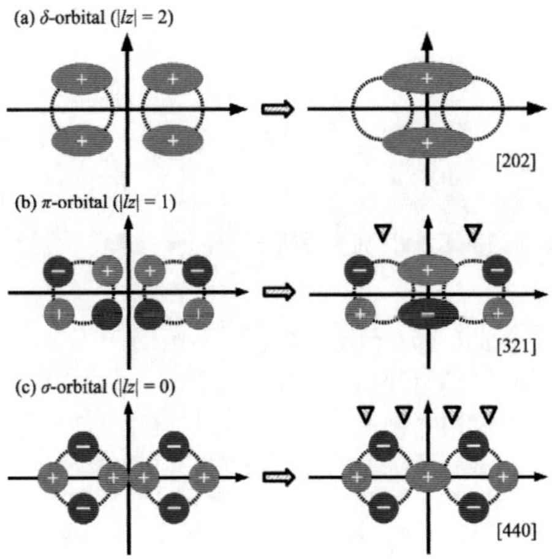

Figure 4. Schematic illustrations of molecular orbitals generated from $0d$ orbits around two ^{16}O cores for (a) δ, (b) π, and (c) σ orbitals. Horizontal axis is z-axis. Inverse triangles show locations of nodes in molecular orbitals in the z direction. Numbers in brackets show Nilsson quanta. This figure is taken from Ref. [12].

orbital, which is a π orbital. As same discussions, The π orbital have two nodes for z-directions, and Nilsson quanta of a π orbital is [321]. Figure 4(c) shows a $|l_z| = 0$ orbital, which is a σ orbital. Nilsson quanta of a σ orbital is [400]. Totally, structures of $K^\pi = 0^+_{\rm SD1}$, $4^-_{\rm SD}$, and $0^+_{\rm SD2}$ are interpreted as ^{16}O + ^{16}O + δ^2, $\delta\pi$, and π^2 structures, respectively, in a cluster picture. It is predicted that $^{33\text{-}36}$S also have positive- and negative-parity SD states with ^{16}O + ^{16}O molecular structure except for negative-parity states in ^{36}S.

4. Conclusions

In conclusions, structures of SD states in ^{34}S by using the AMD and the GCM. In ^{34}S, two positive-parity and one negative-parity SD bands coexist. Those structures are interpreted as ^{16}O + ^{16}O + δ^2, π^2 and $\delta\pi$ molecular orbitals around two ^{16}O cores. $^{33\text{-}36}$S also have SD states except for negative-parity states in ^{36}S. They have also molecular structures. In order to understand largely deformed states in $N \neq Z$ nuclei, clustering and molecular orbital around the cluster core are important.

References

[1] Seya M, Kohno M and Nagata S 1981 *Prog. Theor. Phys.* **65** 204
[2] Kanada-En'yo Y, Horiuchi H and Doté A 1999 *Phys. Rev.* C **60** 064304
[3] Itagaki N and Okabe S 2000 *Phys. Rev.* C **61** 044306
[4] Ito M, Kato K and Ikeda K 2004 *Phys. Lett.* B **588** 43
[5] Ito M 2006 *Phys. Lett.* B **636** 293
[6] Ito M, Itagaki N, Sakurai H and Ikeda K 2008 *Phys. Rev. Lett.* **100** 182502
[7] Ito M, Itagaki N and Ikeda K 2012 *Phys. Rev.* C **85** 014302
[8] Kanada-En'yo Y, Kimura M and Ono A 2012 *Prog. Theor. Exp. Phys.* **2012** 01A202
[9] Ohkubo S and Yamashita K 2002 *Phys. Rev.* C **66** 021301
[10] Kimura M and Horiuchi H 2004 *Phys. Rev.* C **69** 051304
[11] Bisoi A, Sarkar M S, Sarkar S, Ray S, Basu M R, Kanjilal D, Nag S, Selvakumar K, Goswami A, Madhavan N, Muralithar S and Bhowmik R K 2013 *Phys. Rev.* C **88** 034303
[12] Taniguchi Y 2014 *Phys. Rev.* C **90** 054308

IOP Conf. Series: Journal of Physics: Conf. Series **863** (2017) 012051 doi:10.1088/1742-6596/863/1/012051

Search for highly excited states in ^{28}Si

D Montanari[1], S Courtin[1], D G Jenkins[2], C Diget[2], N Yavuzkanat[2], R Neveling[3], J P Mira[3], F Nemulodi[3], F D Smit[3], I Usman[3], P Papka[4], J. A Swartz[4], J J van Zyl[4] and N Orce[5]

[1]IPHC, University of Strasbourg, CNRS-IN2P3, 67000 Strasbourg, France
[2]Department of Physics, University of York, YORK YO10 5DD, UK
[3]iThemba LABS, National Research Foundation, PO Box 722, Somerset West 7129, South Africa
[4]Department of Physics University of Stellenbosch, Stellenbosch 7602, South Africa
[5]Department of Physics, University of Western Cape, Bellville 7535, South Africa

E-mail: daniele.montanari@iphc.cnrs.fr

Abstract. The theoretical and experimental determination of superdeformed states in nuclei in the mass region A≤40 has been since a long time one of the major challenges of nuclear structure studies. Despite the considerable experimental and theoretical work dedicated to this topic, up to now superdeformed bands have been found in only two nuclei, ^{36}Ar and ^{40}Ca. While the experimental signature of the superdeformed nature of those states is irrefutable, their theoretical interpretation is still uncertain. In particular, it is not clear whether clusterisation is responsible of the onset of superdeformation. For this reason, we wanted to investigate an even lighter system, ^{28}Si, where a number of theoretical calculations predict the presence of superdeformation as an effect of the cluster structure of the nucleus.

1. Introduction

The experimental and theoretical study of superdeformation (SD) in lighter nuclei (A≤40) has been at the focus of the interest for decades and still represents one of the most important topics of nuclear structure. It is remarkable that, for ^{40}Ca, an early experimental study [1] showed, more than 40 years ago, the presence of low-spin states associated to $4p-4h$ and $8p-8h$ configurations, which were much strongly populated than the $0p-0h$ ground states. Subsequent experiments [2, 3] confirmed the multiparticle-multihole structure of these states and measured their large transition quadrupole moments, suggesting the presence of superderformation in this nucleus.
However, despite the considerable experimental efforts done in the past, the presence of SD bands has been only recently observed in this mass region for ^{36}Ar [4] and ^{40}Ca [5]. This has been made possible by the coincident detection of particle and gammas by using new-generation high-resolution high-efficiency gamma array detectors.
Under a theoretical point of view, superdeformation in these nuclei has been described by theoretical models in terms of particle-hole shell model excitations [4, 5, 6], cranked Skyrme-Hartree-Fock theory [7] and alpha-clustering configurations by using the Antisymmetrized Molecular Dynamics (AMD) [8, 9, 10] .
Up to now, it remains still uncertain whether the cluster description of superdeformation in the A≤40 mass region reflects a real microscopic structure of the involved nuclei or it is just able to reproduce data with no physical meaning behind.

IOP Conf. Series: Journal of Physics: Conf. Series **863** (2017) 012051 doi:10.1088/1742-6596/863/1/012051

For this reason, it is of particular importance to measure SD states in lighter nuclei where some theoretical calculations based on cluster assumption predict the existence of SD bands at energies higher than the particle break-up threshold [11, 12].

2. Superdeformation in ^{28}Si

The identification of SD bands based on alpha-cluster hypothesis is still an open question for both experiments and theories. In particular for ^{28}Si, a recent study [12] using the AMD model identify a SD band in ^{28}Si with a strong α-^{24}Mg configuration, coexisting with a prolate normal deformed and an oblate ground state rotational band. Moreover, macroscopical-microscopical calculations [13] showed the onset of strongly deformed minima in ^{28}Si nuclear potential at sufficiently high angular momentum and identified possible SD states.
Although some candidate SD states have been identified also in previous experimental [14, 15, 16] and review articles [17], the evidence of their SD nature is still debated.
In particular, the large amount of information available in ref. [15], on the population of excited states of ^{28}Si by means of p-, d- and α-induced reactions, allowed the authors to measure the decay from the 12.86 MeV (6^+_{SD} candidate state) to the 10.945 MeV (4^+_{SD} candidate state). The decay probability was inferred to be quite large, B(E2)$\geq$25 Wu.
For this reason, we decided to study the ^{12}C(^{20}Ne,α)^{28}Si reaction to measure the particle-γ branch and the multipolarity of the (candidate) SD states and to locate higher-lying members of the SD band.

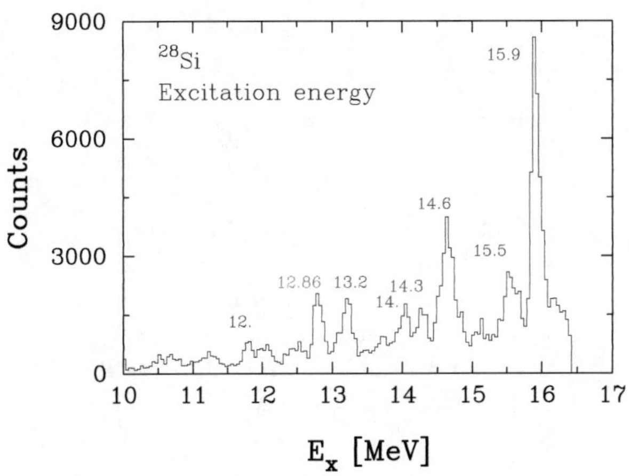

Figure 1. ^{28}Si excitation energy spectrum. Alphas emitted in the ^{20}Ne+^{12}C reaction are detected by the K600 spectrometer positioned at very small angles. The peak corresponding to the excited level of interest for this experiment is labeled in red.

3. The experiment and experimental results

We studied the ^{12}C(^{20}Ne,α)^{28}Si reaction at a beam energy of 50 MeV and used the K600 spectrometer at iThemba LABS (South Africa) to select the population of ^{28}Si states. An accelerated beam of ^{20}Ne, with an average intensity of $\approx$ 50 nA, impinged on a 20μg/cm^2 thick ^{12}C target. The magnetic spectrometer consists of a set of magnetic elements, a sextupole, a quadrupole and two dipoles, used to disperse incoming charged particles around the horizontal plane. Position-sensitive multi-wire drift chambers placed at the focal plane of the spectrometer allow the detection of the position of the light reaction products, in this case α particles emitted by ^{32}S (here for the sake of clarity called α_0). Behind the focal plane detector, a series of two plastic scintillators are used for trigger purposes.

Particles decaying from the remaining ^{28}Si (α_1) have been detected by TIARA-type Si detectors placed around the target position, while the gamma decay branch has been measured by a large NaI detector (300$\times$200 mm^2).

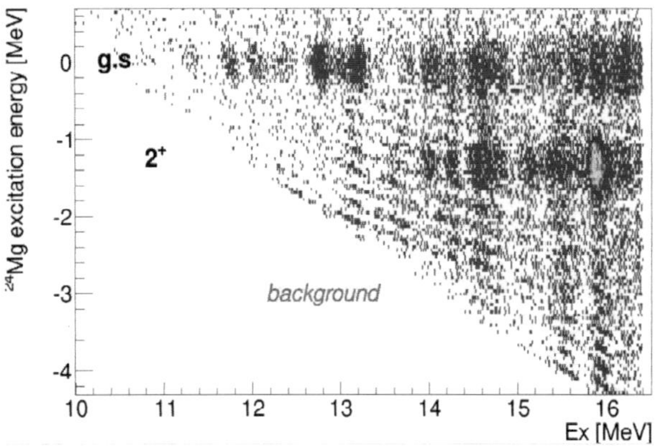

Figure 2. Bi-dimensional matrix showing the correlation between ^{24}Mg and ^{28}Si excitation energies. Despite a quite consistent background level, the population of ^{24}Mg at the ground and 2^+ state is clearly visible.

The measure of the position on the focal plane of α_0 emitted by the compound ^{32}S allows the reconstruction of their kinetic energy and hence of the excitation energy spectrum of ^{28}Si. This is shown in Fig. 1, where it is clearly visible that the 12.86 MeV state (labeled in red in the figure) is quite strongly populated. The fields of the magnetic elements of the K600 spectrometer have

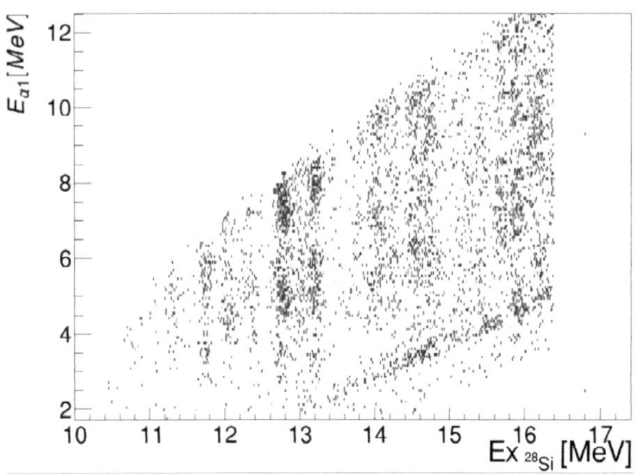

Figure 3. α_1 kinetic energy vs ^{28}Si excitation energy. The bump-like structures visible on the y axis, in correspondence to each excited level of ^{28}Si, reflect the angular distribution of the α_1.

been set in order to allow the detection of the α_0 particles corresponding to an excitation energy window between ≈10-17 MeV. A two dimensional plot of the excitation energy of ^{24}Mg vs the excitation energy of ^{28}Si is represented in Fig. 2. The excitation of ^{24}Mg has been obtained by measuring the energy released by alpha particles emitted by ^{28}Si and by adding it to the energy of alpha particles emitted by ^{32}S plus the kinetic energy of the recoil (^{24}Mg) reconstructed by imposing the conservation of linear momentum. In this figure, two peculiar ridges are clearly visible and they correspond to the population of the ground state and of the first 2^+ state of ^{24}Mg. By selecting the ground state ridge and by putting a gate around the 12.86 MeV state of Fig. 1, one can plot the kinetic energy of alpha particles emitted by ^{28}Si. This kinetic energy, with suitable kinematics calculations, can be transformed into the angular distribution of alpha particles. The bi-dimensional plot of Fig. 3 shows the coincidence between α_0, i.e. ^{28}Si excitation energy, and α_1, by requiring the population of ^{24}Mg at the ground state. It can be noticed that at each excited level of ^{28}Si corresponds a bump-like structure on the y axis. These bumps reflect the angular distribution of the α_1 particles. Through suitable fit procedures, they allow us to extract the spin removed by the α decay of ^{28}Si and therefore to assign spin and multipolarity to the selected excited state of ^{28}Si. An example of such a distribution is shown in Fig. 4 where the kinetic energy of the α_1 emitted by the 12.86 MeV state of ^{28}Si is plotted. This preliminary spectrum is not corrected by the geometrical acceptance of the Si detectors.

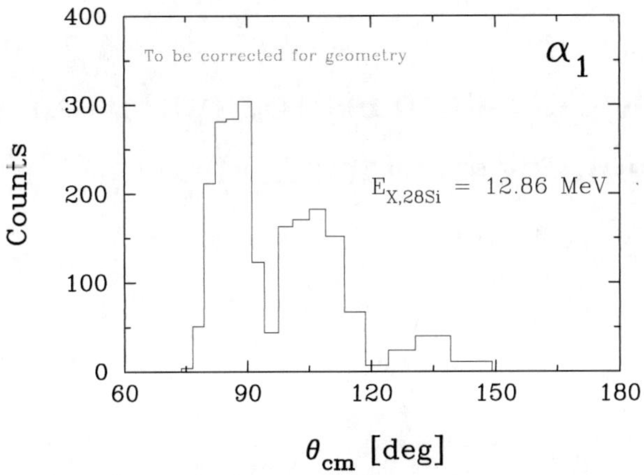

Figure 4. Angular distribution of alphas emitted by ^{28}Si populating the ground state of ^{24}Mg. The angular distribution has been obtained by kinematics calculations. The spectrum is not corrected by the geometry of the detector.

On the gamma detection side, we are able to obtain gamma spectra in coincidence with every peak of Fig. 1. Indeed, it is of particular interest to study the gamma decay associated to the transition from the 12.86 MeV (6^+_{SD} candidate) state to the 10.945 MeV (4^+_{SD} candidate) state, and to compare its intensity to the more probable particle decay branch. Further analysis efforts are being done in order to take out this meaningful result.

References

[1] Middleton R, Garret J D and Fortune H T 1972 *Phys. Lett.* B **39** 339
[2] Endt P M 1990 *Nucl. Phys.* A **521** 1
[3] Wood J L *et al* 1992 *Phys. Rep.* **215** 101
[4] Svensson C E *et al* 2000 *Phys. Rev. Lett.* **85** 2693
[5] Ideguchi E *et al* 2001 *Phys. Rev. Lett.* **87** 222501
[6] Svensson C E *et al* 2001 *Phys. Rev.* C **63** 061301(R)
[7] Inakura T, Yamagami M, Mizutori S and Matsuyanagi K 2002 *Prog. Theor. Phys. Suppl.* **146** 567
[8] Kanada-En'yo Y, Kimura M and Horiuchi H 2002 *AIP Conf. Proc.* **644** 188
[9] Horiuchi H, Y. Kanada-En'yo Y and Kimura M 2003 *Nucl. Phys.* A **722** 80c
[10] Kanada-En'yo Y, Kimura M and Horiuchi H 2003 *C. R. Physique* **4** 497
[11] Kilura M and Horiuchi H 2004 *Phys. Rev.* C **69** 051304(R)
[12] Taniguchi Y, Kanada-En'yo Y and Kimura M 2009 *Phys. Rev.* C **80** 044316
[13] Ichikawa T, Kanada-En'yo Y and Möller P 2011 *Phys. Rev.* C **83** 054319
[14] Kubono S, Morita K, Tanaka M H, Sakaguchi A and Sugitani M 1986 *Nucl. Phys.* A **457** 461
[15] Brenneisen J *et al* 1995 *Zeit für Phys.* A **352** 149/279/403
[16] Tanabe T *et al* 1983 *Nucl. Phys.* A **399** 241
[17] Jenkins D G *et al* 2012 *Phys. Rev.* C **86** 064308

High spin structures in the $A \approx 40$ mass region: from superdeformation to extreme deformation and clusterization (an example of ^{28}Si)

A V Afanasjev and D Ray

Department of Physics and Astronomy, Mississippi State University, Mississippi State, MS 39762, USA

E-mail: `Anatoli.Afanasjev@gmail.com`

Abstract. The search for extremely deformed structures in the yrast and near-yrast region of ^{28}Si has been performed within the cranked relativistic mean field theory up to spin $I = 20\hbar$. The fingerprints of clusterization are seen (well pronounced) in the superdeformed (hyperdeformed) configurations.

1. Introduction

A systematic search for extremely deformed structures in the $N = Z$ and $N = Z + 2$ S, Ar, Ca, Ti and Cr nuclei has been performed for the first time in the framework of covariant density functional theory in [1]. At spin zero such structures are located at high excitation energies which prevents their experimental observation. The rotation acts as a tool to bring these exotic shapes to the yrast line or its vicinity so that their observation could become possible with future generation of $\gamma-$tracking (or similar) detectors such as GRETA and AGATA. Some of the studied nucleonic configurations show the fingerprints of clusterization and nuclear molecules. The best candidates for observation of extremely deformed structures, clusterization and nuclear molecules have been identified in [1]. For example, the addition of several spin units above currently measured maximum spin of $16\hbar$ will inevitably trigger the transition to hyper- and megadeformed nuclear shapes in some of the nuclei (such as ^{36}Ar).

This contribution extends the investigation of [1] to ^{28}Si. Recent experimental studies and throughout analysis of previous experimental data presented in Ref. [2] provide strong indications that the 4^+ and 6^+ states located at 10.946 and 12.865 MeV, respectively, are superdeformed (SD).

2. The details of the calculations and the results for ^{28}Si.

The calculations have been performed in the framework of cranked relativistic mean field (CRMF) theory [4]. The pairing correlations are neglected so the calculations could be considered as fully realistic above $I \sim 8\hbar$. There are several reasons behind this neglect of pairing (see [4] and recent review in [5]); these are the quenching of pairing by rotation (Coriolis anti-pairing effect), substantial shell gaps and blocking effect. Note that only reflection symmetric shapes are considered in these calculations. The CRMF calculations have been performed with the NL3* functional [6] which is state-of-the-art functional for nonlinear meson-nucleon coupling

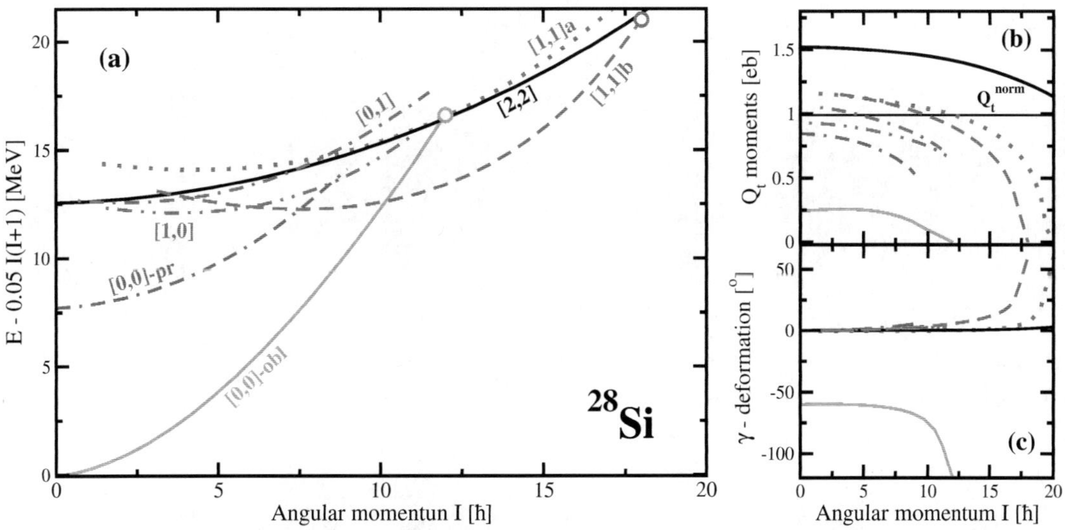

Figure 1. (a) The energies of the calculated configurations in ^{28}Si relative to a smooth liquid drop reference $AI(I+1)$, with the inertia parameter $A = 0.05$. This way of the presentation of the results has clear advantages as compared with the energy versus spin plots, see Sec. 4.1 in Ref. [3] for details. Terminating states are encircled. (panels (b) and (c)) Calculated transition quadrupole moments Q_t and γ-deformations of the yrast and excited SD and HD configurations in ^{40}Ca.

model [7] globally tested for ground state observables in even-even nuclei in [7]. The calculated configurations are labeled by shorthand $[n{,}p]$ labels, where n (p) is the number of neutrons (protons) in the $N = 3$ intruder orbitals.

Fig. 1a shows the calculated configurations in the yrast and near yrast region. At low spin, the [0,0]-obl configuration is yrast up to $I \sim 10\hbar$. It has an oblate shape in agreement with experimental findings [2]. The angular momentum content of this configuration is quite limited and it terminates at $I = 12\hbar$. The yrast line at higher spin is built by superdeformed [1,1]b configuration which terminates at spin $I = 18\hbar$. At higher spin, the hyperdeformed (HD) [2,2] configuration becomes yrast. The transition quadrupole moments Q_t and γ-deformations of the plotted configurations are presented in Figs. 1b and c. The density distributions of the configurations of interest are presented in Fig. 2.

The [0,0]-pr configuration is characterized by a substantial quadrupole and hexadecapole deformations and shows the indications of the development of neck. However, present calculations do not suggest that it could be described as a cluster of two nuclei since the maximum of the density is seen in the central part of nucleus (Fig. 2a). This configuration rather well describes both the excitation energies (compare Fig. 1a with left panel of Fig. 7 in [8]) and the moments of inertia of experimental prolate band; the latter are $J^{(1)} \sim 4\ \hbar^2$/MeV both in experiment [8] and theory.

Based on the comparison of experimental and calculated moments of inertia, the observed 4^+ and 6^+ SD states are most likely associated with the [2,2] configuration. The calculated kinematic moment of inertia of this configuration $J^{(1)} \sim 6.68\ \hbar^2$/MeV, which is nearly constant for large spin range, is only slightly above experimental one ($J^{(1)} \sim 6\ \hbar^2$/MeV [2]). Note that there are large similarities in the predictions of the properties of oblate, prolate and SD bands obtained in present CRMF calculations and the ones within the antisymmetrized molecular dynamics model of [8].

IOP Conf. Series: Journal of Physics: Conf. Series **863** (2017) 012052 doi:10.1088/1742-6596/863/1/012052

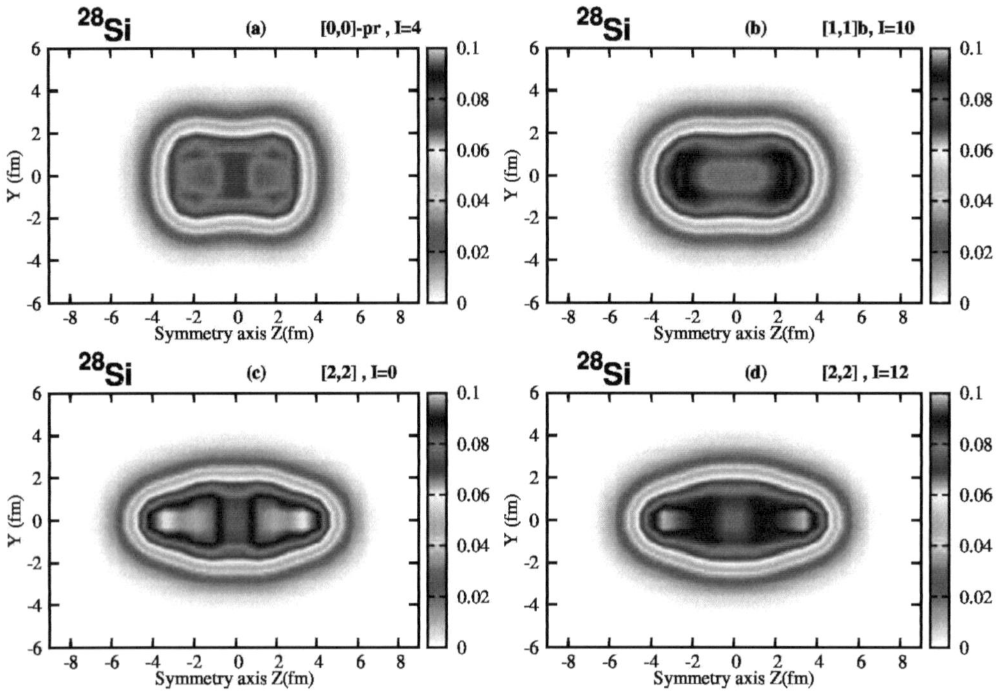

Figure 2. The self-consistent proton density $\rho_p(y,z)$ as a function of y and z coordinates for the indicated configurations in ^{28}Si at specified spin values.

Note that some indications of clusterization are present in the density distribution of the SD [1,1]b configuration (Fig. 2b). The HD [2,2] configuration shows clear signatures of clusterization which is especially pronounced at $I = 0\hbar$ (Fig. 2c). Although the rotation somewhat hinders these signatures (Fig. 2d), they are still present at $I = 12\hbar$.

3. Conclusions

The structure of the configurations in the yrast and near yrast region of ^{28}Si up to $I = 20\hbar$ has been studied in the CRMF theory. The signatures of clusterization are seen in superdeformed [1,1] configurations, but become especially pronounced in the hyperdeformed [2,2] configuration. The observation of such configurations is quite likely with future generation detectors such as AGATA and GRETA. This material is based upon work supported by the U.S. Department of Energy, Office of Science, Office of Nuclear Physics under Award Number DE-SC0013037.

References

[1] Ray D and Afanasjev A V 2016 *Phys. Rev.* C **submitted**
[2] Jenkins D G, Lister C J, Carpenter M P, Chowdury P, Hammond N J, Janssens R V F, Khoo T L, Lauritsen T, Seweryniak D, Davinson T, Woods P J, Jokinen A, Penttila H, Haas F and Courtin S 2012 *Phys. Rev.* C **86** 064308
[3] Afanasjev A V, Fossan D B, Lane G J and Ragnarsson I 1999 *Phys. Rep.* **322** 1
[4] Vretenar D, Afanasjev A V, Lalazissis G A and Ring P 2005 *Phys. Rep.* **409** 101
[5] Afanasjev A V 2016 *"Relativistic Density Functional for Nuclear Structure", (World Scientific Publishing Co), Edited by Jie Meng, Int. Rev. Nucl. Phys.* **10** 305–355
[6] Lalazissis G A, Karatzikos S, Fossion R, Arteaga D P, Afanasjev A V and Ring P 2009 *Phys. Lett.* B **671** 36
[7] Agbemava S E, Afanasjev A V, Ray D and Ring P 2014 *Phys. Rev.* C **89** 054320
[8] Taniguchi Y, Kanada-En'yo Y and Kimura M 2009 *Phys. Rev.* C **80** 044316

IOP Conf. Series: Journal of Physics: Conf. Series **863** (2017) 012053 doi:10.1088/1742-6596/863/1/012053

Asymmetric cluster structure and isoscalar monopole/dipole transitions of ^{28}Si

Y Chiba[1], Y Taniguchi[2] and M Kimura[1]

[1]Department of Physics, Hokkaido University, Sapporo 060-0810, Japan

[2]Department of Medical and General Sciences, Nihon Insitute of Medical Science, Moroyama, Saitama 350-0435, Japan

E-mail: chiba@nucl.sci.hokudai.ac.jp

Abstract. The asymmetric cluster states in ^{28}Si are discussed on the basis of antisymmetrized molecular dynamics. It is found that the inversion doublet bands having the ^{24}Mg+α, ^{20}Ne+^{8}Be and ^{16}O+^{12}C cluster configurations appear in the energy range of $E_x = 10 \sim 15$ MeV. It is demonstrated that the ^{24}Mg+α, ^{20}Ne+^{8}Be cluster states have pronounced isoscalar monopole (IS0) and dipole (IS1) transition strengths from the ground state. This result suggests that the IS0 and IS1 strength are good probe for these cluster states.

1. Introduction

Clustering is one of the important aspects in atomic nuclei and its role in light p-shell nuclei is established well. On the other hand, the clustering of mid sd-shell nuclei is still unknown because of the experimental and theoretical difficulties. One of such examples is ^{24}Mg+α cluster states in ^{28}Si that is of astrophysical interest. The antisymmetrized molecular dynamics (AMD) calculation suggested the positive-parity ^{24}Mg+α cluster bands around $E_x = 10 \sim 15$ MeV [1]. Experimentally, two candidates of ^{24}Mg+α cluster bands were proposed by the resonance scattering [2] and the α-transfer reaction [3]. However, it is unclear if these observed states really corresponds to the AMD result due to the lack of the convincing evidence of the clustering.

An evidence of the parity-asymmetric clustering is the inversion doublet bands [4]. Recently, it was proved that the pair of the 0^+ and 1^- states that constitute the doublet are strongly excited by the isoscalar monopole (IS0) and dipole (IS1) transitions [5, 6]. Indeed, it was observed that the 0^+ and 1^- resonances around $E_x = 10 \sim 15$ MeV are populated by the IS0 and IS1 transitions [7]. Therefore, in this work, we investigate the cluster states and their IS0 and IS1 strengths in ^{28}Si on the basis of AMD [8, 9]. Our aims are (1) the identification of the inversion-doublet of ^{24}Mg+α cluster bands and (2) the prediction of their IS0 and IS1 transition strengths.

2. Theoretical Framework

In this work, we employ the Gogny D1S interaction as an effective interaction. The variational wave function of AMD is given as

$$\Phi_{AMD} = \mathcal{A}\{\varphi_1\varphi_2\cdots\varphi_A\}, \tag{1}$$

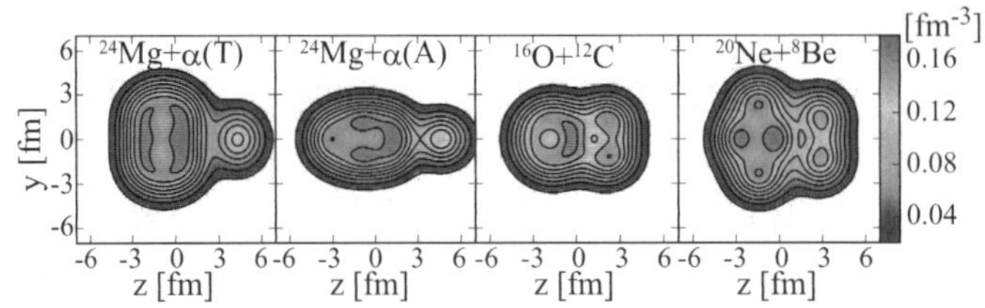

Figure 1. Density distribution of the cluster configurations obtained by the d-constraint.

$$\varphi_i(\boldsymbol{r}) = \exp\left[-\sum_{\sigma=x,y,z}\nu_\sigma\left(r_\sigma - \frac{Z_{i\sigma}}{\sqrt{\nu_\sigma}}\right)^2\right] \otimes (\alpha_i\,|\uparrow\rangle + \beta_i\,|\downarrow\rangle) \otimes \xi_i. \tag{2}$$

The variational parameters are the width $\boldsymbol{\nu}$ and the centroids $\boldsymbol{Z}_i$ of Gaussian wave packet, and the spin α_i and β_i. These parameters are determined by the constrained energy minimization after the projection to the positive- or negative-parity states. The constraint on the quadrupole deformation parameters β and α [10] are imposed. In addition to this, the d-constraint method [11] is introduced to describe the various kinds of cluster configurations. After the energy-minimization, we performed angular-momentum-projection and generator coordinate method (GCM).

3. Results and Discussions

By adopting d-constraint method, we obtained the various kind of cluster configurations as shown in Fig. 1. The obtained cluster basis wave functions can be classified to four groups, ^{24}Mg+α(T), ^{24}Mg+α(A), ^{16}O+^{12}C and ^{20}Ne+^{8}Be cluster configurations. Here, (T) and (A) denote the triaxial and axial deformation of ^{24}Mg+α cluster configurations, respectively.

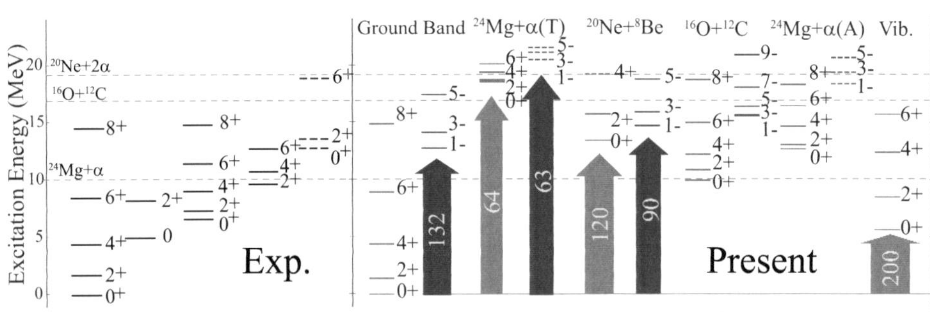

Figure 2. Excitation spectra of ^{28}Si: The observed spectra [2, 3] are plotted in the left hand side. The calculated spectra below 23 MeV are plotted in the right hand side. The dashed lines represents the averaged levels. The numbers in the orange and purple arrows indicate B(IS0) and B(IS1) values in the unit of fm^4 and fm^6 respectively. $B(IS0)$ ($B(IS1)$) values smaller than 1 fm^4 (fm^6) are not displayed.

We superposed these cluster configurations as well as the mean-field configurations, and performed the GCM calculation. Thus-obtained energy spectrum is shown in the Fig. 2, and compared with the observed energy spectra [2, 3]. Although the excitation energies are slightly higher, AMD reproduces the existence of several bands in the energy range of $E_x = 10 \sim 15$ MeV. It is found that each of cluster configurations, ^{24}Mg+α(T), ^{24}Mg+α(A) ^{16}O+^{12}C and ^{20}Ne+^{8}Be, generates the inversion-doublet band owing to their parity-asymmetric structure.

Then, the IS0 and IS1 transition strengths ($B(IS0)$ and $B(IS1)$) of those band-head states are calculated, which is defined as,

$$B(IS0) = |\langle\Phi(0_n^+)||\sum_{i=1}^{A} \boldsymbol{r}_i'^2||\Phi(0_1^+)\rangle|^2, \quad B(IS1) = |\langle\Phi(1_n^-)||\sum_{i=1}^{A} \boldsymbol{r}_i'^2\mathcal{Y}_1(\boldsymbol{r}_i'^2)||\Phi(0_1^+)\rangle|^2. \tag{3}$$

The calculated $B(IS0)$ and $B(IS1)$ values are shown in the Fig. 2. As clearly seen, the IS0 and IS1 transitions are very sensitive to clustering. Their strengths to the ^{24}Mg+α(T) and ^{20}Ne+^{8}Be cluster states are comparable to the single particle estimates $B(IS0)_{s.p.} = 63$ fm^4 and $B(IS1)_{s.p.} = 140$ fm^6. On the other hand, the transitions to the ^{24}Mg+α(A) and ^{16}O+^{12}C cluster states are negligible. This sensitivity can be understood in terms of the dual nature of the ground state. In the Elliot's SU(3) shell model, the ground state of ^{28}Si belongs to the $(\lambda, \mu) = (0, 12)$ irreducible representation. It is easy to show that the ^{24}Mg+α(T) and ^{20}Ne+^{Be} cluster structure become identical to this $(0, 12)$ representation in the zero limit of the inter-cluster distance. This means that ^{24}Mg+α(T) and ^{20}Ne+^{8}Be cluster wave functions are equivalent to the ground state state wave function. In the other words, the degree-of-freedom of ^{24}Mg+α(T) and ^{20}Ne+^{8}Be cluster excitations are embedded in the ground state. The IS0 and IS1 transitions activate the clustering degree-of-freedom embedded in the ground state as discussed in [5, 6] and then developed ^{24}Mg+α(T) and ^{20}Ne+^{8}Be cluster states are populated. On the other hand, ^{24}Mg+α(A) and ^{16}O+^{12}C cluster states belongs to the (12,8) and (12,0) representations which are different from the ground state. Therefore, the IS0 and IS1 transitions from the ground state cannot yield these cluster states. Thus, the IS0 and IS1 transitions are very sensitive and selective probe for clustering.

4. Summary

We investigated the inversion doublets in ^{28}Si using AMD with Gogny D1S interaction. It is found that ^{24}Mg+α(T), ^{24}Mg+α(A), ^{16}O+^{12}C and ^{20}Ne+^{8}Be configurations generates the doublet bands in the energy range of $E_x = 10 \sim 15$ MeV, which are an important evidence of those asymmetric cluster structure. The calculated IS0 and IS1 strengths to 0^+ and 1^- states show that they are very sensitive and selective to cluster structure. The strengths to ^{24}Mg+α(T) and ^{20}Ne+^{8}Be cluster states are enhanced but those to ^{24}Mg+α(A) and ^{16}O+^{12}C cluster states are negligible 0. This sensitivity and selectivity owes to the duality of shell and cluster in the ground state. Our results remarks that IS0 and IS1 transitions are a good probe of asymmetric cluster structure of ^{28}Si.

References

[1] Taniguchi Y, Kanada-En'yo Y and Kimura M 2009 *Phys. Rev.* C **80** 059903
[2] Artemov K P *et al*, 1990 *Sov. J. Nucl. Phys.* **80** 777
[3] Jenkins D G *et al*, 2012 *Phys. Rev.* C **86** 064308
[4] Horiruchi H and Ikeda K, 1968 *Prog. Theor. Phys.* **40** 277
[5] Yamada T, Funaki Y, Horiruchi H, Ikeda K and Tohsaki A 2008 *Prog. Theor. Phys.* **120** 1139
[6] Chiba Y, Taniguchi Y and Kimura M 2016 *Phys. Rev.* C **93** 034319
[7] Peach T *et al* 2016 *Phys. Rev.* C **93** 064325
[8] Kanada-En'yo Y, Kimura M and Ono A 2012 *Prog. Theor. Exp. Phys.* 01A202
[9] Chiba Y and Kimura M 2015 *Phys.Rev.* C **91** 061302
[10] Kimura M, Yoshida R and Masahiro I 2012 *Prog. Theor. Phys.* **127** 287
[11] Taniguchi Y, Kimura M and Horiruchi H 2004 *Prog. Theor. Phys.* **475**

IOP Conf. Series: Journal of Physics: Conf. Series **863** (2017) 012054 doi:10.1088/1742-6596/863/1/012054

Alpha-particle clustering in excited alpha-conjugate nuclei

B Borderie[1], Ad R Raduta[1,2], G Ademard[1], M F Rivet[1], E De Filippo[3], E Geraci[3,4], N Le Neindre[1,5], R Alba[6], F Amorini[6], G Cardella[3], M Chatterjee[7], D Guinet[8], P Lautesse[8], E La Guidara[3] G Lanzalone[6], G Lanzano[3], I Lombardo[6,9], O Lopez[5], C Maiolino[6], A Pagano[3], M Papa[3], S Pirrone[3], G Politi[3,4], F Porto[4,6], F Rizzo[4,6], P Russotto[4,6] and J P Wieleczko[10]

[1]Institut de Physique Nucléaire, CNRS/IN2P3, Univ. Paris-Sud, Université Paris-Saclay, Orsay, France
[2]National Institute for Physics and Nuclear Engineering, Bucharest-Magurele, Romania
[3]INFN, Sezione di Catania, Italy
[4]Dipartimento di Fisica e Astronomia, Università di Catania, Italy
[5]LPC, CNRS/IN2P3, Ensicaen, Université de Caen, Caen, France
[6]INFN, Laboratori Nazionali del Sud, Catania, Italy
[7]Saha Institute of Nuclear Physics, Kolkata, India
[8]Institut de Physique Nucléaire, CNRS/IN2P3, Univ. Claude Bernard Lyon 1, Université de Lyon, Villeurbanne, France
[9]Dipartimento di scienze Fisiche, Università Federico II and INFN, Sezione di Napoli, Italy
[10]GANIL, DSM-CEA, CNRS/IN2P3, Caen, France

E-mail: borderie@ipno.in2p3.fr

Abstract. The nuclear reaction $^{40}Ca+^{12}C$ at 25 MeV per nucleon incident energy was used to produce excited alpha-conjugate fragments from projectile fragmentation mechanism. From a careful selection provided by a complete detection and from comparisons with models of sequential and simultaneous decays, evidence in favor of α-particle clustering from excited light alpha-conjugate nuclei is reported.

1. Introduction

Clustering is a generic phenomenon which can appear in homogeneous matter when density decreases; the formation of galaxies as well as the disintegration of hot dilute heavy nuclei into lighter nuclei are extreme examples occurring in nature. As far as nuclear physics is concerned, the nucleus viewed as a collection of α-particles was very early discussed and in the last forty years both theoretical and experimental efforts were devoted to clustering phenomena in nuclei. Very recently the formation of α-particle clustering from excited expanding self-conjugate nuclei was revealed in two different constrained self consistent mean field calculations [1, 2]. The aim of the present work was to search for experimental evidence of α-particle clustering from very excited and consequently expanding alpha-conjugate nuclei. The chosen experimental strategy was to use the reaction $^{40}Ca+^{12}C$ at an incident energy (25 MeV per nucleon) high enough to possibly produce some hot expanding reaction products, associated with a high granularity,

IOP Conf. Series: Journal of Physics: Conf. Series **863** (2017) 012054 doi:10.1088/1742-6596/863/1/012054

high solid angle particle array (to precisely reconstruct directions of velocity vectors). Then, by selecting the appropriate reaction mechanism and specific events the required information was inferred.

2. Experiment and event selection

The experiment was performed at INFN, Laboratori Nazionali del Sud in Catania, Italy. The beam impinging on a thin carbon target (320 μg/cm^2) was delivered by the Superconducting Cyclotron and the charged reaction products were detected by the CHIMERA 4π multi-detector [3]. The beam intensity was kept around 10^7 ions/s to avoid pile-up events and random coincidences, which is mandatory for high multiplicity studies. CHIMERA consists of 1192 telescopes (ΔE silicon detectors 200-300 μm thick and CsI(Tl) stopping detectors) mounted on 35 rings covering 94% of the solid angle, with very high granularity at forward angles. Details on A and Z identifications and on the quality of energy calibrations can be found in [3, 4, 5, 6]. Energy resolution was better than 1% for silicon detectors and varies between 1.0 and 2.5% for alpha particles stopped in CsI(Tl) crystals.

As a first step in our event selection procedure, we want to exclude poorly-measured events. Without making any hypothesis about the physics of the studied reaction one can measure the total detected charge Z_{tot} (neutrons are not measured). In relation with their cross-sections and with the geometrical efficiency of CHIMERA, the well detected reaction mechanisms correspond to projectile fragmentation (PF) [7, 9] with Z_{tot} = 19-20 (target reaction products not detected) and to incomplete/complete fusion with Z_{tot} = 21-26 [8]. At this stage we can have a first indication of the multiplicity of α-particles, M_α, emitted per event for well identified mechanisms ($Z_{tot} \geq 19$ - see figure 1). M_α extends up to thirteen, which means a deexcitation of the total system into α-particles. Moreover a reasonable number of events exhibit M_α values up to about 6-7.

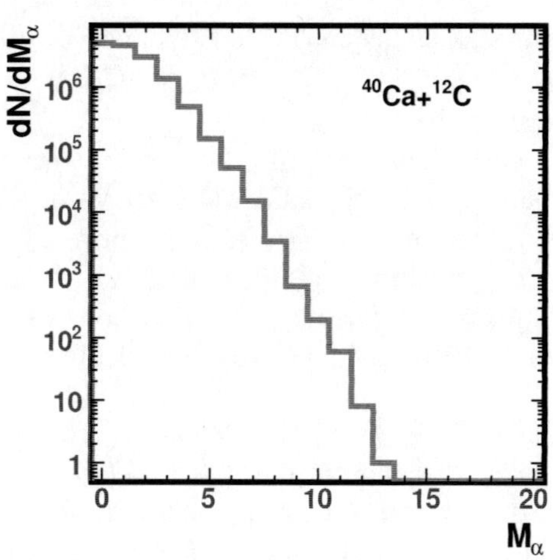

Figure 1. Distribution of α-particle multiplicity, M_α, for well detected events ($Z_{tot} \geq 19$).

The goal is now to tentatively isolate, in events, reaction products emitting α-particles only. References [10, 11] have shown that, at incident energies close to ours, ^{20}Ne or ^{32}S PF is dominated by alpha-conjugate reaction products. Based on this, and expecting the same for ^{40}Ca, we restrict our selection to completely detected PF events (Z_{tot} = 20) composed of one projectile fragment and α-particles. Charge conservation imposes Z_{frag} = 20 - $2M_\alpha$. An example of the mass disribution of the single fragment can be seen in [6].

After this double selection, the question is: from which emission source are the α-particles emitted? Several possible candidates are present and further selections must be done before restricting our study to alpha-sources emitting exclusively the M_α observed (called $N\alpha$ sources in what follows). Possibilities that we must examine are the following:

I) considering the incident energy of the reaction and the forward focusing of reaction products, it is important to identify the possible presence of preequilibrium (PE) α-particles in our selected PF events. With the hypothesis that all the α-particles are emitted from their center-of-mass reference frame, we noted an energy distribution which resembles a thermal one with the presence of a high energy tail starting at 40 MeV, which signs PE emission. To prevent errors on alpha emitter properties, it is necessary to remove events in which such PE emission can be present; an upper energy limit of 40 MeV , found irrespective of M_α was imposed to the α-particle energy.

II) α-particles can be emitted from deexcitation of PF events via unbound states of ^{12}C, ^{16}O, ^{20}Ne and not directly from excited expanding $N\alpha$ sources. We want, for instance, to exclude from the selection an event composed of two fragments (^{24}Mg and ^{12}C*) and one α-particle finally producing one single fragment (^{24}Mg) and four α-particles. Multi-particle correlation functions [12, 5] were used to identify unbound states α-particle emitters and to suppress a small percentage of events (1.6-3.9%).

III) it must be verified that the fragments associated with M_α are not the evaporation residues of excited Ca projectiles having emitted sequentially α-particles only.

As far as the two first items are concerned the effect was to suppress from 8.5 to 12.8% of previously selected events; more details can be found in [6]. The last item will be discussed in the following section.

To conclude on this part, one can also indicate that if excited $N\alpha$ sources have been formed their excitation energy thresholds for total deexcitation into α-particles vary from 20 to 60 MeV when $N\alpha$ moves from 4 to 7. Their mean excitation energy per nucleon is rather constant around 3.3-3.5 MeV which indicates that average lowest densities around 0.7 the normal density may have been reached due to thermal pressure [13, 14]. This density value is a crude estimation.

3. Evidence for alpha-particle clustering

Before discussing different possible deexcitations involved for retained events, information on projectile fragmentation mechanism is needed. Global features of PF events are reproduced by a model of stochastic transfers [9]. Main characteristics for primary events with Z_{tot}=20 are the following: i) excitation energy extends up to about 200 MeV, which allows the large excitation energy domain (20-150 MeV) measured for $N\alpha$ sources when associated to a single fragment and ii) angular momenta extend up to 24 $\hbar$, which gives an upper spin limit for Ca projectiles or $N\alpha$ sources.

Are α-particles emitted sequentially or simultaneously? To answer the question α-energy spectra can be compared to simulations. For excited Ca projectiles and $N\alpha$ sources, experimental velocity and excitation energy distributions as well as distributions for spins are used as inputs. Results of simulations are then filtered by the multi-detector replica including all detection and identification details. Simulated spectra are normalized to the area of experimental spectra.

For sequential emission the GEMINI++ code [15] was used. Before discussing decays of $N\alpha$ sources, we must consider the possible evaporation from Ca projectiles as stated previously. Excitation energy for projectiles is deduced from E^{*}=$E^{*}(N\alpha)$+E_{rel}+Q. E_{rel} is the relative energy between the $N\alpha$ source and the associated fragment (evaporation residue). Comparisons of simulations with experimental energy spectra of α-particles are displayed in figures 2 and 6 of [16, 6] for $M\alpha$ from 4 to 6. They show a rather poor agreement indicating that such an hypothesis seems not correct. Note that no more ^{24}Mg, ^{20}Ne or ^{16}O evaporation residues associated to $M\alpha$ from 4 to 6 are produced in simulations for ^{40}Ca spin distributions centered

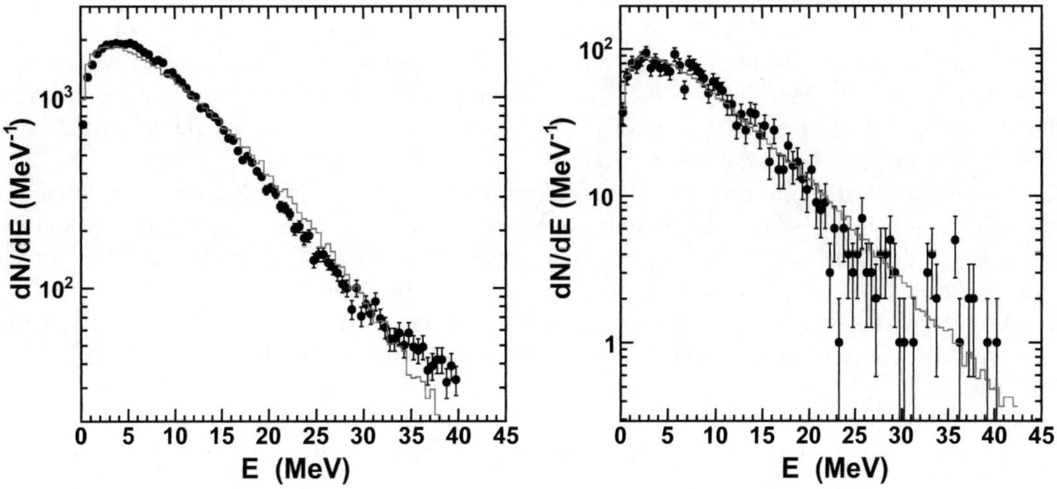

Figure 2. Particle spectra from N_α sources: ^{16}O* (left) and ^{28}Si* (right). Black dots with statistical error bars correspond to experimental data. Histograms superimposed on data correspond to filtered simulations of a simultaneous decay process (see text).

at values larger than $25\hbar$.

Considering now sequential deexcitation of $N\alpha$ sources it appears, as it is shown in figures 3 and 5 of [16, 6], that the agreement between data and simulations becomes poorer and poorer when $N\alpha$ value decreases. Moreover an important disagreement between data and simulations is observed for the percentages of $N\alpha$ sources which deexcite via ^8Be emission [6].

For simultaneous emission from $N\alpha$ sources, a dedicated simulation was done which mimics a situation in which α clusters are early formed when the $N\alpha$ source is expanding [1, 2] due to thermal pressure. By respecting the experimental excitation energy distributions of $N\alpha$ sources, a distribution of $N\alpha$ events is generated as starting point of the simulation. Event by event, the $N\alpha$ source is first split into α's. Then the remaining available energy $(E^* + Q)$ is directly randomly shared among the α-particles such as to conserve energy and linear momentum [17]. Histograms in figure 2 are the results of such a simulation for $N\alpha = 4$ (^{16}O*) and $N\alpha = 7$ (^{28}Si*), which show a good agreement with data even if for the observed ^{28}Si* statistics is low. Similar calculated energy spectra were also obtained with simulations containing an intermediate freeze-out volume stage where α-particles are formed and then propagation of particles in their mutual Coulomb field. In this case angular momentum distributions of $N\alpha$ sources at freeze-out can also be deduced: they exhibit a Maxwell-like shape extending up to $25\hbar$ for $N\alpha = 7$ while mean values vary from 6.7 to $10.4\hbar$ when $N\alpha$ moves from 4 to 7. Note that ^8Be emission is out of the scope of the present simulation.

From these comparisons with both sequential and simultaneous decay simulations it clearly appears that sequential emission is not able to reproduce experimental data whereas a remarkable agreement is obtained when an α-clustering scenario is assumed. Same conclusion is derived for $N\alpha$ equal 5 or 6 [16, 6]. However one cannot exclude that a small percentage of $N\alpha$ sources, those produced with lower excitation energies, sequentially deexcite.

4. Conclusion and perspectives

The reaction ^{40}Ca+^{12}C at 25 MeV per nucleon bombarding energy was used to produce and carefully select specific classes of projectile fragmentation events from which excited $N\alpha$ sources can be unambiguously identified. Their excitation energy distributions are derived with mean values around 3.4 MeV per nucleon and a crude estimation of their mean minimal densities, around 0.7 the normal density, can be deduced.

Their energetic emission properties were compared with two simulations, one involving sequential decays and a second for simultaneous decays. For excited expanding $N\alpha$ sources composed of 4, 5, 6 and 7 α-particles, evidence in favour of simultaneous emission (α-particle clustering) is reported. Those results support mean field calculations of [1, 2].

Work in progress shows that breakup/clustering temperature can be derived from the impressive agreement between α-particle energy spectra and fits with a volume Maxwellian distribution [18]. Finally one can also expect to derive information on the clustering density from the yield of ^{8}Be measured. Indeed within an α-particle clustering picture the production of ^{8}Be can be strongly related to α-α interaction in the freeze-out volume.

References

[1] Girod M and Schuck P 2013 *Phys. Rev. Lett.* **111** 132503
[2] Ebran J P *et al* 2014 *Phys. Rev.* C **89** 031303(R)
[3] Pagano A *et al* 2004 *Nucl. Phys.* A **734** 504
[4] Le Neindre N *et al* 2002 *Nucl. Instr. and Meth. in Phys. Res.* A **490** 251
[5] Raduta Ad R *et al* 2011 *Phys. Lett.* B **705** 65
[6] Borderie B *et al* 2016 *Phys. Lett.* B **755** 475
[7] Borderie B *et al* 1990 *Ann. Phys. Fr.* **15** 287
[8] Eudes P *et al* 2014 *Phys. Rev.* C **90** 034609
[9] Tassan-Got L and Stephan C 1991 *Nucl. Phys.* A **524** 121
[10] Morjean M *et al* 1985 *Nucl. Phys.* A **438** 547
[11] Fuchs H L and Mohring K 1994 *Rep. on Prog. in Phys.* **57** 3
[12] Charity R J *et al* 1995 *Phys. Rev.* C **52** 3126
[13] Friedman W A 1990 *Phys. Rev.* C **42** 667
[14] Borderie B 1997 *Large-Scale Collective Motion of Atomic Nuclei* ed G Giardina G Fazio *et al* (Singapore: World Scientific) pp. 1-14
[15] Charity R J 2010 *Phys. Rev.* C **82** 014610
[16] Borderie B *et al* 2015 *Proc. Int. Conf. on Nucleus-Nucleus collisions (Catania) EPJ Web of Conf.* **117** 07014
[17] Lopez A and Randrup J 1989 *Nucl. Phys.* A **491** 477
[18] Goldhaber A S 1978 *Phys. Rev.* C **17** 2243

From real clusters to experimental nuclear matter phase diagram

L G Moretto and L Phair

Nuclear Science Division, Lawrence Berkeley National Laboratory, Berkeley, CA, USA

E-mail: **lgmoretto@lbl.gov**

Abstract. We have determined the critical properties of infinite uncharged nuclear matter based on cluster emission data. To do this, three obstacles had to be overcome: finite size effects, the Coulomb interaction and appropriate physical picture (not particles coexisting in a box). That lead to the complete liquid to vapor phase diagram of neutral, symmetric, uncharged nuclear matter.

1. Introduction

As virtual clusters exist inside nuclei, real clusters are emitted by excited nuclei, carrying with them information about the nuclear liquid to vapor phase transition. After decades of theoretical and experimental studies, recent papers have published what can be considered a quantitative, credible liquid-vapor phase diagram containing the coexistence line up to the critical temperature [1, 2]. Somewhat unexpectedly, this diagram has not been obtained through the study of caloric curves [3, 4] or anomalous heat capacities [5, 6], but rather through the fitting of the charge distributions in multifragmentation by means of a Coulomb-corrected Fisher's formula [1, 7] giving the cluster composition of a vapor:

$$n_A(T) = q_0 A^{-\tau} \exp[\frac{\Delta\mu A}{T} - \frac{c_0 \varepsilon A^{\sigma}}{T}] \tag{1.1}$$

where q_0 is a normalization constant [7]; τ is the critical exponent giving rise to a power law at criticality; A is the cluster number; $\Delta\mu$ is the difference of chemical potentials between the liquid and the vapor; c_0 is the surface energy coefficient; T is the temperature; $\varepsilon = (T_C - T)/T_C$ is the distance from the critical temperature T_C; σ is another critical exponent (expected to be approximately 2/3). For $\Delta\mu = 0$ the liquid and the vapor are in equilibrium and Eq. (1.1) can be taken to be the equivalent of the coexistence line. More conventionally, one can immediately obtain from Eq. (1.1) the usual p, T and ρ, T phase diagrams by recalling that in Fisher's model, the clusterization is assumed to exhaust all the non-idealities of the gas. It then becomes an ideal gas of clusters. Consequently, the total pressure and density are:

$$p(T) = \sum_A p_A(T) = T \sum_A n_A(T) \tag{1.2}$$

$$\rho = \sum_A A n_A(T) \tag{1.3}$$

as well as the corresponding scaled quantities p/p_c and ρ/ρ_c.

In order to make use of the Fisher description of the cluster concentrations there were three major obstacles to overcome: finite size effects, the problem of the long-range Coulomb interaction, and the (un)physical picture of particles in a box.

2. Finites size effects

Finite size effects are essential to the study of nuclei and other mesoscopic systems. We have developed a general method to deal with finite size effects in phase transitions [8]. In the case of liquid-vapor phase coexistence, a dilute nearly ideal vapor phase is in equilibrium with a dense liquid-like phase. The interesting case of finiteness is realized when the liquid phase is a finite drop[9]. We introduced the concept of the complement (the residual drop which remains after a cluster has been emitted) in order to further quantify finite size effects and to generalize Fisher's theory to better describe the cluster yields from extremely small systems [8]. The complement approach consists of evaluating the change in free energy occurring when a particle or cluster is moved from one (finite) phase to another. In the case of a liquid drop in equilibrium with its vapor, this is done by extracting a vapor particle of any given size from the drop and evaluating the energy and entropy changes associated with both the vapor particle and the residual liquid drop (the complement).

Consider a vapor in equilibrium with a drop of its liquid. The system may be a physical system or the Ising realization of it. For each cluster of the vapor we can make the mental exercise of extracting it from the liquid, determining the change in entropy and energy of the drop and cluster system, and then putting it back in the liquid (the equilibrium condition), as if all other clusters of the vapor did not exist. Fisher's expression can now be written for a drop of size $A0$ in equilibrium with its vapor as follows

$$n_A(T) = q_0[\frac{A(A_0 - A)}{A_0}]\exp\{[A^\sigma + (A_0 - A)^\sigma - A_0^\sigma](\frac{c_0}{T_c} - \frac{c_0}{T})\} \quad (2.1)$$

In other words, we treat the "complement" ($A0 - A$) in the same fashion as a cluster. While different than the standard Fisher expression, Eq. (1) still admits the same T_C as that of the infinite system. This is because the $A0$, A dependence of the surface energy finds its exact counterpart in that of the surface entropy.

We are now prepared to compare the Ising yields of the vapor concentrations to our modified Fisher's droplet model. An example is given in Fig. 1. The yields in the upper panel show the linear Boltzmann behavior associated with a first order phase transition. In the lower panel of Fig. 1 is shown the scaling condition where the ordinate is scaled by the Fisher power law and the abscissa is scaled to reflect the total surface energy over the temperature.

From the Fisher droplet model fits to the yields comes an estimate of the critical temperature of 2.32 ±0.02. This is to be compared to the known analytical solution of $T_c = 2/\ln(1+\sqrt{2}) = 2.269$. Using that information, the coexistence pressure and density can be calculated . The agreement between the Fisher description, the Ising calculation and the known analytic solution of Onsager [16] is excellent.

Thus, we have obtained a practical way of moving from finite to infinite systems with the Fisher droplet model.

3. Coulomb

In ref. [10], we explored the effects of the Coulomb interaction upon the nuclear liquid phase transition. Because large nuclei are metastable objects, phases, phase coexistence, and phase transitions cannot be defined with any generality and the analogy to liquid vapor is ill-posed for these heavy systems. However, it is possible to account for the Coulomb interaction in the decay rates and obtain the coexistence phase diagram for the corresponding uncharged system [1].

4. Physical picture

We assume (just as in compound nuclear decay) that after prompt emission in the initial phase of the collision, the resulting system relaxes in shape and density and thermalizes on a time scale shorter than its thermal decay. At this point, the excited nucleus emits particles in vacuum, according to standard statistical decay rate theory. In this picture, there is no surrounding vapor and no confining box; there is no need for either. By studying the outward flux of the first fragments emitted, we can study the nature of the vapor even when it is absent (the virtual vapor) because of the equivalence of the evaporation and condensation fluxes of a liquid in equilibrium with its saturated vapor. Quantitatively, the concentration $n_A(T)$ of any species A in the vapor is related to the corresponding decay rate $R_A(T)$ (or to the decay width Γ_A) from the nucleus by matching the evaporation and condensation fluxes.

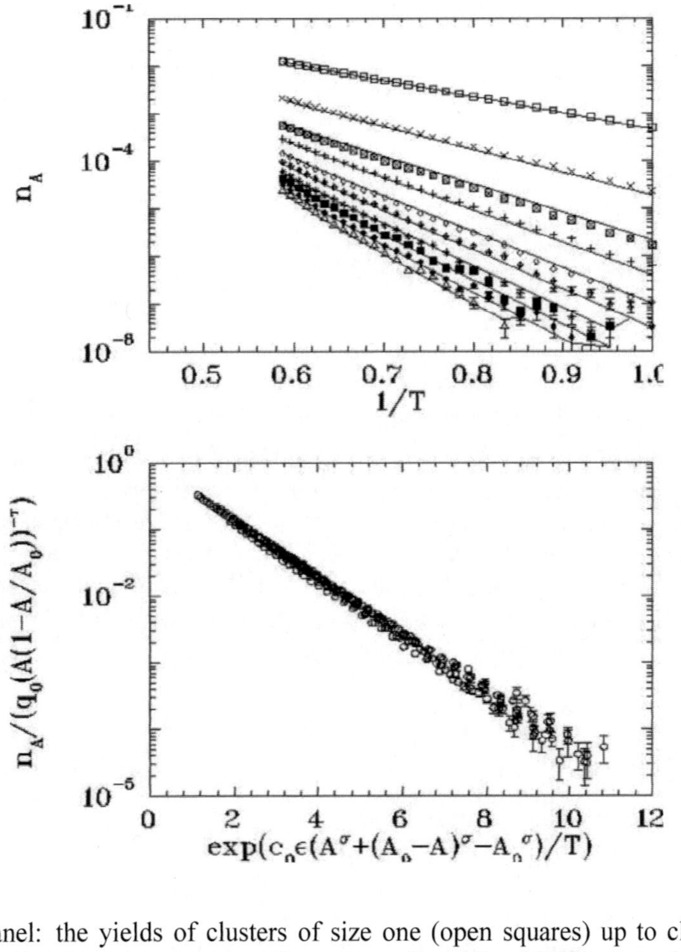

Figure 1: Upper panel: the yields of clusters of size one (open squares) up to clusters of size ten (open triangles) as a function of temperature from a two dimensional lattice gas calculation (lattice of 40 by 40) at constant density ρ = 0.2 (corresponding to a cold liquid drop of A0 = 80) as a function of 1/T. The lines are fits to the yields using the modified Fisher droplet model described in Eq. (2.1). Lower panel: the same yields scaled by the power law on the ordinate and the surface energy over temperature on the abscissa.

This is the fundamental and simple connection between the compound nucleus decay rate, and Eq. (2.1), the modified Fisher droplet model. In the latter, one immediately recognizes in the exponential the canonical expansion of the standard compound nucleus decay rate, namely, the Boltzmann factor. The surface part of the barrier is isolated from all other components, e.g., Coulomb [10], symmetry, and finite size [8]. Thus, the vapor phase in equilibrium can be completely characterized in terms of the decay rate.

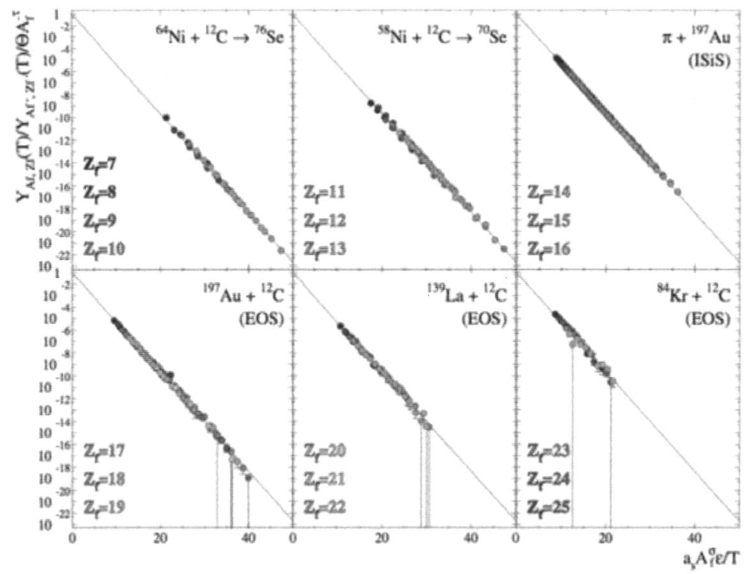

Figure 2: The scaled charge yields for the six indicated reactions. Over 500 points are collapsed onto a single curve which describes the behavior of bulk nuclear matter. The color of the points shows the charge of the fragments. The solid line shows the liquid-vapor coexistence curve of bulk infinite nuclear matter.

Results are shown in Fig. 2 for the indicated reactions. When the yields are scaled by the Coulomb, rotational, symmetry energies, etc. (all terms that are not surface energy), and plotted as a function of the Boltzmann factor associated with surface energy, the different reactions show a similar scaling. This scaled line contains all the information of coexistence between the two phases for chargeless, symmetric, and infinite nuclear matter. The data can be fit to obtain an estimate of T_C and such a procedure yields $T_C = 17.9 \pm 0.4$ MeV. Using the ideal gas approximations(1.2) and (1.3)), estimates can be obtained for the critical density ρ_C and critical pressure p_C. Such calculations give $\rho_C = 0.06 \pm 0.02$ fm^{-3} and $p_C = 0.31 \pm 0.07$ MeV/fm^3.

References

[1] Elliott J B *et al* 2013 *Phys. Rev.* C **87** 054622
[2] Moretto L G *et al* 2011 *J. Phys.* G. **38** 11
[3] Pochodzalla J *et al* 1995 *Phys. Rev. Lett.* **75** , 1040
[4] Chernomoretz *et al* 2001 *Phys. Rev.* C **64** 044605
[5] Gulminelli F *et al* 1999 *Phys. Rev. Lett.* **82** 1402
[6] D'Agostino M *et al* 2000 *Phys. Lett.* B **473** 219
[7] Fisher M E 1969 *Rep. Prog. Phys.* **30** 615
[8] Moretto L G *et al* 2005 *Phys. Rev. Lett.* **94** 202701
[9] Rayleigh L 1917 *Philos. Mag.* **34** 94
[10] Moretto L G *et al* 2003 *Phys. Rev.* C **68** 061602
[11] Fisher M E and Ferdinand A E 1967 *Phys. Rev. Lett.* **19** 169
[12] Landau D P 1976 *Phys. Rev.* B **13** 2997
[13] Coniglio A and Klein W 1980 *J. Phys.* A **13** 2775
[14] Elliott J B *et al* 2005 *Phys. Rev.* C **71** 024607
[15] Mader C M *et al* 2003 *Phys. Rev.* C **68** 064601
[16] Onsager L 1944 *Phys. Rev.* **65** 117
[17] Moretto L G *et al* 2005 *Phys. Rev.* C **72** 064605

Cluster production and nuclear dynamics

M Papa

INFN, Sezione di Catania, Via S. Sofia 64, 95123 Catania, Italy

E-mail: Massimo.Papa@ct.infn.it

Abstract. In this contribution we will briefly illustrate some general aspects related to the cluster production in heavy ion collisions. They concerns the necessary interplay between effective interaction and the description of the cluster formation processes in Molecular dynamics approaches and the time scales separation in Isospin equilibration processes.

1. Introduction

Clustering in nuclei, at the ground state and relatively low excitation energy, is a phenomenon which characterizes the structure of several light nuclei and highlights the role played by the many body correlations and symmetries in nuclear systems [1, 2, 3]. But, clusters and the related formation processes, play a central role also in heavy ion collisions and in astrophysical phenomena in which densities and excitation energies can reach rather high values. In this contribution we want to focus on two aspects in which the mechanism of the reaggregation of nucleons in to cluster, after a violent heavy ion collision, can give information on the behavior of the effective interactions describing and modeling the dynamics in a wide range of densities and excitation energies. In the following section we will try to describe the way in which the necessary correlations introduced in the model, able to produce clusters, affect the parameter values characterizing one of the more widely used effective interaction, the Skyrme one. For this purpose we use the Constrained Molecular Dynamical Model (CoMD)[4, 5]. In section 3 we will show that by looking at the cluster production from an heavy ion collision in a global and "coherent way" it is possible to access in an exclusive way to the early stages of the dynamics of the collision processes. Also in this case others information can be still obtained concerning the effective interaction.

2. Wave-Packets dynamics and Effective interaction.

In heavy ion collision at the Fermi Energy and beyond, the production of light clusters is the most intense cooling mechanism. Every theoretical approach aiming to describe the dynamics of the reactions in a reasonable way has necessarily to describe in some way the nucleon aggregation processes. Moreover, if the description, how desirable, needs to be related directly with the nucleon-nucleon effective interaction it has to be necessarily time dependent on one hand and, on the other, the used approach has to go beyond the mean-field approximation (MFA) just to describe the cluster formation processes. Such approaches due to the extreme complexity of the problem will be necessarily semi-classical. A rather wide class of them is represented by the the so called Molecular Dynamics models. All these approaches, which are different from each other for the way in which quantal effects are eventually introduced, are characterized by the usage of well localized wave-packets to represent the single particles wave functions. It is just this assumption which allows the spontaneous formation of clusters in this kind of models. Concerning the microscopic interaction usually these models use however an effective interaction, like the Skyrme, Gogny, whose strength factors are determined through calculations based on the MFA. Recently [6] this aspect has been investigated in some detail by means of the CoMD model in which Pauli principle is fulfilled at each time step by constraining the N-body equation of motion through a well established numerical algorithm. We start this investigation by studying a relatively simply case: we have studied the Equation of State of the asymmetric Nuclear Matter (NM) around the saturation density by means of the dynamical model, starting from a typical Skyrme interaction containing two and three body interaction plus the iso-vectorial one described by different density functionals. Through an iterative procedure the total energy functional obtained by the model has been determined in such a way to obtain commonly accepted saturation properties $\rho_0 = 0.16 fm^{-3}$, $K_0 \simeq 220MeV$. From this condition, a new set of parameters depending on the degree of stiffness of the iso-vectorial interaction has been found ($T_0 \simeq -159MeV$ and $T_3 \simeq 57MeV$ for $\gamma = 1$). The values of these parameters are rather different from the ones corresponding to the mean-field limit $T_0 \simeq -263MeV$ and $T_3 \simeq 208MeV$ and independent on γ. In figure 1 in panel a) we show the total energy E for symmetric NM as a function of the density computed by the model by using an effective Skyrme interaction with standard parameters values as obtained from the semiclassical MFA approach. Different kind of lines refers to different option concerning the iso-vectorial interaction. It is evident that the nuclear matter properties at the saturation in these cases are not reproduced. In the panel b) we plot the so called E_{bias}. This terms arises just from the iso-vectorial interaction coupled in a self-consistent way with the Pauli principle for which, on average, neutron-proton tend to stay closer with respect to neutron-neutron and proton-proton (deuteron effect). In figure 2 instead we report the obtained results concerning the total energy after the convergency of the iterative procedure which fixes the new set of parameters. A shown in [6] the reasons of this strong differences acting already at density near the saturation one, are related to the spacial correlation introduced by the usage of the Gaussian wave packets. Therefore such results acquire a more general meaning being strictly linked to the cluster formation process and should be investigated also at different densities and in the framework of others approaches.

3. Cluster production and pre-equilibrium effects.

It is well known that that the time related to the heavy ion dynamics extend in a rather wide range of times typically, at the Fermi energy, from $10^{-23}sec$ to $10^{-15}sec$. On the other hand only in the first stages of the reaction the total system experiences the extreme conditions of density and excitation energies. In our scopes the selection of information strictly linked with the first moment of the interaction is therefore of fundamental importance. Usually the complex scenario involving the time evolution is managed by interpreting the obtained results through a weighted sum of quantities obtained from pre-equilibrium models and statistical ones. This last one in general tends to blurry more or less the dynamical effects on the yields of cluster distributions and their correlations. As examples of this unwanted interplay we can quote [7, 8].

In this regard we note that the global and coherent cluster productions can be investigated by looking at the behavior of the following quantity:

$$\langle \vec{D} \rangle = \langle \sum_{i=1}^{m} Z_i(\vec{V}_i - \vec{V}_{c.m.}) \rangle_{\mathcal{K}}. \qquad (1)$$

The brackets indicate the average value over the ensemble $\mathcal{K}$. Z_i, V_i, m are the charges, laboratory velocities, charged particle multiplicity respectively of the produced particles in the selected class of events, respectively. Finally $\vec{V}_{c.m.}$ is the center of mass (c.m.)velocity. The interest on this quantity is triggered by two main reasons:

IOP Conf. Series: Journal of Physics: Conf. Series **863** (2017) 012056 doi:10.1088/1742-6596/863/1/012056

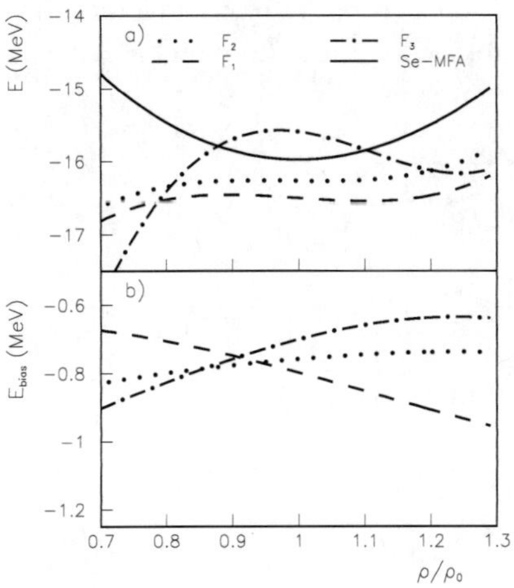

Figure 1. panel a: Total energy per nucleon E for symmetric NM as a function of the reduced density. The solid line represents the results obtained through the semiclassical-MFA. With discontinuous lines we plot the results for CoMD-II calculations corresponding to the first step of the iterative procedure (see the text).panel b:for the same parameter values the values of E_{bias} (see the text) are plotted as a function of the reduced density.

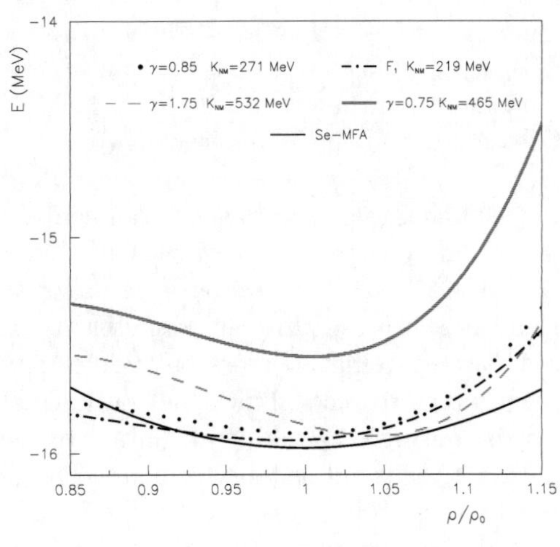

Figure 2. Total energy as a function of the reduced density after the iterative procedure applied to CoMD-II calculations.

-a) because of the symmetries of the statistical decay mode, $\langle\vec{D}\rangle$ is not affected by the statistical emission of all the produced sources in later stages, as it is shown in Ref. [9]. This essentially happens because, due to the vectorial kinematical character of this quantity , for well reconstructed events statistical effects are self-averaged to zero. Therefore $\langle\vec{D}\rangle$ is a rather well suited global variable to selectively evidence dynamical effects related to the Isospin equilibration process;

-b) as shown in Refs [9, 10] this quantity is closely linked with charge/mass equilibration process because it represents the average time derivative of the total dipolar signal in the asymptotic stage (expressed in unit of e).

Recently the experimental investigation on this subject was realized for the first time on

IOP Conf. Series: Journal of Physics: Conf. Series **863** (2017) 012056 doi:10.1088/1742-6596/863/1/012056

the $^{48}Ca +^{27} Al$ system at 40 MeV/A with the multi-detector CHIMERA [11] at the LNS laboratories [12]. In this first stage of investigation we have measured the effective partial signal dipolar signal $\langle D_Z^c \rangle$ along the beam axsis. In this quantity, to eliminate the indetermination associated to systematic errors on the velocities, the true center of mass velocity has been substituted with the ones associated to the subsystem formed by all the charged particles. In

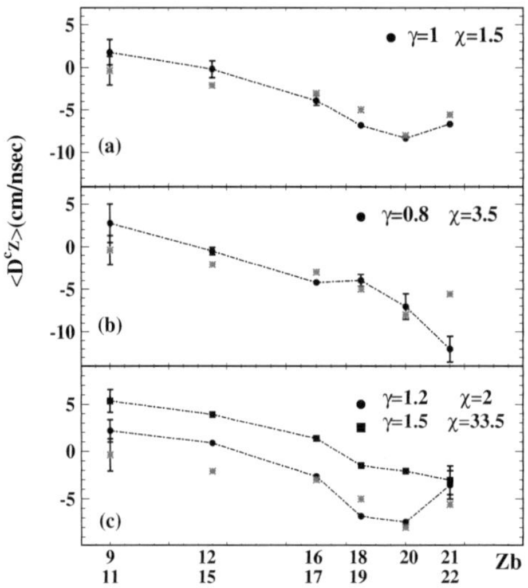

Figure 3. Experimental values of the effective dipolar signal for the $^{48}Ca +^{27} Al$ at 40 MeV compare with CoMD-III calculations for different stiffness values of the symmetry energy.

figure 3 we show with red points $\langle D_Z^c \rangle$ as a function of the charge of the biggest fragment Z_b for a total kinetic energy loss less than 350 MeV. In the different panels we show with black points the calculated value according to CoMD-III [12] calculations. The comparison allows to establish a good agreement with experimental data for an iso-vectorial stiffness parameter $\gamma \simeq 1$. Some efforts will be necessary in the next future to evaluate systematic errors on the velocity measurements by studying symmetrical systems, this will allow to reconstruct the total signal including the effect related to the neutrons. We conclude by noting that taking advantage of the today large efficiency multi-detectors, further investigation on this subject performed for more central events and higher energy aimed will contribute to a clear identification of the isospin dynamical effects and a more accurate study of the related effective interaction.

References

[1] Von Oertzen W 2011 *Int.J.Mod.Phys.* E **10** 765
[2] *Lecture Notes in Physics* vol 818, 2010 *Clusters in Nuclei*, Ed. Beck C (Springer)
[3] Iachello F, Cseh J and Lovay G 1995 *Acta Physica Hungarica New Series Heavy Ion Physics* **1** 91
[4] Papa M, Maruyama T and Bonasera A 2001 *Phys. Rev.* C **64** 024612
[5] Papa M, Giuliani G and Bonasera A 2005 *J. Comp. Phys.* **208** 403
[6] Papa M 2013 *Phys. Rev.* C **87** 014001
[7] Zhou P *et al* 2011 *Phys. Rev.* C **84** 036705
[8] Cussol D *et al* 2002 *Phys. Rev.* C **65** 044604
[9] Papa M *et al* 2005 *Phys. Rev.* C **72** 064608
[10] Giuliani G and Papa M 2006 *Phys. Rev.* C **73** 031601
[11] Pagano A 2012 *Nucl. Phys. News* **22** 25
[12] Papa M *et al* 2015 *Phys. Rev.* C**91** 041601(R)

Pre-equilibrium emission and clustering in medium-mass nuclei: ^{46}Ti from ^{16}O + ^{30}Si, ^{18}O + ^{28}Si, ^{19}F + ^{27}Al

M Cicerchia[1,2], T Marchi[3], F Gramegna[1], M Cinausero[1], J Mabiala[1], D Fabris[4], A Caciolli[2,4], G Collazzuol[2,4], D Mengoni[2,4], M Degerlier[5], L Morelli[6], M Bruno[6], M D'Agostino[6], S Barlini[7], M Bini[7], G Pasquali[7], S Piantelli[7], G Casini[7], G Pastore[7], D Gruyer[7], P Ottanelli[7], S Valdrè[7], N Gelli[7], A Olmi[7], G Poggi[7], E Vardaci[8], I Lombardo[8], D Dell'Aquila[8], S Leoni[9], N Cieplicka[9] and B Fornal[10]

[1]INFN Laboratori Nazionali di Legnaro, Legnaro (PD), Italia.
[2]Dipartimento di Fisica e Astronomia, Univ. di Padova, Padova, Italia.
[3]KU Leuven, Department of Physics and Astronomy Instituut voor Kern- en Stralingsfysica, 3001 Leuven, Belgium.
[4]INFN Sezione di Padova, Padova, Italia.
[5]Science and Art Faculty, Physics Department, Nevsehir Haci Bektas Veli Univ., Nevsehir, Turkey.
[6]INFN Sezione di Bologna e Dipartimento di Fisica e Astronomia, Univ. di Bologna, Bologna, Italia.
[7]INFN Sezione di Firenze e Dipartimento di Fisica e Astronomia, Univ. di Firenze, Firenze, Italia.
[8]INFN Sezione di Napoli e Dipartimento di Fisica, Univ. Federico II Napoli, Napoli, Italia.
[9]INFN Sezione di Milano e Dipartimento di Fisica, Univ. di Milano, Milano, Italia.
[10]Institute of Nuclear Physics, Polish Academy of Sciences Krakow, Poland.
E-mail: cicerchia@lnl.infn.it

Abstract. The study of nuclear cluster states bound by valence neutrons is a field of recent large interest. In particular, it is interesting to study the pre-formation of α-clusters in α-conjugate nuclei and the dynamical condensation of clusters during nuclear reactions. The NUCL-EX collaboration (INFN, Italy) is carrying out a research campaign studying pre-equilibrium emission of light charged particles and cluster properties of light and medium-mass nuclei. For this purpose, a comparative study of the three nuclear reactions: ^{16}O + ^{30}Si, ^{18}O + ^{28}Si and ^{19}F + ^{27}Al, has been recently carried out using the GARFIELD+RCo 4π setup. After a general introduction on the experimental campaign, the preliminary results for the three systems are presented.

1. Introduction

The idea of α-cluster structure in nuclei has been known since a long time [1]. Nowadays, interest in nuclear clustering has gained new momentum due to the study of neutron-rich and exotic weakly-bound nuclei [2]. In light mass nuclei, clustering emission is a good tool to study the interplay between nuclear structure and reaction dynamics. The presence of clusters in medium-mass nuclei is much more complicated. Here, the interplay between preformation and coalescence plays an important role. It is not evident that the emission or capture of α-cluster implies that such an α-cluster is permanently present in the nucleus: it could be formed or dissolved at the instant of emission or capture. Therefore, it is of particular interest to search for α-clustering effects in non-traditional observables, like those deriving from pre-equilibrium process studies. The pre-equilibrium spectra could provide information about the α pre-formation probabilities in compound nucleus [3].

In these past few years, the NUCL-EX collaboration has been engaged in an experimental campaign to study a possible α-cluster structure in α-conjugate nuclei probing pre-equilibrium emission of light charged particles (LCP) at excitation energies close to LCP emission threshold. In particular, the analysis of $^{16}O + ^{116}Sn$ at 8, 12, 16 A MeV [4] brought to the observation of an over-production of α-particles emitted during the non-equilibrium stage of fusion nuclear reactions. One of the hypothesis for the α-particle excess is to be ascribed to the effects induced by the cluster structure in the ^{16}O projectile nucleus. Following these results, a new experiment was proposed to probe the α-cluster structure in ^{16}O in a model independent way: the comparative study of $^{16}O + ^{65}Cu$ and $^{19}F + ^{62}Ni$ at 16 A MeV [5]. Although the analysis is still in progress, evidences of an α-particle fast production excess have been observed for both systems. A slight overproduction is moreover observed in the ^{19}F induced reaction with respect to the ^{16}O one. This can be ascribed to the lower energy of the predicted cluster state of the ^{19}F.

2. Experimental details and preliminary results

To continue the previous campaign, we have performed a comparative study of four nuclear reactions: $^{16}O + ^{30}Si$, $^{18}O + ^{28}Si$ and $^{19}F + ^{27}Al$ at 7 A MeV and $^{16}O + ^{30}Si$ at 8 A MeV. The most important features of these reactions are reported in Table 1.

Table 1. The most important features of the reactions.

Projectile		Target	Grazing	Compound Nucleus					
Ion	E_{lab} (A MeV)	Isotope	θ_{lab} (deg)	Isotope	Mass Asymm.	σ_{fus} (mb)	E* (MeV)	Lab. Vel. (cm/ns)	E.R. Distrib. θ_{lab} (deg)
^{16}O	7.0	^{30}Si	10.1	^{46}Ti	0.30	1081	88.0	1.28	0-30
^{16}O	8.0	^{30}Si	8.8	^{46}Ti	0.30	1070	98.4	1.37	0-30
^{18}O	7.0	^{28}Si	9.0	^{46}Ti	0.22	1110	98.5	1.44	0-28
^{19}F	7.0	^{27}Al	8.9	^{46}Ti	0.17	1100	103.5	1.52	0-28

The main goal of the experiment was to measure and compare the pre-equilibrium component of the particle emission spectra for the three different systems. The ratios between statistical and pre-equilibrium LCP multiplicities and between different exit channels allows to extract information about the influence of α-clustering in these medium-mass systems. In the three reactions at 7 A MeV, the beam velocity was kept constant, as this parameter strongly determines the evolution of pre-equilibrium processes. The $^{16}O + ^{30}Si$ reaction at 8 A MeV was used to populate the ^{46}Ti at the same excitation energy as in the $^{18}O + ^{28}Si$ at 7 A MeV.

The experiment has been performed at the LNL, with the beams provided by the TANDEM-ALPI acceleration system. The GARFIELD array coupled with RCo, fully equipped with digital electronics [6,7], was used to detect LCP and evaporation residues. The performances of the apparatus permit the full event reconstruction and the study of many-body correlations.

Following an initial identification of particles and the energy calibration procedure, a preliminary analysis was performed on an event-by-event basis. The experimental data were initially compared with the simulated events from a statistical model, GEMINI++ [8], where the effects of the filter related to the apparatus replica were taken into account. The GEMINI code can describe only the thermalized emission from a compound nucleus, but it can give information on the differences between the three reactions connected to the major part of the cross section, which is related to the de-excitation of a thermalized source.

The first comparison between the single particle emission spectra and the simulated statistical component of the reactions, shows very similar experimental and simulated proton spectra on the whole angular range. On the contrary, a much larger difference is observed in the α-particle spectra at the most forward angles. Waiting for the calibration procedure to be completed, for a first estimate of the expected fast emission in the four cases, the data have been compared with the predictions of the statistical model simulated with GEMINI++. Figure 1 shown the comparison of proton and α-particles experimental vs. calculated spectra for $^{18}O + {}^{28}Si$ at 7 A MeV. The shape of other reactions spectra are similar.

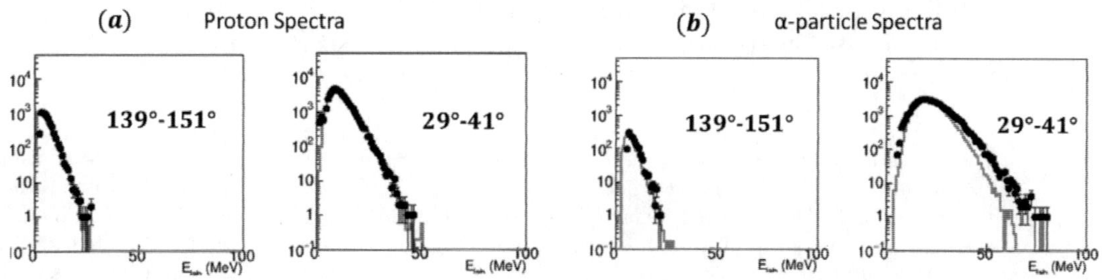

Figure 1. Comparison of laboratory energy spectra (black dots) of protons (*a*) and α-particles (*b*) for the system $^{18}O + {}^{28}Si$ at 7 A MeV with preliminary calculations from GEMINI++ code (red line). The distributions have been normalized to the total area.

3. Conclusion

We have studied the light particle emission from the reactions: $^{16}O + {}^{30}Si$, $^{19}F + {}^{27}Al$ and $^{18}O + {}^{28}Si$ at 7 A MeV and $^{16}O + {}^{30}Si$ at 8 A MeV in order to probe pre-equilibrium and possible α-clustering effects in nuclei. The first preliminary comparison between the experimental and the simulated spectra evidenced that the shape of the proton spectra in the whole angular range and of the α-particle spectra at backward angle are reasonably reproduced by the statistical model. While at forward angle the experimental α-particle spectra show a slightly overproduction of α-particles at higher energies. A more complete analysis is needed to understand the process, looking to pre-equilibrium models which also take into account other types of clusters. This will permit to better disentangle the pre-equilibrium emission. A complete analysis will be performed to correlate the velocity and angular ranges of the heavy fragments and light charged particles. On the other hand, the use of dynamical codes such as the SMF [9] followed by an afterburner decay code and pre-equilibrium models such as the Hybrid model [10] is under study.

References

[1] Von Oertzen W *et al* 2006 *Phys. Rep.* **432** 43
[2] Freer M *et al* 2007 *Rep. Progr. Phys.* **70** 2149
[3] Hodgson P E and Bĕták E 2003 *Phys. Rep.* **374** 1
[4] Kravchuk V L *et al* 2010 *EPJ WoCs* **2** 10006
[5] Fabris D *et al* 2013 *PoS (X LASNPA)* 061.D
[6] Gramegna F *et al* 2004 *Proc. of IEEE Nucl. Symposium* Roma Italy 0-7803-8701-5/04
[7] Bruno M *et al* 2013 *Eur. Phys. Jour.* **A49** 128
[8] Charity R J 2010 *Phys. Rev.* **C82** 014610
[9] Colonna M *et al* 1998 *Nucl. Phys.* **A642** 449
[10] Fotina O V *et al* 2010 *Phys. Atom. Nucl.* **73** 1317

IOP Conf. Series: Journal of Physics: Conf. Series **863** (2017) 012058 doi:10.1088/1742-6596/863/1/012058

Production of exotic nuclei in heavy ion collisions at Fermi energies

R Ogul, N Buyukcizmeci and A Ergun

Selcuk University, Department of Physics, 42075 Kampus, Konya, Turkey

E-mail: rogul@selcuk.edu.tr

Abstract. Isotopic distributions of projectile fragmentation in peripheral heavy ion collisions of ^{86}Kr on ^{124}Sn are calculated within the statistical multifragmentation model. Obtained data are compared to the experimental cross section measurements. We show the enhancement in the production of neutron-rich isotopes close to the projectile, observed in the experiments. The universality of the limitation of the excitation energy induced in the projectile residues is also demonstrated.

1. Introduction

Exotic nuclei displayed in the nuclear landscape can be produced in various types of reactions such as multinucleon transfer, fussion, fission and nuclear fragmentation over a wide range of energy. At low energies near the Coulomb barrier (1-10 MeV/nucleon), neutron-deficient isotopes are produced as a result of the fusion of heavy-ions in central collisions [1]. Similarly, heavier neutron-rich as well as neutron-deficient isotopes are produced as a result of multinucleon transfer reactions [2]. At energies over the Coulomb barrier up to the Fermi regime, however, neutron rich isotopes are populated in Deep Inelastic Transfer (DIT) reactions [2, 3]. At high energies, above the Fermi energy regime, exotic beams over the entire periodic table up to uranium are produced by nuclear fragmentation of relativistic projectiles [4]. Starting from the Fermi energies, nuclear multifragmentation reactions become superior to other reaction channels [5], and one may possibly study the neutron-rich isotope production from projectile fragmentation in Fermi energy regime (20-40 MeV/nucleon). In this short paper, we shall interprete the experimental cross section measurements for the projectile-like fragmentation in the reaction ^{86}Kr+^{124}Sn at 25 MeV/nucleon, given in Ref.[5] with the same parametrization we obtained for the analyses of ALADIN data [6].

2. Isotopic distributions within a statistical approach

In this work the statistical ensemble approach is used for the simulation of projectile fragmentation. In this approach, the formation of the ensembles of excited residual nuclei is determined with the intranuclear cascade (INC) model (the dynamical stage) [7, 8]. Fragmentation of the system and deexitation process are described within the statistical multifragmentation model (SMM) [9]. At low excitation energies ($E^* < 1$ MeV/nucleon), the SMM code can also be used to describe the de-excitation process of primary hot fragments on the basis of standard Weisskopf evaporation and fission scheme (see e.g. Ref.[9]).

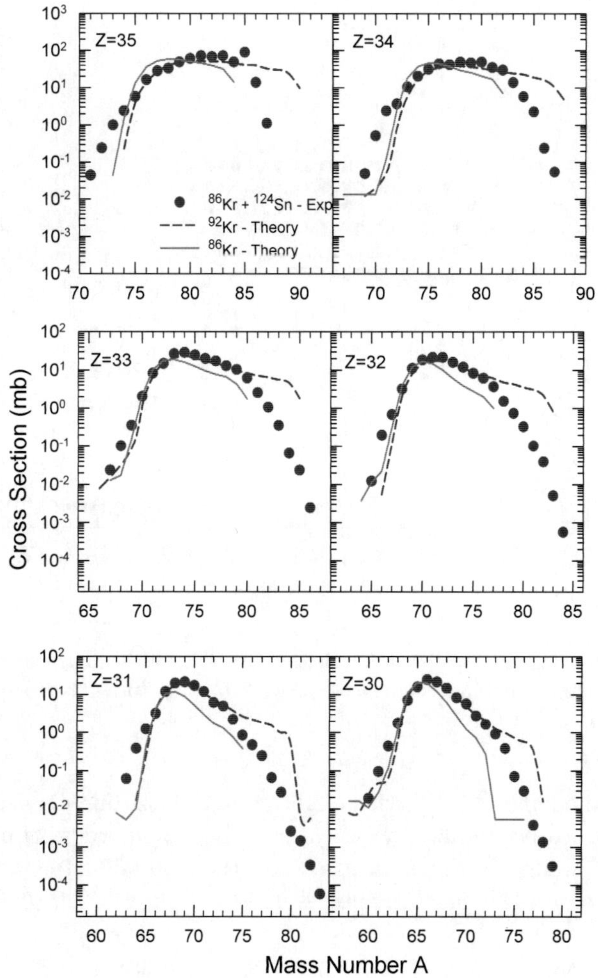

Figure 1. Isotope mass distributions (cross sections) of projectile fragmentation of ^{86}Kr (solid lines), ^{92}Kr (dashed lines), and corresponding experimental data for ^{86}Kr+^{124}Sn (full circles) reactions at 25 MeV/nucleon projectile energies. Here, ^{92}Kr and ^{86}Kr quasiprojectiles are assumed to be formed in ^{86}Kr+^{124}Sn reactions with and without neutron transfer.

It was assumed that neutron rich ^{92}Kr quasiprojectiles are formed in deep inelastic transfer as a result of the interactions between projectile and target nuclei in the studied reaction ^{86}Kr+^{124}Sn at 25 MeV/nucleon. In Fig.1, we show the mass distributions for near-projectile elements $Z = 35 - 30$ produced from the fragmentation of quasiprojectiles as a function of the mass number A of the final fragments. The experimental results (full circles) are taken from Ref. [5]. Choosing the quasiprojectile as ^{92}Kr provided us with the optimum agreement between our predictions and experimental data given in Ref.[5]. We also studied the fragmentation of ^{86}Kr quasiprojectile and it was seen that the produced yields were shortcoming to cover the experimental data. This supports the idea of nucleon exchange at the first stage and the important role of the target in the neutron rich isotope production [10, 11]. The observed agreement between the experimental and theoretical results is satisfactory enough, qualitatively, and considerable differences exist in the yields. Since we did not take the experimental filter into

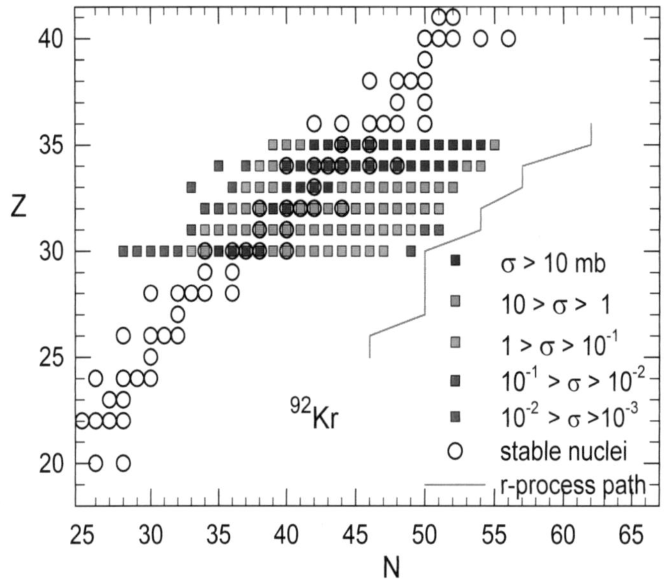

Figure 2. Production cross sections of the fragments $Z = 35 - 30$ from quasiprojectile ^{92}Kr predicted by statistical ensemble calculations and their location with respect to the r-process path (solid line).

account for a more detailed comparison with experimental data, one may see in the neutron rich side in Fig.1 that theoretically predicted yields are at least some orders of magnitude larger with respect to the measured yields. In Fig.2, we also show the results for the location of neutron-rich and proton-rich fragments in the chart of the nuclides.

3. Conclusions

We show the enhancement in the neutron-rich isotope production by using the neutron rich targets at low energies near the Fermi energy regime. This enhancement is associated with peripheral interactions of ^{86}Kr projectile with the neutron rich ^{124}Sn target at 25 MeV/nucleon, where we used the ALADIN parametrization of high energy multifragmentation at 600 MeV/nucleon with the same ensemble approach. This shows the universality of the limitation of the excitation energy induced in the ensemble of projectile residues formed at low and high energy peripheral heavy ion collisions. This kind of analyses may be useful for the studies of exotic nuclei and for the simulation of astrophysical events.

4. References

[1] Reisdorf W 1994 *J. Phys.* G **20** 1297
[2] Devaraja H M *et al* 2015 *Phys. Lett.* B **748** 199
[3] Beliuskina O *et al* 2014 *Eur. Phys. J.* A **50** 161
[4] Geissel H *et al* 1992 *Nucl. Instr. Meth.* B **70** 286
[5] Souliotis G A *et al* 2003 *Phys. Rev. Lett.* **91** 022701
[6] Ogul R, Botvina A S *et al* 2011 *Phys. Rev.* C **83** 024608
[7] Botvina A S, Gudima K K *et al* 1994 *Phys. At. Nucl.* **57** 628
[8] Botvina A S *et al* 1995 *Nucl. Phys.* A **584** 737
[9] Bondorf J P *et al* 1995 *Phys. Rep.* **257** 133
[10] Buyukcizmeci N *et al* 2015 *Nucl. Sci. Tech.* **26** S20507
[11] Kumar S and Ma Y G 2015 *Phys. Rev.* C **91** 034612

Probing the Symmetry Term of the Nuclear Equation of State at High Baryonic Densities

P Russotto[1], L Acosta[2,3], M Adamczyk[4], A Al-Ajlan[5], M Al-Garawi[6], S Al-Homaidhi[5], F Amorini[2], L Auditore[7,8], T Aumann[9,10], Y Ayyad[11], Z Basrak[12], J Benlliure[11], M Boisjoli[13], K Boretzky[10], J Brzychczyk[4], A Budzanowski[14,35], C Caesar[9], G Cardella[1], P Cammarata[15], Z Chajecki[16], M Chartier[17], A Chbihi[10], M Colonna[2], M D Cozma[18], B Czech[14], E De Filippo[1], M Di Toro[2,19], M Famiano[20], S Gannon[17], I Gašparić[9,12], L Grassi[12], C Guazzoni[21,22], P Guazzoni[21,23], M Heil[10], L Heilborn[15], R Introzzi[24], T Isobe[25], K Kezzar[6], M Kiš[10], A Krasznahorkay[26], S Kupny[4], N Kurz[10], E La Guidara[1], G Lanzalone[2,27], P Lasko[4], A Le Fèvre[10], Y Leifels[10], R C Lemmon[28], QF Li[29], I Lombardo[30,31], J Łukasik[14], W G Lynch[16], P Marini[13,15], Z Matthews[17], L May[15], T Minniti[1], M Mostazo[11], A Pagano[1], E V Pagano[2,19], M Papa[1], P Pawłowski[14], S Pirrone[1], G Politi[1,19], F Porto[2,19], W Reviol[32], F Riccio[21,22], F Rizzo[2,19], E Rosato[30,31,35], D Rossi[9,10], S Santoro[7,8],D G Sarantites[32], H Simon[10], I Skwirczynska[14], Z Sosin[4,35], L Stuhl[26], W Trautmann[10], A Trifirò[7,8], M Trimarchi[7,8], M B Tsang[16], G Verde[1], M Veselsky[33], M Vigilante[30,31], Yongjia Wang[29], A Wieloch[4], P Wigg[17], J Winkelbauer[16], H H Wolter[34], P Wu[17], S Yennello[15], P Zambon[21,22], L Zetta[21,23] and M Zoric[12]

[1]INFN-Sezione di Catania, Catania, Italy
[2]INFN-Laboratori Nazionali del Sud, Catania, Italy
[3]Universidad Nacional Autònoma de Mèxico, Mexico
[4]Marian Smoluchowski Institute of Physics, Jagiellonian University, Krakòw, Poland
[5]KACST Riyadh, Riyadh, Saudi Arabia
[6]King Saud University, Riyadh, Saudi Arabia
[7]INFN-Gruppo Collegato di Messina, Messina, Italy
[8]Università di Messina, Messina, Italy
[9]Technische Universität, Darmstadt, Germany
[10]GSI Helmholtzzentrum, Darmstadt, Germany
[11]University of Santiago de Compostela, Santiago de Compostela, Spain
[12]Ruder Boškovič Institute, Zagreb, Croatia
[13]GANIL, Caen, France
[14]H. Niewodniczàski Institute of Nuclear Physics, Krakòw, Poland
[15]Texas A&M University, College Station, USA
[16]NSCL Michigan State University, East Lansing, USA
[17]University of Liverpool, Liverpool, UK
[18]IFIN-HH, Magurele-Bucharest, Romania
[19]Università di Catania, Catania, Italy
[20]Western Michigan University, USA

[35] deceased

IOP Conf. Series: Journal of Physics: Conf. Series **863** (2017) 012059 doi:10.1088/1742-6596/863/1/012059

[21]INFN-Sezione di Milano, Milano, Ialy
[22]Politecnico di Milano, Milano, Italy
[23]Università degli Studi di Milano, Milano, Italy
[24]INFN, Politecnico di Torino, Torino, Italy
[25]RIKEN, Wako, Japan
[26]Institute of Nuclear Research, Debrecen, Hungary
[27]Università Kore, Enna, Italy
[28]STFC Laboratory, Daresbury, UK
[29]Huzhou Teachers College, China
[30]INFN-Sezione di Napoli, Napoli, Italy
[31]Università di Napoli, Napoli, Italy
[32]Washington University, St. Louis, USA
[33]Institute of Physics, Slovak Academy of Sciences, Bratislava, Slovakia
[34]LMU, München, Germany

E-mail: russotto@lns.infn.it

Abstract. In the ASY-EOS experiment flows of neutrons and light charged particles were measured for ^{197}Au+^{197}Au collisions at 400 MeV/nucleon, in order to investigate the strength of the symmetry term of the nuclear equation of state at supra-saturation densities. By comparing the experimental data with the UrQMD transport model predictions, we have extracted a new constraint in agreement with the moderately soft to linear density dependence obtained in the former analysis on FOPI-LAND data, but reducing the associated uncertainty by a factor ~ 2.

1. ASY-EOS experiment results

The neutron-proton elliptic flow ratio has proved to be an effective probe of the high-density behavior of the symmetry energy contribution to nuclear equation of state [1]. By comparing data for the Au+Au system from the FOPI-LAND experiment [2] with calculations performed with the UrQMD transport model [3], an estimate of a moderately soft to linear behavior of the symmetry energy was obtained, although with a considerable statistical uncertainty. The same data set was also compared to calculations performed with the Tübingen QMD model and a constraint compatible with the UrQMD result was obtained [4]. It is worth pointing that in the latter study different parameterizations of the isovector part of the equation of state, the Gogny inspired (momentum dependent) vs. the power law (momentum independent) potential, lead to

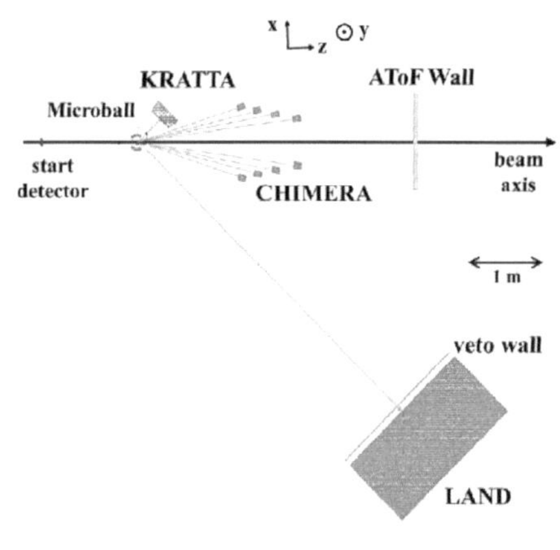

Figure 1. Schematic view of the experimental set-up of the ASY-EOS experiment at GSI.

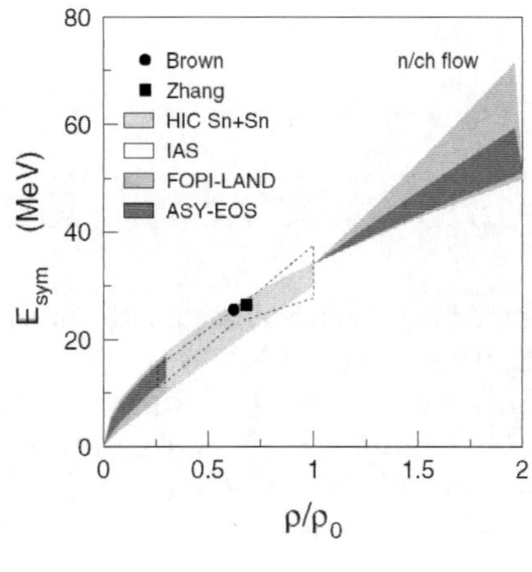

Figure 2. Constraints deduced for the density dependence of the symmetry energy from the ASY-EOS experiment in comparison with the FOPI-LAND result of Ref. [1] as a function of the reduced density ρ/ρ_0. See Ref. [10] for more details.

very similar results, suggesting the possibility to obtain a model independent constraint.
In order to improve the statistical accuracy, the neutron-charged particles elliptic flow ratio in the symmetric collision system ^{197}Au+^{197}Au at 400 MeV/nucleon incident energies was re-measured in the ASY-EOS experiment, carried out at the GSI laboratory in May 2011 (S394 experiment), by coupling the Large Area Neutron Detector (LAND) [5], the Krakòw Triple Telescope Array (KRATTA) [6], the ALADIN Time-of-Flight (AToF) wall [7], 8 rings of the CHIMERA multidetector [8] and 4 rings of the Washington-University Microball array [9], as schematically shown in Fig. 1. More details can be found in Ref. [10].
By comparing the experimental ratio with the UrQMD predictions, a value γ=0.72$\pm$0.19 is obtained for the power-law coefficient of the potential part of the symmetry energy parametrization adopted in the UrQMD model, as shown in Fig. 2 ; the density region probed is $\sim$ 1.5 ρ_0. This new result agrees with the older one from the FOPI-LAND data, but reduces by a factor $\sim$ 2 the total uncertainty.
The results of the ASY-EOS experiment confirms the effectiveness of the elliptic flows ratio in probing the symmetry energy at high-densities. We plan to extend such a kind of measurements at GSI to higher beam energies, up to $\sim$ 1 GeV/nucleon, in order to explore higher densities (up to about 2.5 ρ_0), profiting of the unique opportunity offered by the NeuLAND detector, presently under construction as part of the R3B experimental setup [11] and the availability of very asymmetric radioactive ion beams that will be delivered by the FAIR facility.

References

[1] Russotto P *et al* 2011 *Phys. Lett.* B **697** 471; Russotto P *et al* 2014 *Eur. Phys. Jour.* A **50** 38
[2] Leifels Y *et al* 1993 *Phys. Rev. Lett.* **71** 963
[3] Li Q *et al* 2005 *J. Phys* G **31** 1359
[4] Cozma M D *et al* 2013 *Phys. Rev.* C **88** 044912
[5] Blaich Th *et al* 1992 *Nucl. Instr. and Meth.* A **314** 136
[6] Łukasik J *et al* 2013 *Nucl. Instr. and Meth.* A **709** 120
[7] Schüttauf A *et al* 1996 *Nucl. Phys.* A **607** 457
[8] Pagano A *et al* 2004 *Nucl. Phys.* A **734** 504
[9] Sarantites D G *et al* 1996 *Nucl. Instr. and Meth.* **381** 418
[10] Russotto P *et al* 2016 *Phys. Rev.* C **94** 034608
[11] see *https://www.gsi.de/r3b.*

Dynamical Dipole mode in heavy-ion fusion reactions

C Parascandolo[1], D Pierroutsakou[1], R Alba[2], A Del Zoppo[2], C Maiolino[2], D Santonocito[2], C Agodi[2], V Baran[3,2], A Boiano[1], M Colonna[2], R Coniglione[2], E De Filippo[4], M Di Toro[2,5], U Emanuele[6], F Farinon[7], A Guglielmetti[8], M La Commara[9,1], B Martin[9,1], C Mazzocchi[8], M Mazzocco[10], C Rizzo[2,5], M Romoli[1], C Signorini[10], R Silvestri[9,1], F Soramel[10], E Strano[10], D Torresi[10], A Trifirò[6], M Trimarchi[6]

[1]INFN - Napoli, Napoli, Italy [2]INFN - LNS, Catania, Italy [3]University of Bucharest, Bucharest and NIPNE-HH, Magurele, Romania [4]INFN - Catania, Catania, Italy [5]Dip. di Fisica e Astronomia, Università di Catania, Catania, Italy [6]INFN - Gr. Col. di Messina and Università di Messina, Messina, Italy [7]GSI, Darmstadt, Germany [8]Dip. di Fisica, Università di Milano and INFN - Milano, Milano, Italy [9]Dip. di Fisica, Università di Napoli, Napoli, Italy [10]Dip. di Fisica e Astronomia, Università di Padova and INFN - Padova, Padova, Italy

E-mail: concetta.parascandolo@na.infn.it

Abstract. In this work we give evidence for the first time of the Dynamical Dipole mode in a heavy system in the A$\sim$190 mass region, in both fusion–evaporation and fission events. The ^{40,48}Ca + ^{152,144}Sm reactions at E_{lab}=11(10.1) MeV/nucleon were employed. Our results for evaporation and fission events (preliminary) show that the dynamical dipole mode survives in reactions involving heavier nuclei than those studied previously.

1. The physical problem

The "Dynamical Dipole mode" (DD) is a collective oscillation of protons against neutrons with a dipole spatial pattern inside the atomic nucleus. It is a pre-equilibrium phenomenon, being excited along the fusion path of nuclei with a different ratio of neutrons and protons and decaying through emission of prompt γ-rays [1, 2, 3, 4]. The DD radiation is characterized by i) a centroid energy lower than that of the compound nucleus (CN) GDR in the same mass region due to the high deformation of the emitting source [2, 3] ii) an anisotropic angular distribution with respect to the beam axis since the oscillation is confined in the reaction plane [5] and iii) a γ yield that should depend on both the reaction dynamics and the symmetry term of the EOS [3]. The existence of the DD mode has been probed in deep inelastic and fusion-evaporation heavy-ion collisions [4, 6, 7, 8, 9, 10]. An excess of γ-rays was observed in the GDR energy region for a charge asymmetric reaction, with respect to that of a more charge symmetric one forming the same CN at identical conditions [7, 8, 9] or with respect to statistical model calculations [10]. This γ excess was attributed to the decay of the predicted DD. The emission of DD γ-rays decreases the excitation energy and the initial temperature of the nucleus reaching the statistical phase. This cooling mechanism could be suitable to favour the production of super-heavy elements in hot fusion processes. TDHF calculations [11] showed that the DD γ yield decreases as the mass of colliding ions increases since the reactions with small nuclei are less damped than those involving more nucleons. In order to understand the behavior of the DD in heavier systems than those studied before and to test its usefulness in super-heavy element production, we decided to study the DD in a composite system in the mass region A=190 [12].

2. The experiment: ^{40,48}Ca+^{152,144}Sm at 11 MeV/nucleon

The experiment was performed by using the ^{40}Ca (^{48}Ca) pulsed beam provided by the Superconducting Cyclotron of the Laboratori Nazionali del Sud (LNS, Italy), impinging on a 1 mg/cm^2 thick self-supporting ^{152}Sm (^{144}Sm) target enriched to 98.4%(93.8%) in ^{152}Sm (^{144}Sm) at E_{lab} = 440 (485) MeV. Both entrance channels populate the same CN through a quite different initial dipole moment, 30.6 fm for the ^{40}Ca + ^{152}Sm charge asymmetric reaction and 5.3 fm for the ^{48}Ca + ^{144}Sm more charge symmetric one. The mass asymmetry of the two entrance channels is very similar, namely 0.22(0.18) for the ^{40}Ca + ^{152}Sm (^{48}Ca + ^{144}Sm) system. The formed CN had identical excitation energy in both reactions, E^* = (220±7) MeV, and identical spin distribution: $L_{max} = 74\hbar$ for fusion and $L_{max} = 42\hbar$ for fusion-evaporation, according to PACE2 calculations [13] with a level density parameter a = A/9.5 MeV^{-1}, A being the CN mass. The γ-rays and the light particles were detected by using the MEDEA BaF_2 sphere [14], with a full azimuthal coverage in the polar angular range between 42° and 170°. The fusion-evaporation residues were detected by four position sensitive Parallel Plate Avalanche Counters (PPACs) placed symmetrically around the beam direction at 70 cm from the target at $\theta = 7°$ and subtending 7° in θ. The fission events were selected by detecting the two kinematically coincident fission fragments with position sensitive PPACs, centered at θ = 52.5° symmetrically around the beam axis, at 16 cm from the target covering 22° in both θ and ϕ and allowing the study of γ-ray - fragment angular correlations. Down-scaled single events and coincidence events between at least one fired BaF_2 and a PPAC (two PPACs) for evaporation (fission) events were collected during the experiment. The coincidence request eliminated any cosmic-ray contamination of the γ-ray spectra. The analysis of the light charged particles energy spectra in evaporation events demonstrated that the two reactions lead to the formation of the same CN with the same average excitation energy (see [12]). Therefore any difference in the γ-ray emission between the two reactions can be related to an entrance channel effect.

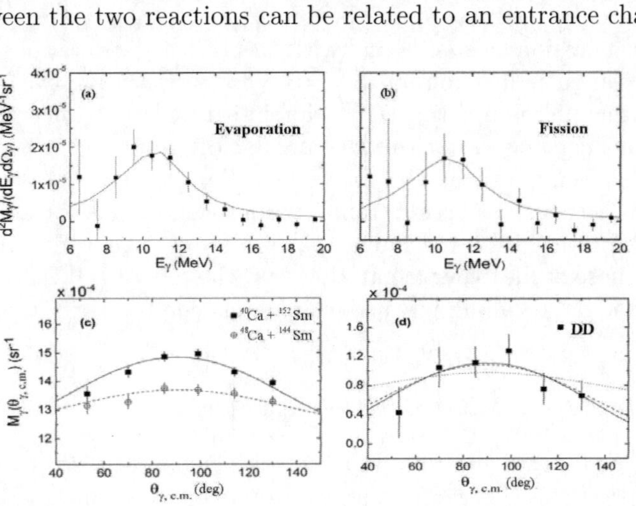

Figure 1. Difference between the charge asymmetric and charge symmetric reaction center-of-mass bremsstrahlung-subtracted γ-ray spectra for fusion-evaporation (a) and fission (b) events. (c) 9 MeV$\leq E_\gamma \leq$15 MeV γ-ray angular distribution of the ^{40}Ca+^{152}Sm (squares) and ^{48}Ca+^{144}Sm (circles) reactions, corrected for the detection efficiency. (d) Angular distribution of the difference between the data of panel (c). The lines are described in the text.

Panels (a) and (b) of Figure 1 show the difference between the center-of-mass bremsstrahlung-subtracted γ-ray spectra of the two reactions in coincidence with the evaporation residues and fission fragments, respectively. The data show an excess of γ-rays in the range E_γ=8-15 MeV in the charge asymmetric channel. This excess can only be related to the DD excitation in the composite system of the ^{40}Ca+^{152}Sm reaction because of its larger charge asymmetry. The DD γ spectrum (symbols in panels (a) and (b)) can be reproduced well by means of a Lorentzian curve folded by the experimental setup response function [15] (lines in the figure), with the DD centroid energy E_{DD} = 11 MeV and the width Γ_{DD} = 3.5 MeV, in both exit channels. E_{DD} is lower than the centroid energy of the GDR built on the ground state of a nucleus of similar mass, E_{GDR} = 14 MeV, confirming the high deformation of the emitting source during the DD γ emission. This result is in agreement with expectations [2, 3] and with previous works [7, 8, 9].

Panels (c) and (d) of Figure 1 display the fusion-evaporation γ-rays angular distribution with respect to the beam axis for the two reactions (c) and for their difference (d), integrated over energy from 9 to 15 MeV and corrected for the detection efficiency. The lines describe the angular distribution of the emitted γ-rays given by the Legendre polynomial expansion $M_\gamma(\theta_\gamma) = M_0[1 + Q_2 a_2 P_2 cos(\theta_\gamma)]$, where a_2 is the anisotropy coefficient and Q_2 is an attenuation factor for the finite γ-ray counter [16] (0.98 in our case). From a best fit to the data, shown with a solid (dashed) line for the ^{40}Ca+^{152}Sm (^{48}Ca+^{144}Sm) reaction, we found a_2 = -0.13 ± 0.03 for the ^{40}Ca+^{152}Sm reaction and a_2 = -0.06 ± 0.02 for the ^{48}Ca+^{144}Sm one. By using the same argument as previously for the spectra, the observed difference in the γ-ray angular distribution of the two systems can only be ascribed to entrance channel effects, namely the DD excitation. Consequently, the experimental angular distribution of the difference between the data of panel (c) (panel (d)) is very anisotropic around 90° and can be reproduced well with a_2 = -1 (solid line) that describes an emission from a dipole oscillation along the beam axis. The dashed line corresponds to a value of a_2 = -0.84 obtained within BNV calculations [3] for evaporation events, while the dotted one shows a more isotropic angular distribution (a_2 = -0.25). The above a_2 values indicate a preferential oscillation axis of the DD along an axis that has not rotated much on the reaction plane during the DD lifetime. Our data therefore suggest that the DD γ-emission time scale is confined at the beginning of the reaction, in agreement with our previous results for evaporation events [9] and with theoretical expectations [5]. By taking into account the DD γ-ray angular distribution (a_2 = -1) for evaporation events and the response function of the experimental setup, the DD yield, integrated over energy and over angle, is (1.2 ± 0.2)*10^{-3} [12]. The analysis for the DD angular distribution for fission events is under way. The experimental results on the DD in ^{40}Ca+^{152}Sm reaction were compared with BNV calculations, based on a collective bremsstrahlung approach of the entrance channel reaction dynamics. These calculations give centroid energy, width and angular distribution of the DD in good agreement with those of the experiment. However, the theoretical DD yield for evaporation events overpredicted the data, calling for further investigation to clarify this aspect.

References

[1] Chomaz P, Di Toro M and Smerzi A 2993 *Nucl. Phys.* A **563** 509
[2] Simenel C, Chomaz P and de France G 2001 *Phys. Rev. Letl.* **86** 2971
[3] Baran V *et al* 2001 *Phys. Rev. Lett.* **86** 182501 and 2001 *Nucl. Phys.* A **679** 373
[4] Papa M *et al* 2005 *Phys. Rev.* C **72** 064608
[5] Baran V *et al* 2009 *Phys. Rev.* C **79** 021603 (R)
[6] Pierroutsakou D *et al* 2003 *Eur. Phys. Jour.* A **16** 423
[7] Flibotte S *et al* 1996 *Phys. Rev. Lett.* **77** 1448; Amorini F *et al* 2004 *Phys. Rev.* C **69** 014608
[8] Pierroutsakou D *et al* 2003 *Eur. Phys. Jour.* A **17** 71 and 2005 *Phys. Rev.* C **71** 54605
[9] Martin B *et al* 2008 *Phys. Lett.* B **664** 47; Pierroutsakou D *et al* 2009 *Phys. Rev.* C **80** 024612
[10] Corsi A *et al* 2009 *Phys. Lett.* B **679** 197; Giaz A *et al* 2014 *Phys. Rev.* C **90** 014609
[11] Simenel C, Chomaz Ph and de France G 2007 *Phys. Rev.* C **76** 024609
[12] Parascandolo C, Pierroutsakou D *et al* 2016 *Phys. Rev.* C **93** 044619
[13] Gavron A 1980 *Phys. Rev.* C **21** 230
[14] Migneco E *et al* 1992 *Nucl. Inst. and Meth.* A **314** 31
[15] Bellia G *et al* 1993 *Nucl. Inst. and Meth.* A **329** 173
[16] Rose M E 1953 *Phys. Rev.* **91** 610

IOP Conf. Series: Journal of Physics: Conf. Series **863** (2017) 012061 doi:10.1088/1742-6596/863/1/012061

In medium fragment break-up of projectile in ^{58}Ar+^{36}Ni central collisions

L Francalanza for INDRA Collaboration

Istituto Nazionale Fisica Nucleare, Sezione di Napoli, Napoli, Italy

E-mail: **francalanza@na.infn.it**

Abstract. Latest results concerning the study of central collisions in 36Ar+58Ni reactions, carried out by using the INDRA 4π detector at the GANIL are presented. Reaction mechanisms involved in very central collisions for such asymmetric system and their evolution as function of incident energy are studied, from 32 to 95 AMeV, suggesting an in-medium “break-up” process of the projectile where the presence of nuclear transparency plays an important role. Such a process, leading to forward emission of fragments and lights particles in the laboratory frame, is found to be more and more relevant at high energies, when the contribution from quasi-fusion processes vanishes. Evidences of emissions from an excited quasi-target (QT*), moving along the beam direction with velocity rising with incident energy under the dragging effect of the in-medium “crumbling” of the projectile are highlighted.

1. Introduction

At intermediate incident energy very central collisions can lead to the formation of a unique source, characterized by excitation energies comparable with the nucleons’ binding energy. This system can meet several decay modes depending on its angular momentum, its density, its excitation energy. Nevertheless, the formation of such a composite excited system in central collisions is subjected to some conditions strictly related to the nuclear stopping, that governs the amount of energy dissipated in the reaction [1].

In the present work we focus on very central collisions (between 0 and 0.3 reduced impact parameter) in ^{36}Ar+^{58}Ni reactions and, after highlighting the not occurrence of single source formation at the highest energies, even in most violent collisions, we concentrate the analysis on the projectile fragmentation and target de-excitation separately. The projectile products, appearing like an anisotropic “crumble” of fragments and light particles, provide interesting evidences about an in-medium fragmentation process that become dominant above 45/50 AMeV, reflecting the strong interplay between mean-field (nuclear degrees of freedom) and n-n collisions (nucleonic degrees of freedom) played in the intermediate energy range [3].

2. Event selection and characterization

We study the so called “complete” events, where the total detected charge and total pseudo-momentum are above 75% of the values they have in the entrance channel of the reaction.

Since we are interested in central collisions, a gate is applied on the global variable $E_{t1,2}$, namely the total transverse energy of light particles (Z=1,2) and concentrate our analysis on collisions with reduced impact parameter b_{red}<0.3.

When the centrality gate is applied most contributions from peripheral events are strongly suppressed and a significant emission at mid-velocity ($v_{par} \approx v_{CM}$) is observed for fragments with charges greater

IOP Conf. Series: Journal of Physics: Conf. Series **863** (2017) 012061 doi:10.1088/1742-6596/863/1/012061

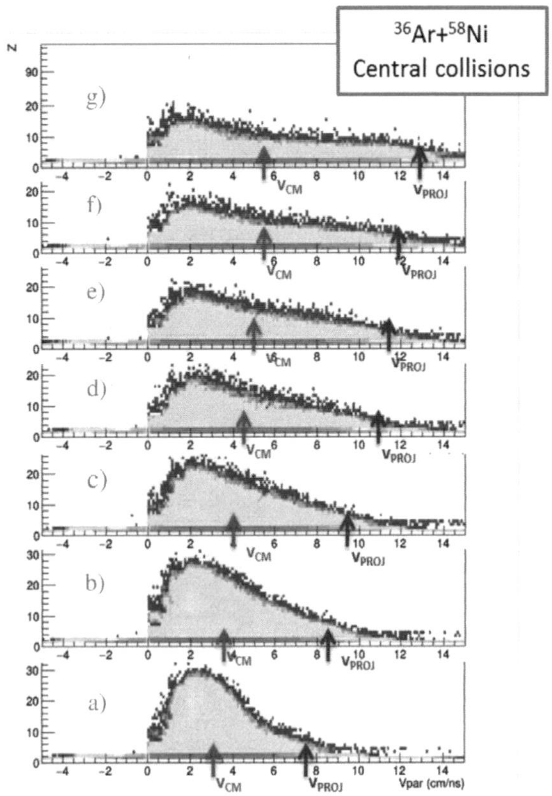

Figure 2.1 vpar-Z correlations at all the E_{inc} range (32, 40, 52, 63, 74, 84, 95 AMeV)

than the charges of the two colliding nuclei at 32 and 45 AMeV (panel a) and b) in fig. 2.1). Such events may correspond to fusion-like process, with a subsequent emission of fragments and light particles through evaporation and multi-fragmentation of a compact source.

Such process becomes less significant and eventually disappears at higher beam energies, indicating a less significant contribution from stopping processes in central collisions. Mechanisms related to transparency phenomena are visible on fig. 2.1 where the v_{par}-Z correlations at all the E_{inc} range is shown: the projectile completely loses its identity after the collision (no QP remnants centered at projectile velocity is observed), undergoing a multiple breakup into fragments with charges from Z=1 to about Z=10-15 (depending on the beam energy).

But, in contrast to what happens at relativistic energies, the reactions products do not show a peak at the projectile velocity, being spread along the v_{par} axis: this aspect seems to suggest that the projectile breakup occurs inside the projectile-target interaction region, where the decay products slow down by dissipating a variable portion of their kinetic energy and then exit the interaction region with parallel velocity values that can span a wide range extending from $v_{par}=v_{CM}$ to about $v_{par}=v_{proj}$. Such results are consistent with a sort of projectile *crumble* breakup occurring in a semi-transparent medium.

Furthermore, an emission of fragments with a large range of charges ($1<Z<Z_{target}$) and parallel velocities centred at low values, approximately v_{par} = 1-2 cm/ns, is observed. Such observations can be associated (see next Section 2.1) to emissions from the excited quasi-target (QT*).

2.1. Reconstruction and characterization of QT and QP sources

In order to better characterize the observed reaction mechanism, we reconstruct the total QT and QP charges by summing the charges of all particles and fragments with v_{par} values lower and greater, respectively, than the center of mass velocity of the reaction: $Z_{QT,QP} = \sum_{(v<v_{CM},v\geq v\_CM)} Z_v$

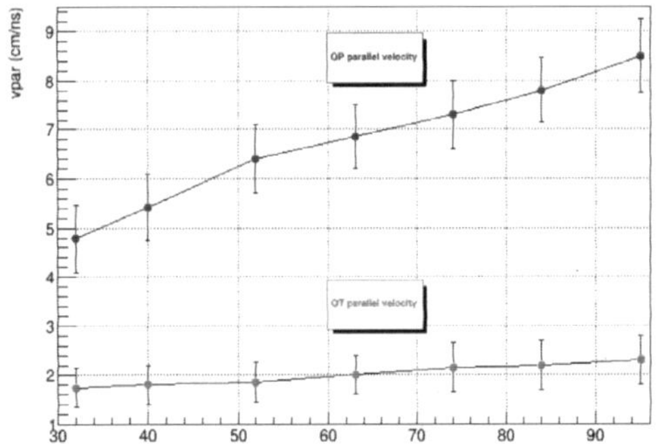

Figura 2.1.1 Center of mass velocity for reconstructed QT (red points) and QP (blu points).

Such reconstruction of the QP and QT sources allows us to also determine their center of mass velocities, shown on fig. 2.1.1.

The v_{CM}^{targ} ranges from 1.9 cm/ns at 42 AMeV to 2.2 cm/ns at 92 AMeV, indicating a sort of gluing and dragging of the target along the beam direction by the projectile crumbling, that become more violent with increasing of beam energy.

Finally, in order to extract some information about the decay mechanism of the QT in coincidence with the in-medium projectile breakup we study the kinetic energy spectra of alpha particles emitted by the QT and measured in its centre of mass reference frame

IOP Conf. Series: Journal of Physics: Conf. Series **863** (2017) 012061 doi:10.1088/1742-6596/863/1/012061

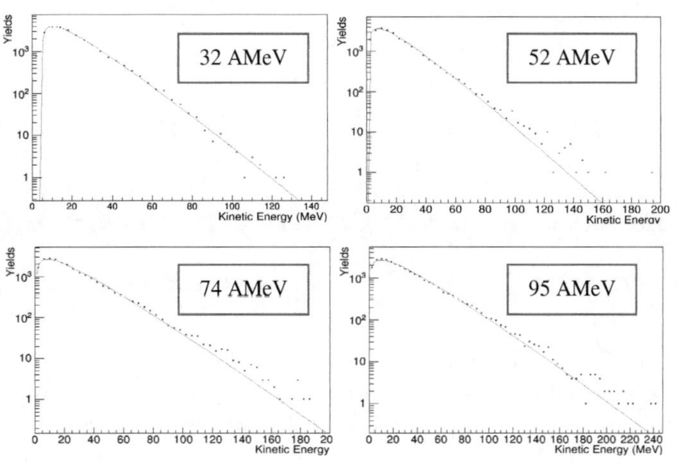

Figure 2.1.2 Energy spectra of alpha particle at 90° in the QT reference fram: solid red lines refer to Maxwellian best-fit.

(fig. 2.1.2). The solid red lines refer to Maxwellian best-fits. To avoid contaminations by other non-QT sources only alpha at 90° in the center of mass of the reconstructed target are selected.
A good agreement between fits and data is observed, indicating that alpha particles are mostly evaporated from the excited target, and temperatures extracted from these fits have values ranging from 11 to 17 MeV.

3. Conclusions and perspectives

The evolution of the reaction mechanisms in central ^{36}Ar+^{58}Ni collisions at E/A=32-95 MeV is studied. The identification of evaporative processes and dynamic multifragmentation and the characterization of fragments emitting sources, together with the study of the kinematics and energy spectra of reaction products emitted in very central collisions, allowed us to isolate a reaction mechanism not yet studied in such kind of reactions, and related to nuclear transparency: a fast "in medium" fragmentation of the projectile that occurs and become dominant with increasing of beam energy in the most violent collisions, where the nuclear stopping saturates (at about 45/50 AMeV) [2]. This mechanisms shows features of a process out of equilibrium, where the mass asymmetry between projectile and target nuclei plays likely an important role. We refer to such phenomenon as an "in-medium crumbling" of the projectile, i.e. a fragmentation of the projectile within the nuclear medium constituted by the whole colliding system in the overlap region. In-medium produced fragments travel through the interaction region and, while quickly escaping from it, they seem to drag the excited quasi-target along the beam direction. The latter finally undergoes an evaporative de-excitation, as it is suggested by the Maxwellian shape kinetic energy spectra of alpha particles in the QT reference frame.
In order to further explore the nature of such reaction mechanism and crumbling breakup phenomenon, it would be important to understand what parameters may favor its observation, namely the total mass of the system and its charge and isospin asymmetry. For this purpose investigations on system with the same mass but different N/Z ratio, or systems with a different mass asymmetry are in progress.

References

[1] Nebauer R, Aichelin J and the INDRA Collaboration 1999 *Nucl. Phys.* A **658** 67 and references therein
[2] Lehaut G *et al* 2010 *Phys. Rev. Lett.* **104** 232701and references therein
[3] Lopez O *et al* 2014 *Phys. Rev.* C **90** 064602 and references therein

IOP Conf. Series: Journal of Physics: Conf. Series **863** (2017) 012062 doi:10.1088/1742-6596/863/1/012062

N/Z effect on reaction mechanisms cross sections in the $^{78}Kr +^{40} Ca$ and $^{86}Kr +^{48} Ca$ collisions at 10 AMeV

B Gnoffo[1,2], S Pirrone[1], G Politi[1,3], M La Commara[4,5], J P Wieleczko[6], E De Filippo[1], P Russotto[1], M Trimarchi[8], M Vigilante[4,5], G Ademard[11], L Auditore[8], C Beck[9], I Bercenau[10], E Bonnet[6], B Borderie[11], G Cardella[1], A Chibihi[6], M Colonna[7], D Dell'Aquila[4,5], S De Luca[8], A D'Onofrio[5,12], J D Frankland[6], G Lanzalone[7,13], P Lautesse[14], D Lebhertz[6], N Le Neidre[15], I Lombardo[4,5], N S Martorana[3,7], K Mazurek[6], S Norella[8], A Pagano[1], E V Pagano[3,7], M Papa[1], E Piasecki[16], F Porto[1,3], L Quattrocchi[1,3], F Rizzo[3,7], G Spadaccini[4,5], A Trifirò[8] and G Verde[11]

[1] INFN Sezione di Catania, Italy
[2] Centro Siciliano Fisica Nucleare e Struttura della Materia, Catania, Italy
[3] Dipartimento di Fisica e Astronomia, Università di Catania, Italy
[4] Dipartimento di Fisica, Università. Federico II Napoli, Italy
[5] INFN Sezione di Napoli, Napoli, Italy
[6] Ganil, CAEN,France
[7] INFN, Laboratori Nazionali del Sud, Catania, Italy
[8] Dipartimento di Fisica, Università di Messina and INFN Sezione di Catania, Italy
[9] IN2P3-IPHC Strasbourg, France
[10] IPNE, Bucarest, Romania
[11] IN2P3-IPN Orsay, France
[12] Dipartimento di Matematica e Fisica, Seconda Università di Napoli, Caserta, Italy
[13] "Kore" Università, Enna, Italy
[14] IN2P3-IPN Lyon, France
[15] IN2P3-LPC CAEN, France
[16] Faculty of Physics and Heavy Ion Laboratory, University of Warsaw, Poland

Abstract. Nuclear reactions between medium-mass nuclei at low energy are characterized by the competition between binary and evaporation process in the compound nucleus de-excitation. A study of the influence of the neutron richness of the entrance channel on the decay paths of the compound nuclei formed in the $^{78}Kr +^{40} Ca$ and $^{86}Kr +^{48} Ca$ at 10 MeV/A is presented. The experiment has been performed at Laboratori Nazionali del Sud by using the CHIMERA 4π multidetector for charged particles. The Kinematical characteristics of the two reactions support the conclusion of a production via long lived system. Besides the results relative to the n-poor system are compared to those obtained at GANIL, performed at 5.5 AMeV, in order to study the energy influence.

1. Introduction

The N/Z ratio, strongly correlated to the isospin degree of freedom, influences the competition among the different reaction mechanisms and have important effects on the aspects characterizing the heavy ion collisions, like for example the production of fragments with an atomic

IOP Conf. Series: Journal of Physics: Conf. Series **863** (2017) 012062 doi:10.1088/1742-6596/863/1/012062

number $3 \leq Z \leq 20$, denoted as the Intermediate Mass Fragments, IMF. The characterization of the production source of the IMFs and the deep comprehension of the effects shown by their emission is one of the goal of nuclear physics in these last years.
In order to study the neutron enrichment effect on the decay modes of the compound nucleus, with a particular attention to the IMFs production, the experiment called ISODEC,[1],[2],[3] has been performed at Laboratori Nazionali del Sud in Catania, by using the beams delivered by the Superconductive Cyclotron and the detection and identification capability of the 4π multi-detector CHIMERA[4],[5]. In fact the reactions studied, $^{78}Kr+^{40}Ca$ (n-poor) and $^{86}Kr+^{48}Ca$ (n-rich), lead to the formation of two systems with N/Z ratios, respectively 1.11 and 1.39, providing the maximum difference available with stable beams.

2. Experimental Results

A deep study of the kinematical characteristic of the reaction products was performed, for details see [3]. The results of the analysis suggest that the IMFs were produced by the two studied systems trough the same mechanism, a fission-like process, but with different probability. In fact the values of the production cross sections of IMFs, plotted versus the atomic number in the fig.1 [3], are higher for the n-poor system compared to n-rich one. Moreover the strong influence of the isopin is also evident in the different oscillation amplitude of the even-odd effect presented by the charge distribution, that is more pronounced for the n-poor system in agreement with other examples in literature [6],[7].This effect is also observed in the plot IMF's yields versus the neutron number, shown in the fig.2, but in this case it is more pronounced for the n-rich system [6]. One should observe that in literature there are few examples of this kind of study, as it is requires an apparatus allowing a good isotopic discrimination; in the ISODEC experiment, the CHIMERA device through the ΔE-E technique provides, with high resolution, the isotopic composition of the fragments with $Z \leq 10$.

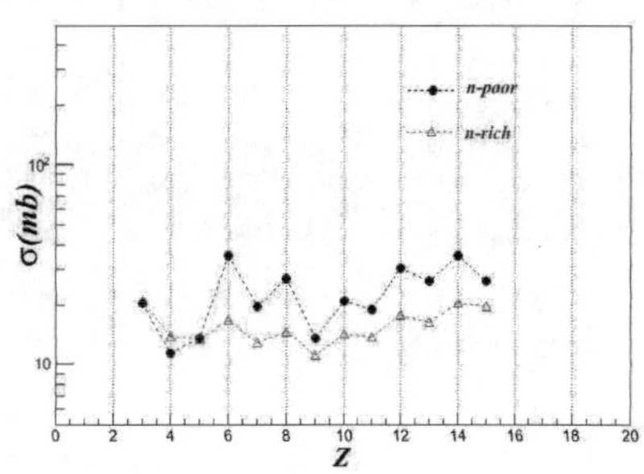

Figure 1. Production cross section for fragments with different atomic numbers for the two reactions $^{78}Kr +^{40} Ca$ and $^{86}Kr +^{48} Ca$. The error bars are inside the graphical symbols.

In order to understand if the IMFs are emitted in an excited state, the relative velocity of an IMF and an α particle emitted in coincidence was studied in the reference frame of the fission fragment. The $v \parallel v \perp$ plots, in the case of very asymmetric, asymmetric and symmetric fission, in the frame of the light fragment and in its complementary partner frame, are shown in the fig.3, compared to those relative of the experiment, realized at GANIL, with the INDRA device, where the $^{78}Kr +^{40} Ca$ reaction was studied at 5.5 AMeV [8]. One observe that in the case of a very asymmetric fission, at higher energy the $v \parallel v \perp$ plot seems to present two coulomb rings, one centered around the velocity of the heavy fission fragment and the other around the velocity of the light one, suggesting that at 10 AMeV, also the light fragment can emit α particles and is thus emitted in excited states. For a better comparison of the results obtained at the two different energies and between these experimental results and the theoretical predictions of statistical and dynamical models the cross sections of the different process will be calculated.

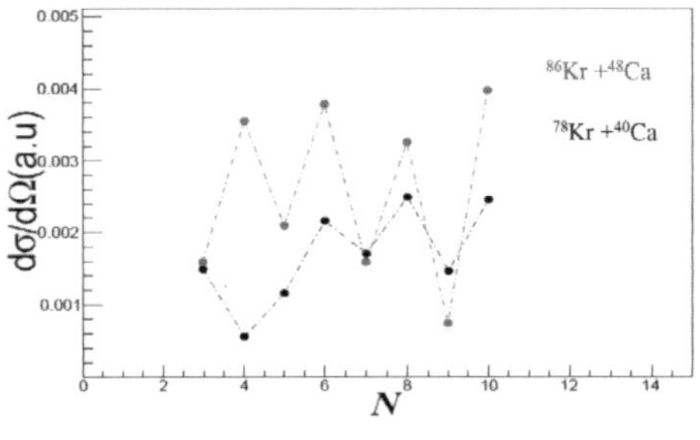

Figure 2. IMFs yields versus the neutron number in black for $^{78}Kr +^{40} Ca$ and in red for $^{86}Kr +^{48} Ca$. The error bars are inside the graphical symbols.

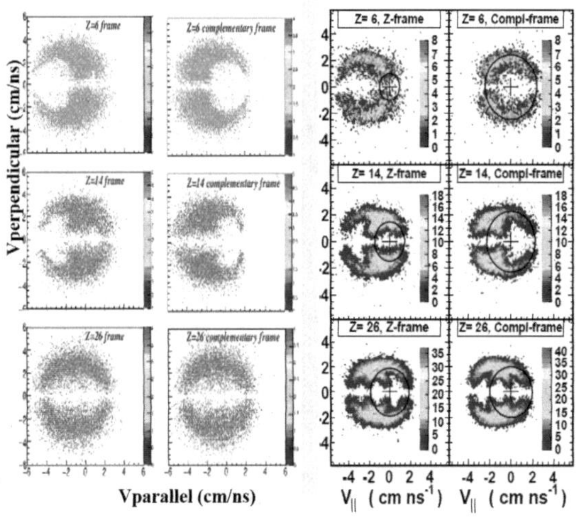

Figure 3. $v \parallel v \perp$ plot for the n-poor system, in the frame of the light fragment and in its complementary partner's referenceframe, on the left for incident energy E= 10AMeV and on the right for E=5.5 AMeV .

References

[1] Pirrone S *et al* 2014 *J. Phys. Conf. Series* **515** 012018
[2] Politi G *et al* 2015 *JPS: Conf. Proc.* **6** 030082
[3] Gnoffo B2016 Nuovo Cimento **39 C** 275
[4] Pagano A *et al* 2012 *Nucl. Phys. news* **22** 25
[5] Politi G *et al* 2005 *IEEE NSS: Conf. Rec.* **N28**
[6] Lombardo I *et al* 2011 *Phys. Rev.* C **84** 024613
[7] Casini G *et al* 2012 *Phys. Rev.* C **86** 011602
[8] Ademard G *et al* 2011 *Phys. Rev.* C **83** 054619

IOP Conf. Series: Journal of Physics: Conf. Series **863** (2017) 012063 doi:10.1088/1742-6596/863/1/012063

α clustering and its connection to the $E1$ response of heavy nuclei

M Spieker[1], **S Pascu**[1,2] **and A Zilges**[1]
[1]Institute for Nuclear Physics, University of Cologne, Zülpicher Strasse 77, 50937 Cologne, Germany
[2]Horia Hulubei National Institute of Physics and Nuclear Engineering, Bucharest, Romania
E-mail: spieker@ikp.uni-koeln.de

Abstract. The generation of the collective nuclear $E1$ response is intimately connected to the breaking of proton-neutron symmetry. In this contribution, we present our recent results on a possible α-cluster dipole mode, which were obtained in the framework of the *spdf* interacting boson model. These extended studies of the low-lying $E1$ response support the general occurence of α-cluster dipole states in atomic nuclei.

1. Introduction

The generation of the collective nuclear $E1$ response is intimately connected to the breaking of proton-neutron symmetry. Especially the low-lying $E1$ strength (LES) or so-called pygmy dipole resonance (PDR), a dipole-strength accumulation below and around the particle-emission thresholds has attracted a lot of interest during the last two decades, see the review articles [1, 2]. Besides other generating mechanisms, the dipole-type oscillation of the neutron-skin has been discussed frequently. Of course, it is interesting to ask the question, whether other mechanisms can account for the experimental systematics. In this contribution, the α-cluster dipole mode will be discussed, which was investigated in the framework of the *spdf* interacting boson model (IBM). First results for the low-lying $E1$ strength in rare-earth nuclei have already been presented and a close connection of the origin of $E1$ strength with the collective p boson was established [3]. The p boson is geometrically identified with a two-body cluster configuration, see, *e.g.*, Refs. [4, 5], where the $\alpha + (A - 4)$ configuration might be considered as the energetically most favored. Therefore, a successful extension of the rare-earth studies to other regions of the nuclear chart might help to establish the α-cluster dipole mode as a possible generator of collective $E1$ strength and, furthermore, contribute to the question as to whether α clustering is a general feature of atomic nuclei.

2. The *spdf* interacting boson model (IBM)

The full one-body $E1$ operator of the *spdf* IBM [6] was adopted to calculate $E1$ transitions.

$$\begin{aligned}\hat{T}(E1) =& e_1[\chi_{sp}(s^\dagger\tilde{p} + p^\dagger\tilde{s})^{(1)} + (p^\dagger\tilde{d} + d^\dagger\tilde{p})^{(1)} \\ &+ \chi_{df}(d^\dagger\tilde{f} + f^\dagger\tilde{d})^{(1)}]\end{aligned} \tag{1}$$

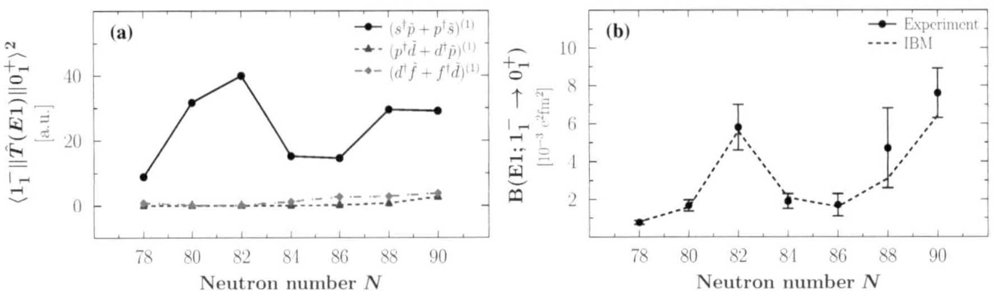

Figure 1. The origin of $E1$ strength in the $N = 78 - 90$ nuclei of the rare-earth region. **(a)** Respective contributions to the $\langle 1_1^-\|\hat{T}(E1)\|0_1^+\rangle^2$ matrix element predicted by the IBM with Eq. (1). **(b)** Experimental data taken from Refs. [7, 8]. Note that the same experimental states have been chosen in **(b)** as in the original figure of Ref. [8].

This operator allows in a very intuitive manner for a clear identification of the most important matrix elements. In the rare-earth nuclei, the $(s^\dagger\tilde{p} + p^\dagger\tilde{s})^{(1)}$ matrix element could be identified as the most collective and accounted in a natural way for the evolution of the experimental $B(E1;1_1^- \rightarrow 0_1^+)$ strength. The other matrix elements evolved as expected for decreasing quadrupole deformation when approaching the $N = 82$ shell closure. The results are shown in Fig. 1. Note, that the *spdf* IBM allows for the discrimination of p boson and *df*-coupled 1^- states in terms of the n_p/n_f ratio, which is a measure for the p-boson and f-boson content of the wave function, respectively. Furthermore, the mixing between these different structures can be accessed. See Ref. [3] for more information.

3. The low-lying $E1$ response of $A < 100$ nuclei

As pointed out in the introduction, it is very interesting to test if the conclusions of the rare-earth region [3] could be extended to other mass regions. For this test, the three nuclei ^{48}Ti, ^{74}Ge, and ^{94}Mo have been chosen. While, on the one hand ^{48}Ti is located between the $N = 20$ and $N = 28$ shell closures with a quadrupole deformation parameter β_2 of about 0.25 [9], ^{74}Ge, on the other hand, is on the edge of the transition from a γ-soft to a γ-rigid nucleus [10, 11]. Furthermore, the vibrational nucleus ^{94}Mo has been the prime laboratory for so-called mixed-symmetry states for decades, see, *e.g.*, the review [12], and an $\alpha + (A - 4)$ structure has already been discussed [13]. For all these nuclei extended information on the $E1$ response up to the particle-emission thresholds is available [7, 14, 15, 16].
The comparison of the experimental data and the IBM predictions for the B(E1) strength is shown in Fig. 2. As in the case of the rare-earth nuclei, the n_p/n_f ratios were also studied to disentangle the different structures of the 1^- states. The general agreement in terms of fragmentation and total strength for experimentally firmly assigned 1^- states is encouraging. Strong p-boson states are observed in all three nuclei. These are states, for which n_p/n_f exceeds one. In contrast to some rare-earth nuclei, very large ratios are observed, which might point at less fragmented α-cluster structures.

4. Conclusion

Our studies of enhanced $E1$ strength due to an α-cluster dipole mode have been extended to $A < 100$ nuclei. In this contribution, the examples of ^{48}Ti, ^{74}Ge, and ^{94}Mo were presented and the *spdf* IBM was able to account for the $E1$ strength distribution as in the case of the rare-earth region. The general applicability of the model to the $E1$ strength of atomic nuclei further strenghtens the claim of Ref. [3], that $E1$ strength could serve as a sensitive measure for α clustering and might, thus, support the general occurrence of α clustering in nuclei. In

IOP Conf. Series: Journal of Physics: Conf. Series **863** (2017) 012063 doi:10.1088/1742-6596/863/1/012063

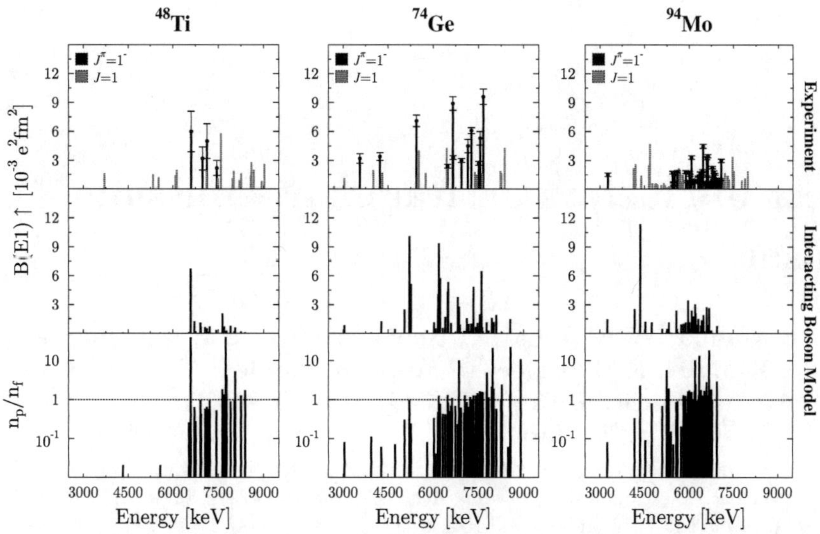

Figure 2. $E1$ distribution in ^{48}Ti, ^{74}Ge, and ^{94}Mo. The experimental data have been taken from Refs. [7, 14, 15, 16]. Firm spin-parity assignments in black, while ambiguous assignments are presented in red.

the future, our studies will be extended and more experimental $E1$ data will be considered. To this end, the isoscalar $E1$ strength constitutes another very promising probe for α clustering [17].

Acknowledgement

This work is supported by the DFG (ZI-510/7-1). M.S. was supported by the Bonn-Cologne Graduate School of Physics and Astronomy. The authors gratefully acknowledge valuable discussions with F. Iachello.

References

[1] Savran D, Aumann T and Zilges A 2013 *Prog. Part. Nucl. Phys.* **70** 210
[2] Bracco A, Crespi F C L and Lanza E G 2015 *Eur. Phys. J.* A **51** 99
[3] Spieker M, Pascu S, Zilges A and Iachello F 2015 *Phys. Rev. Lett.* **114** 192504
[4] Iachello F and Jackson A D 1982 *Phys. Lett.* B **108** 151
[5] Daley H and Iachello F 1983 *Phys. Lett.* B **131** 281
[6] Kusnezov D F 1988 *Nuclear Collective Quadrupole-Octupole Excitations in the U(16) spdf Interacting Boson Model*, PhD Thesis, Princeton University
[7] NNDC Online Data Service, ENSDF database, 2016 `http://www.nndc.bnl.gov/ensdf/`
[8] Fransen C *et al* 1998 *Phys. Rev.* C **57** 129
[9] Pritychenko B, Birch M, Singh B and Horoi M 2016 *At. Data Nucl. Data Tables* **107** 1
[10] Toh Y *et al* 2013 *Phys. Rev.* C **87** 041304
[11] Sun J J 2014 *Phys. Lett.* B **734** 308
[12] Pietralla N, von Brentano P and Litsetskiy A 2005 *Prog. Part. Nucl. Phys.* **60** 225
[13] Souza M A and Miyake H 2015 *Phys. Rev.* C **91** 034320
[14] Degener A *et al* 1990 *Nucl. Phys.* A **513** 29
[15] Jung A *et al* 1995 *Nucl. Phys.* A **584** 103
[16] Romig C *et al* 2013 *Phys. Rev.* C **88** 044331
[17] Chiba Y, Kimura M, Taniguchi Y 2016 *Phys. Rev.* C **93** 034319

The effect of clusters on fragment emission mechanism

S Manna[1], T K Rana[1], C Bhattacharya[1], S Bhattacharya[1], S Kundu[1], K Banerjee[1], P Roy[1], R Pandey[1], V Srivastava[1], A Chaudhuri[1], T Roy[1], T K Ghosh[1], G Mukherjee[1], J K Meena[1], S K Pandit[2], K Mahata[2], A Shrivastava[2] and V Nanal[3]

[1]Variable Energy Cyclotron Centre, 1/AF Bidhan Nagar, Kolkata - 700064, India
[2]Nuclear Physics Division, Bhabha Atomic Research Centre, Mumbai - 400085, India
[3]Tata Institute of Fundamental Research, Mumbai - 400005, India

E-mail: smanna@vecc.gov.in

Abstract. Fragment emission mechanism has been studied for the composites ^{24,25}Mg at same excitation energy (~54 MeV). The inclusive energy distributions of the emitted fragments (Li and Be isotopes) have been measured. It has been observed that the fragments were emitted from a fully energy relaxed composite. The measured yield ratios of the fragments for the two reactions were compared with the statistical model prediction of the same. It was found that unlike Li isotopes, statistical model calculation deviates substantially for the Be isotopes. The observed deviations were due to the boost in the yields whenever there was a known alpha cluster nucleus, either ^{16}O or ^{18}O, as the complementary fragment of the binary decay.

1. Introduction

The presence of cluster structure in nuclei plays a significant role on the fragment emission mechanism for light composite systems, in particular, this would lead to an excess production of nuclei with predominant cluster structure with respect to the prediction of the statistical model. Recently, Morelli et al. [1], have observed some residual deviations from the statistical behaviour in two specific exit channels (carbon with 3α's and oxygen with 2α's channels) for the reaction ^{12}C + ^{12}C at beam energy of 95 MeV, which have tentatively been assigned due to the presence of direct reactions and/or α-clustering effects. Anomalously high branching ratios with respect to the statistical model calculation for these channels have also been observed [1] for the same system. Nevertheless, it is interesting to investigate the signature of cluster-correlation in dissipative binary decay of the same composite system as well as the effect of addition of one valence neutron to the same composite.

2. Experimental Details

The experiment has been carried out at BARC-TIFR Pelletron-Linac facility, Mumbai, using 80 MeV ^{12}C and 78.5 MeV ^{13}C ion beams on ^{12}C target. We used two telescopes, each consisting of ΔE (50 μm) single sided strip detector (SSSD), E (1mm) double sided silicon strip detector (DSSD) and backed by four CsI(Tl) detectors, each of thickness 6 cm. The inclusive energy distributions for various fragments (^{6,7}Li, ^{7,9}Be) have been measured in the angular range of 14°

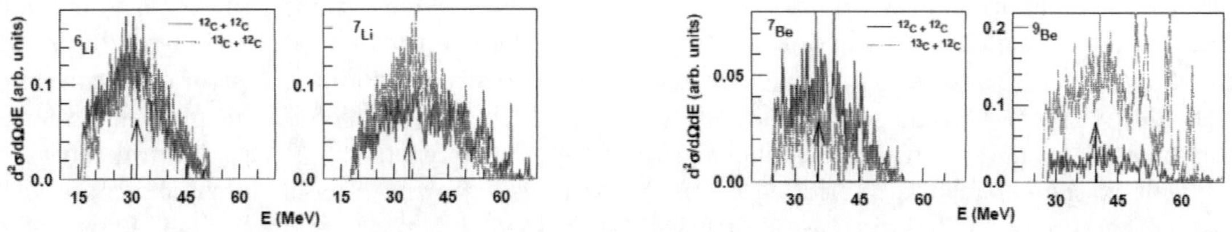

Figure 1. (Color online) (Left) Normalized energy distributions of Li isotopes in two reactions at $\theta_{lab} \sim 15°$. Arrows indicate the mean energies of the fragments predicted by Viola systematics(see text). (Right) Same as Left for Be isotopes.

to 36° in the laboratory. The ^{8}Be spectra were reconstructed from 2α correlation data obtained from the telescopes.

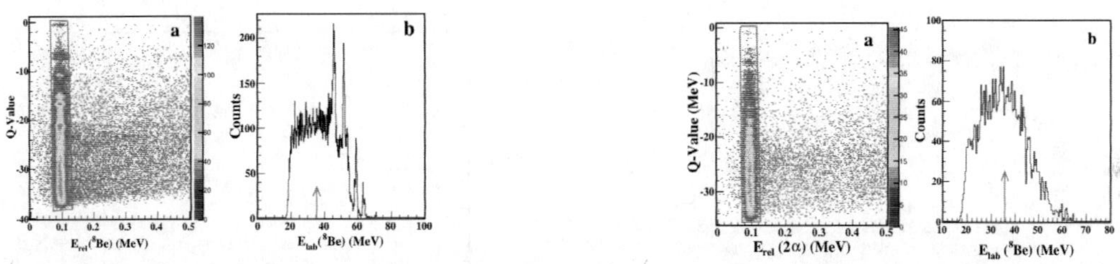

Figure 2. (Color online)(Left)(a)2-d plot between relative energy of 2α and binary beak up Q-value for the ^{12}C + ^{12}C reaction; (b)typical energy spectrum of ^{8}Be at 20°. Arrow indicates the mean energy of the fragment predicted by Viola systematics(see text). (Right)Same as Left for the ^{13}C + ^{12}C reaction.

3. Results and discussion

The normalized energy distributions of different isotopes of Li and Be fragments obtained in the reactions ^{12}C + ^{12}C (blue solid line) and ^{13}C + ^{12}C (red dash-dotted line) have been shown in Figure 1 except for ^{8}Be which is unstable and decays into 2α instantaneously. The ^{8}Be spectra was reconstructed from the measured α-α correlated data for all detected events and the corresponding Q-value was calculated assuming two body reaction and then plotted, as shown in Figure 2(Left)(a) and (Right)(a) for the reactions ^{12}C + ^{12}C and ^{13}C + ^{12}C, respectively. The ground state energy spectrum of ^{8}Be has been obtained by gating on the correlation plots (as shown by red rectangular boxes in Figure 2) and were shown in Figure 2(Left)(b) and (Right)(b). Apart from some sharp transfer channel peaks at high energy ends, the energy distributions for all the fragments (Li and Be) are nearly Gaussian in shape, and their centroid are at the expected kinetic energies for the fission fragments as predicted by the Viola systematics (with correction by corresponding asymmetry factors) [2] (shown by arrows in respective Figures). This indicates that, in all cases, the fragments are emitted from a fully energy relaxed composite as expected for statistical decay of a compound nucleus. The angular distributions of the fragments in centre of mass frame have been shown in Figure 3 and are found to follow $\sim 1/\sin\theta_{cm}$ dependence (shown by solid lines in Figure 3) which further confirms that these fragments have been emitted from a fully energy equilibrated system. The ratios $[R^{exp(Th)} = (Y_{tot}(^{13}C+^{12}C)/Y_{tot}(^{12}C+^{12}C))^{exp(Th)}]$ of yields of each fragment in the two reactions have been compared with the statistical model (CASCADE) [3] prediction for the same as shown in the Figure 4. It is seen that CASCADE predicts quite well for the ^{6,7}Li, but it under predicts for ^{7,9}Be and over predicts for ^{8}Be. This anomalous behaviour can be attributed to the effect of cluster correlation in fragment emission mechanism. In the case of emission of ^{6,7}Li, the complementary binary reaction products are any of the ^{17,18,19}F isotopes which are not known to have any pronounced cluster structure and so the decay is statistical in nature. On the other hand, the emission of any of ^{7,8,9}Be fragments is associated with the emission of one of the isotopes of oxygen ($^{15-18}$O) as complementary fragment in the exit channel. Among them ^{16}O is a well-known alpha-cluster nucleus and the cluster structure of ^{18}O has also been fairly well established [4, 5, 6]. Therefore, yield of ^{7}Be in the ^{13}C + ^{12}C reaction is expected to be higher than the corresponding statistical model prediction if cluster correlation due to ^{18}O contributes to the yield. This additional contribution will not be there in the case of ^{12}C + ^{12}C reaction as the complementary fragment is ^{17}O. Similarly the yield of ^{8}Be and ^{9}Be is associated with the emission of ^{16}O in the reaction channels ^{12}C + ^{12}C and ^{13}C + ^{12}C respectively. Clearly the effect of cluster correlation due to ^{16}O will boost up the yield of the respective decay channels over the statistical emission. This explains the observed trend in Figure 4.

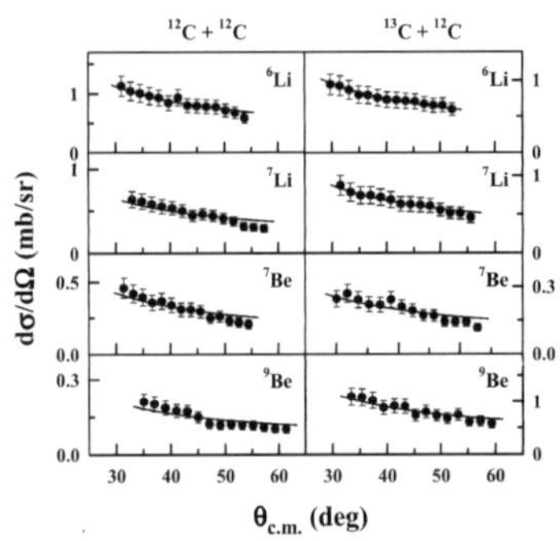

Figure 3. The centre of mass(cm) angular distribution of Li and Be isotopes for ^{12,13}C + ^{12}C reactions. The solid lines correspond to $d\sigma/d\Omega \sim 1/\sin \theta_{cm}$ fit to the data.

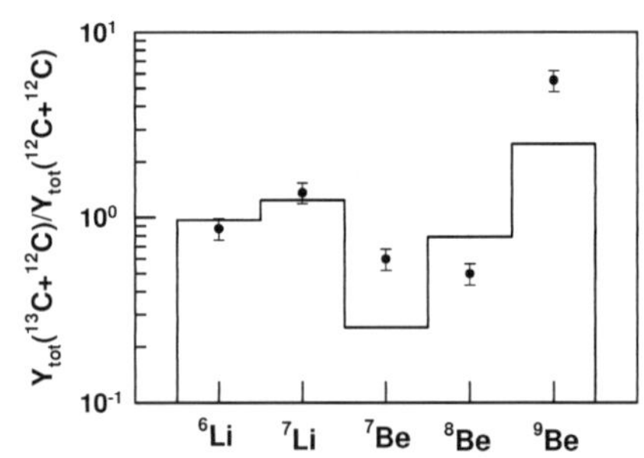

Figure 4. Ratio of fragment yields in ^{13,12}C + ^{12}C reactions. Solid circles represent the data and the histogram shows the corresponding statistical model prediction.

4. Conclusions

The fragment emission mechanism have been studied in $^{13,12}C$+^{12}C reactions. The relative yields of the isotopes of Li and Be emitted from fully energy relaxed composites ^{24,25}Mg* were compared with statistical model prediction. Statistical model prediction agreed well with experimental results for ^{6,7}Li, but deviated significantly for ^{7,8,9}Be. This was due to enhancement in yields for these fragments when the complementary fragments in the exit channel were either of ^{16,18}O isotopes, both being alpha cluster nuclei.

References

[1] Morelli L *et al* 2014 *J. Phys.* G **41** 075107; 2014 *J. Phys.* G **41** 075108
[2] Beck C and Szanto de Toledo A 1996 *Phys. Rev.* C **53** 1989
[3] Pühlhofer F 1977 *Nucl. Phys.* A **280** 267
[4] Curtis N *et al* 2002 *Phys. Rev.* C **66** 024315
[5] Rogachev G V *et al* 2014 *J. Phys. Conf. Series* 569 012004
[6] Johnson E D 2008 *Ph.D. Thesis* Florida State Univ.

IOP Conf. Series: Journal of Physics: Conf. Series **863** (2017) 012065 doi:10.1088/1742-6596/863/1/012065

Tensor correlations in light nuclei with tensor-optimized antisymmetrized molecular dynamics

T Myo[1,2], H Toki[2], K Ikeda[3], H Horiuchi[2] and T Suhara[4]

[1] General Education, Faculty of Engineering, Osaka Institute of Technology, Osaka, Osaka 535-8585, Japan

[2] Research Center for Nuclear Physics (RCNP), Osaka University, Ibaraki, Osaka 567-0047, Japan

[3] RIKEN Nishina Center, Wako, Saitama 351-0198, Japan

[4] Matsue College of Technology, Matsue 690-8518, Japan

E-mail: myo@ge.oit.ac.jp takayuki.myo@oit.ac.jp

Abstract. We investigate the light nuclei using the nucleon-nucleon interaction. We describe the tensor and short-range correlations with the new variational theory named "tensor-optimized antisymmetrized molecular dynamics" (TOAMD). In TOAMD, the correlation functions for the tensor force F_D and the short-range repulsion F_S and their multiples are multiplied to the AMD wave function and the total wave function is expressed as the sum of all the components. We take all the necessary matrix elements of many-body operators coming from the products of the multiple correlation functions and the Hamiltonian. We use the interaction AV8$'$ and discuss the effect of tensor and short-range correlations in the s-shell nuclei, ^{3}H and ^{4}He.

1. Introduction

One of the central issues in nuclear physics is to understand the nuclear structure from the nucleon-nucleon (NN) interaction. The NN interaction has a strong tensor force at long and intermediate distances and a strong repulsion at short distance [1]. It is important to investigate the nuclear structure considering the above characteristics of the NN interaction. The origin of the tensor force is the one-pion exchange interaction, which brings the high momentum components in nuclei. It is necessary to treat the high-momentum components induced by the tensor force in the nuclear wave function. So far, we have included the high-momentum components in nuclei based on the shell model type approach, which we name "tensor-optimized shell model" (TOSM) [2, 3]. In TOSM, we fully optimize the two-particle two-hole (2p2h) states in the wave function. There is no truncation for the particle states in TOSM. In particular, the spatial shrinkage of the particle states with high momentum is essential to achieve convergence of the contributions of tensor force.

The clustering of nucleons is one of the important aspects in the nuclear structure, such as the two-α state in ^{8}Be and the Hoyle state in ^{12}C with the triple-α state. It is difficult to describe the clustering states in the shell model type approach, while the shell model-like states are fairly described in TOSM [3, 4]. Furthermore, the relation between the NN interaction and the clustering mechanism is unclear, although it is known that the α cluster itself contains the large contribution of the tensor force [2, 5]. It is important to understand the nuclear clustering phenomena from the viewpoint of the NN interaction and the tensor force.

In this report, we propose the new theory to investigate the nuclear clustering using the NN interaction. We employ the antisymmetrized molecular dynamics (AMD) [6] as the basis states. We introduce two-kinds of the correlation functions of the tensor-operator type for the tensor force and the scalar-operator type for the short-range repulsion. These correlation functions are multiplied to the AMD basis states. We name this framework "tensor-optimized antisymmetrized molecular dynamics" (TOAMD) [7]. We show how the TOAMD describes the light nuclei with the NN interaction.

2. Tensor Optimized Antisymmetrized Molecular Dynamics (TOAMD)

We use the NN interaction, AV8$'$ in this study [1]. We explain the basic formulation of TOAMD while the detailed part is given in Ref. [7]. We employ the AMD wave function expressed by the Gaussian wave packets of nucleons, which nicely describes the nuclear clustering [6]. We multiply the pair-type correlation functions F_D for tensor force and F_S for short-range repulsion to the AMD wave function, respectively, and superpose them with the AMD wave function $\Phi_{\rm AMD}$. In TOAMD, we can systematically increase the order of the correlation functions by using the products of the correlation functions such as the F_SF_D term. The TOAMD wave function with mass number A is explicitly given as

$$\Phi_{\rm TOAMD} = (1 + F_S + F_D + F_SF_S + F_SF_D + F_DF_S + F_DF_D) \times \Phi_{\rm AMD}, \tag{1}$$

$$\Phi_{\rm AMD} = \mathcal{A}\left\{\prod_{i=1}^{A}\phi_i\right\}, \qquad \phi(\vec{r}) = \left(\frac{2\nu}{\pi}\right)^{3/4} e^{-\nu(\vec{r}-\vec{D})^2}\chi_{\sigma\tau}, \tag{2}$$

$$F_D = \sum_{t=0}^{1}\sum_{i<j}^{A} f_D^t(r_{ij})\, O_{ij}^t\, r_{ij}^2 S_{12}(\hat{r}_{ij})\,, \qquad F_S = \sum_{t=0}^{1}\sum_{s=0}^{1}\sum_{i<j}^{A} f_S^{t,s}(r_{ij})\, O_{ij}^t\, O_{ij}^s\,, \tag{3}$$

with $O_{ij}^t = (\vec{\tau}_i\cdot\vec{\tau}_j)^t$ and $O_{ij}^s = (\vec{\sigma}_i\cdot\vec{\sigma}_j)^s$. The single nucleon wave function $\phi(\vec{r})$ in $\Phi_{\rm AMD}$ has a Gaussian function with a range parameter ν and a centroid position $\vec{D}$ and the spin-isospin part $\chi_{\sigma\tau}$. The function F_D induces the relative D-wave transition via S_{12}. The functions F_D and F_S can excite two nucleons in the AMD state to the high-momentum region corresponding to the 2p2h excitation in the shell model. In Eq. (1) we write terms up to two correlation functions, with which we can show numerical results step by step by adding each term successively. At the same time, we are able to add more correlation terms if necessary.

The energy in TOAMD is given as

$$E_{\rm TOAMD} = \frac{\langle\Phi_{\rm TOAMD}|H|\Phi_{\rm TOAMD}\rangle}{\langle\Phi_{\rm TOAMD}|\Phi_{\rm TOAMD}\rangle} = \frac{\langle\Phi_{\rm AMD}|\widetilde{H}|\Phi_{\rm AMD}\rangle}{\langle\Phi_{\rm AMD}|\widetilde{N}|\Phi_{\rm AMD}\rangle}\,. \tag{4}$$

The correlated Hamiltonian $\widetilde{H}$ and norm $\widetilde{N}$ include the products of correlation functions, such as $F_D^\dagger HF_D$ and $F_D^\dagger F_D$, and become the series of many-body operators. In the case of the two-body interaction V, $F_D^\dagger VF_D$ is expanded from two-body to six-body operators with various combinations of particle coordinates. We classify many-body operators according to the cluster expansion method. We take the matrix elements of all the resulting many-body operators with the AMD wave function.

The radial forms of F_D and F_S are optimized to minimize the energy $E_{\rm TOAMD}$ under the variational principle. We use the Gaussian expansion method to express the relative motion of the pair functions $f_D^t(r)$ and $f_S^{t,s}(r)$ in Eq. (3) for spin-isospin channels, given as

$$f_D^t(r) = \sum_{n=1}^{N} C_n^t \exp(-a_n^t r^2), \qquad f_S^{t,s}(r) = \sum_{n=1}^{N} C_n^{t,s} \exp(-a_n^{t,s} r^2). \tag{5}$$

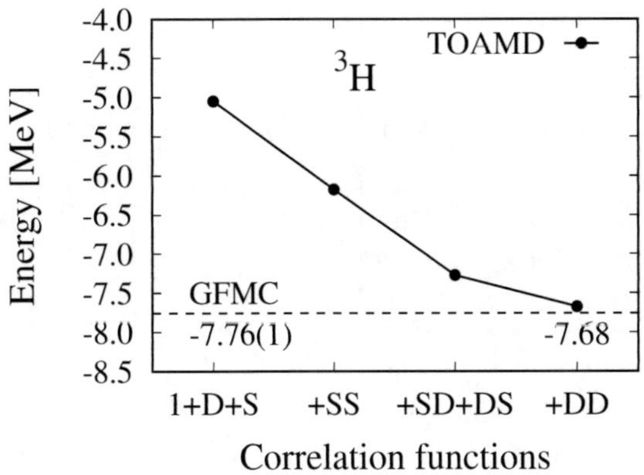

Figure 1. Energy of ^{3}H by adding double correlations successively.

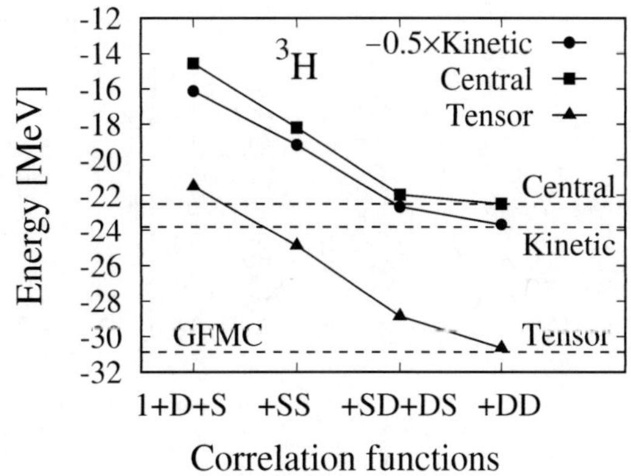

Figure 2. Hamiltonian components of ^{3}H by adding double correlations successively.

Here, a^t_n, $a^{t,s}_n$, C^t_n and $C^{t,s}_n$ are variational parameters. The Gaussian basis number N is taken as seven at most until the solutions are converged. For the ranges a^t_n, $a^{t,s}_n$, we search for the optimized values to gain the total energy. The coefficients C^t_n and $C^{t,s}_n$ are linear parameters in the wave function and determined by diagonalization of the Hamiltonian matrix elements. The positions of the Gaussian wave packets, $\{\vec{D}_i\}$ $(i = 1, \cdots, A)$ are determined variationally. In the calculation of the matrix elements of many-body operators, we adopt the Fourier transformation of the Gaussian forms of the correlation function and the NN interaction [7], which decomposes the many-body operators expressed in terms of the relative coordinates into the separable form in the single particle coordinates.

3. Results of TOAMD

3.1. Single correlation function

We discuss first the TOAMD results for ^{3}H and ^{4}He with the single correlation function in Eq. (1), namely, $\Phi_{\rm TOAMD} = (1+F_S+F_D)\Phi_{\rm AMD}$. In the variation of the AMD wave function, the range ν in Eq. (2) is determined to minimize the energy of each nuclei. The nucleon positions $\vec{D}$ in Eq. (2) become zero for these nuclei and the s-wave configurations are favored in the AMD part. The correlation functions F_D and F_S are optimized for each nuclei. The energies are -5.05 MeV for ^{3}H and -15.37 MeV for ^{4}He, which are underbound in comparison with the Green's function Monte Carlo (GFMC) results; -7.76 MeV for ^{3}H and -25.93 MeV for ^{4}He. We overcome this problem next with the double correlation functions in TOAMD. It is interesting to note that these nuclei are not bound for the case where only the short range correlation F_S is included, namely, $\Phi_{\rm TOAMD} = (1 + F_S)\Phi_{\rm AMD}$.

In the calculation, the many-body operators up to four-body terms are considered in the correlated Hamiltonian $\widetilde{H}$. Up to the two-body terms of $\widetilde{H}$, we cannot obtain the energy minimum and the variational principle is not satisfied. This indicates the inevitable role of the many-body operators in the correlated Hamiltonian. This is related to the spatial ranges of the correlation functions of F_D and F_S, which are extended and increase the contributions of the three- and four-body terms in the correlated Hamiltonian.

3.2. Double correlation function

We include the double correlation functions given in Eq. (1). In Fig. 1, we show the energy of ^{3}H successively by adding various double correlation functions one by one, where $\nu = 0.10$ fm^{-2} variationally. For example, "1+D+S" indicates the calculation within the single correlation

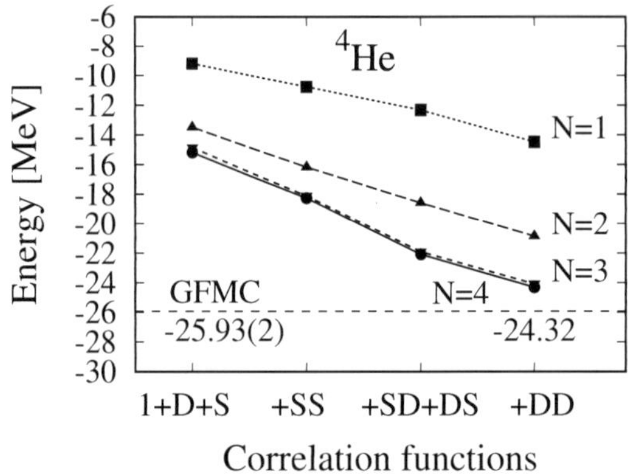

Figure 3. Energy of ^{4}He with the basis number N by adding double correlations successively.

functions and "+SS" indicates the energy by adding $F_S F_S$ component. The components of $F_S F_D$ and $F_D F_S$ are found to give almost identical effect on the energy, and we add two components simultaneously as "+SD+DS". It is found that the final energy up to $F_D F_D$ (+DD) almost reproduces the GFMC results within 80 keV difference. It is also confirmed that the contributions of the kinetic energy, central and tensor forces are reproduced as shown in Fig. 2. From these results, we can learn the accuracy and power of TOAMD and the efficiency of correlation functions F_D and F_S to treat the NN interaction.

For ^{4}He, we show preliminary results in Fig. 3 with the increase of Gaussian basis number N in Eq. (5) from $N = 1$ to $N = 4$ for the case of $\nu = 0.18$ fm^{-2}. The convergence of the energy for N is confirmed at each level of the correlation functions. At $N = 4$ with full components up to $F_D F_D$, the energy is obtained as -24.32 MeV in TOAMD. In order to reach the exact energy, we can increase the order of correlation functions such as the triple component $F_S F_D F_D$. This extension of more correlation functions can be handled systematically in TOAMD.

4. Summary

We propose the new variational theory of "tensor optimized antisymmetrized molecular dynamics" (TOAMD) to describe the nuclear structure with the NN interaction. In TOAMD, the tensor- and central-type correlation functions are multiplied to the AMD wave function. The necessary many-body operators in the correlated Hamiltonian and norm are treated without any truncation. The TOAMD results reproduce the ^{3}H energy using double correlation functions and provide a good binding energy for ^{4}He. Based on the success for the s-shell nuclei, we shall apply TOAMD to the p-shell nuclei with three-body interaction.

Acknowledgments

This work was supported by JSPS Grant: JP15K05091, JP15K17662, JP16K05351.

References

[1] Pieper S C and Wiringa R B 2001 *Annu. Rev. Nucl. Part. Sci* **51** 53
[2] Myo T, Toki H and Ikeda K 2009 *Prog. Theor. Phys.* **121** 511
[3] Myo T, Umeya A, Toki H and Ikeda K 2011 *Phys. Rev.* C **84** 034315
[4] Myo T, Umeya A, Horii K, Toki H and Ikeda K 2014 *Prog. Theor. Exp. Phys.* **2014** 033D01
[5] Kamada H *et al* 2001 *Phys. Rev.* C **64** 044001
[6] Kanada-En'yo Y and Horiuchi H 2001 *Prog. Theor. Phys. Suppl.* **142** 205
[7] Myo T, Toki H, Ikeda K, Horiuchi H and Suhara T 2015 *Prog. Theor. Exp. Phys.* **2015** 073D02

IOP Conf. Series: Journal of Physics: Conf. Series **863** (2017) 012066 doi:10.1088/1742-6596/863/1/012066

Covalent Binding with Neutrons on the Femto-scale

W von Oertzen[1], **Y Kanada-En'yo**[2] **and M Kimura**[3]

[1] Helmholtz-Zentrum Berlin, Hahn-Meitner Platz, 14109 Berlin,Germany.
[2] Department of Physics, Kyoto University, Kyoto, Japan
[3] Hokkaido University, Sapporo 001-0021, Japan.

E-mail: oertzen@helmholtz-berlin.de

Abstract. In light nuclei we have well defined clusters, nuclei with closed shells, which serve as centers for binary molecules with covalent binding by valence neutrons. Single neutron orbitals in light neutron-excess nuclei have well defined shell model quantum numbers. With the combination of two clusters and their neutron valence states, molecular two-center orbitals are defined; in the two-center shell model we can place valence neutrons in a large variety of molecular two-center states, and the formation of Dimers becomes possible. The corresponding rotational bands point with their large moments of inertia and the Coriolis decoupling effect (for K = 1/2 bands) to the internal molecular orbital structure in these states. On the basis of these the neutron rich isotopes allow the formation of a large variety molecular structures on the nuclear scale. An extended Ikeda diagram can be drawn for these cases. Molecular bands in Be and Ne-isotopes are discussed as text-book examples.

1. Introduction

After several decades of work on molecular orbitals in heavy ion collisions (see Ref. [1]), the important step has been taken (Ref. [2]) on defining molecular nuclear states as two body systems with covalent binding with neutron orbitals. This has been initially met with violent objection by the referees, however, the notion of valence neutrons in molecular σ- and π-orbitals soon afterwards has been established by theoretical approaches (with AMD). In the Method of Antisymmetrized Molecular Dynamics (AMD) by H. Horiuchi, Y. Kanada-En'yo, M. Kimura et al [3, 4, 5, 6] the intrinsic density distributions of the valence neutrons in molecular orbits are obtained. For the observation of stable bound molecular states the potential between the two clusters must have "molecular" character, with a repulsion at smaller distances. Best cases (see Ref.[7]) are two α-particles and the α+^{16}O system, shown in Fig. 1. The repulsion at smaller distances is caused by the Pauli-principle acting as soon as a larger overlap of densities occurs.

The recent experimental work has established many molecular rotational bands in light nuclei, see Refs. [8, 9, 10], where many observables for the nuclear structure in terms of two-center molecular states has been found. In work using the (^{7}Li,p)-reactions molecular rotational bands have been established recently in the Oxygen isotopes ^{18,19,20}O. These are intrinsically reflection asymmetric molecules, where the rotational bands appear as *parity inversion doublets.* These are structures consisting of α-particles, with ^{16}O- and ^{14}C-cores. The intrinsically reflection asymmetric molecular states (consisting of α-particles and ^{16}O-cores) have been established in the ^{21,22}Ne-isotopes, see Refs. [12, 14]. With these observations a diagram for nuclear molecular states based on covalent binding between strongly bound cores can be drawn, it is shown in Fig.1. Many of these states have been established in recent [10] experimental works.

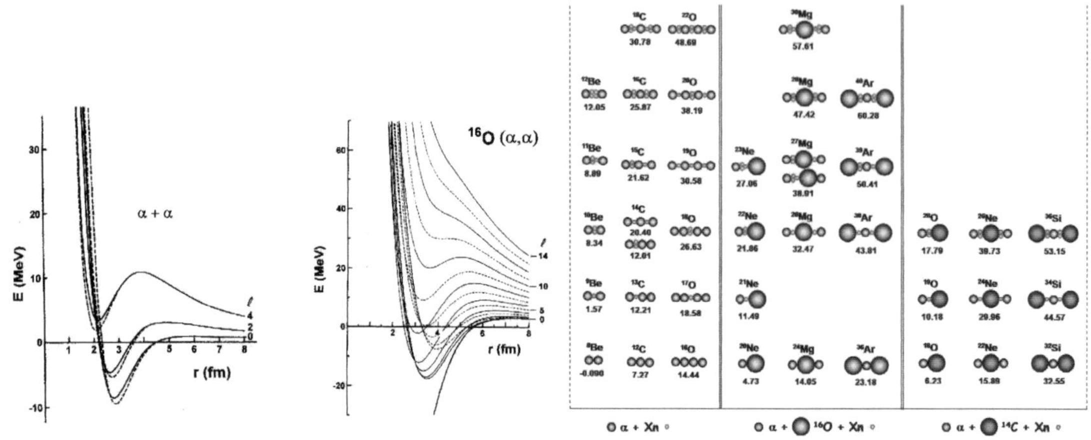

Figure 1. *Left side*: The locally (phase-)equivalent potentials describing the α-α and the α-^{16}O interaction, with "molecular" character, namely the repulsion at small distances. The empirical potentials (full lines) and the theoretical (dashed) are shown, from Ref.[7]. *Right-side*: These potentials allow the formation of various covalently bound states, as suggested in the schematic diagram (the"extended" Ikeda diagram), e.g. the formation of dimers in ^{9,10,11,12}Be as well as in ^{20,21,22}Ne, respectively, see Ref.[10].

2. Dimers, Be- and Ne-isotopes

The molecular structure of the Be-isotopes and Ne-isotopes are text-book examples for physical chemistry on the Femto scale, see the book by Herzberg [11]. We concentrate on these cases in the present contribution. The molecular structures for Be-isotopes can be obtained by inspecting the correlation diagram for valence neutrons and two α-particles, shown in Fig.2. The energy splitting of the levels starting at the larger distances is observed, these are defined by their quantum numbers of molecular structure, as described in Ref.[10]. The rotational bands give clear indications of the σ- and π-orbits, the σ-configurations produce a larger distance between the cores and give larger moments of inertia. The compilation of the rotational bands (excitation energies as function of $j(j+1)$) in the Be-isotopes is shown in Fig.3. This figure illustrates the dependence of the moments of inertia on different configurations. The larger moment of inertia of the $(\sigma)^2$ configuration in the $K = 0_2^*$ - band is clearly visible. The K=0$^-$ band starting with J=1$^-$, is a mixed (σ-π)-configuration.

Another important feature of molecular configurations is the Coriolis-decoupling effect in the rotational bands with $K = 1/2$, which produce the characteristic zig-zag structure seen in Fig.3. The latter appears also in the $K = 1/2$-band in ^{21}Ne, Fig.6. In these cases the spin of the valence particle in a rotational band can be flipped between two projections on the symmetry axis.

It is a case of "complete" spectroscopy, all states in ^{9,10,11,12}Be can be described with valence neutrons in two-center orbitals. We can clearly identify the configurations of the two neutrons occupying σ or π orbits obtained with the valence neutrons in shell model $p_{3/2}$-configurations in the He-isotopes. The mixed orbitals of two neutrons, $(\sigma \times \pi)$ in the 1$^-$ states give the "distorted" wave functions. With this AMD-approach the the internal structure of states of the Be-isotopes are perfectly reproduced.

The method of Antisymmetrized Molecular Dynamics, the AMD [3, 4, 5], is the most powerful tool to illustrate the intrinsic wave functions with the molecular orbitals supporting the concept of molecular orbitals. In fact, the large variety of structures in the Be-isotopes obtained with AMD is shown in Fig. 4.

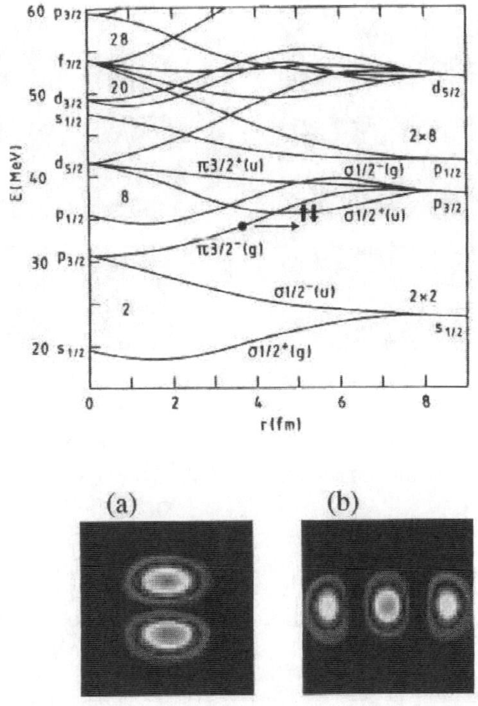

Figure 2. The correlation diagram for two-center neutron orbitals in Be-isotopes. The two orientations of the $p_{3/2}$ states, leading to the formation of four (4) molecular orbitals with the quantum numbers K, symmetry (g,u) and parity, and *gerade/ungerade*-properties as indicated. The possible molecular orbitals are the σ and π orbits, with spatial distributions illustrated in the lower part of the figure.

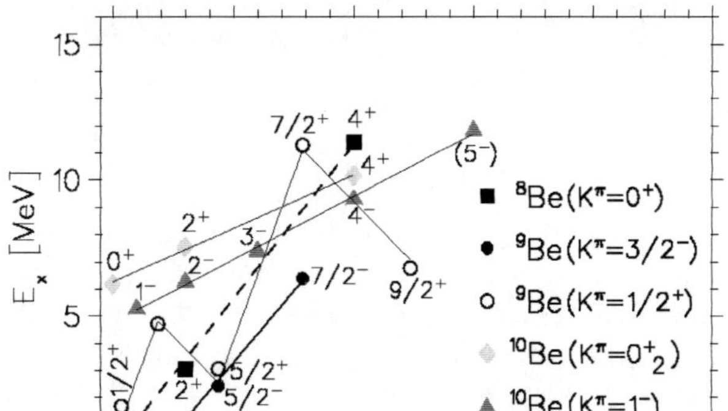

Figure 3. The rotational bands of the isotopes ^{8,9,10}Be. The band with K=0$^+$ based on the excited J=0$^+$ state in ^{10}Be shows a slope corresponding to the larger moment of inertia of the $(\sigma)^2$ configuration. Note the strong *zig-zag* dependence for the $K = 1/2^+$ band in ^{9}Be, due to the Coriolis-decoupling effect. This is also observed in ^{21}Ne as shown in Fig. 6

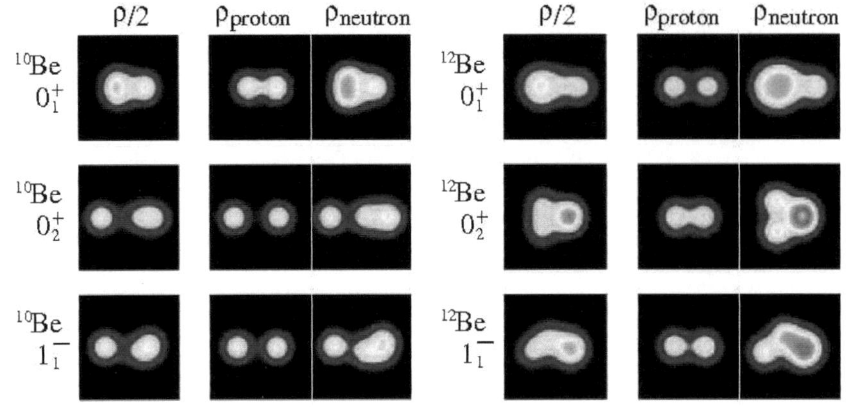

Figure 4. Overview of some intrinsic densities (not projected on parity) for the Be-isotopes ^{10,12}Be. obtained in AMD-calculations [5]. The total densities are shown as well as the cases for protons and neutrons separately, the latter showing the distribution of the valence neutrons.

The very characteristic valence neutron distributions, which give rise to the covalent binding effect in ^{10}Be are further shown in Fig. 5, where in addition the molecular orbital structures of ^{22}Ne are illustrated, again σ- and π-orbits play the dominant role.

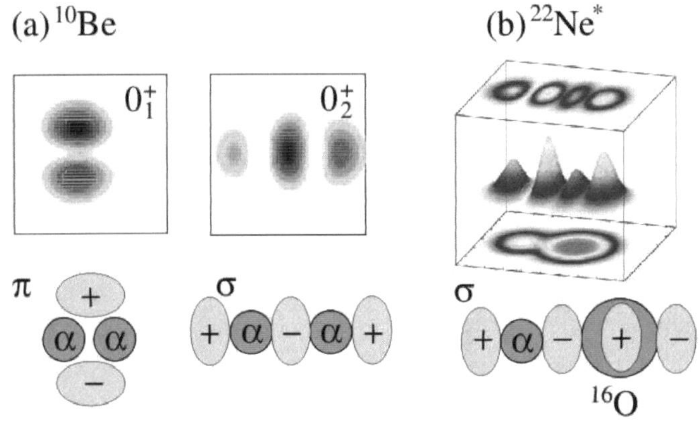

Figure 5. Intrinsic densities of σ- and π-orbits for the Be-isotope ^{10}Be, and for the ^{22}Ne isotope, as obtained in AMD-calculations.

The ^{20,21,22}Ne isotopes are further text-book examples for intrinsically asymmetric binary systems, this was recognized for the cases of $^{20-21}$Ne. The occurrence of parity splitting of rotational bands in ^{20}Ne has already been explained in 1968 [15] by H. Horiuchi. For ^{21}Ne with a single valence neutron unique features are observed [12]. For the π-binding in the $K = 3/2$ band, the splitting is very similar to that of $K = 0^{+-}$-bands in ^{20}Ne, the neutron densities are outside of the symmetry axis, giving space for the exchange as in ^{20}Ne for the $K = 0^{+-}$ states. Then there appears the parity splitting for the $K = 1/2$ bands. The extremely small energy splitting, in this case in ^{21}Ne can be explained in the molecular orbital frame. For the neutrons in the σ-orbits the densities are concentrated on the symmetry axis, preventing an exchange along the axis, in the projection on eigenstates.

Actually in earlier work for ^{21}Ne a systematic survey of the possible rotational bands has been made [13]. In this work the authors used mainly the deformed shell model, however, they were not able to accommodate the many negative parity states ("citation" : to many states with negative parity are observed), because the number of states in the deformed shell model had been exhausted.

The concept of molecular orbitals is a powerful tool for the description of isotopes in light nuclei with neutron excess. Many more nuclei with exotic shapes as in Fig.1 can be anticipated. The chain states based on the configuration with three α-particles (like ^{16}C) have been recently treated explicitly, as reported at this conference.

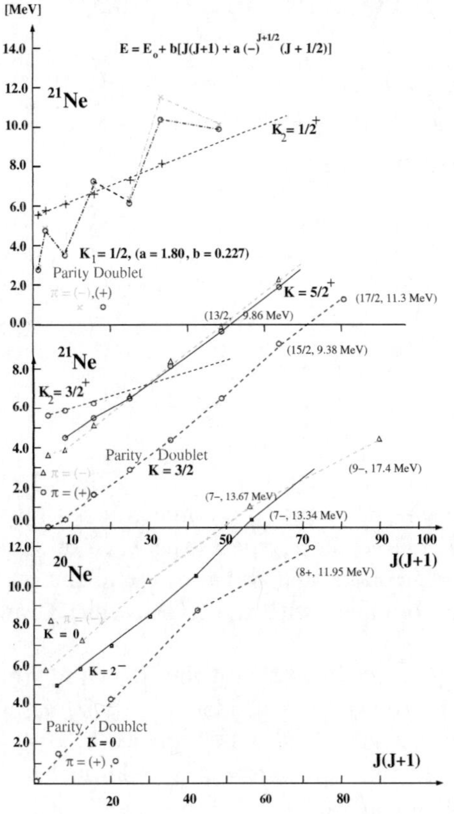

Figure 6. Overview of rotational bands in $^{20-21}$Ne. For ^{20}Ne we observe the well known parity splitting, for the two bands with $K = 0^{+-}$. A similar splitting is observed for the $K = 3/2$ bands in ^{21}Ne, showing a similar splitting as in ^{20}Ne. Note the zig-zag behavior (Coriolis decoupling effect) for the $K = 1/2$ bands. The energy splitting is here only 15-20 KeV, due to the fact that the neutrons in the σ-orbit are concentrated on the symmetry axis.

References

[1] Imanishi B and von Oertzen W 1987 *Phys. Rep.* **155** 29
[2] von Oertzen W 1997 *Z. Phys.* A **357** 355; 1996 *Z. Phys.* A**354** 37
[3] Kanada-En'yo Y and Horiuchi H 1995 *Prog. Theor. Physics (Japan) Suppl.* **93** 115
[4] Kanada-En'yo Y and Horiuchi H 2001 *Prog. Theor. Phys. Suppl.* **142** 206
[5] Kanada-En'yo Y, Kimura M and Ono A 2012 *Prog. Theor. Exp. Phys.* **112** 01A202
[6] Kanada-En'yo Y and Kimura M 2010 *Clusters in Nuclei*, Vol.**1**, ed. C. Beck *Lecture Notes in Physics* **818** 129
[7] Michel F 1989 *Proc. 5th Inter. Conf. on Clustering Aspects*, Kyoto ed. K. Ikeda (Phys. Soc. of Japan) p. 65
[8] Bohlen H G *et al* 2008 *Int.J. Modern Phys.* E **17** 2067
[9] von Oertzen W, Freer M and Kanada-En'yo Y 2006 *Phys. Rep.* **432** 43
[10] von Oertzen W and Milin M 2014 *Clusters in Nuclei*, Vol.**3**, ed. Beck C *Lecture Notes in Physics* **875** 129
[11] Herzberg G 1950 "Molecular spectra and Molecular structure", Vol. **I**, Spectra of Diatomic Molecules, van Nostrand, Princeton.
[12] von Oertzen W 2001 *Eur. Phys. J.* A **11** 403
[13] Roepke H 2000 *Nucl. Phys.* A **674** 95
[14] Kimura M 2007 *Phys. Rev.* C **75** 034512
[15] Horiuchi H and Ikeda K 1969 *Prog. Theor. Physics (Japan) A* **40** 277
[16] von Oertzen W 2010 *et al Eur. Phys. J.* A **43** 17

IOP Conf. Series: Journal of Physics: Conf. Series **863** (2017) 012067 doi:10.1088/1742-6596/863/1/012067

Proton and neutron pair correlations in ^{10}B

H Morita and Y Kanada-En'yo

Department of Physics, Kyoto University, Kyoto 606-8502, Japan

E-mail: morita.hiroyuki.47a@st.kyoto-u.ac.jp

Abstract. We developed an extension of the antisymmetrized molecular dynamics in which we perform energy variation after the projection to isospin eigenstates. In our first application of this method to ^{10}B, we found that low-lying positive-parity states($J^\pi T = 3_1^+0, 1_1^+0$ and 0_1^+1) have a spatially correlated pn pair and 2α structure.

1. Introduction

For a pn pair in nuclei, two types of isospin states exist; the $T = 0$ and $S = 1$ state like a deuteron-like pair and the $T = 1$ and $S = 0$ state analogous to a dineutron pair. In low-lying spectra of $N = Z$ odd-odd nuclei, $T = 0$ and $T = 1$ states compete with each other. To clarify thier structures, we extended the antisymmetrized molecular dynamics.[1] That is the isospin projected and $\beta\gamma$ constraint AMD ($T\beta\gamma$-AMD) combined with the GCM, which is an extended method of the $\beta\gamma$-AMD+GCM.[2]

In our first application of the $T\beta\gamma$-AMD+GCM, we investigate the low-lying states of ^{10}B. ^{10}B has a 2α core and a pn pair [3], and therefore we need to consider spin dynamics and spatial dynamics of the pn pair around the 2α core. In the $T\beta\gamma$-AMD framework, we can see the structures on the energy surfaces by the first principle calculation based on nucleon degrees of freedom and take soft modes into account with the GCM.

2. Formulation

In the framework of the $T\beta\gamma$-AMD, a wavefunction is given by a Slater determinant of Gaussian packets. We performed the energy variation after the isospin and parity projections under the constraint on the quadrupole deformation parameters β and γ:

$$\delta \frac{\left\langle \Phi \middle| H P^J_{MK} P^\pi P^T \middle| \Phi \right\rangle}{\left\langle \Phi \middle| P^J_{MK} P^\pi P^T \middle| \Phi \right\rangle} = 0. \tag{1}$$

After the variation, we superposed the obtained wavefunctions on the $\beta\gamma$ plane to calculate expectation values. Here we used the Volkov No2 force and the spin-orbit part of the G3RS force. See more details in Ref. [4]. The energy variation after the isospin projection is necessary to obtain proper $T = 0$ and $T = 1$ wavefunctions in the whole region of the $\beta\gamma$ planes.

3. Result

In Fig. 1, we show the energy surfaces obtained by the $T\beta\gamma$-AMD and $\beta\gamma$-AMD. We obtained the smooth energy surfaces in the $T\beta\gamma$-AMD results, but we found the serious energy gaps at

IOP Conf. Series: Journal of Physics: Conf. Series **863** (2017) 012067 doi:10.1088/1742-6596/863/1/012067

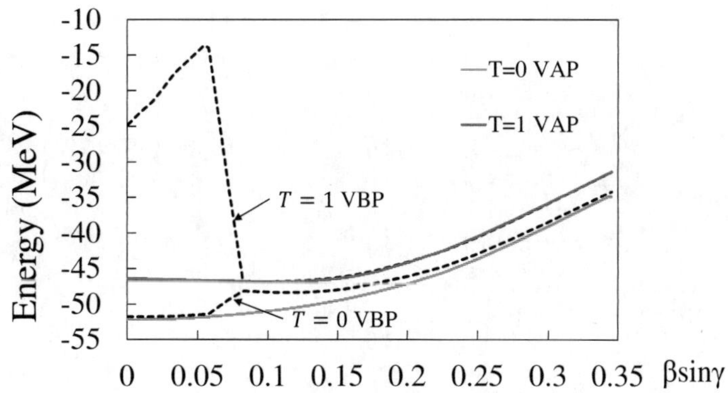

Figure 1. The energy curves at $\beta\cos\gamma = 0.40$ of $T = 0$ and $T = 1$ states for ^{10}B. Results with isospin projection (solid lines) and those without isospin projection (dashed lines) are shown.

$\beta\sin\gamma = 0.07$ in the $\beta\gamma$-AMD results. It indicates that the $T\beta\gamma$-AMD is suitable to describe the low-lying states of $N = Z$ odd-odd nuclei. The energy surface is soft along the $\beta\sin\gamma$ axis, corresponding to the pn pair motion around the 2α core. By superposing these wavefunctions, we can take the motion of the pn pair into account even near the core.

We calculated the energy spectra of ^{10}B and show the results in Fig. 2. We can see that the level ordering and level spacings for each isospin are well reproduced. Moreover, energy gains from the single Slater determinant calculation are remarkably large for all spectra. It implies that the pn pair motion, which is efficiently taken into account by the GCM in the $T\beta\gamma$-AMD, is important in ^{10}B.

We found that spin configuration and spin-orbit force play significant roles in the low-lying states of ^{10}B (see Fig. 3). In the low-lying states, we found a 2α core and a pn pair. For the $T = 0$ spectra, the lowest state ($3^+_1 0$) has the $S = 1$ pn pair moving in the D-wave rotation around the core. $S = 1$ and $L = 2$ of the pn pair are aligned to $J = 3$ to feel strong attractive interaction at the surface of the core. As a result of the alignment, the ^{10}B ground state has the finite spin $J^\pi = 3^+$.

On the other hand, in the first excited state ($1^+_1 0$), the $S = 1$ pn pair moves in S-wave and it is not attracted to the core. For this state, the feature of the pn pair is similar to the $T = 1$ pn pair in the ^{10}B($0^+_1 1$) or the nn pair in the ^{10}Be($0^+_1 1$). For the $1^+_2 0$ state, we found strong $E2$ transition to the $1^+_1 0$ because of the core rotation.

Focusing on the $T = 1$ spectra, the lowest state ($0^+_1 1$) has a pn pair and it moves in the S-wave around the core. This state is the isobaric analogue state of ^{10}Be. We obtained a remarkably large $B(M1)$ from the $0^+_1 1$ to $1^+_1 0$ consistently with the experimental data. These two states, the $0^+_1 1$ ans $1^+_1 0$, have similar spatial structures and show difference in the spin configuration ($S = 0$ and $S = 1$) of the $T = 1$ and $T = 0$ pn pairs. Therefore, these states can be regarded as spin partners.

In the present framework of the $T\beta\gamma$-AMD+GCM, we can treat not only isospin competition but also spin configuration through the isospin projection and antisymmetrization. We can separately describe low-lying levels of $N = Z$ odd-odd nuclei in terms of spin configurations.

4. summary

We extended the AMD framework to that with the isospin projection. In this method, we can treat each isospin eigenstate of the pn pair and control its spin configuration using this method. We investigated the structures of ^{10}B. The reason why the lowest state ($3^+_1 0$) has the finite spin is that the $S = 1$ pn pair rotates around the 2α core in the D-wave so as to gain the spin-orbit

IOP Conf. Series: Journal of Physics: Conf. Series **863** (2017) 012067 doi:10.1088/1742-6596/863/1/012067

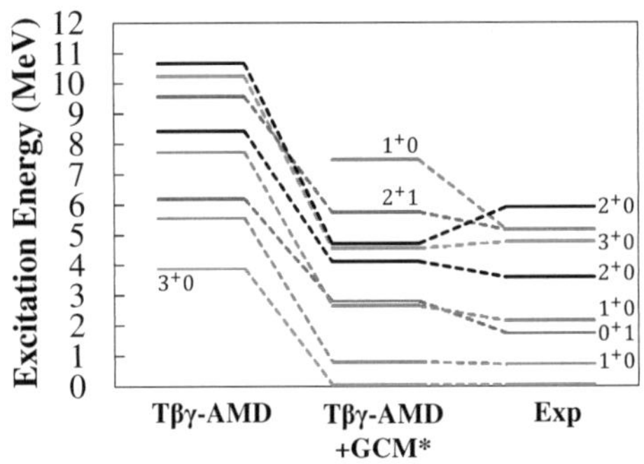

Figure 2. The spectra without GCM (left), those with GCM (middle) and the experimental data (right) of ^{10}B. This figure is taken from Ref. [4] with a slight modification.

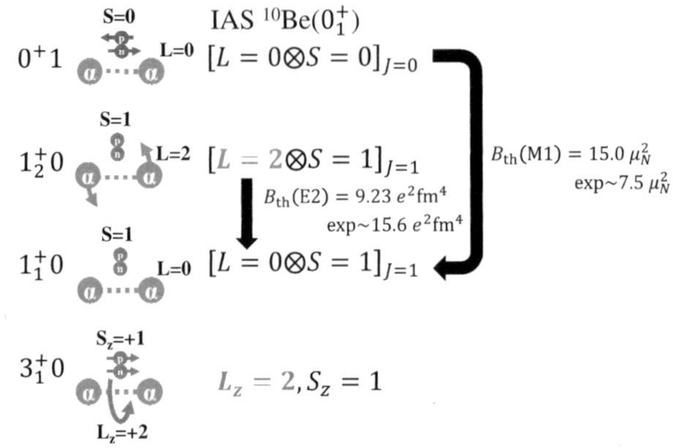

Figure 3. The structures of the low-lying spectra of ^{10}B. $B_{\rm th}$ refers to the reduced transition probability of the present calculation.

potential. The first excited state (1_1^+0) has the $T = 0$ pn pair moving in the S-wave and its spatial structure is similar to the lowest $T = 1$ state (0_1^+1). Accordingly, we obtained the strong $M1$ transition from the 0_1^+1 to the 1_1^+0.

Acknowledgments

The author would like to thank Dr. Kobayashi for fruitful discussions. The computational calculations of this work were performed by using the supercomputer in the Yukawa Institute for theoretical physics, Kyoto University. H. M. was supported by JSPS KAKENHI Grant Number 16J05659. Y. K.-E. was supported by JSPS KAKENHI Grant Number 26400270.

References

[1] Ono A, Horiuchi H, Maruyama T and Onishi A 1992 *Phys. Rev. Lett.* **68** 2898
[2] Suhara T and Kanada-En'yo Y 2010 *Prog. Theor. Phys.* **123** 303
[3] Kanada-En'yo Y, Morita H and Kobayashi F 2015 *Phys. Rev.* C **91** 054323
[4] Morita H and Kanada-Enyo Y, arXiv: 1604.07131

Two-neutron clustering aspects in the transitions induced by the $^{13}C(^{18}O,^{16}O)^{15}C$ reaction at 84 MeV incident energy

D Carbone[1], J L Ferreira[2], F Cappuzzello[1,3], J Lubian[2], C Agodi[1], M Cavallaro[1], A Gargano[4], S M Lenzi[5], R Linares[2] and G Santagati[1]

[1] INFN Laboratori Nazionali del Sud, Catania, Italy
[2] Instituto de Fisica, Universidade Federal Fluminense, Niteroi, RJ, Brazil
[3] Dipartimento di Fisica e Astronomia, Università degli Studi di Catania, Italy
[4] INFN Sezione di Napoli, Napoli, Italy
[5] INFN Sezione di Padova, Padova, Italy

carboned@lns.infn.it

Abstract. The $^{13}C(^{18}O,^{16}O)^{15}C$ reaction is studied at 84 MeV incident energy. The measured cross-section angular distributions are analysed by exact finite-range coupled reaction channel calculations. The two-particle wave functions are extracted using the extreme cluster and the independent coordinate scheme with shell-model derived coupling strengths.

1. Introduction

Direct two-nucleon transfer reactions play an important role in understanding specific properties of the atomic nuclei as for example pairing correlations [1] [2]. Among these, heavy-ion direct transfer reactions at energies close to the Coulomb barrier are useful tools to obtain precise spectroscopic information. Over the last few years, a systematic study on heavy-ion induced one- and two-neutron transfer reactions on different target nuclei was pursued at the INFN-LNS (Italy) by the $(^{18}O,^{17}O)$ and $(^{18}O,^{16}O)$ reactions. Many nuclear systems were explored using the MAGNEX spectrometer [3] to detect the ejectiles. Thanks to its high resolution and large acceptance, high quality inclusive spectra were obtained, even in a largely unexplored region above the two-neutron separation energy in the residual nucleus [4] [5] [6]. New phenomena were unveiled, such as the dominance of the direct one-step transfer of the two neutrons [7] and the presence of broad resonances at high excitation energy in the ^{14}C and ^{15}C spectra. The latter were recently associated to the first experimental signature of the Giant Pairing Vibration [8] [9]. A thorough analysis of the broad structures populated in the $^{13}C(^{18}O,^{16}O)^{15}C$ reaction at high excitation energy, above the two-neutron emission threshold, was presented in ref. [10] and the neutron decay of these structures was investigated in ref. [11]. Moreover, it was demonstrated that the $(^{18}O,^{16}O)$ two-neutron transfer reaction can be used for quantitative spectroscopic studies of pair configurations in nuclear states [12] [13].
From a theoretical point of view, the Coupled Reaction Channel (CRC) approach is necessary to interpret such data. In ref. [12] the experimental absolute cross section of the one- and two- neutron transfer reactions induced by an ^{18}O beam on a ^{12}C target was reproduced for the first time without the need of any scaling factor by means of Exact Finite Range (EFR) CRC calculations. In this manuscript, we analyse the absolute cross section angular distributions for some transitions populated by the $^{13}C(^{18}O,^{16}O)^{15}C$ two-neutron transfer reaction at 84 MeV incident energy in the same framework of our previous work [12].

IOP Conf. Series: Journal of Physics: Conf. Series **863** (2017) 012068 doi:10.1088/1742-6596/863/1/012068

2. The experiment and results

The experiment was performed at the INFN-LNS laboratory in Catania with an $^{18}O^{6+}$ beam at 84 MeV incident energy that impinged on a thin ^{13}C target. Runs with ^{12}C target were also recorded for estimating the background coming from the ^{12}C impurities in the ^{13}C target. The ejectiles were momentum analysed by the MAGNEX spectrometer working in the full acceptance mode [3]. The particle identification and the data reduction technique are the same described in details in refs. [9] [14]. The horizontal and vertical position and angles of the ^{16}O ejectiles, measured by the Focal Plane Detector (FPD) [15] [16], were used as input for a 10th order ray-reconstruction, based on the differential algebraic method implemented in MAGNEX [17] [18].

An example of the obtained energy spectra for the ^{15}C nucleus is shown in Fig. 1 (left panel), in which the ^{14}C background spectrum coming from the ^{12}C impurities in the ^{13}C target is superimposed, after normalization. Several bound and resonant states of the ^{15}C nucleus are recognized, which are almost the same strongly populated in the (t,p) reactions on ^{13}C [19]. In particular, below the one-neutron separation energy (S_n = 1.218 MeV) the only two ^{15}C bound states are identified, i.e. the ground (J^{π} = $1/2^+$) and the $5/2^+$ state at E_x = 0.74 MeV, that are characterized by a dominant single particle configuration [20]. In the region between S_n and the two-neutron separation energy (S_{2n} = 9.39 MeV), narrow resonances at E_x = 3.103 ($1/2^-$), 4.22 ($5/2^-$), 4.66 ($3/2^-$), 6.84 ($9/2^-$, $7/2^-$), 7.35 ($9/2^-$, $7/2^-$) MeV are populated. All of these states are indicated to consist mainly of $2p$-$1h$ configurations (with respect to the $^{14}C_{g.s.}$ vacuum state) [19]. Resonances with a single-particle configuration of a $^{14}C_{g.s.}$ + $1n$ [21] are very weakly populated in the present reaction.

The obtained absolute cross section angular distributions for the ground state and the resonance at 4.22 MeV are shown in Fig. 1 (right panel).

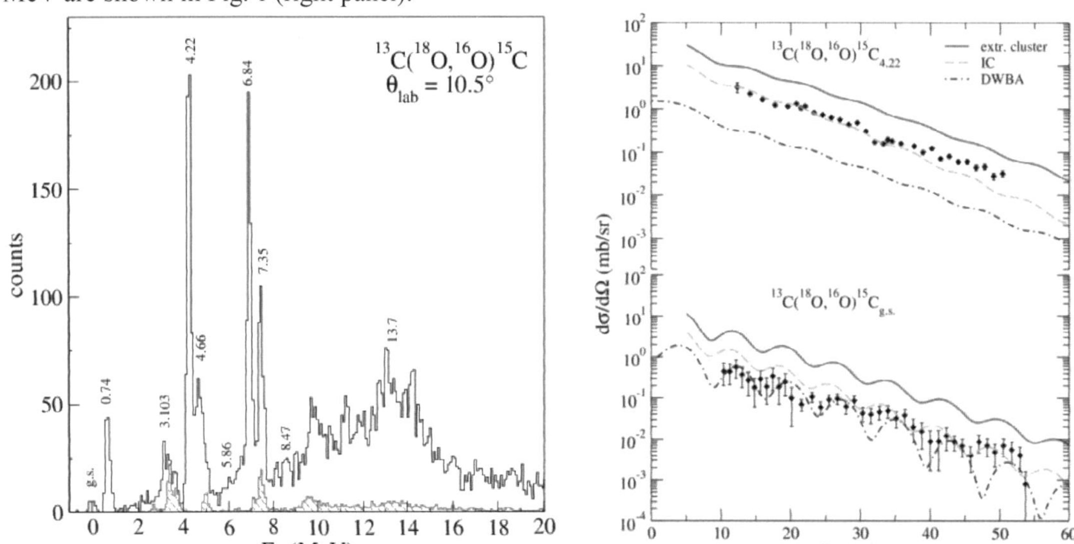

Fig. 1. (left panel) Excitation energy spectrum of the $^{13}C(^{18}O,^{16}O)^{15}C$ reaction for 10° < θ_{lab}< 11°. The red-hatched area corresponds to the background that comes from ^{12}C impurities in the target. (right panel) Experimental cross section angular distributions for the transitions to the g.s. and the state at 4.22 MeV in ^{15}C. Theoretical calculations: extreme cluster calculations (red line), independent coordinate calculations (green line) and two-step sequential DWBA calculations (blue line).

3. Theoretical calculations

Exact finite range coupled reaction channel (CRC) and two-step distorted wave Born approximation (DWBA) calculations for two-neutron transfer reactions were performed to describe the cross section, using the FRESCO code [22], using the same ingredients of ref. [12] [23]. Two different models were adopted for the two-neutron transfer: the first is a direct, simultaneous transfer of the two particles, and the second is a two-step sequential mechanism, which goes through the ^{17}O + ^{14}C intermediate partition. The two-particle wave functions for the one-step two-neutron transfer mechanism were obtained considering two different schemes: i) the extreme cluster model approximation, in which the relative motion between the two transferred neutrons is frozen and separated from the core and the two neutrons are coupled antiparallel to an intrinsic angular momentum $S = 0$, with spectroscopic amplitudes for both target and projectile set to 1.0; ii) the Independent Coordinate (IC) model, in which two-neutron spectroscopic amplitudes determined performing shell-model calculations were used. These calculations assume the ^{12}C as closed core and valence protons and neutrons in the $1p_{1/2}$, $1d_{5/2}$, $2s_{1/2}$ orbits. The effective phenomenological *zbm* interaction [24] was used. Two-step DWBA calculations for the sequential mechanism of two-neutron transfer were also performed in order to check the importance of the two-step mechanism.

The resulting differential cross sections for the direct and the sequential transfer for the ground and the state at 4.22 MeV are shown in Fig. 1 (right panel). The extreme cluster model calculation gives a larger cross section than the data, whereas the IC calculations describe them quite well. The two-step sequential DWBA calculations are much lower than the data for the transition to the 4.22 MeV state, but they are of the same order of the data in the case of the ground state. This difference confirms that the ground state has a dominant single-particle configuration and indeed it is weekly populated in the spectrum (see Fig. 1), whereas the 4.22 MeV state as a dominant $^{13}C_{g.s.} + 2n$ configuration and it is strongly populated in a direct two-neutron transfer mechanism.

References

[1] Guazzoni P *et al* 2011 *Phys. Rev.* C 83 044614
[2] Potel G et al 2011 *Phys. Rev. Lett.* 107 092501
[3] Cappuzzello F, Agodi C, Carbone D and Cavallaro M 2016 *Eur. Phys. J.* A 52 167
[4] Carbone D *et al* 2014 *Phys. Rev.* C 90 064621
[5] Nicolosi D *et al* 2013 *Acta Phys. Pol.* B 44 657
[6] Carbone D *et al* 2014 *Acta Phys. Pol.* B 45 431
[7] D. Carbone *et al* 2011 *J. Phys. Conf. Series* 312 082016
[8] Cappuzzello F *et al* 2015 *Nat. Comm.* 6 6743
[9] Carbone D 2015 *Eur. Phys. J. Plus* 130 143
[10] Cappuzzello F *et al* 2012 *Phys. Lett.* B 711 347
[11] Cavallaro M *et al* 2016 *Phys. Rev.* C 93 064323
[12] Cavallaro M *et al* 2013 *Phys. Rev.* C 88 054601
[13] Ermamatov M J *et al* 2016 *Phys. Rev.* C 94 024610
[14] Cappuzzello F *et al* 2010 *Nucl. Inst. Methods* A 621 419
[15] Cavallaro M *et al* 2012 *Eur. Phys. J.* A 48 59
[16] Carbone D, Cappuzzello F and Cavallaro M 2012 *Eur. Phys. J* A 48 60
[17] Cappuzzello F *et al* 2011 *Nucl. Instr. Methods* A 638 74
[18] Cavallaro M *et al* 2011 *Nucl. Instr. and Meth.* A 637 77
[19] Truong S and Fortune H T 1983 *Phys. Rev.* C 28977
[20] Murillo G, Sen S and Darden S E 1994 *Nucl. Phys.* A 579 125
[21] Goss J D *et al* 1973 *Phys. Rev.* C 8514
[22] Thompson J I http://www.fresco.org.uk/
[23] Pereira D *et al* 2012 *Phys. Lett.* B 710 426
[24] Zuker A P, Buck B and McGrory J B 1968 *Phys. Rev. Lett.* 21 39

Deuteron-Like Correlation for the $T = 0$ Channel in ^{18}F Studied within the Continuum Contributions

H Masui[1] **and M Kimura**[2,3]

[1] Information Processing Center, Kitami Institute of Technology, Kitami 090-8507, Japan
[2] Department of Physics, Faculty of Science, Hokkaido University, Sapporo, 060-0810, Japan
[3] Nuclear Reaction Data Centre, Hokkaido University, Sapporo 060-0810, Japan

E-mail: hgmasui@mail.kitami-it.ac.jp

Abstract. We study the deuteron-like correlation of the valence proton and neutron for the $T = 0$ channel in ^{18}F using the cluster-orbital shell model approach. In this study, the deuteron-like correlation is examined in terms of two key ingredients, cross-shell configuration and continuum contribution. We discuss the cross-shell configuration contributes to the restoration of the symmetry to form the LS-coupling in 1^+, and the continuum contributions plays an important role of the spatial localization of the proton and neutron.

1. Introduction

The proton and neutron correlation in a nucleus has been investigated in the view point of competition of the $T = 1$ and $T = 0$ channels [1, 2, 3]. For the $T = 0\,(S = 1)$ channel, the correlation can be related to the deuteron, since it has the same spin and isospin with the deuteron. In the nucleus, the coupling of the spin-orbit partners $[j_> \otimes j_<]$, which we call the "cross-shell" configuration, is hindered by the the spin-orbit interaction from the core nucleus different from the nucleoens in the deuteron.

We study the effect of the coupling of the spin-orbit partners $[j_> \otimes j_<]^{J=1}_{T=0}$ to the deuteron-like correlation in ^{18}F and also investigate the effect of the continuum contribution to the spatial localization of the proton and neutron. For this purpose, we employ the cluster-orbital shell model (COSM) approach [4] and discuss the deuteron-like correlation in the ^{16}O+p+n three-body model.

2. Model and formalism

The Hamiltonian of the ^{16}O+p+n three-body model in the COSM formalism is formulated as follows:

$$\hat{H} = \sum_{i=1}^{2} \left(\hat{t}_i + \hat{v}_i^{\rm CN}\right) + \left(\hat{T}_{12} + \hat{V}_{12}\right) , \tag{1}$$

where $\hat{T}_{12} = \boldsymbol{p}_1 \cdot \boldsymbol{p}_2 / M_C$ is the recoil term coming from the subtraction of the center of mass motion due to the finite mass M_C of the core nucleus. For $\hat{v}_i^{\rm CN}$, we use the same potential applied in Ref. [5]. For the two-body interaction part $\hat{V}_{12}$, we use an effective nucleon-nucleon interaction, *i.e.* the Minnesota potential [7]. The potential parameters are adjusted to reproduce the $5/2^+$, $1/2^+$ and $3/2^+$ states of ^{17}O and ^{17}F, and 1^+_1 and 0^+_1 state of ^{18}F.

IOP Conf. Series: Journal of Physics: Conf. Series **863** (2017) 012069 doi:10.1088/1742-6596/863/1/012069

The basis set of COSM is constructed in the jj-coupling scheme using basis functions ϕ_α defined in the coordinates of the core+N subsystems as

$$\Phi_m \equiv \mathcal{A}\left\{[\phi_{\alpha_1} \otimes \phi_{\alpha_2}]^{JM}_{TM_T}\right\}_m = \mathcal{A}\{u_1(r_1)\, u_2(r_2) \cdot [j_1 \otimes j_2]_{JM} \cdot [\chi^\tau_1 \otimes \chi^\tau_2]_{TM_T}\}_m \,, \quad (2)$$

where α_i denotes a set of the angular momentum and the isospin of the ith particle. For the radial part of the basis function, we apply the Gaussian expansion method [6]. We take the maximum angular momentum of the basis function as $\ell_{\max} = 6$ in order to include the continuum contribution sufficiently.

3. Deuteron-like correlation in ^{18}F

3.1. Examine the wave function

First, to examine the wave function of ^{18}F for the 1^+ and 0^+ states. We calculate several physical quantities of ^{18}F and obtain the $M1$-transition strength as $B(M1;\, 0^+ \to 1^+) = 18.30\ \mu_N^2$, the magnetic moment as $\mu = 0.811\ \mu_N$ and the r.m.s. radius as $R_{\rm rms} = 2.64$ fm.

For the $B(M1)$-value, other theoretical calculations are obtained as $15.18\ \mu_N^2$ [9] and $18.15\,\mu_N^2$ [3], where the latter is consistent to our result, and experimentally observed value is $19.71\,\mu_N^2$ [8]. Here, the essential difference of the theoretical calculations for Ref. [9] and [3] is the inclusion of the cross-shell configuration to the model space, and the former one does not include the cross-shell. The magnetic moment of other theoretical calculations are also consistent to our result, $0.834\,\mu_N$ [3] and $0.82\,\mu_N$ [10]. Since the magnetic moment is sensitive to the spin-coupling scheme, we decompose it into the orbital angular momentum and spin parts; μ^L and μ^S as $\mu = \langle \sum_i g_l^{(i)} \boldsymbol{l}_i \rangle + \langle \sum_i g_s^{(i)} \boldsymbol{s}_i \rangle \equiv \mu^L + \mu^S$. The angular momentum and spin parts are obtained as $\mu^L = 0.090\,\mu_N$ and $\mu^S = 0.721\,\mu_N$. Therefore, we can confirm that the 1^+ state of ^{18}F has the $S = 1$ dominance in the wave function.

3.2. Configuration dependence of the deuteron-like correlation

Next, we discuss the importance of two ingredients; the cross-term and continuum contributions, to the deuteron-like correlation in ^{18}F. To this end, we classify the set of basis functions into four groups labeled as "CN#". CN1 and CN2 is a set of $(\ell_j)^2$ type basis sets. CN1 consists only the sd-orbits, $(d_{5/2})^2$, $(s_{1/2})^2$, $(d_{3/2})^2$, and CN2 includes higher-orbital $(l_j)^2$ configurations up to $\ell_{\max} = 6$. Since CN1 and CN2 can be applied to the 1^+ and 0^+ states, we can compare the correlation of the valence nucleons within the same model space. CN3 includes the "cross-shell" configurations, $(d_{5/2})(d_{3/2})$ and $(s_{1/2})(d_{3/2})$ in addition to CN2, and CN4 is the full configurations up to $\ell_{\max} = 6$, which corresponds to the inclusion of the continuum contributions.

We calculate the correlation energy and magnetic moment by changing the model space from CN1 to CN4. Here, we define the correlation energy as $E^{n\ell_j}_{\rm Corr}(J^\pi) \equiv (\epsilon^\pi_{0d_{5/2}} + \epsilon^\nu_{0d_{5/2}}) - E(J^\pi)$, where $E(J^\pi)$ is the three-body energy. The obtained wave functions are decomposed into the partial waves. Results for the correlation energy, magnetic moment and the decomposition of the wave function are shown in Figs. 1(a), (b) and (c), respectively.

First, we discuss the importance of the "cross-shell" configurations. We compare the correlation energy of CN1 and CN2 for the 1^+ and 0^+ states. Without the cross-shell configurations, the correlation energy of the 0^+ state is systematically larger than that of 1^+ as shown in Fig. 1(a). Once the cross-shell configuration is added to the model space of the 1^+ state (CN3), the correlation energy is increased by the same amount of the continuum contribution from CN1 to CN2, but the energy is still smaller than that of 0^+. This shows the $T = 1$ channel dominance where the $(\ell_j)^2$-pair is favored in the system.

From Fig. 1(b), the decomposition of the magnetic momentum to the angular momentum and spin part, we find the $(\ell_j)^2$-type model space, CN1 and CN2, give the small magnetic

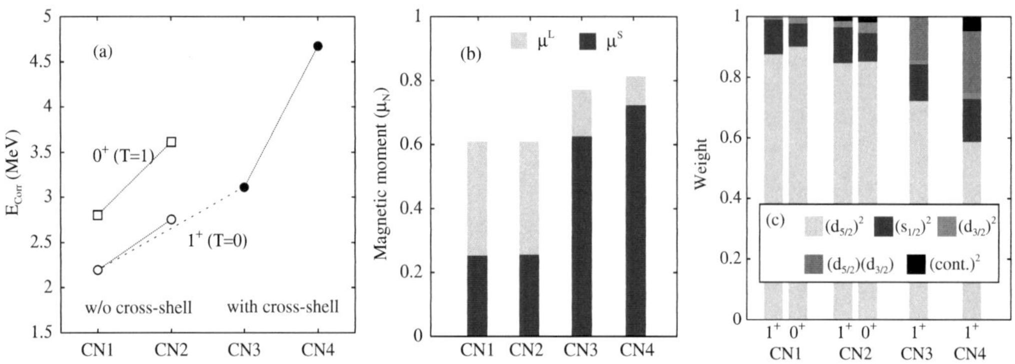

Figure 1. (color online) (a) Correlation energies of the 1^+ and 0^+ states, (b) magnetic moment of the 1^+ state, and (c) decomposition of wave functions of the 1^+ and 0^+ states. The label of the x-axis "CN#" is for the set of configurations. For detail, please see text.

moment and the spin part, Even though the CN2 includes the continuum contribution in the model space. A drastic change of the coupling scheme of the wave function can be achieved by adding the cross-shell configuration to the model space, CN3 and CN4. Since the value of the magnetic momentum for CN4, $\mu = 0.811\mu_N$, is close to the exact value of the sum of the g-factors of the proton and neutron, $0.880\mu_N$, we can consider the component of the LS-coupling with $L = 0$, $S = 1$ channel is dominant in the coupling scheme of the CN4 wave function.

The other important ingredient is the continuum contribution. For the correlation energy, further enhancement can be obtained by adding the continuum to the model space (CN4). The CN4 result of 1^+ gives the correlation energy as $E_{\rm Corr}^{0d_{5/2}}(1^+) = 4.67$ MeV, which is much larger than the 0^+ (CN2) result, 3.61 MeV. The energy changes 1.56 MeV from CN3 to CN4, that is more than 30% of the correlation energy in CN4.

The decomposition of the wave functions are shown in Fig. 1(c). From CN3 to CN4, the cross-shell configuration increases, and on the other hand, the other configurations $(d_{5/2})^2$, $(s_{1/2})^2$, $(d_{3/2})^2$ decrease. This indicates that the cross-shell and continuum configurations are coherently contribute to the wave function of the 1^+ state.

4. Summary

We study the deuteron-like correlation of the valence neutron and proton in ^{18}F using the COSM approach. From the analysis, we showed the essential ingredients of the deuteron-like correlation are the cross-shell configuration, which is the coupling of the spin-orbit partners $j_>$ and $j_<$, and the continuum contributions and its coherence to the correlation energy.

References

[1] Poves A and Martines-Pinedo G 1998 *Phys. Lett.* B **430** 203
[2] Sagawa H, Tanimura Y and Hagino K 2013 *Phys. Rev.* C **87** 034310
[3] Tanimura Y, Sagawa H and Hagino K 2014 *Prog. Theor. Exp. Phys.* **2014** 053D02
[4] Suzuki Y and Ikeda K 1988 *Phys. Rev.* C **38** 410
[5] Masui H, Katō K and Ikeda K 2006 *Phys. Rev.* C **73** 034318
[6] Hiyama E, Kino Y and Kamimura M 2003 *Prog. Part. Nucl. Phys.* **51** 223
[7] Thompson D R, LeMere M and Tang Y C 1977 *Nucl. Phys.* A **286** 53
[8] National Nuclear Data Center, Chart of Nuclides (National Nuclear Data Center, New York, n.d
[9] Lisetskiy A F, Jolos R V, Pietralla N and von Bretano P 1999 *Phys. Rev.* C **60** 064310
[10] Kanada-En'yo Y and Kobayashi F 2014 *Phys. Rev.* C **90** 054332

IOP Conf. Series: Journal of Physics: Conf. Series **863** (2017) 012070 doi:10.1088/1742-6596/863/1/012070

High multiplicity α-particle breakup measurements to study α-condensate states

J Bishop[1], Tz Kokalova[1], M Freer[1], M Assie[3], L Acosta[2,4], S Bailey[1], G Cardella[2], N Curtis[1], E De Filippo[2], D Dell'Aquila[6], S De Luca[5], L Francalanza[6], B Gnoffo[2], G Lanzalone[7,8], I Lombardo[6], N Martorana[7], S Norella[5], A Pagano[2], E V Pagano[7,9], M Papa[2], S Pirrone[2], G Politi[2,9], F Rizzo[7,9], P Russotto[2], L Quattrocchi[5], R Smith[1], I Stefan[3], A Trifirò[5], M Trimarch`ı[5], G Verde[2,3], M Vigilante[6] and C Wheldon[1]

[1] *School of Physics & Astronomy, University of Birmingham, UK*
[2] *INFN, Sezione di Catania, Italy*
[3] *Institut de Physique Nuclaire, CNRS/IN2P3, Université Paris-Sud 11, France*
[4] *Universidad Nacional Autónoma de México, Mexico City, Mexico*
[5] *INFN - Gr. Coll/ Messina and Dipartimento di Fisica - Universitá Messina, Italy*
[6] *INFN Sezione di Napoli & Dipartimento di Fisica, Universitá Federico II, Napoli, Italy*
[7] *INFN LNS, Catania, Italy*
[8] *Universitá Kore, Enna, Italy*
[9] *Dipartimento di Fisica e Astronomia - Universitá di Catania, Italy*

E-mail: `j.bishop.2@pgr.bham.ac.uk`

Abstract.
An experiment was performed to investigate α-condensate states via high α-particle multiplicity breakup. The nucleus of interest was ^{28}Si therefore to measure multiplicity 7 particle breakup events, a highly granular detector with a high solid angle coverage was required. For this purpose, the CHIMERA and FARCOS detectors at INFN LNS were employed. Particle identification was achieved through ΔE-E energy loss. The α-particle multiplicity was measured at three beam energies to investigate different excitation regimes in ^{28}Si. At a beam energy where the energy is sufficient to provide the 7 α-particles with enough energy to be identified using the ΔE-E method, multiplicity 7 events can be seen. Given these high multiplicity events, the particles can be reconstructed to investigate the breakup of α-condensate states. Analysing the decay paths of these states can elucidate whether the state of interest corresponds to a non-cluster, clustered or condensed state.

1. Introduction

In order to investigate to what degree α-particles can form a condensate in light nuclei, an experiment was performed to measure the high α-particle multiplicity break-up of high excitation states in ^{28}Si to measure their decay channels. By measuring the branching ratios corresponding to these different decay paths, information about the wave function of the state can be extracted. The compound nucleus state in ^{28}Si which will decay into 7 α-particles can be populated. The wave functions this level can therefore decay into can be represented as:

$$|\Psi(7\alpha)> = C_1\,|^8\mathrm{Be}\rangle\,|^8\mathrm{Be}\rangle\,|^8\mathrm{Be}\rangle\,|\alpha\rangle \;= C_2\,|^{12}\mathrm{C}\rangle\,|^8\mathrm{Be}\rangle\,|^8\mathrm{Be}\rangle \;= C_3\,|^{16}\mathrm{O}\rangle\,|^{12}\mathrm{C}\rangle = \ldots \qquad (1)$$

The coefficients for each of these decays' wave function can then be calculated from the branching ratios and their values compared [1]. This can be achieved from the proportionality of the decay

IOP Conf. Series: Journal of Physics: Conf. Series **863** (2017) 012070 doi:10.1088/1742-6596/863/1/012070

width to the overlap of the wave functions for the initial $\Psi(7\alpha)$ and final state ψ_i:

$$\Gamma_i(E) \propto |\langle\Psi(7\alpha)|\psi_i\rangle|^2 = |C_i|^2 \prod_{j=1}^{N} P_j(E) \quad (2)$$

with $P_j(E)$ being the penetrability. For an α-condensate state the coefficients C_i should all be equal.

Therefore, to verify the existence of such α-condensate states the various decay channels must be measured; especially the channels comprising of high multiplicity break-ups.

2. Experimental

Measuring a high number of final states products has many difficulties; the primary one being the need for a large angular coverage detector system. Secondly, the beam current must also be sufficiently low such that pileup events are not affecting the detection of these high particle multiplicity events. Finally, the granularity of such a large solid-angular coverage detector must be small enough that while reconstructing events, the resolution is sufficient to distinguish between different excited states.

A good fulfillment of these criteria was available at INFN LNS Catania using the CHIMERA [2] and FARCOS [3] detectors in tandem. The compound nucleus was populated using a ^{16}O beam at 160, 280 and 400 MeV impinging on a carbon target of 58 and 92 $\mu g/cm^2$ with beam currents of $\sim$ 300 pA. These two different detector arrays will briefly be examined.

The CHIMERA detector [2] was originally developed for studying heavy ion collisions, it is however well suited to the current experiment and general nuclear structure studies due to its 1192 telescopes. These telescopes span 94% solid angle coverage with scattering angles from 1° to 176°. These detectors have an angular size ranging from 0.8° for small scattering angles to 8° at large scattering angles. Each telescope comprises of a 290 μm nominal thickness silicon detector followed by a CsI(Tl) crystal which can provide particle identification using ΔE-E. The CsI(Tl) light output is passed through two different readouts which allows for the rise time to be inferred, giving an additional form of particle identification using solely the CsI(Tl) signal. Finally, the timing information of each telescope is also available which can be used for mass identification where the fragment has insufficient energy to pass through the silicon into the CsI(Tl) crystal. For α-particles this corresponds to 24 MeV although the energy required for detection must be higher than this to be above the CsI(Tl) threshold.

The FARCOS detector [3] is an ongoing project to produce a modular detector system with good energy and position information which can be tessellated to provide large solid angle coverage. Each FARCOS module comprises of a 2×2 grid of CsI(Tl) detectors at the rear covered by a DSSD of width 1500 μm. These are then in turn covered by another DSSD of width 300 μm. As with the CHIMERA detector, the primary particle identification is performed by ΔE-E, however, with the FARCOS detector there is a choice between 3 detection stages of performing this identification. The FARCOS detector was employed at low scattering angles to provide improved resolution by virtue of the improved granularity and energy resolution compared to CHIMERA.

3. Analysis

To verify the sufficient solid angular coverage of the detector systems, the multiplicity of α-particles at the three different beam energies can be examined. With the combined detector systems of CHIMERA and FARCOS, and considering any telescopes which were not functionally sufficiently to contribute to the data (i.e. silicon or CsI(Tl) faulty), the overall solid angle coverage was around 80%.

In Figure 1, the α-particle multiplicities can be seen for different beam energies as measured using ΔE-E only. The lowest beam energy of 160 MeV shows a highest multiplicity of 5, this

IOP Conf. Series: Journal of Physics: Conf. Series **863** (2017) 012070 doi:10.1088/1742-6596/863/1/012070

is due to the > 24 MeV needed for an α-particle to punch-through to the CsI(Tl). With only 160 MeV available to share amongst many α-particles, a very equitable distribution of energy is needed for the α-particles to be identified using ΔE-E. It can be seen at 280 MeV, where the available energy is larger, that the required multiplicity of 7 is reached however statistics for this are limited. At 400 MeV, the overall multiplicities are higher than the other beam energies and a reasonable number of multiplicity 7 events are detected. As well as multiplicity 7 events from the ^{16}O+^{12}C reaction with the target, multiplicity 8 events may also be attributed to ^{16}O+^{16}O events with oxygen contaminants in the target. These interactions can be distinguished by looking at the total Q-value of the reaction which can also distinguish from any pile up occurring.

Once α-particles are included which are solely measured in the silicon detector and identified using time-of-flight, the overall multiplicities will increase due to the effective lowering of the α-particle energy threshold. This will particularly affect the 160 MeV beam data where the average α-particle energy is lower and allow for a more unbiased comparison of high multiplicity decays.

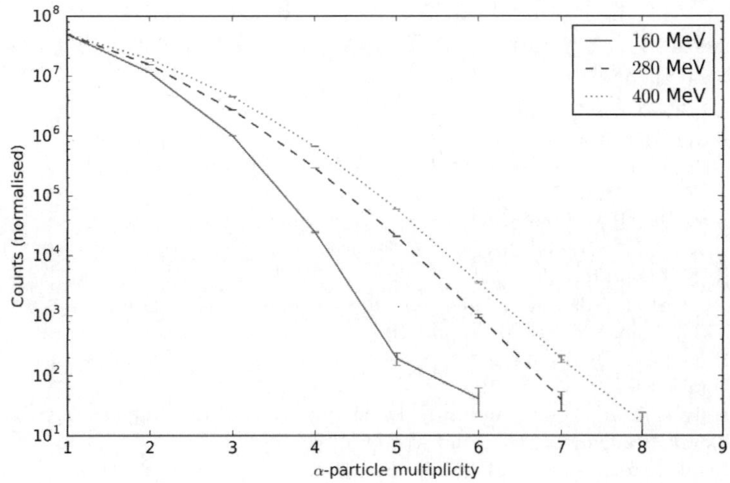

Figure 1. α-particle multiplicities for three different beam energies normalised to multiplicity=1 counts

3.1. Reconstructed spectra

Once an event of high α-particle multiplicity has been identified, all the possible permutations of α-particles can be reconstructed to form an α-α correlation function. In such a statistical decay, a large number of break-up products are expected to proceed via the ^{8}Be ground state. By forming the α-α correlation function, a large contribution at 91.84 keV corresponding to the break-up energy of ^{8}Be can be seen. By further reconstructing the ^{8}Be and remaining α-particles, the breakup of ^{12}C(0_2^+) and ^{16}O* (including the 0_6^+ which is predicted to be a condensate state [4, 5]) can therefore be calculated. It is through these repeated reconstructions that the branching ratios for the decay of the high excitation in ^{28}Si can be determined and therefore the coefficients of the wave function calculated which will provide a distinction between a condensed state, a clustered state and a non-clustered state.

References

[1] Kokalova Tz, Itagaki N, von Oertzen W and Wheldon C 2006 *Phys. Rev. Lett.* **96**, 192502
[2] Pagano A *et al* 2004 *Nucl. Phys.* A **734** 504-11
[3] FARCOS TDR `https://drive.google.com/file/d/0B5CgGWz8Lp00c3pGTWd0cDBoWFE/view`
[4] Tohsaki A, Horiuchi H, Schuck P and Röpke G 2001 *Phys. Rev. Lett.* **87**, 192501
[5] Funaki Y, Yamada T, Tohsaki A, Horiuchi H, Röpke G and Schuck P 2010 *Phys. Rev.* C **82**, 024312

IOP Conf. Series: Journal of Physics: Conf. Series **863** (2017) 012071 doi:10.1088/1742-6596/863/1/012071

Study of resonances produced in light nuclei through two and multi particle correlations

L Quattrocchi[1,2], L Acosta[2,3], F Amorini[4], A Anzalone[4], L Auditore[5], I Berceanu [14], G Cardella[1], A Chbihi[6], E De Filippo[1], S De Luca[5], D Dell' Aquila[9,12], L Francalanza[9], B Gnoffo [1,16], A Grzeszczuk [15], G Lanzalone[4,11], I Lombardo[9,12], I Martel[8], N S Martorana[2,7], T Minniti[17], S Norella[5], A Pagano[1], E V Pagano[2,4], M Papa[1], S Pirrone[1], G Politi[1,2], F Porto[2,7], F Rizzo[2,7], E Rosato [9,12†], P Russotto[1], A Trifirò[5], M Trimarchi[5], G Verde[1,10], M Veselsky[13] and M Vigilante[9,12]

[1] INFN, Sezione di Catania, Catania, Italy
[2] Università di Catania, Dip. di Fisica e Astronomia, Catania, Italy
[3] Instituto de Fisica, Universidad Nacional Autonoma de Mexico, Mexico
[4] INFN, Laboratori Nazionali del Sud, Catania, Italy
[5] Università di Messina e INFN Sezione di Catania, Messina, Italy
[6] GANIL,CEA-IN2P3-CNRS, Caen, France
[7] INFN, Laboratori Nazionali del Sud; Catania, Italy
[8] Departamento de Física Aplicada, Universidad de Huelva, Huelva, Spain
[9] INFN, Sezione di Napoli, Napoli, Italy
[10] IPN Orsay, Orsay, France
[11] Università Kore, Enna, Italy
[12] Università di Napoli Federico II, Dipartimento di Fisica, Napoli, Italy
[13] Slovak Academy of Sciences, Bratislava, Slovakia
[14] Institute for Physics and Nuclear Engineering, Bucharest, Romania
[15] Institute of Physics, University of Silesia, Katowice, Poland
[16] Centro Siciliano di Fisica Nucleare e Struttura della Materia, Catania, Italy
[17] CNR-IPCF, Messina, Italy
† deceased

E-mail: lquattrocchi@unime.it

Abstract. CORRELATION experiment has been performed at INFN-LNS of Catania, using the 4π multi-detector CHIMERA, with the aim of exploring correlations between two and multi light particle produced in $^{12}C+^{24}Mg$ collisions at 35 AMeV. Particular attention has been paid to the decay mechanisms of Hoyle state, an excited resonant state of ^{12}C produced via the triple-α process and characterized by a pronounced molecular like structure with three α particles. The study of the Hoyle state is essential for nucleosynthesis, but it also represents a clearly isolated state that can be studied as a three-α cluster system.

1. Introduction

Heavy Ion collisions at intermediate energies represent an important tool to produce and to study systems in extreme conditions of temperature, density and excitation energies, thus they are the only terrestrial means to access the equation of state of asymmetric nuclear matter. The correlations between particles emitted during these reactions play a crucial role to understand involved dynamics and to investigate some of the spectroscopic properties of several species produced during the evolution of systems. Multi particle correlations have been used to explore the competition between sequential and direct decay mechanisms of resonances produces in these reactions; in particular here three α correlation has been analyzed to study the decay processes of Hoyle state, the second 0^+ excited state of ^{12}C at an energy of 7.65 MeV.

2. Experimental details and results

With the aim of exploring two- and multi-particle correlations in Heavy Ion Collision, an experiment has been performed at LNS-INFN of Catania; ^{12}C+^{24}Mg reaction at 35 AMeV has been investigated using the forward part of CHIMERA 4π array [1]. In order to isolate α particles resulting from decay of excited ^{12}C quasi-projectiles, a criterion has been imposed to restrict velocities of fragments.

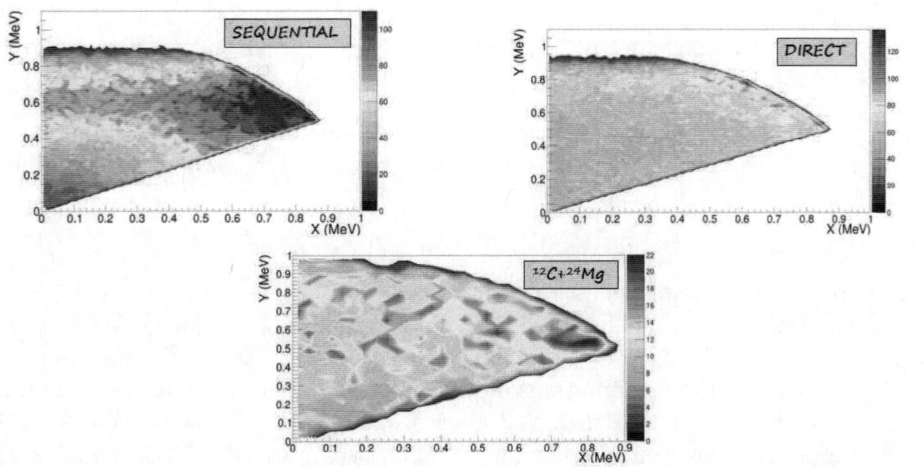

Figure 1. Symmetric Dalitz plots of simulated events for sequential (top left panel) and direct phase-space (top right panel) decay of Hoyle state. Bottom panel: symmetric Dalitz plot, corresponding to decay from Hoyle state region, constructed with experimental data collected in ^{12}C+ ^{24}Mg reaction.

To study the decay mechanisms of Hoyle state, Monte Carlo simulation has been performed for each decay scenario: sequential decay (SD ^{12}C$\rightarrow$^{8}Be-$\alpha$$\rightarrow$3$\alpha$) and direct mechanism (^{12}C$\rightarrow$3α) characterizing a uniform sampling of phase space by the decay products (DDΦ). In order to investigate the various decay modes, energy distributions of decay products have been analyzed adopting the symmetric Dalitz plots [2] constructed using following coordinates:

$$X = \sqrt{3}(\epsilon_j - \epsilon_k), \qquad Y = 2\epsilon_i - \epsilon_j - \epsilon_k, \tag{1}$$

where $\epsilon_{i,j,k} = E_{i,j,k}/(E_i + E_j + E_k)$ are the energies of particles in the center of mass reference frame, normalized to total decay energy. Fig. 1 shows symmetric Dalitz plots for simulated events corresponding to sequential decay (top left panel) and direct phase-space one, DDΦ (top right panel), of the Hoyle state. The same plot constructed using experimental data collected in ^{12}C+ ^{24}Mg reaction is reported in lower panel of Fig. 1; it exhibits a more uniform distribution that does not allow us to exclude any of the two decay mechanisms. For better evaluations, monodimensional ϵ_i distributions [3] have been used. ϵ_i is the highest normalized energy among those of the 3 α particles emitted from the region around the Hoyle state. In the case of the SD mechanism the distribution should have a peak around 0.5. In direct processes instead it should range between 1/3 and 2/3, corresponding to the case of three α particles with an equal energy, and the one of an α particle emitted in a direction opposite to that of the other two, respectively.

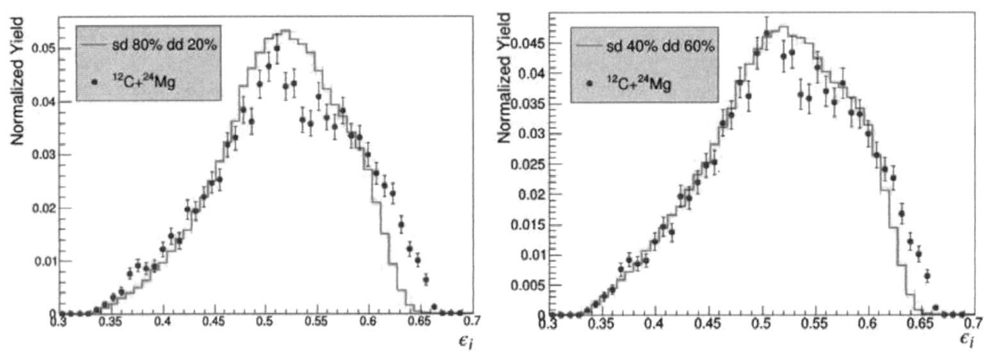

Figure 2. Fits of ϵ_i distribution constructed with experimental data. The different fits have been obtained using two combinations of percentages related to simulated sequential and direct decay mechanisms.

In order to obtain a quantitative estimation for the probability of each decay mechanism, the experimental ϵ_i distribution has been fitted with simulated events in which different percentages of direct and sequential decays have been introduced. Fig. 2 shows the fits of experimental data, relative to two different combinations of percentages related to sequential and direct decay mechanisms. The results reported in Fig. 2 allow us to confirm the presence of an important direct decay component that seems lower than 60% but also larger than 20%. These comparisons can not be used to extract the branching ratio for direct and sequential 3 α decay of Hoyle state, because of the limited energy and angular resolution. However they evidence a significant contribution of direct decay modes for the Hoyle state.

3. Conclusions

Multi-α correlations have been analyzed in peripheral ^{12}C+^{24}Mg reaction at 35 AMeV. The used strategy, based on Dalitz plot techniques, allows us to evaluate the relative contribution of two 3 α decay mechanisms of Hoyle state. The comparison with simulations evidences the contribution of a direct component in Hoyle state decay, already suggested in experiment performed with Heavy Ion collisions [4, 5]. This result could be related to in-medium effects on nuclear structure properties.

References

[1] Pagano A *et al* 2004 *Nucl. Phys.* A **734** 504
[2] Kirsebom O S *et al* 2012 *Phys. Rev. Lett.* **108** 202501
[3] Itoh M *et al* 2014 *Phys. Rev. Lett.* **113** 102501
[4] Raduta Ad *et al* 2011 *Phys. Lett.* B **705** 65
[5] Grenier F *et al* 2008 *Nucl. Phys.* A **811** 233

Clusterization of light nuclei and the Trojan Horse Method

A Tumino[1,2], C Spitaleri[2,3], S Cherubini[2,3], G D'Agata[2,3], L Guardo[2,3], M Gulino[2,3], I Indelicato[2,3], M La Cognata[2], L Lamia[3], R G Pizzone[2], S M R Puglia[2,3], G G Rapisarda[2,3], S Romano[2,3], M L Sergi[2,3] and R Spartá[2,3]

[1]Facoltá di Ingegneria e Architettura, Università degli Studi di Enna "Kore", Enna, Italy
[2]INFN-Laboratori Nazionali del Sud, Catania, Italy
[3]Dipartimento di Fisica e Astronomia, Università degli Studi di Catania, Catania, Italy

E-mail: tumino@lns.infn.it

Abstract. The THM is a unique indirect technique to measure astrophysical rearrangement reactions down to astrophysical relevant energies, where direct experiments are unfeasible. This is done by measuring a suitable three body process in quasi free kinematics. The basic principles are recalled and some applications presented.

1. Introduction

A critical issue in nuclear astrophysics is the Coulomb barrier that makes the study of all nucleosynthesis processes at the relevant energies extremely challenging. The Coulomb repulsion between like charges is indeed responsible for the exponential decrease of the cross section $\sigma(E)$ at those energies. Thus, its behaviour at low energies is quite often extrapolated from higher energies (usually E>100 keV) by means of the astrophysical S(E)-factor

$$S(E) = E\sigma(E)\exp(2\pi\eta), \tag{1}$$

with η the Coulomb parameter of the colliding nuclei, and $\exp(2\pi\eta)$ the inverse of the Gamow factor that removes the Coulomb dependence of $\sigma(E)$. However, extrapolation can be source of additional uncertainties for $\sigma(E)$ due, for instance, to the presence of unexpected resonances.

Another critical issue in the laboratory measurement of nucleosynthesis processes is represented by the electron screening effect that leads to an increased cross section for screened nuclei, $\sigma_s(E)$, compared to the cross section for bare nuclei $\sigma_b(E)$ [1, 2]. Therefore, the so called screening factor, defined as

$$f_{\text{lab}}(E) = \sigma_s(E)/\sigma_b(E) \approx \exp(\pi\eta U_e/E)\,, \tag{2}$$

where U_e is the so-called "electron screening potential"' [1, 2], has to be taken into account to determine the bare nucleus cross section. Indeed, this is the key parameter to determine a reaction rate and the only way to get it measured is via indirect methods ([3, 4] and references therein). They make use of direct reaction mechanisms, such as transfer processes (stripping and pick-up) and quasi-free reactions (knock-out reactions).

In particular, the Trojan Horse Method (THM) ([5, 6, 3] and references therein) has been successfully applied many times in the last two decades to reactions connected with fundamental astrophysical problems. Here we present some of the basic ideas of the THM.

2. Characteristics of the Trojan Horse Method

The THM applies to an appropriate three-body reaction $A + a \to c + C + s$ performed in quasi free (QF) kinematics at energies well above the Coulomb barrier to extract the cross section of a charged particle two-body process $A + x \to c + C$ in the Gamow energy window. This is done making use of the theory for direct reactions, assuming that the nucleus a is described in terms of the $x \oplus s$ cluster structure, e.g. its wave function is required to have a large amplitude for a $x \oplus s$ cluster configuration. In most of the applications performed so far, where $a =$ deuteron, $x =$ proton, $s =$ neutron [7, 8, 9, 10, 11], this is an obvious assumption. The high energy in the $A+a$ entrance channel ensures that the two body interaction takes place inside the nuclear field, without experiencing either Coulomb suppression or electron screening effects. The $A+a$ relative motion is compensated for by the $x - s$ binding energy, determining the so called "quasi-free two-body energy" given by

$$E_{QF} = \frac{m_x}{m_x + m_A} E_A - B_{x-s}. \tag{3}$$

where E_A represents the beam energy, m_x and m_A are the masses of x and A particles respectively, and B_{x-s} is the binding energy for the $x - s$ system. This can be retained as a prescription to determine the best value of the beam energy to populate the relevant E_{QF}. A cutoff in the momentum distribution reflecting the Fermi motion of s inside the Trojan-horse a, defines the range of energies around E_{QF} accessible in the astrophysical relevant reaction. In the Plane Wave Approximation, the three body-cross cross section can be factorized as:

$$\frac{d^3\sigma}{dE_c d\Omega_c d\Omega_C} \propto [KF\, |\varphi_a(\mathbf{p}_{sx})|^2] \left(\frac{d\sigma}{d\Omega_{c.m.}}\right)^{\mathrm{HOES}} \tag{4}$$

where KF is a kinematical factor containing the final state phase-space factor. It is a function of the masses, momenta and angles of the outgoing particles [7]; $\varphi_a(\mathbf{p}_{sx})$ is proportional to the Fourier transform of the radial wave function $\chi(\mathbf{r})$ for the $x - s$ inter-cluster relative motion; $(d\sigma/d\Omega_{c.m.})^{\mathrm{HOES}}$ is the half-off-energy-shell (HOES) differential cross section for the binary reaction at the center of mass energy $E_{c.m.}$ given in post-collision prescription by

$$E_{c.m.} = E_{cC} - Q_{2b}\,. \tag{5}$$

Here, Q_{2b} is the Q-value of the binary reaction and E_{cC} is the relative energy of the outgoing particles c and C. Using the PWA does not change the energy dependence of the two-body cross section but only its absolute magnitude.

In a typical THM experiment the decay products (c and C) of the virtual two-body reaction of interest are detected and identified by means of telescopes (silicon detector or ionization chamber as ΔE step and position sensitive detector as E step) placed at the so called quasi free angles. After the selection of the reaction channel, the most critic point is to disentangle the quasi free mechanism from other reaction mechanisms feeding the same particles in the final state, e.g. sequential decay and direct break-up. An observable which turns out to be very sensitive to the reaction mechanism is the shape of the experimental momentum distribution of the spectator. In order to reconstruct the experimental p_s distribution, the energy sharing method [12] is applied for each pair of coincidence QF angles, selecting c-C relative energy windows of 50 to 100 keV. The extracted experimental momentum distribution is then compared with the theoretical one and further data analysis is limited to the data lying in the region where the agreement between the two distributions exists (usually within few tens of MeV/c).

Therefore, it is possible to derive the HOES $((d\sigma/d\Omega_{cm})^{\text{HOES}}$ from the three-body coincidence yield by simply inverting eq.5. In a final step, the HOES cross section has to be related to the relevant on-energy-shell (OES) cross section by applying the corresponding corrections. In a heuristic approach this consists essentially in replacing the Coulomb suppression in the HOES cross section, by means of the penetrability factor:

$$P_l(k_{Ax}R) = \frac{1}{G_l^2(k_{Ax}R) + F_l^2(k_{Ax}R)} \qquad (6)$$

with F_l and G_l regular and irregular Coulomb wave functions. It was demonstrated that there is no Coulomb barrier in the two-body amplitude extracted from the TH reaction [13, 10, 14] and this is due to the virtuality of particle x. This seems to be the only consequence of off-energy-shell effects as suggested by the agreement between HOES and OES cross-sections for the ^{6}Li(n,α)^{3}H reaction [15].
This procedure does not allow us to extract the absolute value of the two-body cross section. However this is not a real problem since the absolute magnitude can be derived from a scaling to the direct data available at higher energies.
For resonant the two-body reactions, $(d\sigma/d\Omega_{c.m.})^{\text{HOES}}$ has to be worked out to determine the corresponding on-energy-shell (OES) S(E) factor. The corresponding theoretical formalism has been recently developed in a very rigorous way [16] leading to the so called modified R-matrix approach that accounts for HOES effects due to the virtual nature of particle x. By fitting the experimental THM cross section, the reduced width amplitude γ for entrance and exit channels, energy levels and energy shifts can be deduced and used to determine the astrophysical S(E) factor, since these parameters are the same in both direct and THM data. In this way, an exact parameterization of the astrophysical S(E) factor can be obtained overcoming the extrapolation. If the resonance parameters of a single level in the relevant energy region are known, they can be fixed in the fitting procedure to obtain directly the astrophysical S(E) factor in absolute units. Several test studies have been performed in the past years to validate the THM [17], such as the invariance of the two-body reaction amplitude with changing the Trojan Horse nucleus hiding the participant cluster x [18, 19, 20], or the use of momentum distributions from Distorted Wave Born Approximation instead of the simple PWA shape, providing same results within experimental errors [21, 22]. The THM has been applied to many reactions of astrophysical interest connected to fundamental problems in different scenarios, from BBN nucleosynthesis [8, 23, 24, 25, 26, 27, 28, 29, 30, 31, 32, 33] to AGB and more explosive sites [34, 21, 35, 22, 36]. In the last years, reactions involving heavier systems, such as ^{12}C and ^{16}O have been investigated [37, 38].

3. The ^{2}H(d,p)^{3}H and ^{2}H(d,n)^{3}He reactions via the Trojan Horse Method

The Trojan Horse Method was applied for the first time to the ^{2}H(d,p)^{3}H and ^{2}H(d,n)^{3}He reactions by measuring the ^{2}H(^{3}He,p^{3}H)^{1}H and ^{2}H(^{3}He,n^{3}He)^{1}H processes in quasi free kinematics [28, 29, 31]. The ^{3}He+d experiment was performed at 17 and 18 MeV, corresponding to a d-d energy range from 1.5 MeV down to 2 keV. This range overlaps with the relevant region for Standard Big Bang Nucleosynthesis as well as with the thermal energies of future fusion reactors and deuterium burning in the Pre Main Sequence phase of stellar evolution. This is the first pioneering experiment in quasi free regime where the charged spectator is detected. Both the energy dependence and the absolute value of the bare nucleus S(E) factors have been extracted for the first time. They are shown in Figs. 1 and 2 for the for the ^{2}H(d,p)^{3}H and ^{2}H(d,n)^{3}He reactions respectively. THM data are reported as black symbols (full squares and full circles for the 17 and 18 MeV run respectively), with uncertainties accounting for statistical and normalization errors. TH data for the p+^{3}H run at 17 MeV experience a larger scatter with respect to those at 18 MeV, due to the lower statistics in that experiment. Nonetheless, a fair agreement between them is apparent. Direct data from the different sources are shown as colored symbols, as reported in the legend. Their overall trend experience a large scatter, with deviations of more than 15%, in particular for the ^{3}He+n channel, where a new direct study of the cross section over a larger energy region would be desirable. In contrast, the THM S(E) factors exhibit a much smoother behavior, with deviations from the direct data at ultra-low energies, because of the screening, and in the SBBN region.

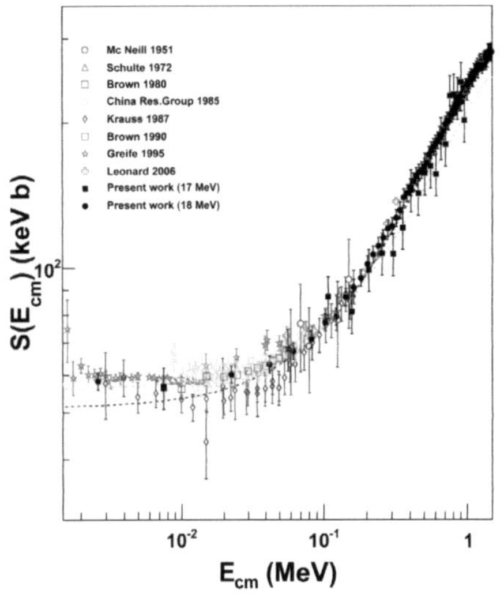

Figure 1. S(E) factor for the ^{2}H(d,p)^{3}H reaction: black solid squares and circles are THM data (17 and 18 MeV respectively); colored symbols represent direct data as described in the legend.

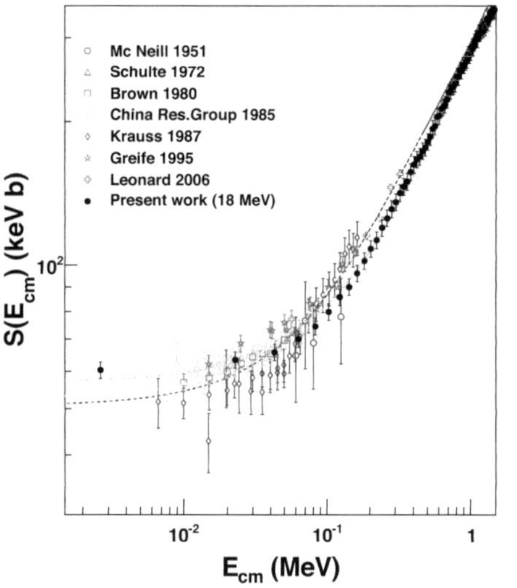

Figure 2. S(E) factor for the ^{2}H(d,n)^{3}He reaction: black solid circles are THM data (18 MeV run only for this channel); colored symbols represent direct data as described in the legend.

The THM S(0) values are of 57.7$\pm$1.8 MeVb for ^{3}H+p and 60.1$\pm$1.9 MeVb for ^{3}He+n. None of the existing fitting curves is able to provide the correct slope of the new data in the full range, thus calling for a revision of the theoretical description. This has consequences in the calculation of the reaction rates with a maximum increase of 20% at the temperatures of the Pre Main Sequence phase and of future fusion reactors.

References

[1] Assenbaum H J *et al* 1987 *Z. Phys.* A **327** 461
[2] Strieder F *et al* 2001 *Naturwissenschaften* **88** 461
[3] Tumino A *et al* 2013 *Few Body Syst.* **54** 869
[4] Tribble R et al 2014 *Rep. Prog. Phys.* **77** 106901
[5] Spitaleri C *et al* 2011 *Phys. At. Nucl.* **74** 1763
[6] Tumino A *et al* 2013 *Few Body Syst.* **54** 745
[7] Spitaleri C *et al* 2004 *Phys. Rev.* C **63** 055806
[8] Tumino A *et al* 2003 *Phys. Rev.* C **67** 065803
[9] La Cognata M *et al* 2006 *Eur. Phys. J.* A **27** 249
[10] Tumino A *et al* 2007 *Phys. Rev. Lett.* **98** 252502
[11] La Cognata M *et al* 2007 *Phys. Rev.* C **76** 065804
[12] Kasagi J *et al* 1975 *Nucl. Phys.* A **239** 233

[13] Mukhamedzhanov A M *et al* 2006 *Eur. Phys. J.* A **27** 205
[14] Tumino A *et al* 2008 *Phys. Rev.* C **78** 064001
[15] Tumino A *et al* 2005 *Eur. Phys. J.* A **25** 649
[16] Mukhamedzhanov A M 2011 *Phys. Rev.* C **84** 044616
[17] Spitaleri C *et al* 2015 *Phys. Rev.* C **91** 024612
[18] Tumino A *et al* 2006 *Eur. Phys. J.* A direct, 1 DOI: 10.1140/epja/i2006-08-038-1
[19] Pizzone R G *et al* 2011 *Phys. Rev.* C **83** 045801
[20] Pizzone R G *et al* 2013 *Phys. Rev.* C **87** 025805
[21] Sergi M L *et al* 2010 *Phys.Rev.* C **82** 032801(R)
[22] La Cognata M *ct al* 2011 *Astrophysical J. Lett.* **739** L54
[23] Pizzone R G *et al* 2003 *Astronomy & Astrophysics* **398** 423
[24] Tumino A *et al* 2004 *Prog. Theor. Phys. Suppl.* **154** 341
[25] Rinollo A *et al* 2005 *Nucl. Phys.* A **758** 146
[26] Qun-Gang Wen *et al* 2008 *Phys. Rev.* C **78** 035805
[27] Qun-Gang Wen *et al* 2011 *J. Phys.* G **38** 085103
[28] Tumino A *et al* 2011 *Phys. Lett.* B **700** 111
[29] Tumino A *et al* 2011 *Phys. Lett.* B **705** 546
[30] Lamia L *et al* 2013 *Astrophysical J.* **768** 65
[31] Tumino A *et al* 2014 *Astrophysical J.* **785** 96
[32] Spitaleri C *et al* 2014 *Phys. Rev.* C **90** 035801
[33] Li Chengbo *et al* 2015 *Phys. Rev.* C **92** 025805
[34] La Cognata M *et al* 2008 *Phys. Rev. Lett.* **101** 152501
[35] La Cognata M *et al* 2010 *Astrophysical J.* **708** 796
[36] La Cognata M *et al* 2013 *Phys. Rev. Lett.* **49** 015106
[37] Tumino A *et al* 2001 *Eur. Phys. J.* A **12** 327
[38] Tumino A *et al* 2016 *Proc. 12th Int. Conf. on Nucleus-Nucleus Collisions 2015* Vol.109 SIF

Measurement of the cross section for the ^{4}He$(\alpha, n)^7$Be reaction as a possible solution to the cosmological lithium problem

T Kawabata[1], T Furuno[1], M Ichikawa[1], N Iwasa[2], Y Kanada-En'yo[1], A Koshikawa[1], S Kubono[3], E Miyawaki[1], T Morimoto[1], M Murata[1], T Nanamura[1], S Nishimura[3], Y Shikata[1], Y Takahashi[1], T Takeda[1], M Tsumura[1], K Watanabe[1]

[1] Department of Physics, Kyoto University, Sakyo, Kyoto 606-8502, Japan
[2] Department of Physics, Tohoku University, Aoba, Sendai, Miyagi 980-8578, Japan
[3] RIKEN Nishina Center, RIKEN, 2-1 Hirosawa, Wako, Saitama 351-0198, Japan

E-mail: `kawabata@scphys.kyoto-u.ac.jp`

Abstract. The cross section for the ^{4}He$(\alpha,n)^7$Be reaction was measured at low energies between E_α = 38.50 and 39.64 MeV motivated by the cosmological lithium problem. On the basis of the detailed balance principle, the cross section for the ^{7}Be$(n,\alpha)^4$He reaction was obtained at $E_{c.m.}$ = 0.20–0.81 MeV close to the Big Bang nucleosynthesis (BBN) energy window for the first time. The obtained cross sections are significantly smaller than the theoretical estimation widely used in the BBN calculations. The present results suggest the ^{7}Be$(n,\alpha)^4$He reaction rate is not large enough to solve the cosmological lithium problem.

1. Introduction

The primordial abundances of the light elements produced in the Big Bang nucleosynthesis (BBN) provide important insights into the early universe. Accurate estimation of the primordial abundances is crucial to test the cosmological theories by comparing the predicted values with the observations. The theoretical predictions of the primordial abundances reasonably well agree with the observations for the helium and deuterium. However, there remains a serious problem that the ^{7}Li abundance does not agree with any theoretical BBN calculations. The primordial abundance of ^{7}Li inferred from observations of metal-poor stars is three times smaller than the calculated values by the BBN theory. This discrepancy is known as the cosmological lithium problem [1], and has been of great interest in recent years.

From a view of nuclear physics, nuclear-reaction rates involved in the BBN theory should be examined. The main process of the ^{7}Li production in the BBN is the electron-capture decay of ^{7}Be, which is synthesized in the ^{3}He$(^4$He$,\gamma)^7$Be reaction. Direct measurements of the cross section for the ^{3}He$(^4$He$,\gamma)^7$Be reaction were extensively carried out in the past by several groups, and uncertainties in this thermonuclear reaction rate are now very small. There is no room to modify the ^{7}Be production rate to solve the lithium problem [2].

Recently, it was pointed out that the ^{7}Li abundance will be greatly reduced in the BBN calculation if the destruction rate of ^{7}Be is enhanced. One of the candidate channels to destruct

^{7}Be is the ^{7}Be(n,α)^{4}He reaction. However, the cross section for the ^{7}Be(n,α)^{4}He reaction at the cosmological energy has been scarcely measured. Since both ^{7}Be and neutron are short-lived nuclei, it is difficult to directly measure the cross section for the ^{7}Be(n,α)^{4}He reaction. Thus, the cross section for the time reverse reaction should be measured. In the present work, we have measured the cross section for the ^{4}He(α,n)^{7}Be reaction at low reaction energies. On the basis of the detailed balance principle, we obtained the cross sections for the ^{7}Be(n,α)^{4}He reaction at $E_{c.m.}$ = 0.20–0.81 MeV close to the BBN energy window for the first time.

2. Experiment and Results

The experiment was carried out at the N0 course in Research Center for Nuclear Physics (RCNP), Osaka University [3]. An α beam accelerated by the AVF cyclotron was transported to the He gas target in the beam swinger magnet. Taking the energy loss in the target into account, the beam energies at the center of the He gas target were determined to be 39.64, 39.30, 38.90, and 38.50 MeV as the nominal beam energies in the present work. The uncertainties of the beam energies were estimated at 40 keV.

The scattered neutrons were detected by a BC-501A liquid scintillation detector located at 13-m away from the target. The sensitive volume of the scintillation detector was a cylindrical shape with the diameter of 200 mm and the depth of 50 mm along the neutron trajectory. A conventional pulse-shape discrimination technique was used to distinguish neutrons from γ rays. The detection efficiency of neutrons by the BC-501 liquid scintillation detector was estimated by using the computer code SCINFUL-CG [4]. We also measured the neutron detection efficiency using the tagged neutrons emitted from the $d + d \to {}^3$He + n reaction. The calculated efficiency agrees with the measurement within the measurement uncertainties.

A He gas target was used in the present work. The He gas was filled at 1 atm in the target cell with the effective length of 6.3 cm. The target cell has the entrance and exit windows with the diameter of 12 mm, and those windows are sealed with the 6-μm aramid films. The mass thicknesses of the He gas and aramid films were 1.0 and 1.7 mg/cm^2, respectively, and the energy loss of the α beam in the He gas target was about 0.5 MeV. The temperature and pressure of the He gas were monitored during the measurement.

The angular distributions of the differential cross sections for the ^{4}He(α,n)^{7}Be reaction are shown in Fig. 1. The solid circles and squares show the differential cross sections for the ground and first excited states in ^{7}Be, respectively. The thick vertical bars present the statistical uncertainties, while the thin vertical bars present the total uncertainties including both the statistical and systematic uncertainties. Unfortunately, the ^{4}He(α,n)^{7}Be events could not be reliably separated from the background events at E_α = 38.50 MeV and $\theta_{c.m.}$ = 77.3°, because the kinematical energy spread of the scattered neutrons is large. Therefore, the upper limit for the differential cross section was given at the 95% confidence level.

Since the angular distribution of the cross section for the two-body scattering of the identical particles must be symmetric with respect to $\theta_{c.m.}$ = 90° in the center of mass frame, the measured cross section can be fitted by a series of the even-odder Legendre polynomials:

$$\sigma(\theta) = \frac{\sigma}{4\pi} \sum_{l=0\,\mathrm{even}}^{l_{max}} \alpha_l P_l(\cos\theta).$$

Because the present measurement was carried out at beam energies close to the $n + {}^7$Be threshold energy, the Legendre polynomial expansion should involve a few of low-oder terms only. We decided l_{max} = 4 for the ground state at $E_\alpha \geq$ 38.90 MeV and 2 for the other states. The fit functions are shown by the solid and dashed lines in Fig. 1. The total cross sections were obtained by integrating the fit functions over the whole solid angle. We also evaluated the total cross sections by using the different l_{max} values in order to estimate the uncertainties stem from the assumption about the l_{max} values.

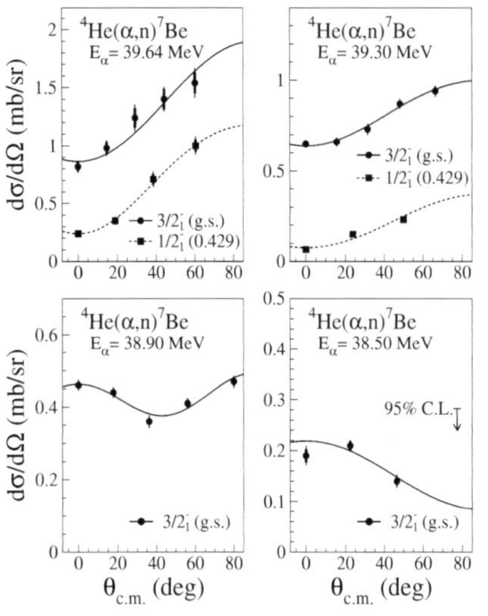

Figure 1. Angular distribution of the differential cross sections for the ^{4}He(α,n)^{7}Be reaction. The solid circles and squares show the differential cross sections for the ground and first excited states in ^{7}Be, respectively. The solid and dashed lines are series of the even-order Legendre polynomials to fit the experimental data.

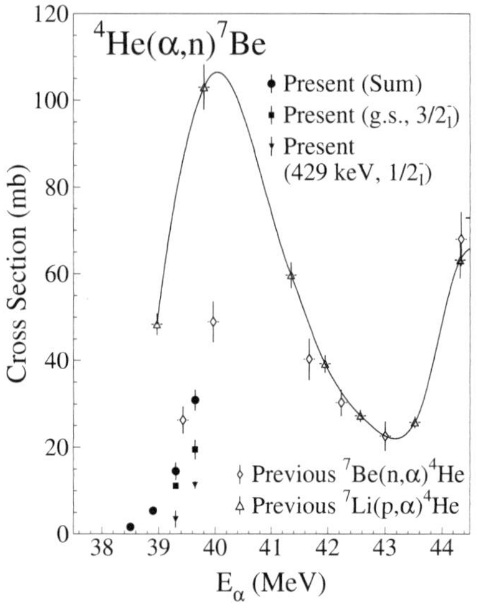

Figure 2. Measured total cross sections of the ^{4}He(α,n)^{7}Be reaction for the ground (solid squares) and first excited (solid triangles) states. The sum of the two states is shown by solid circles. Previous results of the ^{4}He(α,n)^{7}Be and ^{4}He(α,p)^{7}Li reactions [5] are shown by the open diamonds and triangles, respectively. The solid line interpolating the open triangles is drawn for eye-guide.

3. Discussion

The present results are compared with the previous results taken from Ref. [5] in Fig. 2. The solid squares and triangles are the present results for the ground and first excited states. The sum of the two states are shown by the solid circles to directly compare the present results with the previous results from Ref. [5]. The previous results of the ^{4}He(α,n)^{7}Be and ^{4}He(α,p)^{7}Li reactions are shown by the open diamonds and triangles, and the solid line interpolating the open triangles is drawn for eye-guide. The both measurements of the ^{4}He(α,n)^{7}Be reaction overlap around E_α = 39.4 MeV, and they agree within the measurement uncertainties. It should be noted that the present measurement extends the experimental data toward lower energies down to E_α = 38.50 MeV for the first time.

The total cross sections for the ^{4}He(α,n)^{7}Be reaction are similar to those for the mirror ^{4}He(α,p)^{7}Li reaction at higher beam energies, but the two reactions split at lower energies due to the threshold effect because the n + ^{7}Be threshold at E_x = 18.9 MeV in ^{8}Be is higher than the p + ^{7}Li threshold by 1.6 MeV.

From a view of the cosmological lithium problem, the total cross section for the ^{7}Be(n,α)^{4}He reaction is important. The measured ^{4}He(α,n)^{7}Be cross sections for the ground and first excited states in ^{7}Be are separately converted to the cross section for the time reverse reactions on the basis of the detailed balance principle. The solid circles and squares in Fig. 3 show the total cross sections of the (n,α) reaction on the ground and first excited states in ^{7}Be. The cross sections evaluated in Ref. [6] are also plotted by the open triangles together with the cross section from the evaluated nuclear data library ENDF/B-VII.1 [7] shown by the dashed line. The evaluated cross sections are very close to the present data for the ^{7}Be$_{g.s.}$(n,α)^{4}He reaction.

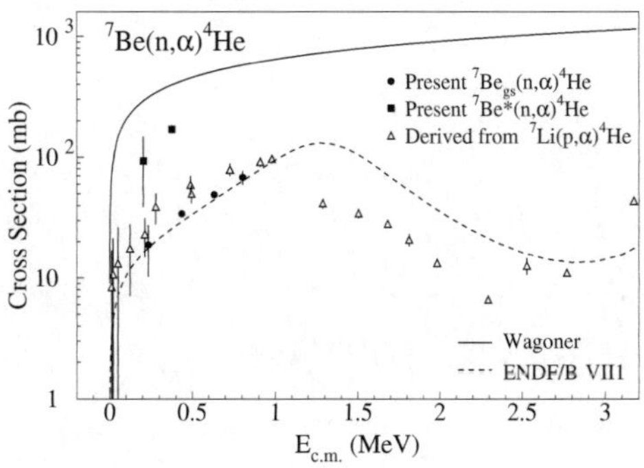

Figure 3. Measured total cross sections for the ^{7}Be(n,α)^{4}He reaction on the ground (solid circles) and first excited (solid squares) states. The open triangles, solid line, and dashed line are the previous evaluations taken from Refs. [6], [8], and [7], respectively.

The cross section for the ^{7}Be(n,α)^{4}He reaction was first estimated by Wagoner in Ref. [8] as shown by the solid line in Fig. 3, which was widely used in the BBN calculations. The present values of the ^{7}Be(n,α)^{4}He cross sections are much smaller than the Wagoner's calculation. The cross sections on the first excited state in ^{7}Be are larger than those on the ground state. Since the low-lying excited states can be thermally populated in the high-temperature environment, the first excited state at $E_x = 429$ keV in ^{7}Be could contribute to the ^{7}Be destruction in principle. However, its excitation energy is too high to make a sizable contribution to the destruction rate at the BBN temperature. Moreover, the cross sections for the ^{7}Be*(n,α)^{4}He reaction are still smaller than the Wagoner's calculation. As a conclusion, the present results suggest the ^{7}Be(n,α)^{4}He reaction does not solve the cosmological lithium problem.

4. Summary

The cross sections for the ^{4}He(α,n)^{7}Be reaction were measured at low energies between $E_\alpha =$ 38.50 and 39.64 MeV. On the basis of the detailed balance principle, the cross sections for the ^{7}Be(n,α)^{4}He reaction were obtained at $E_{c.m.} = 0.20$–0.81 MeV close to the BBN energy window for the first time. The obtained cross sections are significantly smaller than the theoretical estimation widely used in the BBN calculations. The present results suggest the ^{7}Be(n,α)^{4}He reaction rate is not large enough to solve the cosmological lithium problem.

Acknowledgments

The present work was done as a graduation research by the senior students at Kyoto University (KADAIKENKYU P4) over the two years. All the processes of the research project were done by the undergraduate students under the supervision of the faculties. The authors are grateful to RCNP for the deep understanding about the importance of the undergraduate education through the hands-on training at the large accelerator facility. The authors also thank the cyclotron crews at RCNP for their efforts to provide a clean and stable beam. This work was supported by JSPS KAKENHI Grant Numbers JP26287058 and JP15H02091.

References

[1] Fields B D 2011 *Ann. Rev. Nucl. Part. Sci.* **61** 47–68
[2] Cyburt R H and Davids B 2008 *Phys. Rev.* C **78** 064614
[3] Sakai H *et al.* 1996 *Nucl. Instrum. Methods Phys. Res.* A **369** 120–134
[4] Kim E, Endo A and Yamaguchi Y 2002 *J. Nucl. Sci. and Tech.* **39** 693–696
[5] King C H, Austin S M, Rossner H H and Chien W S 1977 *Phys. Rev.* C **16** 1712–1722
[6] Hou S Q, He J J, Kubono S and Chen Y S 2015 *Phys. Rev.* C **91** 055802
[7] Chadwick M B, Herman M, Oblozinsky P *et al.* 2011 *Nucl. Data Sheets* **112** 2887–2996
[8] Wagoner R V 1969 *Astrophys. J. Suppl. Ser.* **18** 247–295

IOP Conf. Series: Journal of Physics: Conf. Series **863** (2017) 012074 doi:10.1088/1742-6596/863/1/012074

Capture reactions into borromean two-proton systems at rp-waiting points

D Hove[1], E Garrido[2], A S Jensen[1], H O U Fynbo[1], N T Zinner[1] and D V Fedorov[1]

[1]Department of Physics and Astronomy, Aarhus University, DK-8000 Aarhus C, Denmark
[2]Instituto de Estructura de la Materia, IEM-CSIC, Serrano 123, E-28006 Madrid, Spain

E-mail: dennish@phys.au.dk

Abstract. We investigate even-even two-proton borromean systems at prominent intermediate heavy waiting points for the rapid proton capture process. The most likely single-particle levels are used to calculate three-body energy and structure as a function of proton-core resonance energy. We establish a linear dependence between two- and three-body energies with the same slope, but the absolute value slightly dependent on partial wave structure. Using these relations we estimate low-lying excited states in the isotones following the critical waiting points. The capture rate for producing a borromean bound state is described based on a full three-body calculation for temperatures about 0.1 – 10 GK. In addition, a simple rate expression, depending only on a single resonance state, is found to comply with the three-body calculation for temperatures between 0.1 and 4 GK. The rate calculations are valid for both direct and sequential capture paths. As a result the relevant path of the radiative capture reactions can be determined.

1. Motivation and purpose

The rapid-proton capture (rp) process is the process whereby nuclei in stellar environments capture an increasing number of protons, and become increasingly heavier [1]. Naturally, these nuclei eventually reach the proton dripline where the process is hindered, and further evolution must wait for the much slower β-dacay. This typically occurs in the accretion disk of a close binary system containing a neutron star [2]. These nuclei along the dripline are called waiting points, and the waiting points with the longest lifetime are called the critical waiting points. Another option, instead of waiting for a natural β-decay, is to skip the gap in the proton dripline by capturing two protons, which would affect the effective lifetime. This effective lifetime of the critical waiting points plays a vital part in determining the overall time scale and relative distributions of elements in the system [3]. The critical waiting points among the relatively heavy nuclei, are ^{64}Ge, ^{68}Se, and ^{72}Kr [2], where ^{68}Se is thought to be most important [4].

These systems are ideal candidates to be studied form a three-body perspective, as adding one proton to the critical waiting point creates an unbound system, but adding two protons would create a bound system, which is the very definition of a borromean system. Another argument in favour of a three-body treatment is that the first excited state in the critical waiting points are 0.9 [5], 0.9[6], and 0.7 [7] MeV for ^{64}Ge, ^{68}Se, and ^{72}Kr respectively, which means they can be considered fairly inert.

This work will focus on studying the proton capture on the critical waiting points from a three-body perspective. More elaborate considerations on the same subject are presented in Ref. [1]. The purpose of this is two-fold; First of all we will examine the relation between two- and three-body energies in the system. Secondly, we will use this in combination with very limited experimental information to estimate the reaction rate of the proton capture process.

2. Theory and method

The few-body formalism applied is the so-called hyperspheric, adiabatic expansion of the three-body Faddeev equations in coordinate space. In the interest of brevity the details of this method will not be explained here, but the interested reader is referred to Refs. [1, 8]. Briefly explained the idea is to introduce one hyperradial, ρ, and five hyperangular, Ω, coordinates, and then solve the angular Faddeev equations for given ρ. This is then used to solve the radial part. The idea is very similar to the idea behind the Born-Oppenhiemer approximation.

Using this formalism the only thing that must be chosen is the three two-body interactions in the system. As we are dealing with a core and two protons, there are only two unique two-body interactions. For the proton-proton interaction a well-established phenomenological interaction, presented in Ref. [9], is used. For the core-proton interaction a Woods-Saxon potential with a central and a spin-orbit part is used,

$$V(r) = V_C(r) + \frac{V_0}{1+e^{(r-R)/a}} + \mathbf{l}\cdot\mathbf{s}\frac{1}{r}\frac{d}{dr}\frac{V_0^{ls}}{1+e^{(r-R_{ls})/a_{ls}}}, \quad (1)$$

where R, R_{ls}, a, and a_{ls} are radial shape and diffuseness parameters kept constant at $R = 7.2$, $R_{ls} = 6.3$, $a = 0.65$ and $a_{ls} = 0.5$ all in fm. The only free parameters are then the strengths V_0 and V_0^{ls}. The intent is to isolate only the partial waves that are most relevant using the free strength parameters. Based on shell model calculations the critical waiting points are known to be situated in the middle of the fpg-shell. However, recent experiments show that the $g_{9/2}$ orbital is not important around $A = 70$ and $N \simeq Z$ [10]. The $f_{5/2}$ and $p_{3/2}$ orbitals are then assumed to be dominating, which is also indicated by the mirror nuclei [11]. To form a negative parity state, and to see the effect of the electric dipole transition ($E1$), an $d_{5/2}$ orbital is at times used instead of $f_{5/2}$.

3. Two- and three-body energies

Using V_0^{ls} to push up the undesired spin-orbit partner, and using V_0 to settle on a specific orbital energy, the relation between two- and three-body energies can be studied. This is seen in Fig. 1 for $^{68}\mathrm{Se} + p + p$.

The single-particle energies of p and f (or p and d) are always identical in Fig. 1. If, for instance, the energy of the f orbital was increased above the energy of p, the black and the green lines would be shifted up by the same amount. The reason for the energy shift being identical is explained in Ref. [1]. However, at some point they would coincide with the corresponding lines with only p, and they would not move further if the energy of f was increased even more. In other words, there is a limit to how much the black and the green line can be moved, and the excitation energy is basically unchanged. The red line is limited in the same way, if the energy of d is increased. But the blue line would move up without limit, if the energy of d was increased, as both p and d is needed to form the negative parity state. As a result the excitation energy for excitations to the 1^- state is not constant, and can increase without limit, depending on the single-particle energies. This has profound consequences for the reaction rate.

As the borromean region of interest is confined by $0 < E_{2b}$ and $E_{3b} < 0$, Fig. (1) places significant constraints on the possible energy levels. Recently, it was possible to measure the proton separation energy of $^{69}\mathrm{Br}$ ($^{68}\mathrm{Se} + p$) to be 0.641 MeV, and the lowest state looked to be an $f_{5/2}$ state [12]. The energy of $p_{3/2}$ is then limited by the borromean region in Fig. 1.

IOP Conf. Series: Journal of Physics: Conf. Series **863** (2017) 012074 doi:10.1088/1742-6596/863/1/012074

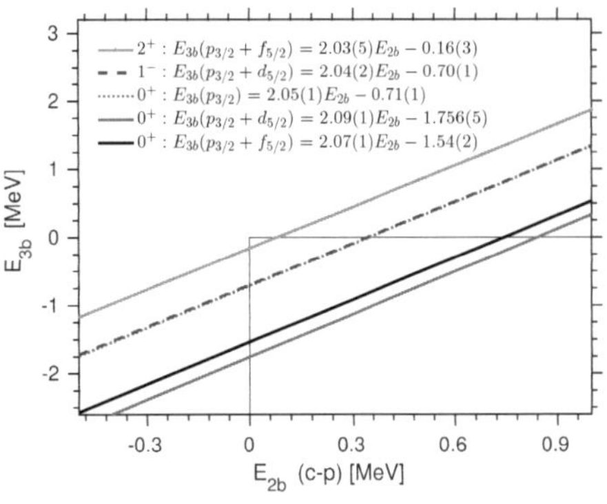

Figure 1. (Color online) The three-body energy as a function of two-body energy for the 0^+ and 2^+ state of ^{68}Se + p + p with $p_{3/2}$ and $f_{5/2}$ waves, as well as for the 0^+ and 1^- state with $p_{3/2}$ and $d_{5/2}$ waves. The single-particle energy of p and f (or p and d) are always identical. The rectangle outlines the borromean region of interest.

4. Reaction rates

As with the three-body formalism, the reaction rate theory will not be explained in detail. The interested reader is referred to Refs. [1, 13]. Seeing that the proton capture takes place in a stellar environment, with temperature T, the energy averaged rate over the corresponding Maxwell-Boltzmann distribution is what is finally calculated,

$$\langle R_{ppc}(E)\rangle = \frac{1}{2T^3}\int E^2 R_{ppc}(E)\exp(-E/T)\,dE, \tag{2}$$

where R_{ppc} is the reaction rate for the one-step γ, three-body transition $p+p+c \rightarrow A+\gamma$. This depends on the photodissociation cross section, which in turn depends on $\langle\Psi_0||\hat{\Theta}_\ell||\psi_\ell^{(i)}\rangle$.

If the resonances are well separated and very narrow, they can be approximated by a Breit-Wigner shape, in which case the integral can be solved analytically. The final, energy average reaction rate would then become

$$\langle R_{ppc}(E)\rangle = \frac{4\pi^3(2\ell+1)\hbar^5}{(\mu_{cp}\mu_{cp,p})^{3/2}}\frac{\Gamma_{eff}(E_R)}{T^3}\exp(-E_R/T), \tag{3}$$

where ℓ is the electric multipole order, while the effective width, Γ_{eff}, is given by the proton, Γ_{ppc}, and the photon, Γ_γ, decay width by $\Gamma_{eff}(E)^{-1} = (\Gamma_{ppc}(E))^{-1} + (\Gamma_\gamma(E))^{-1})$, and the total width is $\Gamma = \Gamma_{ppc} + \Gamma_\gamma$. The proton decay width depends on the long range properties, as the proton has to tunnel through the long-range Coulomb barrier, while the photon decay width depends on the short range properties in particular through the overlap of the final three-body state with the discretized three-body continuum state.

The two-body energy is chosen to be 0.641 MeV for both p and f (or d). The reaction rates are then calculated according to Eqs. (2) and (3). The results are seen in Fig. 2, where the full lines are from the full calculation of Eq. (2), the dashed lines are from the approximation in Eq. (3) using only the lowest resonance state and nothing else, and the dotted line is also from Eq. (3) using the spectrum of resonance states up to 5 MeV. There is only one 1^- resonance state. It should be noted how well the approximation fits with the full calculation using only a single resonance. It should also be noted that the dipole transition has been reduced by a factor of 10^4 to make it fit in the same figure. However, increasing the energy of the $d_{5/2}$ state to just 3 MeV makes the Coulomb barrier unable to support the 1^- resonance, and the rate then drops by many orders of magnitude, making the quadrupole transition the dominating transition.

This result is surprisingly comprehensive, as explained in greater detail in Ref. [1]. The only significant uncertainty is the two-body energy of the lowest $p_{3/2}$ state.

5. Deduced reaction mechanism and adjusted reaction rate

Often, when dealing with three-body capture reactions, it is necessary to assume that the process is either a so-called direct or a sequential process. With the current method not only are both

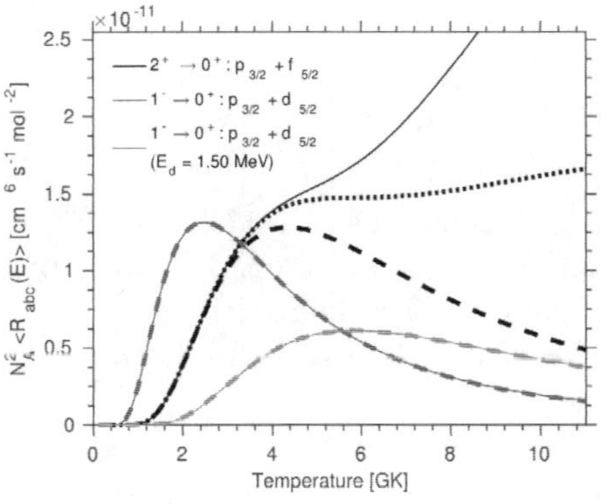

Figure 2. (Color online) The reaction rates for the dipole and quadrupole transition for ^{68}Se $+ p + p \rightarrow$ ^{70}Kr $+ \gamma$ where the allowed single-particle orbitals are adjusted to an energy of 0.641 MeV. The result of the full calculations from Eq. (2) is given by the solid lines. The dashed lines in the same colour are the result of applying Eq. (3) to the lowest resonance. The dotted line is the sum of contributions for spectrum of resonances below 5 MeV using Eq. (3). The $E1$ rates have been scaled down by 10^4 to make the figure readable.

types of reactions included, as well as combinations of the two, it is actually possible to deduce the reaction mechanism from the calculations.

Figure 3 shows the lowest-lying angular part of the wave function, referenced in Sec. 2, as a function of hyperradius and hyperangle. Above it a schematic illustration of the corresponding three-body configuration is included. It shows how, for this particular angular wave function and this particular energy (or two-body potential strengths), one proton would separate from the core-proton system, which equals a sequential decay. Higher-lying angular wave functions could correspond for instance to a direct reaction, where all three particle move apart simultaneously, or to a mixture of a direct and a sequential reaction. However, as long as a sufficient number of angular wave functions are included, all possible reaction paths are also included. In this case the most like reaction mechanism, corresponding to the lowest angular wave function, is the sequential reaction.

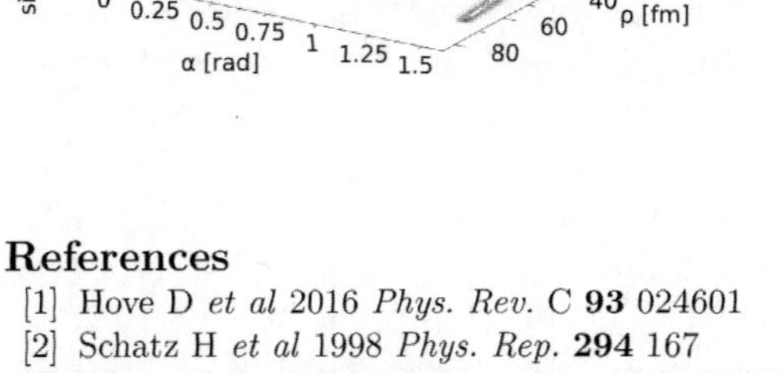

Figure 3. (Color online) The square of the angular wavefunction mentioned in Sec. 2, multiplied by the phase factor $\cos^2(\alpha)\sin^2(\alpha)$ and integrated over Ω_x and Ω_y, as a function of hyperradius ρ and hyperangle α for first angular wave function of the 2^+ state in the set of Jacobi coordinates with first (x) coordinate between core and proton.

References

[1] Hove D *et al* 2016 *Phys. Rev.* C **93** 024601
[2] Schatz H *et al* 1998 *Phys. Rep.* **294** 167
[3] Brown B A *et al* 2002 *Phys. Rev.* C **65** 045802
[4] Tu X L *et al* 2001 *Phys. Rev. Lett.* **106** 112501
[5] Singh B 2007 *Nucl. Data Sheets* **108** 197
[6] Mccutchan E A 2012 *Nucl. Data Sheets* **113** 1735
[7] Abriola D *et al* 2010 *Nucl. Data Sheets* **111** 1
[8] Nielsen E *et al* 2001 *Phys. Rep.* **347** 373
[9] Garrido E *et al* 2004 *Phys. Rev.* C **69** 024002
[10] Nicholas A J *et al* 2014 *Phys. Lett.* B **733** 52
[11] Nesaraja C D 2014 *Nucl. Data Sheets* **115** 1
[12] Del Santo M *et al* 2014 *Phys. Lett.* B **738** 453
[13] de Diego R *et al* 2011 *Phys. Lett.* B **695** 324

Search for the rare γ-decay mode in ^{12}C

M Tsumura[1], T Kawabata[1], T Furuno[1], A Koshikawa[1], M Murata[1], T Morimoto[1], S Adachi[2], A Tamii[2], K Hatanaka[2], T Ito[2], J Zenihiro[3], S Kubono[3], M Itoh[4], Y Matsuda[4], Y Maeda[5], S Sakaguchi[6], H Akimune[7], H Fujimura[8], I Ou[9], T Hashimoto[10] and C Iwamoto[11]

[1] Department of Physics, Kyoto University, Sakyo, Kyoto, 506-8502, Japan.
[2] Reserch Center for Nuclear Physics, Osaka University, Ibaraki, Osaka, 567-0047, Japan.
[3] RIKEN, Wako, Saitama, 351-0198, Japan.
[4] Cyclotron and Radioisotope Center, Tohoku University, Sendai, Miyagi, 980-8578, Japan.
[5] Miyazaki University, Miyazaki, 889-2192, Japan.
[6] Kyusyu University, Fukuoka, 812-8581, Japan.
[7] Konan University, Kobe, Hyogo, 658-8501, Japan.
[8] Wakayama University, Wakayama, 640-8441, Japan.
[9] Okayama University, Okayama, 700-0082, Japan.
[10] Institute for Basic Science, Daejeon, 34047, Korea.
[11] Center for Nuclear Study, the University of Tokyo, Wako, Saitama, 351-0198, Japan.

E-mail: tsumura@scphys.kyoto-u.ac.jp

Abstract. Excited states in ^{12}C nuclei play a very important role in the nucleosynthesis in the universe. ^{12}C nuclei are synthesized by the triple α reaction. Normally the triple α process proceeds via the 0_2^+ state (Hoyle states) at $E_x = 7.65$ MeV in ^{12}C. However, at high temperature $T > 10^9$, the highly excited 3α resonance states such as the 3_1^- state at $E_x = 9.64$ MeV might play a part of the triple α reaction. Unfortunately, the γ-decay probability of the 3_1^- state has not been determined, therefore we planned to measure ^{1}H(^{12}C, ^{12}Cp) reaction without measuring γ-rays. The test experiment using the solid hydrogen target and GAGG($Gd_3Al_2Ga_3O_{12}$) was recently carried out at the cyclotron facility in RCNP. The results of the test experiment are reported in this paper.

1. Introduction

1.1. ^{12}C *synthesis in the universe*

Excited states in ^{12}C nuclei play a very important role in the nucleosynthesis in the universe. ^{12}C nuclei are synthesized by the triple α reaction as shown in Fig.1. Firstly, excited states of ^{12}C are produced as 3α resonance states. Most of these excited states immediately decay into three α particles, but a tiny fraction of these excited states decays into the ground state of ^{12}C by emitting γ rays. Normally this process proceeds via the 0_2^+ state (Hoyle states) at $E_x = 7.65$ MeV in ^{12}C. The radiative decay width of the 0_2^+ state is known to be $\Gamma_\gamma = 3.7 \pm 0.5$ meV and the total width for this state is $\Gamma = 8.5 \pm 1.0$ eV [1]. However, at high temperature $T > 10^9$, the highly excited 3α resonance states such as the 3_1^- state at $E_x = 9.64$ MeV and the 2_2^+ state at $E_x = 9.84$ MeV might play a part of the triple α reaction [2]. The total width of the 3_1^- state is known to be $\Gamma = 34 \pm 5$ keV [1], but the radiative decay width of this state has not been determined. Therefore, we have planned to determine the radiative width.

1.2. *γ-decay probability of the 3_1^- state*

It is difficult to measure the γ-decay probability of the 3_1^- state because it is very small. The present knowledge about the total decay width Γ, γ-decay width Γ_γ and γ-decay probability

IOP Conf. Series: Journal of Physics: Conf. Series **863** (2017) 012075 doi:10.1088/1742-6596/863/1/012075

Γ_γ/Γ of the 3_1^- state is summarized in Table 1 [3]. The direct γ-decay width to the ground state ($\Gamma_{\gamma 0}$) was obtained from the inelastic electron scattering. Since total γ-decay width Γ_γ must be larger than $\Gamma_{\gamma 0}$, the lower limit of Γ_γ of the 3_1^- state is $\Gamma_{\gamma 0} = 0.31(4)$ meV. The upper limit was reported from the measurement of the ^{12}C(α, α' ^{12}C) reaction [4]. In this previous measurement, only the upper limit of Γ_γ was determined because the ^{13}C contaminants in the isotopically enriched ^{12}C target caused serious background. Therefore, we decided to use the inverse kinematic reaction ^{1}H(^{12}C, ^{12}Cp), because no ^{13}C contaminants are included in a ^{12}C beam.

Table 1. Total decay width Γ, γ-decay width Γ_γ and γ-decay probability Γ_γ/Γ of the 3_1^- state [3].

Γ	Γ_γ	Γ_γ/Γ
34(5) keV	$> 0.31(4)$ meV < 14 meV(1σ C.L.)	$> 9.1\times10^{-9}$ $< 4.1\times10^{-7}$

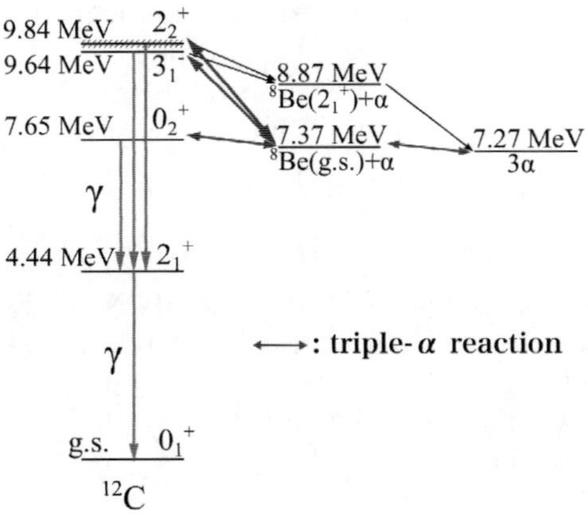

Figure 1. Level diagram related to the triple α reaction.

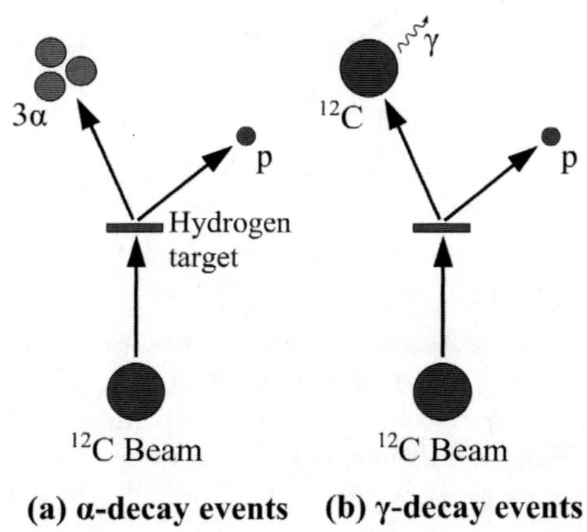

Figure 2. (a) α-decay events and (b) γ-decay events in the ^{1}H(^{12}C, ^{12}Cp) reaction.

2. Experimental consideration

2.1. Experimental procedure

A hydrogen target is bombarded with a ^{12}C beam as shown in Fig. 2. The excitation energy of the scattered ^{12}C is determined from the energy and angle of the recoil proton. Most of the ^{12}C nuclei excited to the 3_1^- state decay into 3α as shown in Fig. 2(a), but a tiny fraction of them decays into the ground states by emitting γ-rays as shown in Fig. 2(b). Therefore, if the scattered ^{12}C is detected in coincidence with the recoil proton, the γ-decay event is identified without the γ-ray detection. The γ-decay probability Γ_γ/Γ is obtained from the ratio of the number of all the inelastic events to the number of the γ-decay events.

2.2. Setup

The test experiment was performed at the cyclotron facility in Research Center for Nuclear Physics, Osaka University in April, 2016. A solid hydrogen target [5] with a thickness of 0.5 mm is bombarded with a 250-MeV ^{12}C beam as shown in Fig. 3. The recoil proton is detected and particle-identified by GION (Si+GAGG telescope) in the scattering chamber. The recoil angle of the proton is obtained from the double-sided Si strip detector. The excitation energy of the scattered ^{12}C is determined by the energy and angle of the recoil proton.

IOP Conf. Series: Journal of Physics: Conf. Series **863** (2017) 012075 doi:10.1088/1742-6596/863/1/012075

The scattered ^{12}C is analyzed by the Grand Raiden spectrometer and detected by the two multi-wires drift chambers (VDC1 and VDC2) and two plastic scintillators at the focal plane of the spectrometer. The thicknesses of the plastic scintillators are 1 mm at the upstream and 10 mm at the downstream. The scattered ^{12}C cannot penetrate the 1-mm scintillator, while α particles can penetrate and stop in the 10-mm scintillator. Therefore, the trigger signal for the ^{12}C detection was generated using the anti-coincidence technique between the two scintillators.

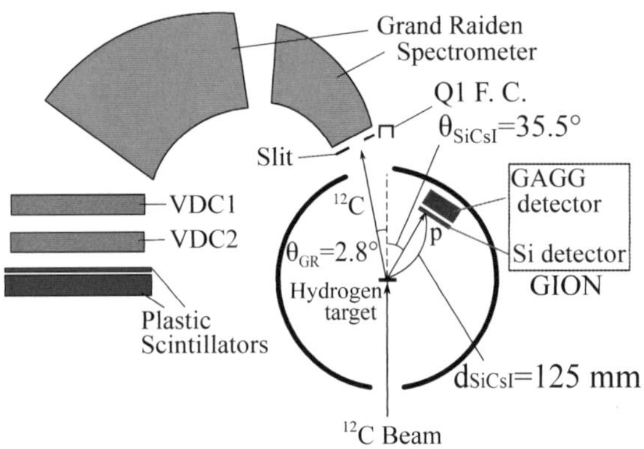

Figure 3. Schematic view of the experimental setup.

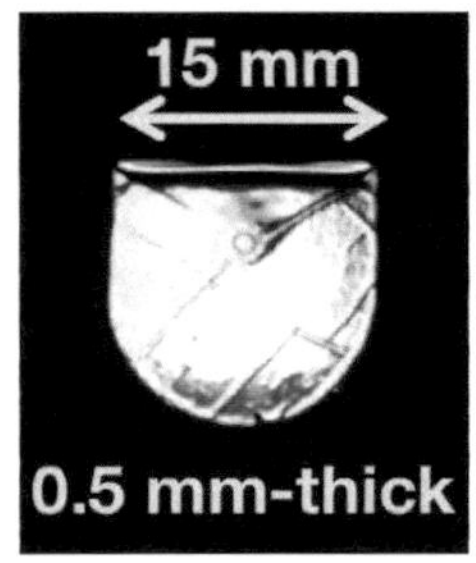

Figure 4. Solid hydrogen target used in this test experiment.

Figure 5. GAGG crystals wrapped in ESR films (a part of GION telescope array).

2.3. Improvement of S/N
A detailed Monte Carlo simulation was performed to estimate the signal to noise (S/N) ratio and it was found that the S/N ratio was less than 1/250 if no special treatment to remove backgrounds was done. Therefore we introduced some tricks to improve the S/N ratio, for example, a thin solid hydrogen target for impurities reduction, an event-tagging recoil proton detector for removal of accidental coincidence events, and so on.

2.3.1. The first test experiment The first test experiment was carried out in October, 2013. In the first test experiment, we used a CH_2 film as a target and a Si+CsI telescope as a recoil proton detector. However, the continuous background for the γ-decay events limits the sensitivity to the γ-decay probability Γ_γ/Γ in the first test experiment. As a result, the sensitivity to the γ-decay probability Γ_γ/Γ in the measurement was $2.0(5) \times 10^{-5}$, and it was not good enough to determine the γ-decay probability Γ_γ/Γ of the 3_1^- state.

2.3.2. Solid hydrogen target(SHT) A solid hydrogen target system was introduced for impurities reduction. The thin target is needed for the high resolution measurement because the energy loss in the target makes the energy resolution worse. However it is difficult to make large thin SHT because of its low thermal conductivity. Therefore we introduced ortho-para converter to the SHT system [5] in order to use para-H_2, whose thermal conductivity is 10 times larger than normal-H_2. As a result, we succeeded in making SHT as shown in Fig. 4.

2.3.3. GION In the first test experiment, the pile-up events of CsI detector caused serious background. Therefore we introduced the GION telescope array instead of the Si+CsI telescope. GION [**G**AGG($Gd_3Al_2Ga_3O_{12}$) based light **ion** counter telescope] consists of a double-sided Si strip detector and 24 GAGG crystals. GAGG is new scintillating crystal [6] and its decay time is much faster than CsI's as shown in Table 2. The GAGG crystals are wrapped in ESR films as shown in Fig. 5, and mounted on avalanche photodiodes (APD). The temperature of GION was monitored during the measurement. The temperature was stable and the fluctuation of the gain of the APDs was acceptable.

IOP Conf. Series: Journal of Physics: Conf. Series **863** (2017) 012075 doi:10.1088/1742-6596/863/1/012075

Table 2. Property of CsI and GAGG [6].

	Density (g/cm^3)	Energy resolution (FWHM)@662 keV	Decay time (ns)	Light output (photon/MeV)
CsI(Tl)	4.51	6%	1000	56000
GAGG(Ce)	6.63	5–6%	88	65000

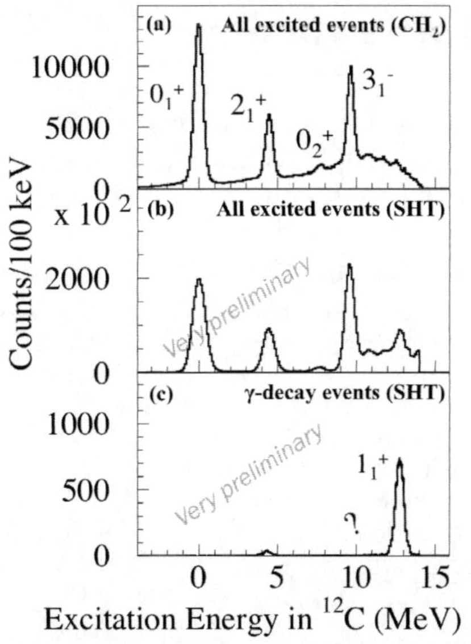

Figure 6. Excitation energy spectra including (a) all the excited events using CH_2 target, (b) all the excited events using SHT and (c) the γ-decay events only.

3. Results of the test experiment

This test experiment using SHT and GION was carried out in April, 2016. Fig. 6(a) and Fig. 6(b) show the excitation energy spectra of ^{12}C. Fig. 6(a) was obtained using the CH_2 target instead of SHT from a 8-hours measurement in the first test experiment, while Fig. 6(b) was taken by using the SHT from a half-hour measurement in this test experiment. The prominent peaks of the ground state, the 2_1^+ state at $E_x = 4.44$ MeV, the 0_2^+ state at $E_x = 7.65$ MeV, and the 3_1^- state at $E_x = 9.64$ MeV are clearly observed in both Fig. 6(a) and 6(b). However, the backgrounds in Fig. 6(b) are drastically reduced compared with Fig. 6(a). Fig. 6(c) shows the excitation energy spectrum of the γ-decay events. The peaks of the 2_1^+ state at $E_x = 4.44$ MeV and the 1_1^+ state at $E_x = 12.71$ MeV are clearly observed. Further analysis is still on going. After we confirm the feasibility of the present measurement, we are going to acquire the data for 10 days.

References

[1] Itoh M, Akimune H, Fujiwara M, Garg U, Hashimoto N, Kawabata T, Kawase K, Kishi S, Murakami T, Nakanishi K, Nakatsugawa Y, Nayak B, Okumura S, Sakaguchi H, Takeda H, Terashima S, Uchida M, Yasuda Y, Yosoi M and Zenihiro J 2011 *Phys. Rev.* C **84** 1
[2] Hoyle F and Fowler W A 1960 *Astrophys. J.* **132** 565
[3] Anzenberg-Selove F 1990 *Nucl. Phys.* A **506** 158
[4] Chamberlin D, Bodansky D, Jacobs W W and Oberg D L 1974 *Phys. Rev.* C **10** 909
[5] Matsuda Y, Tsumura M, Sakaguchi H, Terashima S, Zenihiro J, Kawabata T, Otsu H, Beaumel D, Tian Z, Maeda Y, Gotanda S, Kawakami S, Aoi N and Yamamoto T 2015 *J. Rad. Nucl. Chem.* **305** 897
[6] Kurosawa S, Shoji Y, Yokota Y, Kamada K, Valery I. Chani and Yoshikawa A 2014 *J. Crys. Growth* **393** 134

Active target MAIKo to investigate cluster structures in unstable nuclei

T Furuno[1], T Kawabata[1], M Murata[1], S Adachi[2], Y Ayyad[3], H Baba[4], T Baba[1], M Cavallaro[5], T Hashimoto[6], Y Ishii[1], R Kobayakawa[1], Y Matsuda[7], Y Matsuoka[1], T Morimoto[1], T Nanamura[1], S Beceilo-Novo[3], H J Ong[2], A Sakaue[1], J Tanaka[8,2], I Tanihata[2], D T Trong[2], M Tsumura[1] and H D Watanabe[1]

[1] Department of Physics, Kyoto University, Kyoto 606-8502, Japan
[2] Research Center for Nuclear Physics, Osaka University, Osaka 567-0047, Japan
[3] National Superconducting Cyclotron Laboratory, Michigan State University, MI 48824, USA
[4] RIKEN Nishina Center, Saitama 351-0198, Japan
[5] INFN Laboratori Nazionali del Sud, Catania 95125, Italy
[6] Rare Isotope Science Project, Institute for Basic Science, Daejeon 305-811, Korea
[7] Cyclotron and Radioisotope Center, Tohoku University, Miyagi 980-8578, Japan
[8] Faculty of Science and Engineering, Konan University, Hyogo 658-8501, Japan

E-mail: `furuno@scphys.kyoto-u.ac.jp`

Abstract. Study on clustering of nucleons in nuclei is recently focusing on unstable nuclei where new kinds of structures, namely molecular structures with excess nucleons, are predicted. The Coulomb shift of energy in the mirror system is suggested to reflect the size of these structures. Although the missing mass spectroscopy is expected to give access to these structures even beyond particle decay thresholds without any biases in excitation energy spectra but the detection of very low energy particles is challenging. To satisfy the requirement, a new active target system MAIKo has been developed at RCNP. The detector was commissioned using a ^{13}C beam under the same kinematical condition as that of RI beam experiments.

1. Introduction

The clustering phenomena are observed in the various hierarchies from the subatomic scale to the cosmic scale. In order to understand the essential nature of each physics hierarchy, it is important to find out why the clustering structures appear. Investigation of the clustering of nucleons is thus necessary to understand many-body problems of nucleons. Especially the alpha clustering has been widely studied among self-conjugate $4n$ nuclei over the past fifty years. Recently, research interests in alpha clustering are extending from stable nuclei to unstable neutron-rich nuclei. In such nuclei, excess neutrons are speculated to occupy the particle orbits between alpha cores and exhibit a variety of new cluster structures, namely molecular structures. For example, rotational bands of alpha molecular structures in ^{10}Be were discussed by Freer *et al* [1].

It is a natural question whether proton-rich molecular structures with valence protons are also be formed. Moreover, it is recently pointed out that energy shifts between isobaric analogous states unveil the dynamics of the constituent clusters [2]. Usually, energy levels in proton-rich

IOP Conf. Series: Journal of Physics: Conf. Series **863** (2017) 012076 doi:10.1088/1742-6596/863/1/012076

sides are higher than those in the mirror nuclei due to the repulsive Coulomb force. However, if the relative wave function between the clusters contains a large s-wave component, the energy shifts will become smaller because the wave functions of the clusters are spatially extended. In case of the ^{10}C-^{10}Be system, the 0_1^+ and 0_2^+ states are considered to be $2\alpha + 2N$ molecular structures where two valence nucleons move among two alpha cores. The 0_3^+ and 0_4^+ states are considered to be well-developed α+^{6}Be or α+^{6}He structures. The energy shift of the 0_2^+ and 0_3^+ states are expected to be smaller than those of the 0_1^+ and 0_4^+ states, respectively. This is because the 0_2^+ and 0_3^+ states mainly contain the s-wave components while the 0_1^+ and 0_4^+ states contain the p-wave and the d-wave components, respectively. This can be regarded as cluster Thomas-Ehrman shift. In order to confirm the cluster Thomas-Ehrman shift, it is mandatory to perform experiments on both of the mirror system. However, experimental information on ^{10}C is scarce. Therefore, we propose a novel experiment to search for alpha cluster states in ^{10}C to compare with ^{10}Be data.

Alpha inelastic scattering is the best probe to systematically search for the 0^+ states discussed above because it has a selectivity to excite the natural parity states and the reaction is described by a simple mechanism. The 0_3^+ and 0_4^+ states in ^{10}C will decay into the 2α+2p channels. Invariant mass spectroscopy on such states is difficult because all of the decay fragments must be detected. On the other hand, missing mass spectroscopy is useful because only the recoil alpha particle needs to be detected. The differential cross sections calculated by the distorted-wave Born approximation (DWBA) with different transferred angular momenta are shown in Fig. 1. The measurement should cover $\theta_{\rm CM} \sim 3^\circ$ to search for the 0^+ states because the cross section for the monopole transition becomes dominant. As shown in Fig. 2, the recoil alpha particles are emitted from the target with very small energy ($\sim$ 500 keV) at $\theta_{\rm CM} \sim 3^\circ$. If one tries to perform a measurement with a conventional setup using a external detectors, the He gas target should be extraordinary thin which makes the luminosity irredeemable.

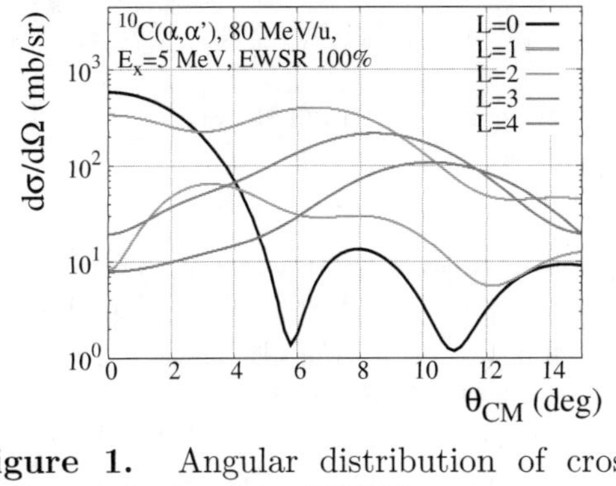

Figure 1. Angular distribution of cross sections calculated by DWBA.

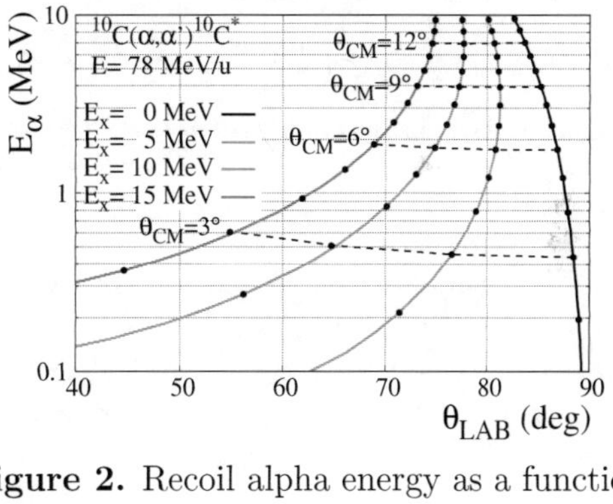

Figure 2. Recoil alpha energy as a function of recoil angle.

2. Design of MAIKo

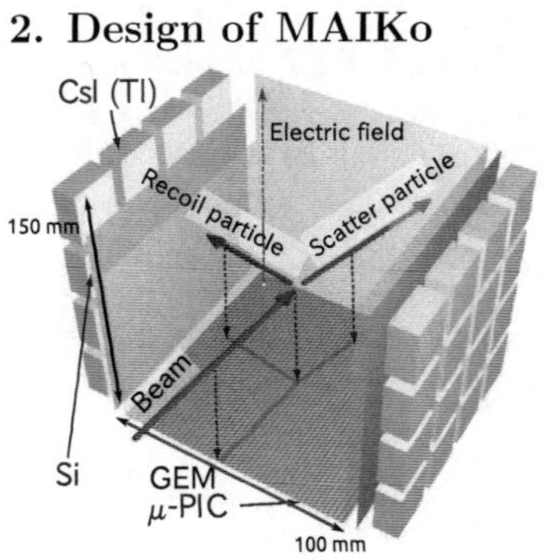

Figure 3. Schematic view of MAIKo.

To overcome the problem stem from the low momentum transfer kinematics, a new active target system MAIKo (μ-PIC based Active target for Inverse Kinematics.) has been recently developed at Research Center for Nuclear Physics (RCNP), Osaka University.

A schematic view of MAIKo system is shown in Fig. 3. The system consists of a time projection chamber (TPC) which can provide the three-dimensional information of charged particle trajectories. The main feature of the active target is to employ the detection gas of the TPC as the target gas. In particular, He gas is filled in the TPC chamber to perform alpha inelastic scattering. Since the scattering is measured inside the sensitive volume of the TPC, the detection threshold for the recoil particle is lowered. Moreover, the thick gas target and large solid angle of the TPC increase the statistics.

The TPC has a sensitive volume of $100\times100\times150$ mm^3. For the amplification of the ionized electron, Gas Electron Multiplier (GEM) and Micro-PIxel Chamber (μ-PIC) [3] are utilized. The drift electrons are first amplified by GEM and then further multiplied around the anode pixels of μ-PIC. The electron drift time and position are recorded through anode and cathode strips of μ-PIC with a pitch of 400 μm. The anode and cathode strips are arranged orthogonally and they provide two-dimensional projections of the particle trajectories (Fig. 4). The two-stage amplification achieves larger total gas gain than the single amplification. It suppresses the breakdown and enables the stable operation of the TPC. This also helps to reduce the amount of quench gas which can be a background source. The structure and the readout system of the TPC are described in detail in Refs. [4][5].

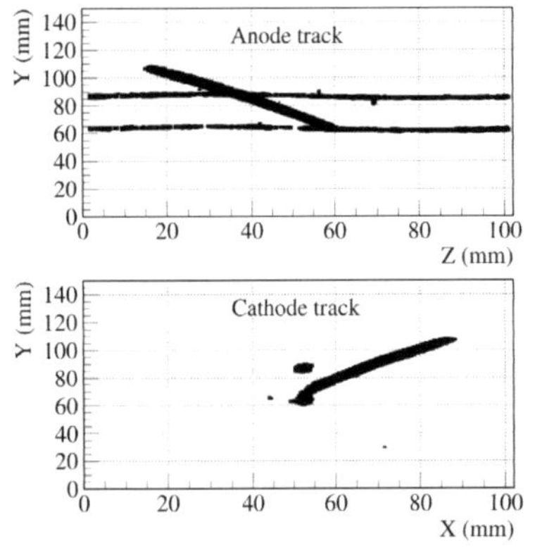

Figure 4. Track example of a scattering event.

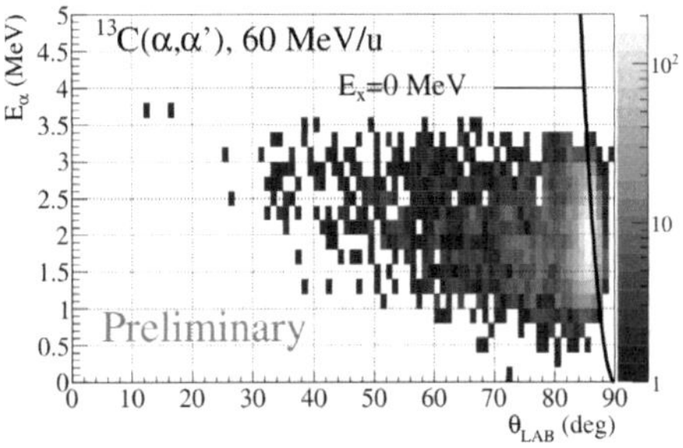

Figure 5. Reconstructed energy versus angle of recoil alpha particle.

3. Test experiment

A test experiment was carried out at RCNP EN beam line using a 60-MeV/u ^{13}C beam. The kinematic condition of the experiment was almost same as that of the ^{10}C experiment. The TPC chamber was filled with He(98%)+CF_4(2%) at 100 kPa to measure alpha inelastic scattering. A typical track of a scattering event is shown in Fig. 4. Each trajectory in the anode and cathode images was identified by an algorithm based on Hough transform method [6]. The track of the recoil alpha particle was fitted by a straight line to reconstruct the scattering event. Energy of the recoil alpha particle was calculated from the range in the TPC gas. Figure 5 shows the correlation between the reconstructed energy and angle of the recoil alpha particle. A clear locus corresponding to the elastic scattering was observed. The detection threshold was successfully lowered to less than 1 MeV. A further test experiment on ^{10}C will be performed in November 2016. The detection threshold will be further lowered to less than 500 keV by reducing the gas pressure to 50 kPa.

References

[1] Freer M *et al* 2001 *Phys. Rev.* C **63** 034301
[2] Ito M 2016 *EPJ Web Conf.* **117** 06014
[3] Ochi A *et al* 2001 *Nucl. Inst. Meth. Phys. Res.* A **471** 264
[4] Furuno T *et al* 2014 *Jour. Phys.: Conf. Series* **569** 012042
[5] Mizumoto T *et al* 2015 *Nucl. Inst. Meth. Phys. Res.* A **800** 40
[6] Furuno T *et al* 2016 *RCNP annual report*

IOP Conf. Series: Journal of Physics: Conf. Series **863** (2017) 012077 doi:10.1088/1742-6596/863/1/012077

The measurement of ^{4}He photodisintegration with MAIKo active target

M Murata[1], S Adachi[2], H Akimune[3], T Furuno[1], T Kawabata[1] and M Tsumura[1]

[1] Department of Physics, Kyoto University, Kyoto, Kyoto 606-8502, Japan
[2] Research Center for Nuclear Physics (RCNP), Osaka University, Ibaraki, Osaka 567-0047, Japan
[3] Facluty of Science and Engeneering, Konan University, Kobe, Hyogo 658-8501, Japan

E-mail: m.murata@scphys.kyoto-u.ac.jp

Abstract. The phtodisintegration reactions of ^{4}He nuclei *i.e.*, ^{4}He(γ,pt) and ^{4}He(γ,^{3}He)n, play crucial roles for the nucleosynthesis in the universe, but the result of previous experimental studys show sizable discrepancies. We performed the measurement of these reactions with the TPC-based MAIKo active target. The MAIKo active target enable us to detect, the sensitivity for the low energy decay particles emitted to full solid angle. The mearsurement were carried out at the New SUBARU synchrotron radiation facility, and the qusi-monoquromatic pulsed photon beam were irradiated on MAIKo. The data analysis has been partially done. The number of true events were counted by tracking of recorded images, and the total flux of photon beams was determined by the analysis of the spectrum recorded by the NaI beam monitor.

1. Introduction

^{4}He is the self-con ugate nucleus with doubly closed-shell sturucture and it plays crucial role not only in cluster phenomena, but also nucleosynthesis in the universe. It is pointed out that the cross sections of the ^{4}He photodisintegration reactions ^{4}He(γ,pt) and ^{4}He(γ,^{3}He)n are important to clarify the scenarios of nucleosynthesis.

irstly, the corss section of this reaction is inportant from a view of nucleosynthesis driven by neutrinos in a suparnova[1]. Two neutral current reactions ^{4}He(ν,$\nu'p$)^{3}H and ^{4}He(ν,$\nu'n$)^{3}He are the startig point of the neutrino driven process. Although it is di cult to measure these reactions directly, they are analogicaly understood by the photodisintegration reaction.

Secondly, this reaction may give us an insight into the big bang nucleosynthesis[2]. In order to solve a problem in which the promodial abundance of ^{6} i is significantly underestimated by the standard big bang nucleosynthesis (BBN) theory, the possibility of the unconsiderd cotribution from heavy relic particles produced before BBN era is pointed out. In this scenario, the radiative decay of relic particles plays a role of extra photon source, and these photons induce the photodisintegration reaction of ^{4}He to produce the seeds of ^{6} i. The validity of this scenario depends on the cross section of the ^{4}He photodisintegration.

The ^{4}He photodisiontegration has long been caught a greate deal of attention and intensively studied by various researchers[3, 4], but the measured values of its cross sections have not been settled at the relaible values, yet. or both the ^{4}He(γ,pt) and ^{4}(γ,^{3}He)n reactions, almost all the data show agreement in the higher energy region but discrepancy in the lower energy domain

IOP Conf. Series: Journal of Physics: Conf. Series **863** (2017) 012077 doi:10.1088/1742-6596/863/1/012077

between the decay threshold at $E_\gamma \sim 20$ Me and the giant dipole resonance (R) peak region around $E_\gamma \sim 26$ Me . Especially, the data reported by T.Shima *et al.* shows the remarkably de erent trend[5], which suggests the peak of R is located at energies higher than 30 Me . This astonishing result are obtained from the measurement with time pro ection chamber. It is necessary to verify this reslut by the same approach with more improved detector system.

In the present study, we measured the cross section of ^{4}He photodistintegration in the energy regions between E_γ 22.3 − 33.3 Me . e used the qusi-monochromatic photon beam generated by laser Compton scattering technic and 4π acceptance gas based tracking detector, MAIKo active target.

2. Experimental Setup

The measurement was performed with the MAIKo active target shown in ig.1.[6] MAIKo has a great advantage for the decay particle measurement near the threshold. Basically, the operation principle of MAIKo is the same way as time pro ection chamber. The signals induced by the drift electrons were read out via the strip shaped electrods on μ-PIC.[7] The detection gas filled in MAIKo consists of 98 natural He and 2 C $_4$ with a total pressure 2 atm. Since the detection gas also plays a role of target of photon induced reactions simlutaneously, all the photodisintegration reactions were occured inside the sensitive region. MAIKo, therefore enable us to measure low energy charged particles over full-solid angle.

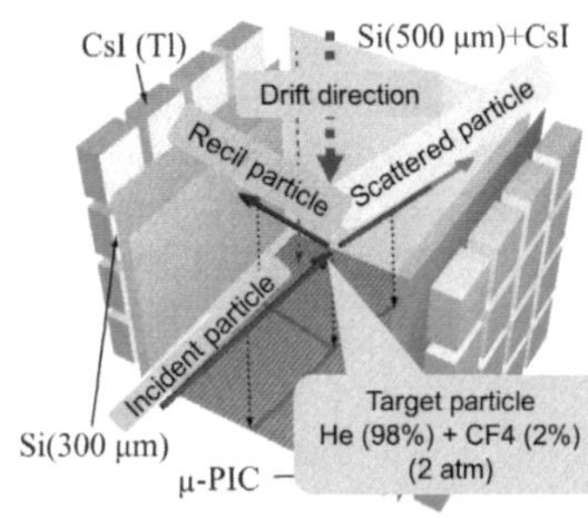

Figure 1. MAIKo active target. (Auxiliary CsI detectors were not used.)

The experiment was carryied out at the synchrotron radiation facility New S BAR [8]. MAIKo was installed at the beam line 01 in this facility as shown in ig.2. This beam line is designed for the experiment with the laser-Compton-scattering (CS) photons[9]. A CS photon was produced by the head-on collision of linearly polarized Nd Y $_4$ laser in the 1st and 2nd harmonics (λ 1064, 532 nm) and a relativistic electrons (E_e 792 − 1147 Me) orbiting inside the storage ring. The lead collimator limits the scattering angle of photons, and only photons scatterd to the very backword angle are transported to the MAIKo active target. The photon energy beam can be changed simply by changing the energy of laser and electrons, and the qusi-mochromaticity of beams are ensured by the collimation of scattering angle. or the beam flux determination, a NaI scintilation detector was installed at the downstream of MAIKo, and the photons passing through the MAIKo were measured.

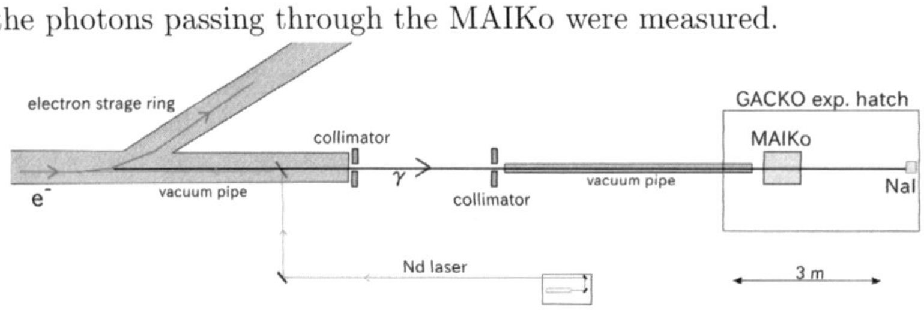

Figure 2. Setup of beam line 01 in New S BAR .

3. Data Analysis

All the events recorded by MAIKo were classified by the shape of particle tra ectories, and the number of ^{4}He photodisintegration events was counted. A typical image of the ^{4}He(γ,pt) event were shown in ig.3. The background events caused by the photodisintegration of heavy nuclei (C,) and a Compton scattered high energy electron were also observed. In order to eliminate background events, firstly, all these events were classified in accordance with the number of linear tra ectories, and electron events which normaly consists of zigzag and thin tra ectories were excluded. The qualities of classification procedure were assured event by event. Secondly, linear tra ectories were fit by linear functions, and three dimensional tra ectories were reconstructed. inally, ^{4}He events were selected by the cut conditions required by kinematics. Thus the background-free event set were obtained.

Total flux of beam photons during the measurement was monitored with the NaI scintilation detector. A typical spectrum of the NaI detector is shown in ig.4. Since the response time of the NaI scintillator is much shorter than the time interval between beam bunches but much longer than time spread of the beam bunch, photons in each bunch are observed as a pile-up event. Therefore, the pulse height of the NaI detector is discretized by the the number of photons in each beam bunch as seen in ig.4. A response function of the NaI detector for one photon was measured with a low intensity photon beam, and response functions for multi photons were obtained by convolutiong the single-photon response function. The average number of photons was determined by fitting the pulse height spectrum in ig.4 with these response functions. inally total photon flux was determined by the product of the measurement time, laser frequency, and average multiplicity.

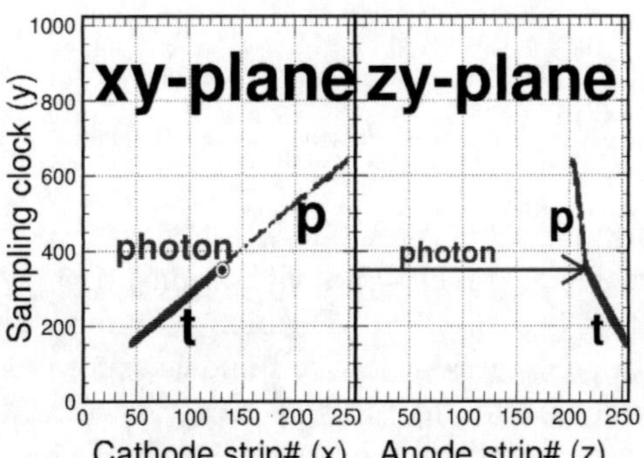

Figure 3. ^{4}He(γ,pt) image.

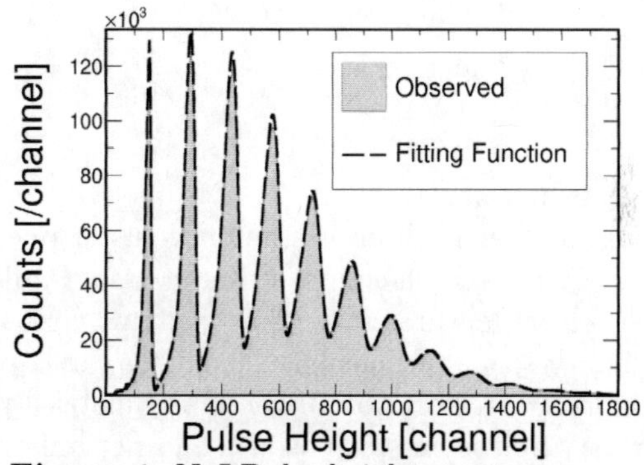

Figure 4. NaI Pulse height spectrum.

The photodisintegration cross section are obtained from the number of events, total beam flux, and target thickness. The result at seven beam energies will be reported soon else where.

4. Summary

e carried out the measurement of ^{4}He photodistintegration ^{4}He(γ,pt), ^{4}He(γ,^{3}He)n cross section at seven energies between E_γ 22.3 – 33.3 Me . The measurement was performed with the CS photon beams and MAIKo active target. The result will be reported soon else where.

References

[1] Suzuki T and Kajino T 2013 *J. Phys. G: Nucl. Part. Phys.* **40** 083101
[2] Kusakabe M, Kajino T, Yoshida T, Shima T, Nagai Y and Kii T 2009 *Phys. Rev. D* **79** 123513
[3] Raut R, Tornow W, Ahmed M W, Crowell A S, Kelley J H, Rusev G, Stave S C and Tonchev A P 2012 *Phys. Rev. Lett.* **108** 042502
[4] Tornow W, Kelley J H, Raut R, Rusev G, Tonchev A P, Ahmed M W, Crowell A S and Stave S C 2012 *Phys. Rev.* C **85** 061001
[5] Shima T, Naito S, Nagai Y, Baba T, Tamura K, Takahashi T, Kii T, Ohgaki H and Toyokawa H 2005 *Phys. Rev.* C **72** 044004
[6] Furuno T *et al.* 2014 *3 rd Int. Workshop on "State of the Art in Nuclear Cluster Physics"* (Yokohama: Kanto Gakuin University) 012042
[7] Ochi A, Nagayoshi T, Koishi S, Tanimori T, Nagae T and Nakamura M 2001 *Nucl. Inst. Math. Phys. Res.* A **478** 196
[8] Ando A, Amano S, Hashimoto S, Kinosita H, Miyamoto S, Mochizuki T, Niibe M, Shoji Y, Terasawa M, Watanabe T and Kumagai N 1998 *J. Synchrotron Radiation* **5** 342
[9] Miyamoto S, Asano Y, Amano S, Li D, Imasaki K, Kinugasa H, Shoji Y, Takagi T and Mochizuki T 2006 *Radiation Measurements* **41** S179

Closing remarks from a theoretical perspective

Y Kanada-En'yo
Department of Physics, Kyoto University, Kyoto 606-8502, Japan
E-mail: yenyo@ruby.scphys.kyoto-u.ac.jp

Abstract. This paper is based on a summary talk for theoretical works presented in the 11*th* International Conference on "Clustering Aspects of Nuclear Structure and Dynamics", Naples (Italy), May 23-26, 2016.

1. Introduction

Among fundamental physics behind cluster phenomena, symmetry is one of the most important physics in cluster phenomena. Of course, Pauli principle, duality and saturation property are also essential features in nuclear systems consisting of two species of Fermions. Symmetries in coordinate space of nuclei are rotational and parity symmetries. As well known, spontaneous breaking and restoration of these symmetries provide specific features of energy spectra. The symmetry in isospin-spin space is another important symmetry. For example, an alpha cluster is the lightest scalar particle in the isospin-spin space. The Pauli principle, duality, and saturation property also play important roles in formation and dynamics of clusters. In particular, the Pauli blocking strongly influences on cluster formation and cluster motion. It also gives a major contribution to density dependence of clustering. Moreover, the duality and saturation property are responsible for coexistence of mean-field and cluster states.

2. Algebraic approaches

The symmetry breaking and restoration in the coordinate space are essential to understand low-lying spectra of light $Z = N$ nuclei. An uncorrelated state has the rotational symmetry. However in realistic system, clusters are formed because of many-body correlations and the system has periodic density as a standing wave on the surface. As a result, the rotational symmetry is broken into a discrete point-group symmetry. Once clusters are formed, inter-cluster motion can be activated. With decrease of density, the symmetry is restored and the system goes to a spherical cluster gas in a low density limit.

Respecting symmetries, algebraic approaches are developing to describe energy spectra and transitions in light nuclei (see Refs. [1, 2, 3] and references therein). Iachello and Bijker discussed symmetries in 2α, 3α, and 4α dynamics, and successfully described energy spectra and transitions of the ground and excited bands in ^{8}Be,^{12}C, and ^{16}O with Z_2, D_{3h}, T_d symmetries and vibration modes built on them. Theoretical spectra correspond well to the recently observed experimental data of ^{12}C as presented in many talks in this conference. It means that the realistic ^{12}C has such the discrete symmetry as a leading component. Cseh and Lévai have applied algebraic approaches to $Z = N$ *sd*-shell nuclei such as ^{28}Si, and neutron-rich nuclei [2, 3].

The algebraic approach has been also adopted in *ab initio* calculation to reveal the hidden symmetry as discussed by Draayer [4]. In the symplectic no-core shell model, the large scale full model space is decomposed into physically relevant subspaces. It has been shown that the leading symplectic symmetry accounts for the dominant components of low-lying states of ^{6}Li, ^{12}C, ^{16}O, ^{20}Ne, and so on. This success indicates again that the cluster, namely, the symmetry is not a model assumption but it is a fundamental degree of freedom and actually exists in realistic nuclear systems.

3. Localization and non localization

As mentioned previously, because of many-body correlations, clusters are formed at the nuclear surface. In a compact state with normal density, clusters are localized at the nuclear surface because the Pauli blocking effect is rather strong in the inner region. With decrease of density, clusters become free from the Pauli blocking effect and can move freely like a gas at low density. This is nothing but non-localization of clusters. Thus, the cluster feature changes depending on the density from the localized cluster at normal density and the non-localized cluster at low density. It should be stressed that there exist tow kinds of clustering, the localized and non-localized clusters.

Horiuchi and his collaborators proposed a new idea of the container picture which unifies these two kinds of clustering [5]. They pointed out that clusters are localized at the nuclear surface because of the Pauli blocking effect, whereas clusters far from a core are non-localized by a dynamical effect. In other words, the non-localization of clusters is a natural consequence of the symmetry restoration at low density, namely, the quantum fluctuation of the cluster center of mass motion. With the container picture, Horiuchi and Funaki described the localization and the non-localization of clusters in the ground and excited states of ^{12}C, ^{16}O, ^{20}Ne, ^{10}Be, and so on. Funaki discussed rotation of cluster gas states starting from the Hoyle state in ^{12}C using the THSR wave functions. More general discussions of the THSR wave functions and α condensation were given by Schuck. Suhara presented how the α-breaking component affects 3α dynamics in ^{12}C[6]. A new description of cluster gas states in medium-mass nuclei was proposed by Imai [7].

What is the answer to ^{12}C? Our main concern is the cluster structure in the Hoyle band. Is it a localized 3α clusters with a triangle shape or not? My answer is as follows. We have two different answers. If we look at the inner part of the Hoyle state we find the localized 3α as a dominant component because of the strong Pauli bocking effect. However, if we look at the outer tail region of the Hoyle state, we see a non-localized clusters like a gas at a low density. This is nothing but one of the multifacet aspects of ^{12}C.

4. Cluster formation

In the cluster formation, the Pauli blocking, i.e., Fermi surface plays an important role. As already mentioned, the localized clusters are formed at nuclear surface or inside nuclei at normal density, whereas clusters are non-localized at low density. It should be emphasized that the localized clustering can be described within mean-field approaches because it is caused by many-body correlation at the Fermi surface. However, the non-localized clustering is beyond the mean-field picture because there is no Fermi surface.

Khan discussed the transition between quantum liquid and crystal in nuclear systems [8]. Here, the crystal is a kind of localized clustering. He showed clusterization in Z=N=even nuclei and neutron-rich Be isotopes with a relativistic mean-field calculation. Sambataro and Lasseri investigated the 4-body correlation with a quarteting model, in which correlation between two pairs is taken into account [9]. The isospin symmetry restoration in correlating two pairs plays a significant role to form quarteting α-type correlation. Horiuchi *et al.* achieved full 5-body calculation of ^{12}C+4N and showed the α cluster formation and spectroscopic amplitudes in excited states of ^{16}O.

In heavy nuclei, the α-cluster formation has been a long-standing problem discussed in relation to α-decay lifetimes. Two kinds of clustering contribute to the α-decay width; the α-cluster formation at the surface (the localized cluster) and the α-decay dynamics (the non-localized cluster). A theoretical problem is how to combine these two kinds of clustering. To this end, hybrid approaches have been adopted to study α decay and α spectroscopic factor in heavy nuclei. In particular, the theoretical description of α decays from ^{212}Po is one of the revival hot topics as discussed by Roepke and Lovas [10, 11]. It was pointed out that the effect of Fermi surface (Pauli blocking), i.e., the antisymmetrization effect should be carefully taken into account to calculate the α-decay life time.

Two-nucleon correlation is another type of many-body correlation in nuclear systems. The proton-neutron pair formation in Z=N=odd nuclei such as ^{10}B and ^{18}F has been studied by Morita and Masui [12, 13]. These works may give a hint to deuteron-type and quarteting condensations in nuclear systems.

Let me emphasize recent remarkable progress of *ab initio* calculations, which gave a great impact to cluster physics. In these years, many efforts have been made to describe the cluster structure of excited states of ^{12}C with *ab initio* calculations. The recent results of *ab initio* Monte Carlo shell model [14] and those of simplectic no-core shell model [4] were presented by Otsuka and Draayer, respectively. It was proved that clusters are evidently built in many nucleon dynamics from realistic nuclear force in excited states of ^{12}C and also in neutron-rich Be isotopes. Now we are ready to go to further fundamental questions: What is the roles of the tensor and 3-body forces in nuclear clustering? Answers will come soon in *ab initio*-type calculations. Myo proposed a new method of semi *ab initio* approaches called tensor optimized AMD [15].

5. Cluster phenomena in isospin asymmetric systems and heavy systems

Cluster physics is being expanding widely toward the isospin asymmetric and heavy-mass regions in nuclear chart. In such systems, sub systems are not simple rigid clusters but they are themselves complex objects. A key problem is smooth connection between one-center and two-center limits. A one-center system changes gradually into a two-center system through a largely deformed state, a strong-coupling two-center state, and to a weak-coupling two-center state. In the weak-coupling limit, the essential degree of freedom is the relative motion between clusters, which is well decoupled from internal degrees of freedom (DOF) of clusters. As two clusters approach to each other, the internal DOF (excitations) of clusters become more important and two systems merge into a strong-coupling state. A theoretical problem is how to access to the transitional region, in which DOF are not clearly separated.

Let me first mention about cluster features in neutron-rich nuclei. von Oertzen discussed cluster structure of neutron-rich nuclei with the molecular orbital model, which is successful to describe energy spectra of neutron-rich Be and Ne isotopes [16]. When valence neutrons occupy the longitudinal σ-orbital sticking two clusters, the state has a strong-coupling cluster structure with a large deformation and constructs rotational bands. On the other hand, when valence neutrons occupy atomic orbitals instead of molecular orbitals, the state has a weak-coupling cluster structure and corresponds to a cluster resonance, which usually appears in an energy region higher than molecular orbital states. Further rich cluster phenomena are expected in neutron-rich nuclei. For instance, a three-center structure with a linear-chain configuration was predicted in neutron-rich C by von Oertzen, Suhara, and Baba [16, 17, 18]. Valence neutrons play an important role to stabilize the linear chain structure. Experimental measurements of band members of the linear chain structure were reported by Yamaguchi [19].

With increase of the mass number, systems become more complex, and internal excitations give significant effects to resonances and low-energy reactions via coexistence of different cluster channels, neck formation, multi-center phenomena, breakup process in reactions, and so on. In *sd*-shell and medium-mass nuclei, different cluster channels degenerate in excited states near threshold energies. For instance, α-cluster states and strong-coupling two-center cluster states appear almost in the same energy region near threshold energies. Because of duality, those cluster states correspond to superdeformed states. In highly excited states, molecular resonances have been observed. Usually observed molecular resonances are weak-coupling cluster resonances near or above the Coulomb barrier. A key problem is how to access to the transitional region near the threshold much below the Coulomb barrier. In a two-center cluster picture, key problems are internal excitations, channel coupling, cluster dissociations. Many attempts from the theoretical and experimental sides have been made to reveal cluster states in the transitional region. The coexistence of different cluster channels such as α-cluster states, 2α-cluster states, and ^{12}C-cluster states in ^{24}Mg and ^{28}Si have been studied theoretically and experimentally as discussed by Kimura, Royer, Chiba, and Kravvaris (Refs. [20, 21, 22] and references therein). Rotational bands of the superdeformation and its relation to O+O cluster cores with valence neutrons in ^{34}S were discussed by Taniguchi based on AMD calculation [23] and Afanasjev with Cranking RMF calculation [24].

Uegaki investigated C+C resonances constructed by two oblately deformed clusters and compared them with Si+Si cluster resonances [25]. In talks by Spieker, Kimura and Chiba, it was addressed that isovector dipole, isoscalar dipole, and monopole transitions are useful probes for nuclear clustering [21, 26, 27, 28].

Fusion and fission phenomena are contributed by further complex dynamics in the transitional region between one-center and two-center systems. Itkis gave a excellent review on fission of superheavy nuclei induced by light- and medium-mass nuclei and discussed clustering effects [29]. Carjan investigated spontaneous fission of very heavy nuclei by Cassinian oval models, and showed important roles of the compact fission mode, ^{132}Sn-like fragments, and octupole DOF in competition of (super)symmetric and asymmetric fission [30]. Dynamics of three fragments fission such as collinear cluster tri-partition and clustered chain-like prescission has been discussed by Pyatkov and Kamanin [31].

6. Advances in nuclear reactions

An extreme case of internal excitation of a cluster is breakup process in nuclear reaction. breakup of a cluster play an important role in low-energy reaction of weakly bound nuclei, which has been often discussed with astrophysical interests. I should remark recent developments of theoretical approaches for 3-body decay such as Hyperspherical approaches, 3-body CSM, and 4-body CDCC, in which 3-body continuum states are properly taken into account. These approaches enable us to directly connect theoretical calculations with experimental data with reliable reaction theories. Using these approaches, dynamics of 2,3-body breakup of weakly-bound systems and its effect to reaction cross sections have been intensively investigated. Descouvemont presented significant effects of breakup in ^{6}He scattering on heavy targets with the microscopic 4-body CDCC [32]. Watanabe *et al.* investigated breakup effects in ^{6}Li scattering with 4-body CDCC[33]. 3-body Coulomb breakup of ^{22}C at the neutron-drip line has been studied with 4-body Coulomb corrected eikonal model by Pinilla *et al.* [34]. Two-body breakup and incomplete fusion of weakly bound nuclei and ^{16}O have been attracting a great interest as discussed in Hussein's review talk and Samarin's talk [35, 36].

The low-energy resonances and reactions have been intensively studied theoretically and experimentally in particular with astrophysical interests. Many theoretical and experimental works concerning this subject have been presented in this conference: $2N$ emission [37], $2p$ capture [38], photodisintegration of ^{9}Be [39], radiative α capture, α-decay, scattering and quasi-molecule [40]. $2N$ decays, in particular, $2p$ decays from proton-rich nuclei beyond the drip line are attracting a great interest in physics of unstable nuclei. Grigorenko investigated dynamics of $2p$ decays from resonances with 3-body calculations while paying attention to true and sequential decays [37]. Hove calculated reaction rate of $2p$ capture at rapid-proton process waiting points in $Z \sim 70$ region and showed dominant sequential and direct $2p$ capture at high and low temperature, respectively [38].

7. Clustering in heavy ion collisions and nuclear matter

Clustering in finite temperature medium is a key to understand fragmentation in heavy ion collision (HIC). Emitting particles reflect nuclear matter information such as the EOS and symmetry energy though formation and dissolution of clusters in dynamical process of HIC. Recently, isospin asymmetric collision data have been used to obtain asymmetric matter information with molecular dynamics (MD) and statistical approaches [41, 42]. For example, Papa found that dipole degree of freedom relates to the symmetry energy through isospin equilibration process in HIC by investigating interaction dependence of average dipole in the Constrained MD calculation [41]. In discussions of phase diagram of nuclear matter, Moretto showed the liquid-vapor line obtained from HIC fragments assuming saturated vapor in a cluster gas [43]. Clustering in finite temperature medium also gives contribution to stellar matter as discussed by Typel [44]. He demonstrated model dependence on mass fraction in neutron star matter.

8. How to probe clustering

How one can probe clustering from observed data? Ito proposed that sizes of excited states can be determined by inelastic scattering [45]. He obtained enhanced spatial size of $^{12}C(2_2^+)$ from scattering data. Fukui discussed α-transfer reactions as a probe for α-cluster probability at surface [46].

9. Summary

In summary, I give my perspectives to the following questions: What we learn so far? What are the keys? Where we are going? Key words for theoretical study in cluster physics could be: N-body calculations with continuum. Microscopic treatment of nucleonic degrees of freedom. Calculations based on realistic nuclear force. Challenges to heavier systems and heavier clusters. Density, isospin, temperature dependences of clustering. Experimental probes for clustering. Our research subject is now being expanding widely toward the large mass number, isospin asymmetry, high excitation energy regions. By changing the excitation energy and the isospin asymmetry, we can access to low-density systems. In such new areas, we will see rich physics and encounter exotic phenomena related to cluster physics, because clustering is an essential DOF, in particular, at sub normal density. One might wonder a question "to be clustering or not to be clustering". However, it may not be the problem in a sense because cluster aspects arise everywhere in nuclear systems because of duality and symmetry. In general, different aspects in two limits can be smoothly connected with each other as seen in the connection between shell-model and cluster structures, the connection between one-center and two-center systems, and that between light-mass and heavy-mass regions, low energy and high energy regions, and so on. In many cases, the most essential and difficult problems exist in the transitional region between two limits. What we should do is to approach to the transitional region from both limits and find analogies and differences between two limits. Then we obtain a unified understanding of nuclear phenomena. Of course, we should not forget about other connections, the one between realistic and effective nuclear forces and that between theoretical and experimental studies. I would like to ask a question to myself. Is "Cluster physics" progressing in nuclear physics. The word "cluster" is useful but sometimes confusing. I would say, in the progress of nuclear physics, cluster concept is separating from and merging with other concepts again and again following decoupling and coupling of DOF in nuclear many-body systems.

I would like to greatly thank the organizers for their hospitality and many supports. The conference was well organized and done successfully. I believe that every participants enjoyed the conference and fruitful discussions in a beautiful and friendly place, Naples.

References

[1] Iachello F 2016 Naples conference; Bijker R 2016 Naples conference; Bijker R and Iachello F 2014 *Phys. Rev. Lett.* **112** 152501
[2] Cseh J 2016 Naples conference; Cseh J and Riczu G 2016 *Phys. Lett.* B **757** 312
[3] Lévai G 2016 Naples conference; Lévai G 2013 *Phys. Rev.* C **88** 014328
[4] Draayer J P 2016 Naples conference; Launey K D et al. 2016 *Prog. Part. Nucl. Phys.* **89** 101
[5] Horiuchi H 2016 Naples conference; Funaki Y 2016 Naples conference; Funaki Y, Horiuchi H and Tohsaki A 2015 *Prog. Part. Nucl. Phys.* **82** 78
[6] Suhara T 2016 Naples conference; Suhara T and Kanada-En'yo Y 2015 *Phys. Rev.* C **91** 024315
[7] Imai R 2016 Naples conference
[8] Khan E 2016 Naples conference; Ebran J P, Khan E, Niksic T and Vretenar D 2014 *Phys. Rev.* **90** 054329
[9] Sambataro M 2016 Naples conference; Sambataro M and Sandulescu N 2016 *Phys. Rev.* C **93** 054320
[10] Röpke G 2016 Naples conference; Xu C *et al* 2016 *Phys. Rev.* C **93** 011306
[11] Lovas R G 2016 Naples conference; Lovas R G *et al* 1998 *Phys. Rep.* **294** 265
[12] Morita H 2016 Naples conference; Morita H and Kanada-En'yo Y 2016 *Preprint* arXiv:1604.07131
[13] Masui H 2016 Naples conference; Masui H and Kimura M 2016 *Prog. Theor. Exp. Phys.* **2016** 053D01
[14] Otsuka T 2016 Naples conference; Yoshida T *et al* 2015 *JPS Conf. Proc.* **6** 030028
[15] Myo T 2016 Naples conference; Myo T *et al* 2015 *Prog. Theor. Exp. Phys.* **2015** 073D02
[16] Oertzen W 2016 Naples conference; Oertzen W, Freer M and Kanada-En'yo Y 2006 *Phys. Rep.* **432** 43
[17] Suhara T and Kanada-En'yo Y 2010 *Phys. Rev.* C **82** 044301
[18] Baba T 2016 Naples conference; Baba T, Chiba Y and Kimura M 2014 *Phys. Rev.* C **90** 064319
[19] . Yamaguchi H 2016 Naples conference
[20] Royer G 2016 Naples conference; Royer G, Ramasamy G and Eudes P 2015 *Phys. Rev.* C **92** 054308
[21] Kimura M 2016 Naples conference; Chiba Y 2016 Naples conference; Chiba Y and Kimura M 2015 *Phys. Rev.* C **91** 061302
[22] Kravvaris 2016 Naples conference; Volya A and Tchuvil'sky Y M 2015 *Phys. Rev.* C **91** 044319
[23] Taniguchi Y 2016 Naples conference; Taniguchi Y 2016 *EPJ Web Conf.* **117** 04009
[24] Afanasjev A V 2016 Naples conference; Ray D and Afanasjev A V 2016 *Phys. Rev.* C **94** 014310
[25] Uegaki E 2016 Naples conference; Uegaki E and Abe Y 2016 *EPJ Web Conf.* **117** 07024
[26] Yamada T *et al* 2012 *Phys. Rev.* C **85** 034315
[27] Chiba Y, Kimura M and Taniguchi Y 2016 *Phys. Rev.* C **93** 034319
[28] Spieker M 2016 Naples conference; Spieker M *et al* 2015 *Phys. Rev. Lett.* **114** 192504
[29] Itkis M 2016 Naples conference
[30] Carjan N 2016 Naples conference; Carjan N *et al* 2015 *Nucl. Phys.* A **942** 97
[31] Pyatkov Yu V and Kamanin D V 2016 Naples conference
[32] Descouvemont P 2016 Naples conference; Descouvemont P 2016 *Phys. Rev.* C **93** 034616
[33] Watanabe S 2016 Naples conference; Watanabe S *et al* 2012 *Phys. Rev.* C **86** 031601
[34] Pinilla E C 2016 Naples conference; Pinilla E C and Descouvemont P 2016 *Preprint* arXiv:1602.06477
[35] Hussein M S 2016 Naples conference
[36] Samarin V V 2016 Naples conference
[37] Grigorenko L V 2016 Naples conference; Grigorenko L V *et al* 2015 *Phys. Rev.* C **91** 024325
[38] Hove D 2016 Naples conference; Hove D *et al* 2016 *Phys. Rev.* C **93** 024601
[39] Katō K 2016 Naples conference; Odsuren M 2016 Naples conference; Kikuchi Y, Odsuren M, Myo T and Katō K 2016 *Phys. Rev.* C **93** 054605
[40] Solovyev A S 2016 Naples conference; Tchuvil'sky Y M 2016 Naples conference; Solovyev A S, Igashov S Y and Tchuvil'sky Y M 2016 *EPJ Web Conf.* **117** 09017
[41] Papa M 2016 Naples conference; Papa M *et al* 2015 *Phys. Rev.* C **91** 041601(R)
[42] Ogul R 2016 Naples conference; Imal H *et al* 2015 *Phys. Rev.* C **91** 034605
[43] Moretto L G 2016 Naples conference
[44] Typel S 2016 Naples conference
[45] Ito M 2016 Naples conference
[46] Fukui T 2016 Naples conference; Fukui T *et al.* 2016 *Phys. Rev.* C **93** 034606

IOP Conf. Series: Journal of Physics: Conf. Series **863** (2017) 012079 doi:10.1088/1742-6596/863/1/012079

Future Challenges for Nuclear Clustering

M Freer

School of Physics and Astronomy, University of Birmingham, Birmingham, B15 2TT, UK

E-mail: M.Freer@bham.ac.uk

Abstract. This contribution will focus on some of the advances that have been made since the last Cluster Conference in Debrecen, Hungary, and the challenges that remain for the subject before the next conference in Brazil. It addresses the question of cluster structure above and below the cluster decay threshold, dynamical symmetries, molecular structures and clustering at the neutron drip-line.

1. Introduction

The subject of nuclear clustering is almost as old as nuclear physics itself, and for those reasons and the fact that it has its own conference series, it is often forgotten why it is important to understand *nuclear* clustering. After all, clustering appears in many systems from biological, chemical and physical and can result in structures which are highly appealing to the human sense of organization and symmetry. However, to think only of the cluster arrangements would make nuclear clustering no more than a curio. Clustering reveals much about the nature of the force through which the constituent components interact and the symmetries that result. It is important to remember this connection in order to maintain the relevance of the subject to nuclear physics. The nuclear strong interaction is clearly complex and this is revealed in the detail of the unbound and bound light nuclei. The α-particle is one of the most highly bound light nuclei with a very high-lying, $\sim$20 MeV, first excited state, and here the array of correlations, that include not only $n-n$ and $p-p$ but $n-p$, maximize the binding energy. The tendency of other nuclei to optimize their own binding by generating spatial and momentum correlations induces the formation of clusters. This is responsible for clustering in α-conjugate nuclei, Borromean and molecular systems, alike.

How the clusters arrange themselves in part derives from the minimization of the total energy, but is also strongly influenced by the fact that the cluster structures grow from the single-particle nature of the nucleus, driven by the mean-field. The mean-field generates magic numbers and symmetries - which may or may not be those of α-particles. This tension is seen in approaches such as the no core shell model [1] where a realistic nucleon-nucleon interaction is employed within a harmonic oscillator basis. In some instances, the cluster structure emerges naturally from the available, low $\hbar\omega$, valence space, for example ^{8}Be. The contrasting case is the Hoyle-state in ^{12}C, which explores much more fully the entire $\hbar\omega$ spectrum of the harmonic oscillator. Herein lies the signature that the clustering degree of freedom has taken over and the correlations outweigh the mean-field. Understanding the interplay between the limiting cases requires experimental intervention and in particular precision measurements. The following sections explore some outstanding challenges for the field which might be addressed along the path to the next Cluster Conference.

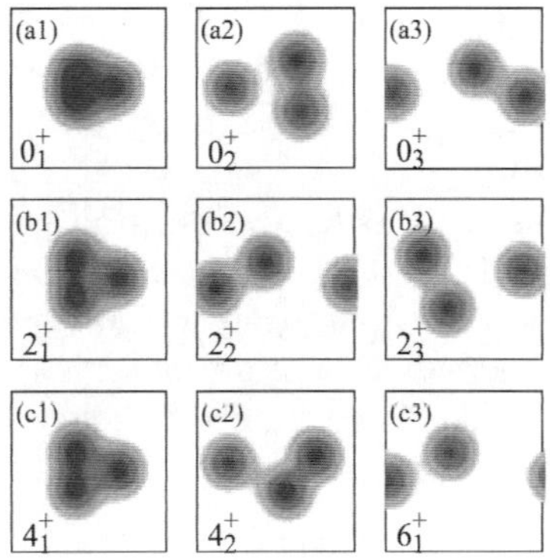

Figure 1. AMD densities of the ground state band, 0_1^+: a1, b1 and c1, Hoyle-band, 0_2^+ a2, b2, c2 and 0_3^+ band a3, b3, c3, from [2].

2. Clustering above and below the decay threshold and dynamical symmetries

It is recognised and accepted that clustering plays a crucial role above the cluster-decay threshold. Here this is well illustrated by the AMD calculations for ^{12}C (Fig. 1). The densities associated with the structures above the 3α decay threshold demonstrate that the nucleus precipitates into 3 clusters. However, below the decay threshold, and even in the ground-state, clustering also appears to have a structural influence. Demonstrating such a structure exists, experimentally, is not simple. The ground-state is compact, though still reveals the 3α structure, which is also strongly influenced by the ^{8}Be+α substructure. The collective excitations of this state (b1) and (c1) 2^+ and 4^+ states. Ultimately, measurements of the electromagnetic transition strengths, $B(E2)$, between the states above the α-decay threshold are the litmus test of which states may be strongly structurally linked.

Such measurements have been performed for ^{8}Be between the 2α 4^+ and 2^+ cluster states [3]. In that instance the states are separated by close to 8 MeV and hence the transition probability is strongly amplified by the E^5 dependence of the $B(E2)$ reduced transition amplitude. Hence, the competition with the α-decay is enhanced and a measurement is possible. The measurement demonstrates the collective nature of the excitations of ^{8}Be. This raises an interesting question [4]. The excited states of ^{8}Be are associated with resonances with large widths (2^+ 1.5 MeV, 4^+ ~3.5 MeV) which imply lifetimes of 10^{-22} seconds. This is the timescale for a nucleon to cross a nucleus and it is anticipated that the time scale for collective correlations and rotations to develop is significantly longer. It is intriguing to then understand why the 0^+, 2^+ and 4^+ states have a sequence of energies which display a rotational behaviour $E(4^+)/E(2^+)$=3.7. Is the key feature, the underlying cluster structure and the resulting symmetry rather than rotational characteristics and the states are symmetry-linked rather than rotationally-linked?

Understanding such issues may be improved by examining the structure of ^{12}C.

Here the dynamical symmetries of 3α-system correspond to a spinning top with a triangular point symmetry. The rotational properties of these states are given by

$$E_{J,K} = \frac{\hbar J(J+1)}{2\mathcal{I}_{Be}} - \frac{\hbar K^2}{4\mathcal{I}_{Be}} \quad (1)$$

where $\mathcal{I}_{Be}$ is the moment of inertia corresponding to two touching α-particles, which can be determined from the ^{8}Be ground-state rotational band [5]. One would expect that based on this structure there should be a number of rotational bands with different values of K. For K^π=0^+, the rotations will be around an axis which lies in the plane of the three α-particles, generating a series of states 0^+, 2^+, 4^+ .... These correspond to the rotation of a ^{8}Be nucleus - the rotation axis passing through the centre of the third α-particle. The next set of rotations correspond to the rotation around an axis perpendicular to the plane of the triangle, with each α-particle having one unit on angular momentum - giving $L = 3 \times 1\hbar$; $K^{\pi}=3^{-}$. Rotations around this axis and that parallel to the plane combine to give a series of states 3^{-}, 4^{-}, 5^{-}.... Such an arrangement possesses a D_{3h} point group symmetry. The experimental quest to observe this structure has resulted in the observation of the 4^{-} and 5^{-} members of the $K^{\pi} = 3^{-}$ rotational band [6], which has been taken as confirmation of the D_{3h} structure. However, as observed in the present conference, the arrangements of the experimentally observed states into rotational-like bands is not unique and equally it is possible to construct a sequence that has $SU(3)$ symmetry [7]. Experimental measurements of the electromagnetic transition rates is likely to be important in resolving this open question.

Similarly, the 4α system should be described by the tetrahedral symmetry group; T_d. Here the properties are those of a spherical top, with equal moments of inertia. If one assumes the separation of the α-particles is that which is associated with the ^{8}Be ground state, $\mathcal{I}_{Be}$, then the rotational energies are given by

$$E_J = \hbar^2 \frac{J(J+1)}{4\mathcal{I}_{Be}} \tag{2}$$

The rotations of the tetrahedral structure corresponds to the equivalent rotation of two ^{8}Be nuclei around their symmetry axis and hence the $4\mathcal{I}_{Be}$ in the denominator. The symmetry then dictates that all values of J are permitted except J=1, 2 and 5; states with J= 0, 4 and 8 have even parity and J=3, 7 and 11 have negative parity. A key feature of this structure would be degenerate 6^{+} and 6^{-} states. A similar conclusion can be found in the recent work of Bijker and Iachello [8]. The experimentally observed states at 6.130 MeV, 3^{-}; 10.356, 4^{+} and 21.052 MeV 6^{+} have been linked in this latter work to the collective excitations of the tetrahedral structure. These same calculations predicted states at 6.132, 10.220 and 21.462 MeV and electromagnetic transition strengths $B(E3)$ and $B(E4)$ of 181 and 338 e^{2}fm^{2L} compared with experimental values of 205(10) and 378(133) e^{2}fm^{2L}. The comparison between experiment and theory is compelling.

An alternative theoretical approach is provided by the Alpha Cluster Model (ACM) calculations [9]. These calculations identify a number of cluster structures, including a tetrahedral arrangement of the four α-particles in the ground-state. In addition, a planar arrangement of α-particles is found for the first excited 0^{+} state. These structures gives rise to rotational bands. The main difference between the ACM and Algebraic Cluster Model (ACM') of Ref. [8] is evident in the assignment of the 10.356 MeV 4^{+} state. The ACM assigns it to the planar rotational structure, whereas the ACM' links it to the tetrahedral ground-state. What is clear from measurements of the α-decay branching ratios for decay to the ^{12}C ground state and first excited states is that the states in the ACM planar band, above the alpha-decay threshold, all have very similar decay properties - they predominantly decay to the ^{12}C ground state [10]. This similar structure conflicts with the tetrahedral interpretation and indicates a collective excitation built around a ^{12}C+α cluster structure where the total angular momentum of the state is generated by the orbital motion of the α-particle around the ^{12}C core. To arrive at a better understanding of the cluster symmetries of ^{16}O further electromagnetic transition strengths need to be determined. These include states above the α-decay threshold, where small branching ratios ($< 10^{-5}$) make such studies very challenging.

A further outstanding puzzle is the question over the existence of the 4α-particle chain state. This dates back to some ground-breaking measurements performed by Chevallier [11] of the ^{12}C(^{4}He,^{8}Be)^{8}Be reaction which showed a number of resonances that decayed into two ^{8}Be nuclei. Given the structure of the final state and that the resonances appeared to lie on a $J(J+1)$ trajectory with a moment of inertia consistent with a 4α linear arrangement, a chain state was conjectured. These data have stood without proper test for many decades. The theoretical study using a cranked Hartree-Fock approach [12] indicated that, unlike the experimental data, the chain-like structure should *not* appear at low spins, but is rather stabilised by the rotations and hence wold exist at higher spins. These are spins beyond when the experimental data indicated sharp resonances. A subsequent experimental study, repeating the measurement of the same reaction used by Chevallier came to a different conclusion regarding the nature, J^{π}, of the resonances [13]. As such, there remains no definitive evidence for the existence of chain-like states in ^{16}O. A study at higher spins would help in this exploration.

A significant contribution to the Cluster Conference continues to be Condensate, or gas-like, α-particle like states [14]. In ^{12}C these are associated with the Hoyle-state and in ^{16}O the 15.2 MeV 0^+ state. The relationship between the Hoyle-state and the 2^+ excitation at 10 MeV has yet to be definitively resolved as has the existence of the proposed 4^+ state close to 13 MeV. The theoretical behaviour in terms of the predicted energies of these 2^+ and 4^+ states does not match that observed experimentally and this remains a puzzle. Similarly, it has not been demonstrated that the 15.2 MeV state in ^{16}O does indeed have a 4α structure. As such there remains much work to be done in the characterisation of the states in both these nuclei above the α-decay thresholds.

3. Molecular states

The story behind molecular structures was told by von Oertzen at the present conference (see the proceedings for the detail). The experimentally observed behaviour of the beryllium isotopes is remarkable in terms of how it may be characterised in terms of the covalent exchange of one or more valence neutrons. This behaviour is also captured in the no core shell model [15] and AMD calculations [16]. The extension of these ideas to the 3-centre systems where neutrons are delocalised beyond two-cores was also explored. Here the experimental situation is not so clear. Measurements presented at the conference indicated the original interpretation of molecular bands [16] may need some revision. The conference heard about measurements of ^{10}Be+α resonant scattering [17] populating resonances in ^{14}C. Together with other recent measurements of the same reaction [18, 19] it is possible that a consistent understanding of the structure of molecular structures will emerge in these more complex systems.

For both the two-centre and three-centre molecular systems it is important to go beyond the present description of their molecular behaviour in terms of experimental states matching rotational-like character. The Coriolis-decoupling of the $1/2^+$ band in ^{9}Be provides some confidence of the underlying structure, but precision measurements are now required to really make progress in addressing questions such as "Do these states really have a molecular structure, or are they equally well described by a mean-field type approach?" Measurements such as those performed by for the $A = 10$ analogues ^{10}C, ^{10}B* and ^{10}Be are called for [20].

4. The drip-lines

A long-standing prediction, of which the Conference was reminded [21], was that made at the Rab Cluster Conference. This was the prediction that at the neutron drip-line the large neutron excess and the energetic advantage of maximising the $n-p$ interaction is accommodated by clusterising the core, which is shown as a cartoon in Fig. 2.

Such a prediction remains to be verified, or otherwise. Nevertheless, the recent measurements of the potential $4n$-decay of ^{8}He, which were presented [22], appear to show evidence for $4n$-correlations does indicate that the drip-line may yet hold some surprises. The Conference is keen to learn more from the ongoing programme at RIKEN [23].

5. Summary and Acknowledgements

The contribution to the Conference from the experimental community was rich and varied. This continues to drive an understanding of the structure of light nuclei and the role of cluster

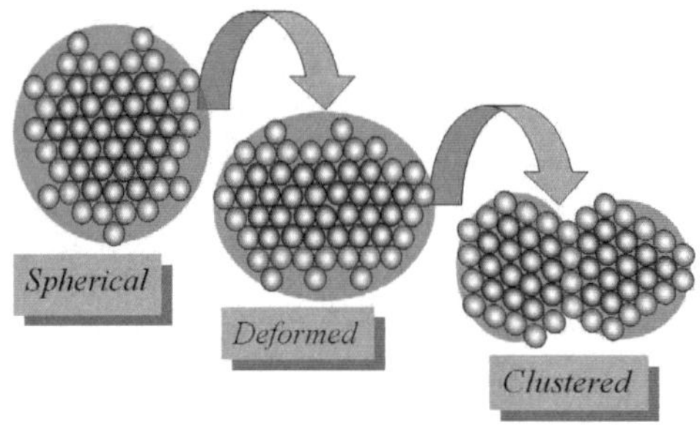

Figure 2. A schematic of the evolution of cluster structure in the ground-states of drip-line nuclei. The blue spheres represent nucleons associated with nuclear matter of *normal* isospin, whereas the red spheres are the excess valence neutrons. The transition from spherical through deformed to clustered permits a more even distribution of the valence neutrons.

correlations. However, to make further and substantial progress there is a strong need for comprehensive and precision measurements. Comprehensive in the sense that for key states all of their characteristics are measured and with a *precision* which allows the testing of models of the structure and interaction. In order to make progress there is a strong need for greater coordination in the experimental and indeed theory communities, to define the key objectives for future experimental measurements, much as is done in communities such as particle physics. This should be a key objective on the journey to Brazil.

The Conference would like to acknowledge the professional efforts of the local organising committee and further to recognise the contribution of Paulo Gomez, whose research into nuclear reactions at low energies often took advantage of the clustered structure of light nuclei. He will be greatly missed.

References

[1] Draayer J, contribution to the present conference
[2] Kanada En'yo Y 2007 *Prog. Theor. Phys.* 117 655
[3] Datar V M *et al* 2013 *Phys. Rev. Lett.* 111 062502
[4] Courtesy of Nazarewicz W
[5] Hafstad L R and Teller E 1938 *Phys. Rev.* **54** 681
[6] Marín-Lambarri D J *et al* 2014 *Phys. Rev. Lett.* **113** 012502
[7] Cseh J, contribution to the present conference
[8] Bijker R and Iachello F 2014 *Phys. Rev. Lett.* **112** 152501. See also the contribution from M Gai
[9] Bauhoff W, *et al* 1984 *Phys. Rev.* C **29** 1046
[10] Wheldon C, *et al* 2011 *Phys. Rev.* C **83** 064324
[11] Chevallier P *et al* 1967 *Phys. Rev.* **160** 827
[12] Ichikawa T, Maruhn J A, Itagaki N and Ohkubo S 2011 *Phys. Rev. Lett.* **107** 112501
[13] Curtis N, *et al* 2013 *Phys. Rev.* C **88** 064309
[14] See contributions from Itoh, Funaki and Schuck
[15] Maris P, Caprio M A and Vary J P 2015 *Phys. Rev.* C **91** 014310
[16] von Oertzen W, Freer M and Kanada-Enyo Y 2016 *Phys. Rep.* **432** 43
[17] See contribution Yamaguchi
[18] Freer M *et al* 2014 *Phys. Rev.* C **90** 054324
[19] Fritsch A *et al* 2016 *Phys. Rev.* C **93** 014321
[20] Lister C J 2012 *J. Phys.: Conf. Ser.* **381** 012010
[21] Horiuchi H, conference dinner presentation
[22] Shimoura S, contribution to the present conference
[23] Marques M, contribution to the present conference